# International Conference E-Learning and Digital Learning (ELDL2024), and Sustainability, Technology and Education (STE2024)



Held at the Multi-Conference on Computer Science and Information Systems (MCCSIS 2024)

Budapest, Hungary
13-15 July 2024

ISBN: 979-8-3313-0304-4

**Printed from e-media with permission by:**

Curran Associates, Inc.
57 Morehouse Lane
Red Hook, NY 12571

**Some format issues inherent in the e-media version may also appear in this print version.**



International Association for Development of the Information Society (IADIS)
Rua Sao Sebastiao Da Pedreira
No. 100, 30, 1050-209
Lisbon, Portugal

Phone: 351 21 3151373
Fax: 351 21 3151244

secretariat@iadis.org

**Additional copies of this publication are available from:**

Curran Associates, Inc.
57 Morehouse Lane
Red Hook, NY 12571 USA
Phone: 845-758-0400
Fax: 845-758-2633
Email: curran@proceedings.com
Web: www.proceedings.com

# PROCEEDINGS
# OF THE
# INTERNATIONAL CONFERENCES

# E-LEARNING AND DIGITAL LEARNING 2024

# AND

# SUSTAINABILITY, TECHNOLOGY AND EDUCATION 2024

**Budapest, Hungary**
**JULY 13 - 15, 2024**

Organised by

iadis

international association for development of the information society

As a member of Crossref (a non-profit membership organization for scholarly publishing working with the purpose to make content easy to find, link, cite and assess) each published paper in this book of proceedings will be allocated a DOI (Digital Object Identifier) number for its fast and easy citation and indexation.

Volume Editors:
Miguel Baptista Nunes, Pedro Isaías, Tomayess Issa and Theodora Issa

Computer Science and Information Systems Series Editors:
Piet Kommers and Pedro Isaias

Associate Editor: Luís Rodrigues

# TABLE OF CONTENTS

FOREWORD ix

PROGRAM COMMITTEE xiii

KEYNOTE LECTURE xvii

SPECIAL TALK xix

TUTORIAL xxi

## FULL PAPERS

***E-LEARNING AND DIGITAL LEARNING***

EXAMINING ACCESSES TO OPEN ACCESS SIMULATOR OBJECTS IN ENGLISH AND IN PORTUGUESE 3
*Ana M. B. Pavani and William S. Barbosa*

TEACHING ACADEMIC WRITING IN THE ERA OF LARGE LANGUAGE MODELS: A CASE STUDY 11
*Andrea Molinari and Elena Molinari*

IMPLEMENTING BLENDED LEARNING AS A CULTURALLY RESPONSIVE TEACHING METHOD TO THE NEEDS OF STUDENTS WITH CULTURAL AND LINGUISTIC DIVERSITY 19
*Vaso Anastasiou, Charoula Angeli and Yiota H. Diakou*

EXTENDING COMMON USABILITY HEURISTICS LIST WITH LEARNER-CENTERED STRATEGIES FOR AN EXHAUSTIVE E-LEARNING DESIGN ASSESSMENT 27
*Nadia Menad*

THE OBSTACLES TO DIGITAL TRANSFORMATION AND THEIR IMPACT ON CYBER THREATS IN MOROCCAN HIGHER EDUCATION: A MULTI-CRITERIA APPROACH 35
*Abdelilah Chahid, Souad Ahriz, Kamal El Guemmat and Khalifa Mansouri*

USING AN ESCAPE ROOM GAME TO EXPLORE THE RELATIONSHIPS BETWEEN GAMEPLAY SELF-EFFICACY, EXPERIENTIAL VALUE AND LEARNING PROGRESS IN A PERFORMING ARTS COURSE 43
*Jon-Chao Hong, Yi-An Chien and Chih-Hsuan Ho*

WELCOME: TURN ON YOUR CAMERA, YOU HAVE A LEAD ROLE 51
*Eduardo Hamuy, Ricardo Abarca, Rolando Cori and José Garrote*

PLAY2ESCAPE - A DIGITAL ESCAPE ROOM IN MATH CLASSES FOR UNDERGRADUATE TOURISM STUDENTS - CASE STUDY 59
*Sónia Pais, Ana Elisa Sousa and Maria Sofia Lopes*

FORMATION OF BACHELOR'S SKILLS IN THE FIELD OF MACHINE LEARNING AND INTELLIGENT DATA ANALYSIS 67
*Marina Lapenok, Sergei Patrushev and Lidiya Shestakova*

STUDENT-CENTERED PERSONALIZATION OF INDIVIDUAL EDUCATION THROUGH REUSABLE AND AUTONOMOUS LEARNING UNITS - THE SPIRAL MODEL 75
*Sorana Cimpan, Francesco Floris, Marina Marchisio Conte and Sergio Rabellino*

DIGITAL FOR WELFARE. STORIES OF "THIRD SECTOR" REALITIES FACING THE CHALLENGES OF DIGITAL TRANSFORMATION 83
*Dalia Gallico*

TINKERCAD AS A VIRTUAL LABORATORY 91
*József Udvaros and Norbert Forman*

LEARNING ANALYTICS VISUALISATIONS OF SCHOOL DROPOUT IN VOCATIONAL EDUCATION 98
*Christos Samaras, Anna Mavroudi and Vassilios S. Verykios*

DIGITAL TECHNOLOGIES AND THEIR IMPACT ON METACOGNITIVE ASPECTS IN LEARNING RECOVERY IN SCIENTIFIC AND HUMANISTIC SUBJECTS 105
*Andrea Balbo, Alice Barana, Giulia Boetti, Marina Marchisio Conte and Sara Omegna*

TAILORING EDUCATION TO DATABASE TRENDS: CHALLENGES IN CURRICULUM DESIGN FOR ADVANCED DATABASE TOPICS 113
*Rosa Maria Reis, Bertil P. Marques and Ricardo Almeida*

RATING UNIVERSITIES IN HONG KONG, MACAO AND SINGAPORE BY GENERATIVE ARTIFICIAL INTELLIGENCE (AI): ARE VARIOUS AI MODELS CONSISTENT? 121
*Victor K. Y. Chan*

DIGITAL EDUCATION IN THE TEACHING AND LEARNING OF EXPONENTIAL AND LOGARITHMIC FUNCTIONS 129
*Alberto Codeço, Pedro Miguel Faria and Isabel Araújo*

DETECTING BLENDED LEARNING, COLLABORATIVE LEARNING AND INTERNATIONALISATION IN MILITARY EDUCATION: A SYSTEMATIC LITERATURE REVIEW 137
*Alice Barana, Giulia Boetti, Antonino Cambria, Marina Marchisio Conte and Enrico Spinello*

STUDENTS' PERCEPTION OF INTERNATIONAL COLLABORATION AND CULTURAL UNDERSTANDING IN DIGITALLY DRIVEN EDUCATION 145
*Marina Marchisio Conte, Fabio Roman, Matteo Sacchet, Enrico Spinello, Anna Molnár, Shay Attias, Paul Tudorache, Jaroslav Kompan and Andrew Dolan*

A STUDY ON THE PROMOTION OF E-LEARNING TO THE PROFESSIONAL DEVELOPMENT OF RURAL TEACHERS IN GUANGXI, CHINA 154
*Chenfei Xiang, Jiachao Wei, Yuchen Gou, Shan Liu and Liqing Huang*

INTEGRATION OF ARTIFICIAL INTELLIGENCE IN TEACHER TRAINING: TRAINING TO INNOVATE IN THE DIGITAL AGE 161
*Bruno F. Gonçalves*

JUDOMATHS: AN INTELLIGENT SEMANTIC SYSTEM TO SUPPORT SKILLS-BASED LEARNING OF FRENCH THIRD-CYCLE MATHEMATICS 169
*Clément Aubeuf*

GATEWAY TO SUCCESS: IMPROVING ACADEMIC ACHIEVEMENT THROUGH ONLINE MODULES 176
*Linda van Ryneveld*

PARTICIPATORY PROCESSES IN THE CONTEXT OF THE DEVELOPMENT AND IMPLEMENTATION OF (DIGITAL) LEARNING OFFERINGS: A CASE STUDY ON THE SUPPORT OF INTERNSHIP PHASES IN BUSINESS MANAGEMENT STUDIES 183
*Yvonne Bönninger, Anne Florence Merkle, Elke Mählitz-Galler, Gunnar Voß, Beate von Velsen-Zerweck and Michael A. Herzog*

PREDICTING MOBILE LEARNING APPS' USER RATING BASED ON THEIR USABILITY EVALUATION BY GENERATIVE ARTIFICIAL INTELLIGENCE (AI) 191
*Victor K. Y. Chan*

EVALUATION OF THE IMPACT OF ACTIVE LEARNING USING AUGMENTED REALITY ON CADETS' KNOWLEDGE 199
*Jadranko Tuta and Ljerka Luic*

***SUSTAINABILITY, TECHNOLOGY AND EDUCATION***

GREEN IT ADOPTION BY SOFTWARE DEVELOPERS: KEY FACTORS FROM THE THEORY OF PLANNED BEHAVIOR PERSPECTIVE 210
*Yassine Talas and Hajer Rabii*

STUDYING THE ENERGY CONSUMPTION OF INSTAGRAM FUNCTIONALITIES ON PC AND TABLET 219
*Elisa Jiménez, Carlos Pulido, Coral Calero, Mª Ángeles Moraga, Félix García and Alberto Gordillo*

ASSESSMENT OF HYGROTHERMAL PROPERTIES OF ADOBE BRICKS. RESEARCH EXPERIENCE AT KEENE STATE COLLEGE 227
*Fernando del Ama Gonzalo, Paul Fowler and Zachary Punch*

RESEARCH ON AN ABSORPTION AIR-CONDITIONING SYSTEM MADE FROM SUSTAINABLE MATERIALS 235
*Yuxian Zhou, Wenquan Wang and Xiuwei Li*

EFFECTS OF LEARNING ENVIRONMENTS ON STUDENTS' SKILLS DEVELOPMENT: A COMPARATIVE VIEW OF EIGHT EUROPEAN COUNTRIES 243
*Ana Tecilazić and Biljana Birač*

GENERATIONAL PERSPECTIVE ON ENERGY-EFFICIENCY AWARENESS: A COMPARATIVE STUDY 253
*Gabriella Francesca Amalia Pernice, Valeria Orso and Luciano Gamberini*

A SUSTAINABILITY CAMPUS WALK AS A GAMIFIED TYPE OF BLENDED LEARNING 261
*Anna-Marie Tuomala and Alexandra Maksheeva*

## SHORT PAPERS

***E-LEARNING AND DIGITAL LEARNING***

THE IMPACT OF THE MAKER MOVEMENT ON ENTREPRENEURSHIP FROM A RESOURCE PERSPECTIVE 273
*Tsai-Hsin Chen, Tsz-Yu Huang and Chun-Wang Wei*

THE PDSA APPROACH TO UNDERSTANDING SELF-REGULATED LEARNING STRATEGIES 277
*Yassine Safsouf*

SHIFTING PERCEPTIONS OF AI: AN INTEGRATED LEARNING PROGRAM IN THE MIDDLE SCHOOL SETTING 281
*Won Jung Kim and Sangjin An*

COLLABORATIVE ONLINE INTERNATIONAL LEARNING (COIL): SWOT ANALYSIS FROM A TECHNICAL ENGINEERING COURSE PERSPECTIVE 286
*Jaime Garbanzo-León, Luís Jorge Gonçalves and Jorge Medina*

COGNITIVE LOAD MEASUREMENT WITH ELECTROENCEPHALOGRAPHY BASED ON MACHINE LEARNING ALGORITHM 291
*Yi-Hsuan Wang, Hsing-Yi Wu and Chun-Wang Wei*

ZINES AND DIGITAL TASK-BASED PROJECTS: CONTENT-BASED INSTRUCTION AND TASK-BASED LEARNING 295
*Peter Gobel and Makimi Kano*

## REFLECTION PAPERS

### *E-LEARNING AND DIGITAL LEARNING*

ADAPTING ASSESSMENT DESIGN: NAVIGATING TECHNOLOGICAL ADVANCES AND ACCESS TO INFORMATION 303
*Fran Greyling and Linda van Ryneveld*

ADVANCING EARLY CHILDHOOD EDUCATION: A CONCEPTUAL FRAMEWORK FOR ARTIFICIAL INTELLIGENCE APPLICATIONS IN EARLY CHILDHOOD EDUCATION 307
*Hesham Alomyan*

## POSTERS

### *E-LEARNING AND DIGITAL LEARNING*

APPLYING VIRTUAL LEARNING ENVIRONMENTS ASSISTS LANGUAGE LEARNERS TO STUDY ENGLISH 315
*Pei-Lin Liu and Chiu-Jung Chen*

DESIGN OF STACKABLE CERTIFICATE COURSE MODULES IN AN ONLINE PROGRAM 318
*Parviz Ghandforoush*

LEARNING ANALYTICS WITH MAPPER: INVESTIGATING PARAMETER CONFIGURATION 321
*Hitoshi Inoue, Koichi Yasutake, Osamu Yamakawa, Takahiro Tagawa, Takahiro Sumiya and Kiyomi Okamoto*

# FOREWORD

These proceedings contain the papers and posters of the: 18th International Conference on e-Learning and Digital Learning (ELDL 2024) and 12th International Conference on Sustainability, Technology and Education (STE 2024), which was organised by the International Association for Development of the Information Society, 13 - 15 July, 2024. These conferences are part of the 18th Multi Conference on Computer Science and Information Systems 2024, 13 - 15 July, which had a total of 603 submissions.

The e-Learning and Digital Learning (ELDL) 2024 conference aims to address the main issues of concern within e-Learning. This conference covers both technical as well as the non-technical aspects of e-Learning.

The conference accepted submissions in the following seven main areas: Organisational Strategy and Management Issues; Technological Issues; e-Learning Curriculum Development Issues; Instructional Design Issues; e-Learning Delivery Issues; e-Learning Research Methods and Approaches; e-Skills and Information Literacy for Learning.

The above referred main submission's areas are detailed below:

**Organisational Strategy and Management Issues**
- Higher and Further Education
- Primary and Secondary Education
- Workplace Learning
- Vocational Training
- Home Schooling
- Distance Learning
- Blended Learning
- Change Management
- Educational Management
- Continuous Professional Development (CPD) for Educational and Training Staff
- Return on e-Learning Investments (ROI)

**Technological Issues**
- Learning Management Systems (LMS)
- Managed Learning Environments (MLEs)
- Virtual Learning Environments (VLEs)
- Computer-Mediated Communication (CMC) Tools
- Social Support Software
- Architecture of Educational Information Systems Infrastructure
- Security and Data Protection
- Learning Objects
- XML Schemas and the Semantic Web
- Web 2.0 Applications

**e-Learning Curriculum Development Issues**
- Philosophies and Epistemologies for e-learning
- Learning Theories and Approaches for e-learning
- e-Learning Models
- Conceptual Representations
- Pedagogical Models
- e-Learning Pedagogical Strategies
- e-Learning Tactics
- Developing e-Learning for Specific Subject Domains

**Instructional Design Issues**
- Designing e-Learning Settings
- Developing e-Learning Pilots and Prototypes
- Creating e-Learning Courses
- Collaborative Learning
- Problem-based Learning
- Inquiry-based Learning
- Blended Learning
- Distance Learning
- Designing e-Learning Tasks
- E-learning Activities
- Online Groupwork
- Experiential Learning
- Simulations and Modelling
- Gaming and Edutainment
- Creativity and Design Activities
- Exploratory Programming
- Learning Analytics

**e-Learning Delivery Issues**
- e-Delivery in different contexts
    - Higher and Further Education
    - Primary and Secondary Schools
    - Workplace Learning
    - Vocational Training
    - Distance Learning
- Online Assessment
- Innovations in e-Assessment
- e-Moderating
- e-Tutoring
- e-Facilitating
- Leadership in e-Learning Delivery
- Networked Information and Communication Literacy Skills
- Participation and Motivation in e-Learning

**e-Learning Research Methods and Approaches**
- Action Research
- Design Research
- Course and Programme Evaluations
- Systematic Literature Reviews

- Historical Analysis
- Case Studies
- Meta-analysis of Case Studies
- Effectiveness and Impact Studies
- Evaluation of e-Learning Technologies
- Evaluation of Student and Tutor Satisfaction
- Learning and Cognitive Styles
- Ethical Issues in e-Learning

**e-Skills and Information Literacy for Learning**
- Teaching Information Literacy
- Electronic Library and Information Search Skills
- ICT Skills Education
    - in schools and colleges
    - for business, industry and the public sector
    - in adult, community, home and prison education
    - informal methods (peer groups, family)
- Education for Computer-mediated Communication Skills
    - Netiquette
    - Online safety for children and vulnerable users
    - Cybercrime awareness and personal prevention
- Student Production of Online Media
    - Web design
    - Digital storytelling
    - Web 2.0 tools
    - etc.
- Digital Media Studies

The International Conference on Sustainability, Technology and Education 2024 (STE 2024) has the purpose to address the main issues which occur by evaluating the relationship between Sustainability, Education and Technology. The main areas of interest for this conference are: Sustainability and Leadership, Sustainability and Green IT, Sustainability and Education.

These events received 189 submissions from more than 21 countries. Each submission has been anonymously reviewed by an average of four independent reviewers, to ensure that accepted submissions were of a high standard. Consequently, only 33 full papers were approved, which meant an acceptance rate of 18%. A few more papers were accepted as short and reflection papers and posters. An extended version of the best papers will be selected for publishing in the Interactive Technology and Smart Education (ITSE) journal (ISSN:1741-5659), IADIS International Journal on WWW/Internet (ISSN: 1645-7641) and also in the IADIS International Journal on Computer Science and Information Systems (ISSN: 1646-3692). Other outlets may also receive extended versions of the best papers.

Besides the presentations, the conference also included one keynote speech from an internationally distinguished researcher. We would therefore like to express our gratitude to Professor Helen Crompton, Executive Director of the Research Institute of Digital Innovation in Learning (RIDIL), Old Dominion University, USA, for being the keynote speaker. The conference also featured one special talk by Professor Cathie Norris, Regents Professor, University of North Texas, USA, Professor Elliot Soloway, Arthur F. Thurnau Professor, University of Michigan, USA, and Professor Anne Tapp, Saginaw Valley State University, USA, and a tutorial by Veronica Ruberti, Lorenzo Benussi, Chiara Ciociola and Flavio Renga, LINKS Foundation, Italy.

A successful conference requires the effort of many individuals. We would like to thank the members of the Program Committee for their hard work in reviewing and selecting the papers that appear in this book. We are especially grateful to the authors who submitted their papers to this conference and to the presenters who provided the substance of the meeting. We wish to thank all members of our organizing committee.

Last but not least, we hope that everybody enjoyed the presentations, and we invite all participants for next years′ edition of these conferences.

Miguel Baptista Nunes, Xi'an Jiaotong-Liverpool University, China
Pedro Isaias, Universidade Aberta (Portuguese Open University), Portugal and
The University of New South Wales (UNSW – Sydney), Australia
*e-Learning and Digital Learning 2024 Conference Program Co-Chairs*

Tomayess Issa, Curtin University, Perth, Australia
Theodora Issa, Curtin University, Perth, Australia
*Sustainability, Technology and Education 2024 Conference Program Co-Chairs*

Piet Kommers, University of Twente, The Netherlands
*MCCSIS 2024 General Conference Chair*

July 2024

# PROGRAM COMMITTEE

**E-LEARNING AND DIGITAL LEARNING 2024**
**CONFERENCE PROGRAM CO-CHAIRS**

Miguel Baptista Nunes, Xi'an Jiaotong-Liverpool University, China
Pedro Isaias, Universidade Aberta (Portuguese Open University), Portugal
& The University of New South Wales (UNSW – Sydney), Australia

**SUSTAINABILITY, TECHNOLOGY AND EDUCATION 2024**
**CONFERENCE PROGRAM CO-CHAIRS**

Tomayess Issa, Curtin University, Perth, Australia
Pedro Isaías, Universidade Aberta (Portuguese Open University), Portugal
& The University of New South Wales, Australia
Theodora Issa, Curtin University, Perth, Australia

**MCCSIS 2024 GENERAL CONFERENCE CO-CHAIRS**

Piet Kommers, University of Twente, The Netherlands
Pedro Isaias, The University of New South Wales (UNSW – Sydney), Australia

**E-LEARNING AND DIGITAL LEARNING 2024**

**CONFERENCE COMMITTEE MEMBERS**

Adamantios Koumpis, University Hospital Cologne, Germany
Airina Volungeviciene, Vytautas Magnus University, Lithuania
Alexandru Vulpe, University Politehnica of Bucharest, Romania
Ali Fawaz Shareef, The Maldives National University, Maldives
Ana Barata, ISEP-GILT, Portugal
Andreas Bollin, Klagenfurt University, Austria
Angélica De Antonio, Universidad Politecnica de Madrid, Spain
Antoanela Naaji, Vasile Goldis West University of Arad, Romania
Antonella Nuzzaci, University of Messina, Italy
Antonio Hervás-Jorge, Universitat Politecnica de Valencia, Spain
Antonio Navarro, Universidad Complutense de Madrid, Spain
Apostolos Gkamas, University of Ioannina, Greece
Bertil P. Marques, Instituto Superior de Engenharia do Porto (ISEP), Portugal
Bruno F. Gonçalves, Instituto Politécnico de Bragança, Portugal
Carolina Lorenzo, University of Alicante, Spain
Charalampos Karagiannidis, University of Thessaly, Greece
Charoula Angeli, University of Cyprus, Cyprus
Christos Bouras, University of Patras, Greece
Christos Troussas, University of West Attica, Greece
Demetrios Sampson, University of Piraeus, Greece
Dessislava Vassileva, Sofia University "st. Kliment Ohridski", Bulgaria
Dimitra Pappa, National Center for Scientific Research - Demokrit, Greece
Egle Butkeviciene, Kaunas University of Technology, Lithuania
Eliza Stefanova, Sofia University, Bulgaria
Elliot Soloway, University of Michigan, USA
Esma Aimeur, University of Montréal, Canada

## SUSTAINABILITY, TECHNOLOGY AND EDUCATION 2024
## CONFERENCE COMMITTEE MEMBERS

# KEYNOTE LECTURE

## GENERATIVE AI: AFFORDANCES AND CHALLENGES FOR EDUCATION

**Professor Helen Crompton**
**Executive Director of the Research Institute of Digital Innovation in Learning (RIDIL), Old Dominion University, USA**

## ABSTRACT

As ChatGPT and other generative AI tools permeate the educational landscape, educators find themselves at a crossroads, reexamining their teaching strategies and assessment methods. This presentation offers a thorough examination of ChatGPT's impact on teaching and learning, highlighting its potential to enhance instructional delivery and the overall learning journey. It underscores the imperative for educators to recognize and tackle the inherent limitations and potential misuses of these tools, promoting a balanced and responsible integration into the educational framework.

# SPECIAL TALK

## EVERY TEACHER HAS THEIR OWN VOICEBOT: IS IT WORTH THE COSTS?

**Cathie Norris, Regents Professor, University of North Texas, USA**
**Elliot Soloway, Arthur F. Thurnau Professor, University of Michigan, USA**
**&**
**Anne Tapp, Professor, Dept. of Teacher Education, College of Education, Saginaw Valley State University, USA**

## ABSTRACT

In the evolving landscape of educational technology, tools that enhance personalized learning experiences are pivotal. Current technologies adeptly address individual student needs within lessons, yet there is a notable gap in personalizing pedagogical practices. We are beta testing a pioneering tool designed to bridge this gap: a personalized VoiceBot for each teacher. The goal is to use AI to amplify the classroom teacher's role rather than to diminish it.

The VoiceBot generates feedback and instructional content in the teacher's own voice, enhancing the connection and continuity between a teacher and their students. This tool supports diverse applications:

- **Essay Feedback**: The VoiceBot provides written feedback on student essays, maintaining the distinctive voice, style, and tone of the teacher's feedback.
- **Reading Assistance**: For students facing reading challenges, the VoiceBot articulates task instructions (verbal) aloud in the teacher's voice, making instructions more accessible and, it appears, personally relevant.
- **Individual Tutoring**: In complex problem-solving scenarios (e.g., math), the VoiceBot offers step-by-step tutoring in the teacher's (verbal, written, or both) voice, fostering a personalized learning and personalized teaching environment.

Once the VoiceBot is created for an individual teacher, it is deployed automatically by our next-generation teaching and learning environment, the Roadmap Platform. For example, rather than a teacher spending considerable time recording the instructions and the content in a typical, K-5 PDF document, the Roadmap Platform uses the teacher's VoiceBot to automatically record all the text! Our VoiceBot leverages "few-shot learning" to effectively learn and replicate a teacher's "voice" from minimal samples—requiring only 3-4 instances of written feedback for essay grading, and approximately 90 seconds of verbal recording. Preliminary results indicate a strong preference among students for feedback that uses their teacher's authentic voice – written or verbal, as opposed to a generic or alternate instructor's voice.

However, there are real costs to using the VoiceBot. For example, we estimate that deploying the VoiceBot for a typical "Roadmap curricular lesson" in social studies, English, math, or science may cost approximately $.02 (U.S.) per lesson. That cost won't scale! And, while there is some evidence that "personalized teaching" does lead to increased achievement, there is virtually no data on using a VoiceBot to produce such "personalized teaching." Another question for which research is needed: what happens if (when?) students come to realize that the voice they are reading/hearing isn't really their teacher's voice, but an AI-version? Will students still find such "AI-based, personalized teaching" an enhancement?

While still in the preliminary stages, the VoiceBot has demonstrated significant potential to revolutionize teacher-student interactions and instruction delivery. As we continue to refine these technologies, further research will explore its broader impacts on teaching efficacy and student engagement – and cost. We welcome your collaboration – please, join us!

# TUTORIAL

## HOW TO DESIGN ONLINE STEM COURSES WITH CREATIVE LEARNING

**Veronica Ruberti, Lorenzo Benussi, Chiara Ciociola and Flavio Renga**
**LINKS Foundation, Italy**

## ABSTRACT

For instructional designers and teachers, few things are more challenging than designing tasks for online, collaborative and experiential learning. Learner characteristics, previous knowledge and contexts might vary very much, resulting in different degrees of engagement, interaction and collaboration between students. Given the growing demand for Computer-Supported Collaborative Learning (CSCL), the design of collaborative scripts emerges as a solution. Collaborative scripts are defined as adaptable instructional guidelines that support participation, coordination as a group towards a common goal, individual knowledge acquisition, and metacognition (Weinberger 2011).

In this tutorial, participants are involved in the process of designing a collaborative script based on the Creative Learning (CL) approach (Resnick 2017). According to this approach, the learning task should focus on creating a project that students are passionate about and interested in. Furthermore, the design process should support peer collaboration and playful experimentation as a learning pathway. By interacting and cooperating in groups, participants in this tutorial explore and apply CL to design a CSCL module from a case study. Imagining themselves at the beginning of the school year, participants co-design the first course unit for high school students. Depending on their preference, participants choose a subject (science, physics or mathematics) and co-design an online learning task that should meet a twofold objective. On one hand, to promote a positive attitude towards the target discipline by stimulating students' interest, curiosity, motivation and involvement. On the other, to make the relevance of the subject visible.

Given the varied learner characteristics and contexts, the tutorial focuses on how to include inclusive strategies in the online setting. Firstly, participants discover how to welcome students' preferences by providing an online wide-wall learning environment. For example, facilitators should provide learners with a combination of inspirational working themes and creative tools to include students' different interests and attitudes into the learning process (Resnick 2017). Secondly, participants reflect on how to adapt the learning task to novice-expert differences by designing a low-floor and high-ceiling environment. In other words, facilitators should select suitable tools so that every student can start and complete their project without frustration. At the same time, every student, if wished, can work on progressively more complex projects, experiencing what is known as "hard fun" (Papert 1993) in a high-ceiling environment (Resnick 2017). Then, participants discover how to address different learning styles by including a project demo with a modeling function at the beginning of the script and scaffolding materials to guide students along the whole learning task. Last but not least, in this tutorial we explore the fundamental role of facilitators. Facilitators work as mentors and coaches. They create an inclusive, constructive and stimulating environment for all the participants. They encourage students to explore and collaborate. Hence, they coach the online learning process by providing resources, support and feedback just in time. Moreover, they support participants to overcome difficulties, especially by encouraging peer support.

To conclude, participants elaborate how to assess the achievement of the learning objectives. A collective logbook is suggested as a useful assessment tool. The logbook can be used as a pre and post-assessment tool by including guiding questions aligned with the learning objectives. According to the learning goals of the given case, the pre-intervention questions could be: What do you expect from today's module? Given today's topic, what are you most passionate about? Why do you think it is useful to learn more about this topic? As well, the post-intervention questions could be: How did it go today? What difficulties did you have and how did you overcome them? What are you most passionate about? Why do you think it is useful to learn more about this topic? The logbook could provide valuable feedback for future implementation improvements and interesting insights about learning needs to be included in the following modules to design.
To sum up, the tutorial's output will be: a CL task for an online module, and the relative script and set of guidelines. While applying CL to script design, participants have the opportunity to experience and live the approach first-hand, to ponder the degree of collaboration, motivation and learning, as well as the feasibility of the CL approach for CSCL.

**KEYWORDS**

Instructional Design Issues, Designing Online Settings, Collaborative Learning, Experiential Learning, Creativity and Design Activities, Creative Learning

# Full Papers

# EXAMINING ACCESSES TO OPEN ACCESS SIMULATOR OBJECTS IN ENGLISH AND IN PORTUGUESE

Ana M. B. Pavani and William S. Barbosa
*Pontifícia Universidade Católica do Rio de Janeiro*
*Rio de Janeiro, RJ, Brazil*

**ABSTRACT**

This work focuses on analyzing access patterns to a collection of simulators within an Open Educational Resources project. The simulators are originally developed in Portuguese and then translated into English. The investigation compares the numbers of accesses and the countries accessing each pair of simulators over a six-year period (2018-2023). The data utilized for this analysis are sourced from an openly available dataset in the references. Two distinct analytical approaches are employed. The first relies on calculating averages, while the second involves analyzing access patterns under specific conditions using yearly data. The findings reveal significant differences in access patterns, attributed not only to language variations but also to the thematic contents addressed by the simulators.

**KEYWORDS**

Open Educational Resources, Simulator Objects, Accesses, Simulators in Portuguese, Simulators in English

## 1. INTRODUCTION

Online materials play a pivotal role in e-learning and digital education, particularly within the framework of Open Educational Resources (OER). Various designations are associated with such materials, including Learning Object (LO) as defined by IEEE (2002), Shareable Content Object (SCO) according to the ADL Guide (2004), Reusable Chunks of Instructional Media proposed by Wiley (2000), and Reusable Learning Objects as discussed by Alsubaie and Alshawi (2009).

This work focuses on examining accesses to a select set of OERs that are available at Open Educational Resources @PUC-Rio (https://www.maxwell.vrac.puc rio.br/projetosEspeciais/OER/Home.php). The OERs under consideration are simulators, though the collection hosts others of different natures. They have special characteristics that led to their choice for observation. The simulators and their special characteristics are presented in section 2. The objective of this analysis is to discern access patterns, including the numbers of countries accessing each OER and the total numbers of accesses, for both the Portuguese (pt) and English (en) versions. The primary research question guiding this study is:

**Research Question (RQ): Are the patterns of accesses (numbers of countries and of accesses) different for simulators in Portuguese (pt) and their English (en) versions?**

This work is an assessment of this project that started in May 2015 with simulators in Portuguese and the first English version made available in April 2016. The objective of the English versions is to support the sharing of online materials whose implementations are resource and time consuming, Since February 2023, Spanish (es) versions of the simulators have been developed and published with the same objective. The authors of the OERs all agree with a Creative Commons License (https://creativecommons.org/) in order to contribute to the worldwide effort based on Create-Share-Reuse.

It is important to remark that The Most Spoken Languages Worldwide in 2023 (Statista, 2023), lists English (1$^{st}$), Spanish (3$^{rd}$), French (4$^{th}$) and Portuguese (8$^{th}$) among the 10 most spoken languages in the world; the other 6 are Russian and Eastern languages. Wang and Towey (2017) list Language and Localization as one of the challenges for the use of OERs in higher education. They also cite the existence of the Universia Digital Library that offers Learning Materials in Portuguese and Spanish. So, language is a wider concern.

This work is divided in 4 sections besides this Introduction. Section 2 describes the context at the university and the available data. Section 3 presents the data to be analyzed, the analysis and the results. Section 4 comments the results.

# 2. CONTEXT AT THE UNIVERSITY, AVAILABLE DATA AND THE ANALYSIS PERFORMED

## 2.1 The OER Collection

Pontifícia Universidade Católica do Rio de Janeiro (PUC-Rio) operates an IR – Institutional Repository (Lynch, 2003) that hosts thousands of digital contents created at the university. It is the Maxwell System (https://www.maxwell.vrac.puc-rio.br) that publishes Electronic Theses and Dissertations (ETDs), Senior Projects (TFCs), Monographs, Research Data (RD), journals, research & technical reports, restricted courseware and OERs. In order to make OERs easier to find, an aggregator, Open Educational Resources @PUC-Rio (https://www.maxwell.vrac.puc-rio.br/projetosEspeciais/OER/Home.php), with all interfaces in English was deployed. There are aggregators for ETDs, TFCS and RD too.

Since the focus of this work is OERs, some additional information about them is necessary. The characteristics of the OER collection are:

01. The collection has 493 titles with 1,567 objects – many titles have multiple objects;
02. 453 titles are in Electrical Engineering (EE) and Science & Technology (S&T) topics while 40 are in Portuguese as a Second Language (PSL);
03. OERs in EE and S&T are of various natures – hypermedia objects, videos, podcasts and simulators;
04. The topics addressed by the EE and S&T OERs are part of the courses in the curricula of EE was well as Control Engineering and Computer Engineering in institutions all over the world;
05. Simulators are the focus of this work because they are the most interactive in the collection and stimulate students being proactive;
06. Simulators have been developed by faculty, undergraduate students, graduate students and technical staff; the participation of students was twofold – they came with the suggestions of topics they wanted to learn better and then became the authors of the simulators;
07. Simulators are first implemented in Portuguese and, afterwards, translated into English and Spanish;
08. Currently, there are 43 simulators in Portuguese, 22 in English and 22 in Spanish. Others are under development;
09. Simulators are powered by Scilab (https://www.scilab.org/) which is installed on a server and accessed by the IR;
10. Each simulator has a theoretical introduction with help and instructions of use – the interfaces are in HTML and they communicate with Scilab via PHP programs;
11. Users do not see the Scilab interface and do not have to learn how to use it, they just have to fill the boxes with functions and/or parameters.

In the previous section, it was stated that the usage of the simulators in English when compared to the originals in Portuguese could serve as an indication if the effort to translate is worth. It is important to mention that only the interfaces in HTML are translated. For this reason, the effort is small and the results presented later on in this work indicate it is worth. A second benefit of this analysis is to guide the authors in deciding which new simulators to develop; this has already been happening as mentioned later on in this work.

The first simulator was published in May.2015 and the first English version in April.2016. This time frame yields an important consequence for this work – data are available for e very limited number of years. Since the authors are comparing accesses to pairs of simulators – the original in Portuguese and the corresponding English version, the limitation grows bigger because it is necessary to examine when both are published.

It is important to mention that all OERs on the collection are included on the database of MERLOT – Multimedia Educational Resources for Learning and Online Teaching (https://www.merlot.org).

A final comment in this section respects the student's participation, which was in a voluntary basis when the students suggest, in opposition to the work developed by Andone et al. (2020) where the creation of OERs were "renewable" assignments in contrast to "disposable" tasks. Their work addresses different types of OERs - multimedia objects, while this work focuses on simulators.

## 2.2 Available Data

The IR runs on a relational Database Management System (DBMS). A unique sequential number identifies each and all contents. The webserver is Apache that records accesses to the id numbers of the contents with the time stamps and the IP addresses where they come from. Every hour, the Apache server log is scanned to identify the numbers and the IP addresses, that are converted to countries using GeoIP Free. This information is stored on a database table that is organized by year-month-content id-country-number of accesses; accesses are added every time the log is scanned. This table contains information for all digital contents on the IR. Data on accesses are publicly available on the IR.

To organize data, a dataset was created compiling numbers of accesses for each original in Portuguese and the corresponding English version. The condition to be included in the dataset is that both had to had been available from Jan 01 to Dec.31 of the year under consideration. In order to have data on more simulators, the starting year was decided to be 2018. The dataset is available on the IR – Pavani and Temporão (2024). Data on the dataset start on Jan.01.2018 for the first 10 sets (Portuguese and English) and end on Dec.31.2023. Each set is in a sheet. The remaining 10 sheets have later starting dates due to the dates of publication of their English versions. Only the first 10 sets are examined in this work. The 10 sets under consideration are:

- Set 01 – Nyquist Stability Criterion;
- Set 02 – Root Locus Plot;
- Set 03 – Bode Plot for First and Second Order Systems;
- Set 04 – Fourier Transform of a Continuous Time Pulse Function;
- Set 05 – Inverted Pendulum;
- Set 06 – RL Series Circuit;
- Set 07 – RC Series Circuit;
- Set 08 – First Order Systems;
- Set 09 – Combinations of Continuous Time Functions;
- Set 10 – Combinations of Discrete Time Functions.

The 10 sets are clearly related to courses on Signals & Systems (S&S), Electric & Electronic Circuits (E&EC) and Control Systems (CS). The courses are core of the three engineering curricula mentioned before.

Table 1 classifies the sets according to the courses whose students are expected to use them.

Table 1. Classification of simulators according to the courses they support

| Set | S&S | E&EC | CS |
|---|---|---|---|
| 1 | | | ✓ |
| 2 | | | ✓ |
| 3 | | ✓ | ✓ |
| 4 | ✓ | | |
| 5 | | | ✓ |
| 6 | ✓ | ✓ | ✓ |
| 7 | ✓ | ✓ | ✓ |
| 8 | ✓ | ✓ | ✓ |
| 9 | ✓ | | ✓ |
| 10 | ✓ | ✓ | ✓ |

The analysis of the Table 1 delineates several discernible patterns: (1)The numbers of simulators across the three courses are different; (2) Noteworthy, is the cross-applicability of specific topics, initially tailored for the E&EC course, such as RC Series Circuit and RL Series Circuits, which serve a pedagogical purpose in the CS and S&S domains, facilitating the comprehension of 1$^{st}$ Order Linear Systems dynamics; and (3) These topics can have recurrent utilization; this is evident when parameters and/or functions undergo modifications.

## 3. DATA ANALYSIS

Data were extracted from the dataset and organized in 10 tables, one for each set. Since all sets were observed from 2018 to 2023, all tables have the same numbers of rows. Each table is divided in two vertical blocks – one for the numbers of countries that accessed the two versions and the other for the numbers of accesses. Objects in Portuguese are treated separately from their English versions – one column for each in each block. The tables reorganize data in the first 10 data sheets of the dataset. The last row of each table contains the averages of the numbers in the other rows in the same columns of the table.

Tables 2 and 3. Numbers of Countries and Numbers of Accesses Sets 1 (Nyquist Stability Criterion) and 2 (Root Locus Plot)

| | Countries | | Accesses | |
|---|---|---|---|---|
| | pt | en | pt | en |
| **2018** | 8 | 14 | 359 | 148 |
| **2019** | 14 | 28 | 468 | 255 |
| **2020** | 9 | 26 | 761 | 205 |
| **2021** | 11 | 18 | 687 | 229 |
| **2022** | 8 | 18 | 392 | 130 |
| **2023** | 10 | 14 | 494 | 196 |
| **μ** | **10.00** | **19.67** | **526.83** | **193.83** |

| | Countries | | Accesses | |
|---|---|---|---|---|
| | pt | en | pt | en |
| **2018** | 8 | 62 | 1,026 | 3,491 |
| **2019** | 7 | 65 | 1,353 | 6,611 |
| **2020** | 8 | 77 | 1,272 | 7,884 |
| **2021** | 19 | 61 | 2,141 | 2,365 |
| **2022** | 10 | 50 | 605 | 1,313 |
| **2023** | 8 | 56 | 907 | 1,904 |
| **μ** | **10.00** | **61.83** | **1,217.33** | **3,928.00** |

Tables 4 and 5. Numbers of Countries and Numbers of Accesses
ts 3 (Bode Plot for 1st and 2nd Order Systems) and 4 (Fourier Transform of a Continuous Time Pulse Function)

| | Countries | | Accesses | |
|---|---|---|---|---|
| | pt | en | pt | en |
| **2018** | 15 | 19 | 2,127 | 99 |
| **2019** | 21 | 32 | 3,415 | 247 |
| **2020** | 25 | 23 | 2,556 | 175 |
| **2021** | 28 | 42 | 2,457 | 332 |
| **2022** | 25 | 66 | 1,754 | 1,167 |
| **2023** | 28 | 81 | 2,653 | 3,464 |
| **μ** | **23.67** | **43.83** | **2,493.67** | **914.00** |

| | Countries | | Accesses | |
|---|---|---|---|---|
| | pt | en | pt | en |
| **2018** | 14 | 6 | 315 | 44 |
| **2019** | 6 | 8 | 320 | 68 |
| **2020** | 5 | 23 | 233 | 84 |
| **2021** | 8 | 25 | 139 | 99 |
| **2022** | 9 | 27 | 282 | 135 |
| **2023** | 10 | 12 | 161 | 117 |
| **μ** | **8.67** | **16.83** | **241.67** | **91.17** |

Tables 6 and 7. Numbers of Countries and Numbers of Accesses Sets 5 (Inverted Pendulum) and 6 (RL Series Circuit)

| | Countries | | Accesses | |
|---|---|---|---|---|
| | pt | en | pt | en |
| **2018** | 14 | 18 | 3,374 | 149 |
| **2019** | 16 | 12 | 3,515 | 91 |
| **2020** | 16 | 17 | 4,401 | 323 |
| **2021** | 16 | 7 | 4,374 | 173 |
| **2022** | 12 | 21 | 2,863 | 203 |
| **2023** | 15 | 28 | 3,029 | 600 |
| **μ** | **14.83** | **17.17** | **3,592.67** | **256.50** |

| | Countries | | Accesses | |
|---|---|---|---|---|
| | pt | en | pt | en |
| **2018** | 8 | 9 | 609 | 23 |
| **2019** | 13 | 5 | 990 | 40 |
| **2020** | 7 | 14 | 1,174 | 63 |
| **2021** | 16 | 16 | 1,200 | 50 |
| **2022** | 14 | 10 | 1,100 | 67 |
| **2023** | 16 | 26 | 1,053 | 160 |
| **μ** | **12.33** | **13.33** | **1,021.00** | **67.17** |

Tables 8 and 9. Numbers of Countries and Numbers of Accesses Sets 7 (RC Series Circuit) and 8 (First Order Systems)

| | Countries | | Accesses | |
|---|---|---|---|---|
| | pt | en | pt | en |
| **2018** | 13 | 6 | 485 | 32 |
| **2019** | 9 | 4 | 1,219 | 42 |
| **2020** | 10 | 6 | 980 | 45 |
| **2021** | 15 | 5 | 845 | 19 |
| **2022** | 13 | 4 | 961 | 21 |
| **2023** | 11 | 25 | 740 | 182 |
| **μ** | **11.83** | **8.33** | **871.67** | **56.83** |

| | Countries | | Accesses | |
|---|---|---|---|---|
| | pt | en | pt | en |
| **2018** | 6 | 3 | 185 | 25 |
| **2019** | 7 | 3 | 286 | 28 |
| **2020** | 9 | 4 | 457 | 47 |
| **2021** | 14 | 6 | 691 | 37 |
| **2022** | 11 | 7 | 516 | 45 |
| **2023** | 12 | 20 | 453 | 89 |
| **μ** | **9.83** | **7.17** | **431.33** | **45.17** |

Tables 10 and 11. Numbers of Countries and Numbers of Accesses Sets 9 (Combinations of Continuous Time Functions) and 10 (Combinations of Discrete Time Functions)

| | Countries | | Accesses | |
|---|---|---|---|---|
| | pt | en | pt | en |
| **2018** | 5 | 3 | 1,384 | 105 |
| **2019** | 9 | 4 | 3,246 | 97 |
| **2020** | 6 | 4 | 981 | 47 |
| **2021** | 13 | 5 | 1,753 | 40 |
| **2022** | 8 | 7 | 1,404 | 79 |
| **2023** | 5 | 5 | 1,335 | 55 |
| **μ** | **7.67** | **4.67** | **1,683.83** | **70.50** |

| | Countries | | Accesses | |
|---|---|---|---|---|
| | pt | en | pt | en |
| **2018** | 6 | 2 | 734 | 25 |
| **2019** | 6 | 3 | 1,415 | 98 |
| **2020** | 5 | 4 | 93 | 40 |
| **2021** | 7 | 6 | 127 | 54 |
| **2022** | 6 | 7 | 416 | 79 |
| **2023** | 9 | 5 | 213 | 71 |
| **μ** | **6.50** | **4.50** | **499.67** | **61.17** |

A quick look at the 10 previous tables indicates that the numbers are quite different from one set to the others, between the two languages in the same set and even in different years of the time series of one object. As it can be observed, tables 2 to 11 show larger numbers of accesses to objects in Portuguese compared to those in English (an exception is table 3); similar numbers of countries in both languages, except for tables 2 and 3, where the numbers of countries that accessed simulators in English surpass the ones in Portuguese. The subsections that follow analyze some characteristics of the data.

## 3.1 Analysis Based on the Averages

This subsection examines data based on the averages of numbers of accesses and numbers of countries where they came from for each set. These data are presented in Table 12 consolidating the last rows of Tables 2 - 11.

Table 12. Averages of Numbers of Countries and Numbers of Accesses

| | | μ Countries | | μ Accesses | |
|---|---|---|---|---|---|
| | | pt | en | pt | en |
| **Set 1** | **Nyquist Stability Criterion** | 10.00 | 19.67 | 526.83 | 193.83 |
| **Set 2** | **Root Locus Plot** | 10.00 | 61.83 | 1,217.83 | 3,928.00 |
| **Set 3** | **Bode Plot for 1st and 2nd Order Systems** | 23.67 | 43.83 | 2,493.67 | 914.00 |
| **Set 4** | **Fourier Transform of a Continuous Time Pulse Function** | 8.67 | 16.83 | 241.67 | 91.17 |
| **Set 5** | **Inverted Pendulum** | 14.83 | 17.17 | 3,592.67 | 256.50 |
| **Set 6** | **RL Series Circuit** | 12.33 | 13.33 | 1,021.00 | 67.17 |
| **Set 7** | **RC Series Circuit** | 11.83 | 8.33 | 871.67 | 56.83 |
| **Set 8** | **First Order Systems** | 9.83 | 7.17 | 431.33 | 45.17 |
| **Set 9** | **Combinations of Continuous Time Functions** | 7.67 | 4.67 | 1,683.83 | 76.50 |
| **Set 10** | **Combinations of Discrete Time Functions** | 6.50 | 4.50 | 499.67 | 61.17 |

The inspection of Table 12 yields interesting information:

01. Across six instances (sets 1 – 6), the averages of the numbers of countries accessing the English versions exceed these of the originals in Portuguese. Notably, this disparity reaches up to six times higher in one case and nearly double in three others.

02. The sets with highest average access numbers, both in Portuguese and in English, predominantly belong to the six aforementioned cases, with the exception of set 4, which typically pertains to S&S courses.

03. Four out of the six sets are utilized in Control Systems courses at PUC-Rio and globally.

04. Sets 6 and 7 are used in topics of the E&EC courses at PUC-Rio and all over the world.

05. Sets 4, 9 and 10 are used in S&S courses at PUC-Rio and all over the world

06. The English version of the Root Locus Plot, as depicted in Table 2, garnered accesses from 50 or more countries over the six-year period under scrutiny.

07. Except for the CS simulators, the others have significantly higher numbers of accesses to the Portuguese originals when compared to the English version.

08. As can be observed, sets 1-3, which exhibited greater diversity in access across countries, are centered on topics related to the knowledge base of CS.

09. Consistency in Accesses per Country: Except for Sets 2 and 3, where the numbers of countries accessing content in English exceed those accessing content in Portuguese, there generally is consistency in the number of accesses per country between the two languages.

10. Temporal Trends: While there are fluctuations in the numbers of accesses over the years, there does not seem to be a clear upward or downward trend across sets. However, some sets show an increase in accesses over time (e.g. Sets 1 and 2), while others show more variability (e.g. Sets 5 and 6).

11. Impact of Content Type: Different types of content (e.g. Nyquist Stability Criterion, Fourier Transform) seem to attract different levels of interest, as evidenced by variations in access numbers across sets.

The observed patterns in access numbers appear to be closely tied to the topics addressed by the simulators. Notably, CS emerges as the subject area with the highest access numbers, manifesting in some cases through the Portuguese versions and in others through their English counterparts.

## 3.2 Analysis based on Specific Accesses to Some Sets

This subsection examines sets in which one or both objects garnered a minimum average of 1,200 accesses, denoting at least 100 average accesses per month. This choice was to be able to analyze more data since the time frame of this work is small due to the dates of the creation of the simulators.

Four sets are under examination – Root Locus Plot, Bode Plot for 1st and 2nd Order Systems, Inverted Pendulum, and Combinations of Continuous Time Functions. To conduct this analysis, the tables in section 2 and the dataset are examined.

For a more comprehensive understanding of the relationships between the numbers of countries and of accesses across both versions of each simulator, a bi-dimensional graphic was generated for each set. Within each graph, the horizontal axis denotes the number of countries, while the vertical axis represents the number of accesses. Bullets indicating the Portuguese original are depicted as round, while those representing the English version are triangular. Control Systems emerges as the topic with the highest number of accesses for certain instances, observed in both the Portuguese and English versions.

### 3.2.1 Set 2 – Root Locus Plot and Set 3 – Bode Plot for 1st and 2nd Order Systems

Figures 1 and 2 are the bi-dimensional graphics for sets 2 and 3.

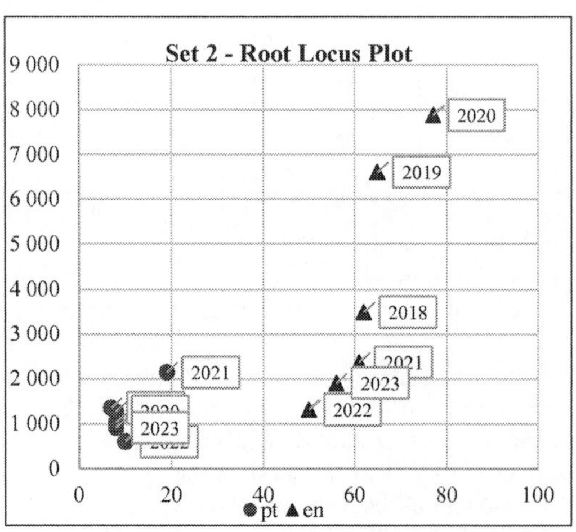


Figure 1. Numbers of countries and of accesses to the Root Locus Plot objects in pt and en 2018-2023

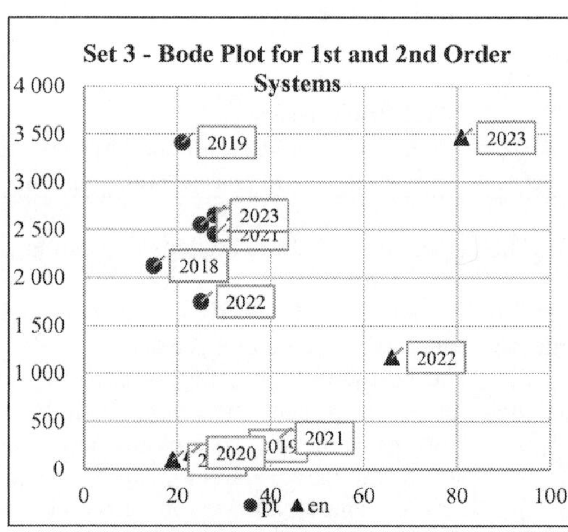


Figure 2. Numbers of countries and of accesses to the Bode Plot for 1st and 2nd Order Systems objects in pt and en 2018-2023

The first observation about Set 2 – Root Locus Plot is that average number of countries that accessed the English version is six times higher than the corresponding average to the Portuguese original. The average number of accesses is three times as high. In order to examine the numbers with more detail, Table 2 and the dataset are considered. Figure 1 shows the numbers of countries and corresponding numbers of accesses for each of the years 2018-2023 as presented in Table 2. Figure 1 yields a better perception of the relations among numbers of countries and of accesses. A comment can be made: the bullets of countries and of accesses are very close together in the pt original and far apart from the bullets in the en version; bullets of the en version are far apart from each other too. This characteristic indicates that both the numbers of countries and of accesses are higher for the en version. Concerning Set 3 – Bode Plot for 1st and 2nd Order Systems, it is easily seen that the numbers of accesses to the Portuguese original are higher than the ones to the English version. On the other hand, the numbers of countries are either similar or higher for the English version.

Both sets are used in CS courses and the Bode Plot is used in E&EC too.

### 3.2.2 Set 5 – Inverted Pendulum and Set 9 – Combinations of Continuous Time Functions

Figures 3 and 4 are the bi-dimensional graphics for sets 5 and 9.

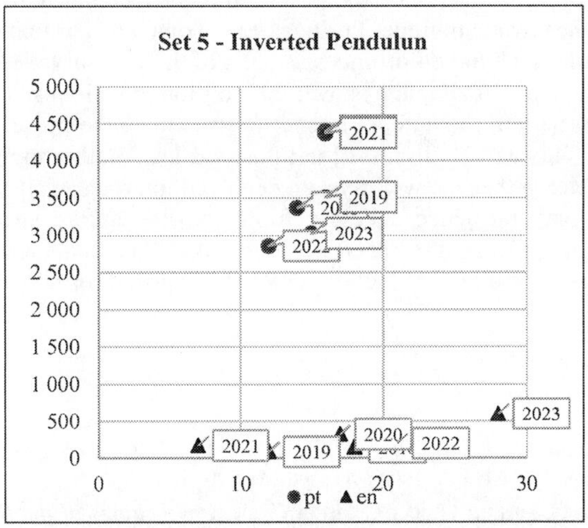


Figure 3. Numbers of countries and of accesses to the Inverted Pendulum objects in pt and en 2018-2023

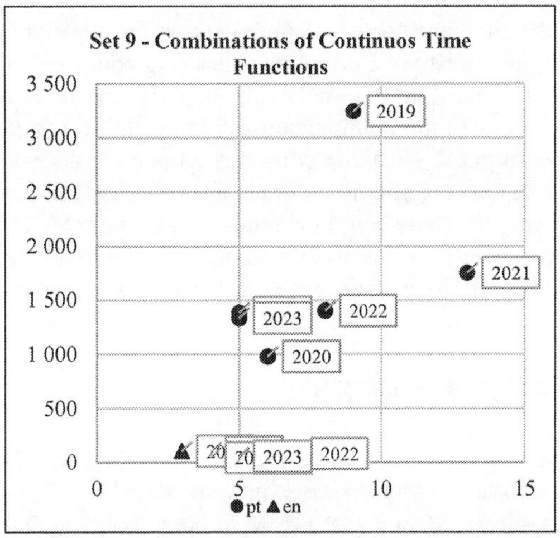


Figure 4. Numbers of countries and of accesses to the Combinations of Continuous Time Functions objects in pt and en 2018-2023

The first observation about Set 5 – Inverted Pendulum is that the numbers of accesses to the English version are much lower than the ones to the Portuguese original – the average number of accesses for Portuguese is 14 times the average for English. The high numbers of accesses in the six years under consideration are due to Brazil. This can easily be seen by averaging the number of accesses from Brazil as presented in the dataset; this number is 3,367.33. On the other hand, the average number of countries that accessed the objects is slightly higher for the English version. A possible explanation for these numbers is that the Inverted Pendulum, in general, is addressed in Control Systems courses in Brazil. Concerning Set 9, it is clear that both the numbers of countries and the numbers of accesses are higher for the Portuguese original. Once more, the higher numbers are due to Brazil. Combinations of functions, though necessary in many courses, is mainly focused in Systems & Signals courses.

### 3.2.3 Comments on the Four Sets

When the four sets are examined together, it is clear that the numbers of accesses to the CS simulators are higher regardless of where they originated from. If the numbers of countries are compared, the differences are not significant except in Set 9.

Accesses from Brazil to the English versions happen in significant numbers. This can easily be verified examining the dataset.

## 4. FINAL ANALYSIS, COMMENTS AND CONCLUSION

The numbers presented and discussed in this work were gathered from the Maxwell System using Open Access applications. The dataset was organized and made available in Open Access to make analysis and understanding easier.

The data present a complex picture of user behavior and preferences. While Portuguese content seems to be more popular overall, there are variations across sets and languages that warrant further investigation. Additionally, factors such as the use in undergraduate courses and impact factor of the repository over the world may influence these patterns and should be considered in a more comprehensive analysis.

Another factor that may have influenced data is the COVID pandemic. From 2020 to 2021 or 2022, many universities switched to online learning and online materials may have become more useful.

The Research Question seems to be answered, revealing differences in access patterns, both in terms of the number of accesses and the countries of origin, between simulators in Portuguese and their English versions. Additionally, variations in access numbers are noted across different topics addressed by the simulators, with CS emerging as a favored subject.

The CPLP – Community of Portuguese Language Countries (https://www.cplp.org/) includes Angola, Brazil, Cape Verde, Guinea-Bissau, Equatorial Guinea, Mozambique, Portugal, São Tomé and Príncipe, and Timor-Leste as Portuguese-speaking countries, all of which have Portuguese as one of the official languages. It is important to remark it is the only official language of Brazil that has over 200 million inhabitants.

Given the abundance of data available for examination, the authors intend to investigate the presence of Portuguese-speaking countries among the accesses. This will be discussed in future works. Another topic that will deserve attention comes from the fact that last March the Maxwell System deployed interfaces that can be switched between Portuguese and English. This may influence accesses from non-Portuguese speaking countries. Further work will analyze this. And last but not least, it is necessary to add more data to the analysis. This will be accomplished as time goes by, since it is necessary that the Simulators be exposed for more time.

## REFERENCES


ADL. (2004). *ADL Guidelines for Creating Reusable Content with SCORM 2004"*, July 2008, available https://adlnet.gov/assets/uploads/ADLGuide2004_Final_073108.pdf. Last accessed May 2022.

Alsubaie, M. and Alshawi, M. (2009). Reusable Objects: Learning Object Creation Cycle, *Proceedings of the Second International Conference on Developments in eSystems Engineering*, 2009, pp. 321-325, doi: 10.1109/DeSE.2009.63. Last accessed May 2022.

Andone, D., Mihaescu, V., Vert, S. Ternauciuc A. and Vasiu, R. (2020). Students as OERs (Open Educational Resources) co-creators, 2020 IEEE 20th International Conference on Advanced Learning Technologies (ICALT), Tartu, Estonia, 2020, pp. 34-38, doi: 10.1109/ICALT49669.2020.00017.

Institute of Electrical and Elesctronics Engineers. (2002). *IEEE Standard for Learning Object Metadata*, in *IEEE Std 1484.12.1-2002*, vol., no., pp.1-40, 6 Sept. 2002, doi: 10.1109/IEEESTD.2002.94128. Last accessed in May 2022.

Lynch, C. (2003). Institutional Repositories: essential infrastructure for scholarship in the digital age, ARL Bimonthly Report, nº 226, USA, 2003.

Pavani, A.M.N and Temporão, G.P. (2024). Comparative Accesses to Simulator OERs in Portuguese and their English Versions Available from Open Educational Resources@PUC-Rio – Extended Set and Time Frame, doi.org/10.17771/PUCRio.ResearchData.65826.

Statista. (2023). The Most Spoken Languages Worldwide in 2023, https://www.statista.com/statistics/266808/the-most-spoken-languages-worldwide/. Last accessed in January 2024.

Wang, T. and Towey, D. (2017). Open educational resource (OER) adoption in higher education: Challenges and strategies, 2017 IEEE 6th International Conference on Teaching, Assessment, and Learning for Engineering (TALE), Hong Kong, China, 2017, pp. 317-319, doi: 10.1109/TALE.2017.8252355. keywords: {Open Educational Resources;Conferences;Training;Government;Production;Technological innovation;Open Educational Resources (OERs);barriers to change;higher education}

Wiley, II, D.A. (2000). *Learning Object Design and Sequencing Theory*, PhD dissertation presented at Brigham Young University, United States, June 2000, available https://opencontent.org/docs/dissertation.pdf. Last accessed May 2022.

# TEACHING ACADEMIC WRITING IN THE ERA OF LARGE LANGUAGE MODELS: A CASE STUDY

Andrea Molinari[1] and Elena Molinari[2]
[1]*Dept. of Industrial Engineering University of Trento, Trento 38122, Italy*
[2]*TUI Musement, Milan, 20139, Italy*

**ABSTRACT**

This study delves into incorporating Large Language Models (LLMs) into academic writing education, focusing on a case study involving Bachelor's degree students. In an era where artificial intelligence (AI) significantly influences educational methodologies, our paper explores the effective use of LLMs in advancing students' academic writing abilities. A study was conducted with undergraduates in an academic writing course spanning three academic years. The primary assignment was to create a minimum three-page paper, including a literature review on a self-selected topic with high-quality references. The experimental approach involved comparing students' papers (n=24) with versions generated by LLMs (OpenAI ChatGPT-4 and Jenni), followed by a blind review to assess the quality of both the student-written and LLM-generated papers, using identical evaluation criteria. Preliminary findings suggest that in the specific domain of literature review and academic writing, (a) papers created by LLMs scored better in most if not all, the evaluation areas compared to their human-written counterparts and (b) in the review process, LLM-generated papers were indistinguishable from those written by humans. The study also discusses challenges like maintaining academic integrity and nurturing critical thinking skills in an AI-assisted learning environment.

**KEYWORDS**

Artificial Intelligence, Teaching, Academic Writing

## 1. INTRODUCTION

This article presents the results of an experiment grounded on specific insights regarding the teaching and evaluation methodologies of a traditional academic writing course within scientific and economic faculties and the impact of artificial intelligence tools, specifically Large Language Models and Generative Pretrained Transformers, on generating academic essays. Integrating advanced tools such as Large Language Models (LLMs) presents opportunities and challenges in the rapidly evolving educational technology landscape. There is at least two decades-long research on implementing technology in the classroom, where results have been equivocal, with examples of technology-based interventions reporting significant, positive effects (Englert et al., 2007; Warren et al., 2008) and others reporting minor or adverse effects (e.g., Rowley & Meyer, 2003; Goldenberg et al., 2011). Several studies have focused on AI-assisted language learning tools to improve English language learners' writing skills (Little et al., 2018), and this is an immediate usage of these technologies due to their excellent performance (especially in the English language) in providing better and richer text than average students of any order (Liu et al., 2021, Wu et al., 2021, Fitria, 2023).

A wealth of research has delved into the impact of AI on improving English language learning outcomes (Sun et al., 2021) or in technical disciplines (Huang A. Y. et al., 2023). For instance, Huang X. et al. (2023) conducted a study comparing students' academic performance and engagement levels in an AI-enhanced course versus a traditional one. Results indicated that those in the AI course achieved higher academic success and engaged more actively in their learning tasks than their peers in the non-AI course. Other studies follow the same line by examining the role of AI-assisted language learning tools in enhancing the writing skills of English language learners (Liu et al., 2021; Seufert et al., 2021; Hsiao and Chang, 2023; Yan, 2023). These studies explore how AI-supported language learning influences the writing abilities of English as a Foreign Language (EFL) learners, finding a substantial positive effect. Yan (2023) assessed the impact of ChatGPT, an AI-based language learning tool, on EFL learners' writing proficiency and observed notable enhancements in their writing due to AI-assisted learning. In today's digital era, characterized by the widespread accessibility of technology and online resources, learners can effortlessly practice their language skills, mainly writing, at their convenience and from any location (Yan, 2023). This practice is increasingly supported by sophisticated artificial intelligence (AI)-enabled computer and mobile applications. These applications provide interactive and tailored resources for refining writing skills and improved motivation (Jiang, 2022; Meunier et al., 2022; Yan, 2023).

Other approaches involve incorporating AI tools alongside traditional teaching practices, such as discussions, drafting, revising, and providing formative feedback. By incorporating AI tools into these existing practices, instructors can support students in developing linked thinking and writing skills while leveraging the benefits of AI assistance in areas such as generating alternative versions of the text and assessing the AI output for relevance, accuracy, completeness, bias, timeliness, and writing quality (Stanford University, 2023). The sources also highlight the need for students to learn how to effectively use AI tools in academic writing tasks, including setting directions and prompts and assessing the AI output (Song & Song, 2023). Instructors are expected to provide explicit instruction on the appropriate application of AI tools in academic writing tasks and continuously monitor their impact on students' writing skills. Chong (2021) highlights the importance of balancing AI assistance and students' independent thinking and engagement in the writing process. The findings are in line with other studies that suggest an integration of AI tools, specifically the use of the equipped AI-powered writing tool, to positively affect English second postgraduate students' academic writing performance, thus contributing to their overall improvement in English writing proficiency (Song & Song, 2023).

However, it is essential to note that a complete reliance on AI for writing tasks may lead to a reduction in accuracy and in writing autonomous abilities. Therefore, instructors must balance AI assistance and students' critical thinking and engagement in the writing process. The studies mentioned emphasize the need for a structured learning environment incorporating formative feedback, drafts, and revisions to support students' writing development. By adopting these essential teaching practices, instructors can effectively navigate the integration of AI tools into their academic writing courses, promoting students' critical thinking skills while harnessing the benefits of AI technology (Stanford University, 2023), and enhancing their pedagogical strategies. Students must learn to set directions, prompt the AI effectively, assess the AI outputs, and make necessary revisions and adjustments to improve their writing.

This paper aligns with the previously presented research lines. However, it approaches the research questions with a qualitative analysis of academic texts produced by bachelor students for an academic writing course. It compares these texts with texts produced by AI tools with the same title and abstract. It represents the summary of almost one year of comparison between human-generated academic texts and AI-generated academic texts. It explores the effectiveness of LLMs in a practical setting, focusing on a case study involving Bachelor's degree students enrolled in an academic writing course. We critically examine the role of LLMs, specifically OpenAI ChatGPT-4 and Jenni, in enhancing academic writing skills among undergraduate students. The study spanned three academic years, offering a substantial temporal frame to gauge the impact of LLMs on the academic writing prowess of these undergraduates. Central to the research is an experimental approach where the outputs of top-performing students (n=20) are juxtaposed with those generated by LLMs. This comparison, grounded in a blind review process using uniform evaluation criteria, seeks to assess the quality and distinguishability of student-written and LLM-generated academic papers, particularly in literature review. The paper is organized as follows: chapter 2 illustrates the methods used to collect and analyze academic texts produced by students and then generated with AI tools. Chapter 3 summarizes the most relevant results, considering the qualitative nature of this research. In Chapter 4, we present some preliminary considerations on the collected data, and finally, in Chapter 5 we draw some conclusions to be further investigated in future research.

## 2. METHODS

The study included 24 students from different European countries enrolled in three courses related to Academic Writing, in scientific (Computer Science / Engineering) and economic (Economy and Management) Bachelor's degree programs in two different Italian Universities. The course encompasses instruction in the techniques for composing the main components of a scientific paper: abstract, introduction, research methods, results, discussion, and conclusions. For each of the main components of an article, the course introduces construction methods without delving specifically into linguistic aspects, as it is a "technical" course rather than an English language course. For instance, the CARS model (Swales, 1990, Sánta, 2015). is employed for writing the introduction, dividing the writing process into various logical steps that culminate in the final product.

The final exam involved creating a literature review on a topic the student chose, composed of at least three pages according to a standard conference/journal template, with no figures, tables or any other graphical element. The evaluation focused not so much on the quality of the chosen scientific content but instead on the alignment between the academic writing techniques taught in the course and the work produced by the student. The students' final work was meticulously checked using the university's anti-plagiarism tools, which, until November 2022, were very effective in verifying the correctness and originality of the paper and the absence of plagiarism.

The evaluation criteria for the submitted paper are represented in Table 1, with the weights expressed in thirtieths. The student passed the exam with a score of 18/30. The evaluation was focused on two main elements: the techniques used to write the various sections (title, abstract, introduction, conclusions and references) and the use of academic language in the paper (vocabulary, grammar, cohesion, and coherence of text). The sum of weights is 31 to provide a possibility of the "cum Laude" score.

Table 1. Evaluation criteria for the papers

| Title | Abstract | Intro | Conclusions | References | Vocabulary | Grammar | Cohesion and coherence | Total |
|---|---|---|---|---|---|---|---|---|
| 1 | 3 | 8 | 3 | 2 | 4 | 4 | 6 | 31 |

The final results, obtained from the sum of the scores, are rounded up to the nearest whole number, and the grade is expressed in multiples of 0.5.

Because of the nature of the requested paper (a literature review) and in order to allow students to exert an effort proportional to the number of credits acquired (3 or 6 ECTS), sections like "methods", "results" and "discussions" were excluded from the final paper. Papers for the experiment have been collected in the past 3 years before the disruptive advent of LLM on the market to not condition students in their writing and to have reasonable assurance that the papers were not produced using ChatGPT, Jenni, or similar tools. The papers (n=24) have been selected in order to represent three classes of results, with eight papers for each class: a) fair (21 <= grade <=24), b) good (24 <= grade <=27), c) very good (28 <= grade <=31). Grades below 21 have not been considered, as these results are usually represented by students whose final essays are often of low quality and easily improvable. Many students of this type choose to retake the exam to enhance their score, so most of the type of paper is an "approach" to the exam rather than their final essay, and for this reason, they have not been included in the analysis.

For each grading category within the class, papers are distributed equally among the four different scores in the interval. This approach ensured a comprehensive analysis encompassing the full spectrum of student performance, from those achieving the least to those achieving the most and those representing the typical or average level of achievement within each grade category. An initial group of 80 students was selected and divided into the three mentioned classes. Students' papers have been associated with the respective scores in the various sections assigned by the teacher. Authors' names were removed from their articles, and from these 80 anonymized articles, the three classes were randomly composed to prevent any association between a student and their paper.

After the composition of the sample of papers, we submitted the paper's title and abstract to the AI tools. Through a series of interactions with the AI tools using a standard prompt, "enlarge and rephrase the previous text," we have generated a new paper of approximately the same length and content as the original one. Twenty-four new papers have been produced and blindly evaluated with the same criteria for the students' papers. Regarding time dedicated to recreating the original paper using LLMs, one-twentieth of the formally expected time was allocated for such an article. The ECTS mechanisms stipulate a student effort of 25 hours for each ECTS, so the anticipated workload corresponds to approximately 100 hours. Deducting the hours spent on lectures from this workload, a student has roughly sixty hours left to dedicate to composing the paper. In creating the new paper using artificial intelligence tools, about three hours were allocated for generating each article's title, abstract, introduction, conclusions, and references, thereby limiting the number of interactions for subsequent refinements within this time frame. The result produced by the artificial intelligence engines was then evaluated, and we present the results of the comparison between the grade obtained by the student and the grade that would have been obtained had the student used an AI engine and one-twentieth of the anticipated time.

## 3. RESULTS

Table 2 below is a summary of the most relevant data we collected. Data is presented for each student cohort concerning the various components of the required academic article, encompassing both the technical and linguistic aspects. A final column is included, showcasing the average outcomes achieved by the students in their articles alongside the results derived from the evaluation of articles generated by AI. Cells, where the AI yielded results inferior to student-generated texts are highlighted with color.

Table 2. summary of results of student-/AI-generated texts (C3=Very Good, C2 = Good, C1 = Fair)

| | Title | | Abstract | | Intro | | Conclusions | | References | | Vocabulary | | Grammar | | Cohesion coherence | | Class Students Avg Score | Class AI avg score |
|---|---|---|---|---|---|---|---|---|---|---|---|---|---|---|---|---|---|---|
| C3 | 0,9 | 0,7 | 3,0 | 2,8 | 7,4 | 7,9 | 2,8 | 2,9 | 2,0 | 1,9 | 4,0 | 3,8 | 3,9 | 3,8 | 5,6 | 5,9 | 29,8 | 29,9 |
| C2 | 0,9 | 0,7 | 1,8 | 2,8 | 6,3 | 7,9 | 2,8 | 2,6 | 1,7 | 1,7 | 3,6 | 3,8 | 3,4 | 3,9 | 5,3 | 5,8 | 26,0 | 29,4 |
| C1 | 0,8 | 0,8 | 1,8 | 2,6 | 4,1 | 7,8 | 2,6 | 2,7 | 1,3 | 1,6 | 3,3 | 3,8 | 3,3 | 3,8 | 5,3 | 5,8 | 22,5 | 29,0 |

It is observed that the texts generated by AI are of inferior quality compared to those produced by the students in the title section across all three student categories. The inferiority of the AI-generated texts is relatively consistent (albeit marginally) in the texts produced compared to the excellent students. At the same time, the AI generates superior texts in virtually all other sections for all other student categories.

The most relevant difference between human-generated and AI-generated texts in favor of humans is in the quality and appropriateness of the title and the references. Here, students' final scores are superior or similar to those of AI-generated text. The techniques and tips taught in academic writing contexts appear to be sufficient to maintain a minimal difference. Due to the uniqueness of the results compared to the other sections, we manually analyzed these two sections of the papers. For titles, this is probably a matter of the evaluators' taste and the time devoted to the AI-generated text. Three hours were devoted to generating AI texts, and a relatively minimal portion of the time was devoted to the composition of the title, evidently in proportion to its impact on both the grade and the structure of the scientific essay. Conversely, students, recognizing the title's significance as a tool for the academic communication and promotion of their work, almost certainly allocated more attention to it. This heightened focus could account for the markedly superior outcome achieved by human effort as compared to AI.

Concerning references, those generated by students were generally more appropriate and connected to the sentences they referred to. This aspect, as expected, mainly happened in the top-scored papers, where evidently, the quality of the student's research and writing and the consequent overall quality of the article were closely linked to the quality of the bibliographic references analyzed and cited. The citation style was without mistakes from AI, and the references generated by AI were appropriate. However, from a scientific point of view, students' generated references (in top papers) were more connected to the context and more interesting for the reader. Due to the sample size constraints, in Table 3, we present a few simple statistical elements, with a preference for a more detailed and qualitative interpretation. A more substantial statistical significance is deferred to subsequent, more in-depth analyses.

Table 3. simple stats on final results

| | AI Avg improvement | AI Avg % improvement | AI Min Improvement | AI Max Improvement |
|---|---|---|---|---|
| Class 3 - Very good | **0,1** | **0,65%** | **-6,45%** | **10,71%** |
| Class 2 - Good | **3,4** | **13,08%** | **7,41%** | **20,00%** |
| Class 1 - Fair | **6,5** | **29,18%** | **20,83%** | **42,86%** |

Here, we can readily corroborate the findings presented in Table 2, noting a significant increase in improvement as the quality of the student-generated papers decreases. As hypothesized, the students with the poorest performance would benefit tremendously from using AI-generated texts in terms of their final grade, increasing their scores by approximately 30%, with peaks nearing 43%. This essentially amounts to an advancement of two grade categories.

## 4. DISCUSSION

The research questions addressed in this study were predominantly qualitative, bordering on foundational inquiries about the essence of teaching specific disciplines and evaluating outcomes using traditional methodologies. A primary question centers not exclusively on the performance of AI engines in generating academic texts but on the level of text creation required for Bachelor's level students unaided by additional tools. The quality of the outcomes produced by generative pre-trained transformers (GPTs) models, trained explicitly for academic writing, is becoming increasingly evident and impressive. The main goal, however, was to identify the areas where LLMs provide significant improvements and understand and enhance educational processes.

For this, it is primarily evident that students with lower performance scores benefit the most from using artificial intelligence tools. This is predominantly true for the lengthier sections of an academic paper, such as the introduction, abstract, and conclusions. The longer the text, the greater the advantage compared to what a student produces natively. As highlighted by previous studies, the linguistic aspect is a significant area of improvement. The term 'linguistic area' does not solely refer to the use of the English language but also encompasses vocabulary, grammatical structures, and coherence and consistency in academic language. Generative models, also trained using scientific articles, produce results in this particular language and mode of expressing scientific thought that is decidedly superior to what a student can learn in a traditional academic course.

Another intriguing finding from our sample pertains to the distinct net improvement across the three categories. The sole instances of performance deterioration, relative to the grades assigned by the instructor, involve the three top students. We have recorded only one case of identical grading (albeit with slightly varied weighting), whereas, in the remainder of the sample, the outcomes of the AI-generated texts are invariably superior, if not significantly so. The inference here is relatively straightforward: those who benefited most in terms of grading are the ones who initially received the lowest scores. However, examining the specifics of the areas where the AI underperformed is particularly noteworthy compared to the "human-generated" papers. This phenomenon was observed among the students in the C1 category (grades from 28 to 31), and what is intriguing is the detailed analysis of the aspects where AI fell short.

Given the lengthy process of recreating a document of equivalent length for each selected paper, we know that quality issues may compromise these data. Nevertheless, the fact that the AI system has, on average, penalized top-performing students (<10%), and subsequently demonstrated equal or vastly superior performance compared to over 90% of the students corroborates our foundational hypothesis. With a large student cohort, the assessment of their prowess in academic writing is impracticable in an offline setting. This element does not diminish the validity of AI tools in supporting educational activities, nor does it significantly augment the considerable apprehensions surrounding this synergy. It can be asserted that, within this sample and context, all indications suggest a radical shift in the perception of learning assessment through producing written material non-face-to-face. The impact is particularly pronounced on traditional education, homework assignments, and educational approaches involving content creation processes. Our findings reveal that AI performs these tasks in a remarkably superior fashion compared to human efforts, too, in a fraction of the time. Notably, this enhancement is substantial for most participants, culminating in an improvement for students with lower scores ranging from 20% to over 40%.

Another line of inquiry concerned the validity of student evaluation methods used at various educational levels, from elementary school to Bachelor's degree programs. An additional expected contribution was to contemplate whether it still makes sense to allow students to work unsupervised on creating an essay, which could be more effectively accomplished by artificial intelligence. The results of our limited research are self-evident. Suppose we aim to ensure a fair evaluation that certifies a student's knowledge and skills in a specific subject. In that case, the methods of content production not supervised by an instructor do not guarantee this outcome. Despite being under public pressure to provide methods of differentiation between human-generated and AI-generated content, even the developers of LLMs have not been able to guarantee such results. Even with some cases of "hallucinations" by AI, it is impossible to distinguish a text generated by a human from one generated by artificial intelligence with certainty in areas such as scientific production linked to university courses, leading one to wonder if it will ever be feasible due to the continuous progress of these AI tools.

Various educational institutions at different levels have underestimated this phenomenon, primarily due to some educators' unawareness of the potential use of these tools for generating content. It is still common to find course syllabi where the final grade is derived from projects, papers, or texts that students can independently create alone or in groups. In some contexts, this task becomes essentially replicable by an AI engine, producing results inconceivable within the time and effort constraints set by the credit system. This leads to invalidated results and can discourage students who choose not to use such tools and instead wish to gauge their capabilities, causing widespread demotivation. The established "hourglass" structure for academic essays encompasses several sections where AI can substantially contribute, thereby rendering even long academic essays (like theses) highly susceptible to the technical and ethical dilemmas highlighted in this brief study. Given the number of ECTS credits allocated to theses and considering the inevitable prolonged timeframe typically involved in thesis production, we question whether the traditional form of a thesis, as known for decades, still holds relevance. The technical answer is somewhat predictable: it is nonsensical to envision a thesis where 40 to 80% of the content can be AI-generated with minimal student involvement. This involvement is certainly not commensurate with the expectations set by current evaluation mechanisms, particularly concerning the "originality of contribution" expected by supervisors and thesis committees. This issue is, as of now, insufficiently acknowledged.

Therefore, this research's preliminary findings illuminate our perspective and research questions. We were able, quite simply, to produce scientific papers of a lower-medium level in one-twentieth of the theoretical time that consistently outperformed those produced by humans. Moreover, it was mainly the less effective and efficient students who showed the most significant improvement. Concerns were raised about demotivation and, on the other hand, about continuing their growth if an AI tool can do this with practically no effort, suggesting that LLMs match and, in some instances, surpass human efforts in generating academic texts. Furthermore, the study reveals that in the blind review process, papers crafted by LLMs were mainly superior in terms of academic quality to their human-written counterparts. This revelation opens up a discourse on the implications of such technologies in academic settings, particularly in maintaining academic integrity and fostering critical thinking skills.

Moreover, we must delve into educators' evolving role in an AI-assisted educational landscape. Generative AI technologies are altering educational methodologies, content creation, and evaluation processes, especially in disciplines profoundly affected by these advancements. We need more contributions to understanding AI's impact on education and a profound re-evaluation of pedagogical strategies in the face of these disruptive technologies.

## 5. CONCLUSIONS

This paper examined the effectiveness of integrating AI-assisted tools in improving learners' writing performance and efficiency in academic contexts. The study also addressed teachers' concerns about potential adverse effects on academic writing skills. It emphasized the importance of providing instruction on the appropriate application of AI tools for academic writing tasks. The study found that using AI-assisted tools could significantly impact learners' writing performance and efficiency, as AI-generated texts can provide many suggestions for improvement. At the same time, as generated text has been demonstrated to be remarkably better than students' text, this invalidates any previous model of examining students on an autonomously produced paper.

The study highlights the importance of integrating AI tools within a structured learning environment that includes discussion, drafting, and revision opportunities to support students' development of linked thinking and writing skills. Furthermore, we suggest that teaching practices incorporate formative feedback from instructors and peers, encouraging students to further engage in drafts and revisions to enhance their writing skills. Teaching academic writing with the integration of AI tools can improve students' writing skills while increasing their interest in the subject. The study also acknowledges the concerns raised by authors regarding possible detrimental effects on students' academic writing skills over time. To address these concerns, the study recommends providing explicit instruction on the appropriate use of AI tools and continuously monitoring and assessing their impact on learners' writing skills over an extended period. The case study demonstrated the potential benefits of using large language models and AI-assisted tools in teaching academic writing.

Future developments will primarily aim at an extended involvement of students in producing, comparing, and commenting on AI-generated text, expanding the sample size used to strengthen the results presented in this article statistically. Regrettably, the creation of academic texts related to literature reviews no longer seems to be a teaching possibility for evaluation. Large Language Models, especially for English non-native students, generate texts of good, if not excellent, linguistic quality, rich in appropriate references suitable for a bachelor's level. In the future, the aim is to elevate the level of analysis for master's/PhD students on more specific topics and parts of scientific articles that are not merely a literature review.

## REFERENCES

Chong, D. (2021) Research on Artificial Intelligence-based English Writing Blended Teaching Mode. Available at: https://doi.org/10.1088/1742-6596/1852/3/032018.

Englert, C. S., Zhao, Y., Dunsmore, K., Collings, N. Y. & Wolbers, K. (2007) Scaffolding the writing of students with disabilities through procedural facilitation: Using an Internet-based technology to improve performance, Learning Disability Quarterly, 30(1), 9–29.

Fitria, T. N. (2021). Grammarly as AI-powered English writing assistant: students' alternative for writing English. Metathesis J. Eng. Lang. Liter. Teach. 5, 65–78. doi: 10.31002/metathesis.v5i1.3519

Goldenberg, L., Meade, T., Cooperman, N. & Midouhas, E. (2011) Impact of a technology-infused middle school writing program on sixth-grade students' writing ability and engagement, Middle Grades Research Journal, 6(2), 75–97.

Hsiao, J. C., and Chang, J. S. (2023). Enhancing EFL reading and writing through AI-powered tools: design, implementation, and evaluation of an online course. Interact. Learn. Environ. 1-16, 1–15. doi: 10.1080/10494820.2023.2212716

Huang, A. Y., Lu, O. H., & Yang, S. J. (2023). Effects of artificial Intelligence–Enabled personalized recommendations on learners' learning engagement, motivation, and outcomes in a flipped classroom. Computers & Education, 194, 104684.

Jiang, R. (2022). How does artificial intelligence empower EFL teaching and learning nowadays? A review on artificial intelligence in the EFL context. Front. Psychol. 13:1049401. doi: 10.3389/fpsyg.2022.1049401

Little, C.W., Clark, J.C., Tani, N.E. and Connor, C.M. (2018), Context and Implications Document for: Improving writing skills through technology-based instruction: a meta-analysis. Rev Educ, 6: 202-204. https://doi.org/10.1002/rev3.3115

Liu, C., Hou, J., Tu, Y. F., Wang, Y., & Hwang, G. J. (2023). Incorporating a reflective thinking promoting mechanism into artificial intelligence-supported English writing environments. Interactive Learning Environments, 31(9), 5614-5632.

Meunier, F., Pikhart, M., and Klimova, B. (2022). New perspectives of L2 acquisition related to human-computer interaction (HCI). Front. Psychol. 13:1098208. doi: 10.3389/fpsyg.2022.1098208

Rowley, K. & Meyer, N. (2003) The effect of a computer tutor for writers on student writing achievement, Journal of Educational Computing Research, 29(2), 169–187.

Sánta, B. (2015). Introducing the Literary Critic: The CARS Model in the Introductions of Academic Papers in Literary Criticism. International Journal of English Language & Translation Studies. 3(2), 84-93. Retrieved from http://www.eltsjournal.org

Seufert, S., Burkhard, M., & Handschuh, S. (2021, November). Fostering Students' Academic Writing Skills: Feedback Model for an AI-enabled Support Environment. In Innovate Learning Summit (pp. 49-58). Association for the Advancement of Computing in Education (AACE).

Song, C., & Song, Y. (2023). Enhancing academic writing skills and motivation: assessing the efficacy of ChatGPT in AI-assisted language learning for EFL students. Frontiers in Psychology, 14.

Stanforf University (2023) AI Tools in Teaching and Learning. Available at: https://teachingcommons.stanford.edu/news/ai-tools-teaching-and-learning.

Sun, Z., Anbarasan, M., & Praveen Kumar, D. J. C. I. (2021). Design of online intelligent English teaching platform based on artificial intelligence techniques. Computational Intelligence, 37(3), 1166-1180.

Swales, J. (1990) "Genre Analysis". Cambridge University Press.

Warren, S. J., Dondlinger, M. J. & Barab, S. A. (2008) A MUVE towards PBL writing: Effects of a digital learning environment designed to improve elementary student writing, Journal of Research on Technology in Education, 41(1), 113–140.

Wu, L., Wu, Y., and Zhang, X. (2021). L2 learner cognitive psychological factors about artificial intelligence writing corrective feedback. Engl. Lang. Teach. 14, 70–83. doi: 10.5539/elt.v14n10p70

Yan, D. (2023). Impact of ChatGPT on learners in a L2 writing practicum: an exploratory investigation. Educ. Inf. Technol. 28, 13943–13967. doi: 10.1007/s10639-023-11742-4

# IMPLEMENTING BLENDED LEARNING AS A CULTURALLY RESPONSIVE TEACHING METHOD TO THE NEEDS OF STUDENTS WITH CULTURAL AND LINGUISTIC DIVERSITY

Vaso Anastasiou, Charoula Angeli and Yiota H. Diakou
*University of Cyprus, Cyprus*

**ABSTRACT**

The present action research analyzed the learning environment during the implementation of Blended Learning (BL) in fourth-grade students at a public school in Cyprus, with an emphasis on the needs of children with cultural and linguistic diversities. The research spanned three academic years and included a quantitative analysis of academic performance and a qualitative analysis based on the educator's reflective comments and artifacts from the use of digital tools. Although the initial adoption of BL created disorganization, it ultimately highlighted the potential of the method, offering a flexible and supportive learning environment for students from marginalized groups. The researcher observed that the changes in the learning environment affected learning based on the opportunities offered for interaction, for creating a sense of collectivity, and for personal development and improvement.

**KEYWORDS**

Blended Learning, Cultural and Linguistic Diversity, COVID-19

## 1. INTRODUCTION

The COVID-19 pandemic rapidly transformed the world, coming at a time of demographic change and social upheaval. These changes significantly affected educational practices, increasing the demand for technology integration more than ever (Singh et al., 2021; European Commission, n.d.). Additionally, migration waves have created schools and classes with cultural and linguistic diversity. Equal participation of all students necessitates the adoption of multicultural education that embraces appropriate social learning environments (Parkhouse et al., 2022; Bennett et al., 2018). This research focuses on BL as a teaching method integrating technology in a classroom setting with students of diverse cultural backgrounds.

BL combines face-to-face teaching with online activities (Staker & Horn, 2012). It has gained ground in the educational field due to the dynamism it offers, combining tried-and-tested traditional methods with innovative approaches (Dziuban et al., 2018; Rasheed et al., 2020). Although it offers advantages such as flexibility and autonomy in learning (Li & Wang, 2022; Rasheed et al., 2020), challenges are also highlighted, such as the need for self-regulation and digital proficiency (Van Laer & Elen, 2017; Boelens et al., 2017).

Culturally Responsive Education (CRE), on the other hand, focuses on integrating cultural diversity into teaching practices. It centers on all those children 'bring with them' into the classroom, not on the 'deficit' and what they cannot do (Parkhouse et al., 2022; Bennett et al., 2018). This approach encourages equal participation of students from various cultural backgrounds, increasing academic performance and empowerment of all students, particularly those from historically marginalized groups. However, there is a need for more extensive training of educators in CRE practices and for them to acquire the necessary knowledge and skills to utilize and integrate the cultural capital of their students (Parkhouse et al., 2022).

The present research studies the learning environment during the implementation of BL in fourth-grade students (aged 9-10 years) in a public urban school of Cyprus, examining how this environment responded to the needs of students with cultural and linguistic differences. Specifically, the research questions are as follows:

- How did BL affect the learning environment of fourth-grade elementary students during the Covid-19 pandemic?

- Did the implementation of BL differently affect the performance of fourth-grade elementary students with diverse cultural/linguistic backgrounds compared to the performance of students with the dominant cultural/linguistic background during the Covid-19 pandemic?

The existence of a research gap in investigating the impact of BL within the context of learning environments in Cypriot primary education underscores the significance and necessity of this research. The findings of the research have both practical and theoretical importance. Practically, they provide guidelines for the successful implementation of BL in elementary schools, especially in classes characterized by cultural and linguistic diversity. Theoretically, the research seeks to explore the challenges and outcomes of BL, contributing to the clarification of research gaps and understanding BL's response to the needs of all students, including those with diverse cultural and linguistic backgrounds.

# 2. LITERATURE REVIEW

## 2.1 Blended Learning

BL is defined as the combined use of traditional face-to-face teaching with physical presence in educational buildings and online learning. A fundamental principle of BL is that these two modes of teaching are interconnected and complementary, ultimately offering students some degree of control over the time, place, path, and/or pace of their learning (Staker and Horn, 2012; Christensen et al., 2013).

Through the study of existing BL programs, Staker and Horn (2012) categorized four primary models: Rotation, Flex, Self-Blend, and Enriched-Virtual. These models reflect different approaches to implementing BL, with each category addressing different educational needs (Staker and Horn, 2012; Christensen et al., 2013). For example, in the Self-Blend model, students individually alternate between fully online courses and face-to-face teaching, while in the Enriched Virtual model, the rotational program involves all school students combining classroom attendance with online teaching.

During the COVID-19 pandemic, BL was adopted as an emergency measure, with many educators adapting to the new reality and new methods of teaching and learning (Singh et al., 2021). This experience led to the utilization of tools for ensuring the quality of BL, such as frameworks and rubrics. These tools are useful at all stages of BL courses, from planning to implementation, execution, and redesign. An example is the grade-level rubrics offered by the organization Quality Matters (Quality Matters, n.d.). Consequently, the pandemic provided an opportunity to highlight the advantages of BL, paving the way for further digital integration.

## 2.2 Education in a Multicultural Society

Efforts to support educators with appropriate pedagogical practices for multicultural classes date back thirty years, with rich research material in the field of multicultural education and various theoretical approaches (Bennett et al., 2018). Regardless of the theoretical approach, all aim to respond to the needs of each student, incorporating their cultural background and experiences into the learning process. The literature review focuses on the range of these approaches, from the perspective of CRE to theories that emphasize the social dimension of learning, highlighting the complexity of the sociocultural dimension of education.

### 2.2.1 Culturally Responsive Education

CRE has been influenced by many researchers and theorists who have contributed to the development and investigation of this educational approach. Gloria Ladson-Billings (1995) is among them, identifying both the empowerment of collectivity and the creation of a caring school climate as cornerstones of Culturally Relevant Pedagogy. These elements support not only students' academic performance but also their overall well-being Geneva Gay (2002) outlines the fundamental characteristics of culturally responsive teaching. Additionally, Additionally, Villegas and Lucas (2002) identify the essential characteristics that culturally responsive educators must possess. Other researchers (Bennett et al., 2018; Gunn et al., 2021) discuss the fundamental characteristics and practices that interconnect to support multicultural classes for young learners. Integrating theories from the field of CRE with other disciplines, de Groot et al. (2023) highlight an approach focused on creating bonds within the school environment. This approach recognizes the importance of fostering a close connection between individuals and their surroundings.

### 2.2.2 Learning as a Social Experience

Examining the micro-level of education, each classroom is defined as a unique community with its own culture, socially constructed by its members. This culture emerges from the teaching practices, values, rules, and interpersonal relationships formed within the classroom (Levy, 2007).

Thus, within the school, each student, as a member of a class 'community,' has the right to integrate and participate in its activities and practices. Wenger (1998, 2009) defines these communities as 'Communities of Practice,' within which members share a common interest and engage in a process of collective learning. Three levels of participation are recognized: the core, comprising members with leadership roles; the active group, engaging in regular participation; and the peripheral group, participating passively.

The perception and involvement of students in 'community' practices and interactions determine their social positions and influences within the classroom. The different social positions formed by student interaction within the classroom's social environment require educators to be sensitive and consider the imbalances that may arise during teaching. Interaction with classmates can either be beneficial for learning or, conversely, a hindrance, underscoring that learning opportunities are not equal for all students (Brown, 2007).

### 2.2.3 Towards a Learning Environment for All

In conclusion, it is evident that the unique learning environment of each class significantly affects the quality of learning and the achievement of educational goals. The class community functions as a learning community where the participation and interaction of its members are crucial for learning. Thus, these approaches identify common perceptions about the learning environment of the class:

a) The learning environment is influenced by the opportunities for interaction among members, as mentioned by Wenger (1998) regarding Communities of Practice, and by de Groot et al. (2023) in creating bonds within the class. Gay (2002) emphasizes the importance of creating communication channels with all students.

b) Opportunities for personal improvement and development are also crucial. Wenger (1998) highlights the relationship between contribution to knowledge and professional progress in Communities of Practice. Ladson-Billings (1995) emphasizes the importance of academic achievements and the well-being of children.

c) The learning environment is influenced by the creation of a sense of collectivity in the class community. Gay (2002) talks about creating learning communities, while Ladson-Billings (1995) refers to the empowerment of collectivity, as do Bennett et al. (2018). Wenger (1998, 2009) underscores the importance of knowledge contribution in Communities of Practice.

## 3. METHODOLOGY

An action research approach was employed for this study, involving three cycles. Each cycle's implementation was revised based on the findings from the previous one.

## 3.1 Participants

The researcher, a female elementary school teacher, was employed at a public school during the 2019–2020 school year, marking her 18th year of service in elementary education. Although she had experience integrating technology into teaching, she had not encountered remote online education before the implementation of the first research cycle.

Table 1. Research Participants by School Year

| | School year 2019-2020 | | School year 2020-2021 | | School year 2021-2022 | |
|---|---|---|---|---|---|---|
| | Total | non-native speaking families | Total | non-native speaking families | Total | non-native speaking families |
| Boys | 11 | 3 | 10 | 1 | 10 | 1 |
| Girls | 11 | 0 | 11 | 4 | 10 | 0 |
| Total | 22 | 3 | 21 | 5 | 20 | 1 |

As indicated in Table 1, the research involved a total of 63 students from three different 4th-grade classes at an elementary school, with the educator overseeing these classes over three successive school years, from the 2019–2020 to the 2021–2022 school years. Regarding the number of children with different cultural and linguistic backgrounds: in the 2019–2020 school year, there were three such children (S1, S2, and S3); in the 2020–2021 school year, there were five (S4, S5, S6, S7, and S8); and in the 2021–2022 school year, there was one child (S9).

## 3.2 Data Collection Methods

For the quantitative data collection on performance during the face-to-face teaching phase, the educator prepared identical written evaluation tests in Greek language and mathematics for all three classes. In the online teaching phase, grades from digital activities, also prepared by the educator, were recorded. Additionally, observations and reflective comments were noted in the 'Educator's Diary.' Material from the digital tools used, specifically ClassDojo, Microsoft Teams, and Google Classroom, was also collected.

## 3.3 Digital Tools

Throughout the three research cycles, the ClassDojo application facilitated communication between educators and parents and served portfolio functions.

In the first cycle, students engaged in practical exercises using school computers and tablets in small groups. For the online phase, personal devices were utilized alongside tools like Microsoft Forms, Google Forms, Kahoot challenge, Padlet, and Learning Apps.

In the second cycle, face-to-face instruction employed tablets on a 1:1 basis, incorporating additional practice tools such as WORDWALL, Quizizz, Liveworksheets, Mindmeister, and Pixton. Microsoft TEAMS was the platform of choice during the online phase.

The third cycle saw the integration of Google Classroom for face-to-face teaching as a learning management system, complemented by tools like Google Docs, Jamboard, and Scratch. Unlike previous cycles, online learning was personalized, not engaging the entire class. Instructions and materials, aligned with in-class activities, were made available to individual students through Padlet.

## 3.4 Research Procedures

The research included three investigative cycles, starting and ending with face-to-face teaching, while following different trajectories. During the first cycle, the school year began without an emergency plan, but with the onset of the COVID-19 pandemic, teaching continued online through asynchronous remote education. Enriched Virtual and Self-Blend models were used. In the second cycle, the year began with face-to-face teaching and continued online due to the pandemic following the Enriched Virtual model. During the third cycle, there was no school closure due to the pandemic. Self-Blend model was used.

## 3.5 Data Analysis

For the data analysis in this research, both quantitative and qualitative methods were utilized. The quantitative analysis aimed to assess the impact of BL on the academic performance of students from diverse cultural and linguistic backgrounds in comparison to their peers from the dominant cultural/linguistic group. The qualitative analysis explored the effects of BL on the learning environment amidst the COVID-19 pandemic and examined the teaching practices of the educator. This examination was based on entries in the educator's diary and the review of artifacts from the digital tools employed.

# 4. RESULTS

## 4.1 Statistical Analyses

The ANOVA analysis revealed no significant statistical differences in overall performance based on the academic year. However, the performance of students from Greek-speaking family environments ($\bar{X}$= 88.58, SD = 10.04) was higher compared to students from non-Greek-speaking environments ($\bar{X}$= 60.75, SD = 18.54). The use of an independent samples t-test revealed that this difference was statistically significant at $\alpha$= 0.05 ($t = 6.90$, $df = 61$, $p < .001$).

Table 2. Descriptive Statistics of Student Performance

| | Face to face teaching | | | | Online teaching/teaching with digital tools | | | |
|---|---|---|---|---|---|---|---|---|
| Student | $\bar{X}_g$ | SDg | $\bar{X}_m$ | SDm | $\bar{X}_g$ | SDg | $\bar{X}_m$ | SDm |
| S1 | 67.2 | 14.5 | 79.2 | 16.7 | - | - | - | - |
| S2 | 53.6 | 14.2 | 70.4 | 21.1 | - | - | - | - |
| S3 | 56.2 | 9.2 | 57.2 | 22.6 | - | - | - | - |
| $\bar{X}_{cycle\ 1}$ | 80.4 | 2.9 | 82 | 9.87 | - | - | - | - |
| S4 | 35.6 | 16.1 | 43.2 | 8.1 | 56.5 | 55.5 | 100 | 0 |
| S5 | 58.4 | 16.8 | 68.1 | 19.9 | 61 | 24.1 | 85 | 12.9 |
| S6 | 73.8 | 5.8 | 68.3 | 13.4 | 76.2 | 21.8 | 68 | 31.3 |
| S7 | 91.7 | 4.6 | 86.9 | 11.2 | 79.1 | 23.8 | 60 | 60 |
| S8 | 46.7 | 15.2 | 57.5 | 17 | 67.5 | 24.6 | 92 | 10.9 |
| $\bar{X}_{cycle\ 2}$ | 85 | 2.8 | 85.3 | 3.4 | 83.9 | 8.9 | 89 | 6.66 |
| S9 | 38.6 | 19.35 | 42.1 | 26.4 | 54.7 | 26.8 | 74.9 | 30.9 |
| $\bar{X}_{cycle\ 3}$ | 86.3 | 3 | 85.3 | 4.9 | 84.6 | 10.7 | 83.01 | 13.8 |

Note: g: Greek language, m: Maths

During the first cycle, students S1, S2, and S3 demonstrated below-average performance in the face-to-face teaching phase. Specifically, in written assessments for Greek and mathematics, their performances consistently fell below the class average, as detailed in Table 2. In the absence of written evaluations during the online teaching phase, student engagement was inferred from portfolio posts. S1 posted 39 times, aligning with the class average, whereas S2 and S3 posted only 7 and 11 times, respectively, indicating incomplete task completion. The content of their posts primarily consisted of photographs of their work in books and notebooks, lacking the production of additional digital content that could have fostered their creative and technological skills, unlike many of their peers.

In the second cycle, students S4, S5, S6, S7, and S8 generally had lower performances in Greek and mathematics, except for S7 who performed better in Greek. During the online phase, the students showed an improved picture with several performances above the class average, as indicated in Table 2

During the third cycle, student S9 performed below the class average in both Greek and mathematics. The quarantine period highlighted his challenges with self-regulation, as he failed to complete his assigned tasks. However, the integration of technology in face-to-face teaching proved beneficial for him, allowing his performance, particularly in mathematics, to approach the class average.

## 4.2 Analysis of Qualitative Data

The qualitative analyses presented next focus on analyzing the learning environment, considering opportunities for interaction, personal improvement and development, and collectivity in the class.

During the first cycle's face-to-face teaching phase, efforts were made to foster a learning community that supported interaction, personal development and improvement, and a sense of collectivity for all students, despite language barriers. This was achieved through student-centered activities, enriching the curriculum with non-formal activities, arranging desks in groups, and assigning specific roles in collaborative tasks, which all promoted open dialogue and idea sharing. The forced transition to online teaching due to the pandemic significantly limited opportunities for interaction and collectivity. Although many children maintained their engagement through tasks in digital portfolios, the opportunity for personal development was limited, especially for children with different family linguistic backgrounds. The return to face-to-face teaching in the classroom occurred under unprecedented conditions due to health restrictions and social distancing. Opportunities for interaction, collectivity, personal improvement, and development were significantly reduced.

In the second cycle teachers and students were prepared for a potential transition to online teaching. Opportunities for interaction in face-to-face teaching remained limited, with desks placed frontally and without collaborative activities. The digital skills of the children were enhanced using tablets, providing opportunities for personal development. The online teaching of this cycle included structured lessons, teaching objectives, the use of a quality assurance framework of Quality Matters (Quality Matters, n.d.), and assessment through digital tools. Active participation and self-regulation of students were sought. In this environment, students with different cultural and linguistic backgrounds benefited more, and some students, such as S4, S5, S6, and S8, showed improved performances and increased their learning engagement using digital tools. The return to face-to-face teaching was marked by an increased sense of collectivity and recognition of the value of school as a space for socialization.

In the third cycle, face-to-face teaching leveraged technology based on learning theories and management systems to foster personal development and improvement. Despite ongoing health protocols limiting physical interaction, technology effectively 'bridged' these distances, adding significant value to teaching. Collaborative activities utilized tools from a sociocultural approach, encouraging active engagement with technology. This approach notably improved participation and performance, particularly for students like S9. Online teaching was implemented in instances of illness or exposure to COVID-19, with materials on Padlet mirroring those in Google Classroom. This continuity allowed students to maintain their learning from home and seamlessly reintegrate into the classroom upon their return.

# 5. CONCLUSION

Throughout the three cycles, the learning environments encountered were significantly shaped by health protocols due to the COVID-19 pandemic. These environments presented diverse opportunities for interaction, personal development, improvement, and fostering a sense of collectivity. Notably, children from less privileged backgrounds found themselves more adversely affected by these changes. This observation aligns with findings from other studies examining the pandemic's impacts on education (Herrmann et al., 2021; Singh et al., 2021; Pittman et al., 2020; Casimir et al., 2022; Popyk, 2021).

Drawing on Wenger's Communities of Practice theory (1998, 2009), the study observed varying levels of participation among students within the classroom community. Notably, students from culturally diverse backgrounds often found themselves in roles of peripheral participation. This diminished engagement and the corresponding lower academic performance among these students underscore a broader trend toward academic marginalization faced by children from immigrant backgrounds, as documented in existing literature (Brown, 2007; Watson, 2011; Levy, 2007).

In the first phase of the first cycle before the pandemic, there was a positive learning environment during face-to-face teaching, where students had the opportunity for interaction, a sense of collectivity, and personal development and improvement. The use of various teaching techniques and student-centered approaches led to enhanced personal improvement and social participation (Watson, 2011; Brown, 2007). However, during the pandemic, the unpreparedness for transitioning to online teaching from face-to-face teaching caused an increase in social inequalities (Herrmann et al., 2021; Singh et al., 2021; Pittman et al., 2020; Casimir et al.,

2022; Popyk, 2021). Vulnerable children with inadequate self-regulation skills faced difficulties in online learning (Li & Wang, 2021; Van Laer & Elen, 2016).

The thorough preparation and training for online teaching in the second cycle resulted in a more effective and seamless transition from traditional to online methods. The transition from traditional to online teaching was smooth, following criteria for quality implementation and effectiveness. Students from diverse cultural and linguistic backgrounds benefited, as BL improved their academic results and participation, offering a flexible and supportive learning environment (Li & Wang, 2021; Staker & Horn, 2012).

In the face-to-face teaching under restrictive health protocols and social distancing, the lack of flexibility and personalized approaches did not benefit all students, especially those from disadvantaged groups. Limited interaction and teacher-centered methods restricted opportunities for personal development and collectivity. This contrasts with findings supporting the need for flexibility and cultural diversity in learning environments to support all children (Watson, 2011; Gay, 2002; Villegas & Lucas, 2002). Children with linguistic diversities performed better in the online phase of BL, which offered a more favorable learning environment.

In the third cycle's face-to-face teaching, the employment of a learning management system and the integration of technology, grounded in learning theories, notably enhanced student interaction through digital collaborative activities, effectively 'bridging' the gap imposed by social distancing. This enriched learning environment was the result of cumulative experiences and training from the preceding years. Research indicates that such technological integration and system use not only bolster academic performance but also enhance students' motivation for learning (Rachmadtullah et al., 2020).

The integration of technology in the classroom, despite social distancing constraints, facilitated interaction and led to improved academic outcomes. However, the need for further research into the challenges and possibilities of BL remains, as the various phases of teaching present differences in many factors, such as teaching methods, the use of digital tools, and health protocols.

## 5.1 Future Applications and Suggestions

This research reveals that BL can be effectively incorporated within the framework of culturally responsive education, favoring all students, especially those from different cultural and linguistic backgrounds. The effectiveness of BL hinges on a seamless blend of face-to-face teaching and online learning, which are interconnected. The enhanced academic performance among students from diverse backgrounds underscores BL's potential in accommodating the educational needs of such learners. Future research should build on this study's findings to develop BL approaches that optimize the beneficial aspects of each phase. This includes:

- Face-to-face teaching in a learning environment that provides opportunities for interactions, personal development, and improvement, and the creation of a sense of collectivity.
- Online teaching using learning management systems and reliable standards.
- Integration of technology based on learning theories.

Additionally, it would be valuable to investigate the impact of the duration and specific models of BL implementation on student performance. Understanding these variables can provide deeper insights into optimizing BL strategies for diverse educational settings.

## 5.2 Limitations

This research is subject to limitations that should be acknowledged. The quantitative and qualitative data collected do not support broad generalizations to other schools, participants, or contexts beyond Cyprus due to the small and non-representative sample size. Additionally, the implementation of BL was constrained by health protocols and social distancing measures necessitated by the COVID-19 pandemic, which may limit the applicability of the findings in different conditions or future scenarios. Despite these constraints, the insights garnered offer valuable perspectives on educational practices during the pandemic crisis.

## REFERENCES


Bennett, S. V., Gunn, A. A., Gayle-Evans, G., Barrera, E. S., & Leung, C. B. (2018). Culturally responsive literacy practices in an early childhood community. *Early Childhood Education Journal*, 46, 241-248.

Brown, R. (2007). Exploring the social positions that students construct within a classroom community of practice. *International Journal of Educational Research*, 46(3-4), 116-128.

Casimir, O. A., Blake, S. C., Klosky, J. V., & Gazmararian, J. A. (2023). Adaptations to the Learning Environment for Elementary School Children in Georgia during the COVID-19 Pandemic. *Journal of Child and Family Studies*, 1-14.

Christensen, C. M., Horn, M. B., & Staker, H. (2013). *Is K-12 Blended Learning Disruptive? An Introduction to the Theory of Hybrids*. Clayton Christensen Institute for Disruptive Innovation.

de Groot, I., Leijgraaf, M., & van Dalen, A. (2023). Towards culturally responsive and bonding-oriented teacher education. *Teaching and Teacher Education*, 121, 103953.

European Commission. (n.d.). *A Europe fit for the digital age*. Retrieved from: https://commission.europa.eu/strategy-and-policy/priorities-2019-2024/europe-fit-digital-age_en

Gay, G. (2002). Preparing for culturally responsive teaching. *Journal of teacher education*, 53(2), 106-116.

Gunn, A. A., Bennett, S. V., Alley, K. M., Barrera IV, E. S., Cantrell, S. C., Moore, L., & Welsh, J. L. (2021). Revisiting culturally responsive teaching practices for early childhood preservice teachers. *Journal of Early Childhood Teacher Education*, 42(3), 265-280.

Herrmann, L., Nielsen, B. L., & Aguilar-Raab, C. (2021, June). The Impact of COVID-19 on interpersonal aspects in elementary school. *In Frontiers in Education* (Vol. 6, p. 635180). Frontiers Media SA.

Ladson-Billings, G. (1995). But that's just good teaching! The case for culturally relevant pedagogy. *Theory into practice*, 34(3), 159-165.

Levy, M. (2007). Culture, culture learning and new technologies: Towards a pedagogical framework. *Language Learning and Technology*, 11(2), 104–127.

Li, S., & Wang, W. (2022). Effect of blended learning on student performance in K-12 settings: A meta-analysis. *Journal of Computer Assisted Learning,* 38(5), 1254–1272.

Parkhouse, H., Bennett, E., Pandey, T., Lee, K., & Johnson Wilson, J. (2022). Culturally Relevant Education as a Professional Responsibility. *Educational Researcher*, 51(7), 474–480. https://doi.org/10.3102/0013189X221092390

Pittman, J., Severino, L., DeCarlo-Tecce, M. J., & Kiosoglous, C. (2021). An action research case study: Digital equity and educational inclusion during an emergent COVID-19 divide. *Journal for Multicultural Education*, 15(1), 68-84.

Popyk, A. (2021). The impact of distance learning on the social practices of schoolchildren during the COVID-19 pandemic: reconstructing values of migrant children in Poland. *European Societies*, 23(supl), S530-S544

Quality Matters. (n.d.). Course Design Rubric Standards. Retrieved from: https://www.qualitymatters.org/sites/default/files/PDFs/StandardsfromtheK-12RubricFifthEdition.pdf

Rachmadtullah, R., Marianus Subandowo, R., Humaira, M. A., Aliyyah, R. R., Samsudin, A., & Nurtanto, M. (2020). Use of blended learning with moodle: Study effectiveness in elementary school teacher education students during the COVID-19 pandemic. *International journal of advanced science and technology*, 29(7), 3272-3277.

Rasheed, R. A., Kamsin, A., & Abdullah, N. A. (2020). Challenges in the online component of blended learning: A systematic review. *Computers & Education*, 144, 103701.

Singh, J., Steele, K., & Singh, L. (2021). Combining the best of online and face-to-face learning: Hybrid and blended learning approach for COVID-19, post vaccine, & post-pandemic world. *Journal of Educational Technology Systems*, 50(2), 140-171.

Staker, H., & Horn, M. B. (2012). *Classifying K–12 blended learning*. Retrieved from: http://www.christenseninstitute.org/wp-content/uploads/2013/04/Classifying-K-12-blended-learning.pdf

Van Laer, S., & Elen, J. (2017). In search of attributes that support self-regulation in blended learning environments. *Education and Information Technologies*, 22, 1395-1454.

Villegas, A. M., & Lucas, T. (2002). Preparing culturally responsive teachers: Rethinking the curriculum. *Journal of teacher education,* 53(1), 20-32.

Watson, S. L. (2011). Somebody's gotta fight for them: A disadvantaged and marginalized alternative school's learner-centered culture of learning. *Urban Education*, 46(6), 1496-1525.

Wenger, E. (1998). Communities of practice: Learning as a social system. *Systems thinker*, 9(5), 2-3.

Wenger, E. (2009). A social theory of learning. *Contemporary theories of learning*, 209-218.

# EXTENDING COMMON USABILITY HEURISTICS LIST WITH LEARNER-CENTERED STRATEGIES FOR AN EXHAUSTIVE E-LEARNING DESIGN ASSESSMENT

Nadia Menad
*University of Science and Technology of Oran, Algeria*

**ABSTRACT**

The learner-centered approach to e-learning is an emerging philosophy that has been winning prominence in the educational space in nowadays. This approach emphasizes creating a more engaging e-learning environments by putting the needs of the learner first and creating an adapted situation where the learner can succeed. Learners are allowed to learn autonomously in a positive manner and be enough motivated. As motivation is tied to autonomy and the manner of designing interfaces have a direct impact on the learner. For this purpose, e-learning application designers should adapt this way of thinking as a part of their conception. Our contribution aims to extend an existing usability evaluation interface heuristics. These common list has been enriched in our proposal with a learner-centered strategies to propose a new exhaustive e-learning interface design assessment questionnaire.

**KEYWORDS**

User Experience, Educational Platforms, Interface Evaluation, Online Educational Environment, Learner-Centered Strategies, Remote Teaching Quality

## 1. INTRODUCTION

From the learner-centered approach to e-learning, the learner-centered approach to e-learning education focuses on creating a more engaging learning space with a set of activities and immersive experiences, where the learners have a high level of engagement by placing the needs of the learner first. Each time a learner encounters a tough situation, their brain is forced to work harder and think about past knowledge, process new information, reflect, and then make a decision. It is not enough for a learner to simply recall knowledge when they are confronted with dealing circumstances they must apply it as well. However, understanding your own learning preferences is a good place to start when considering the benefits of a learner-centered e-learning approach. Psychological researchers (Marius et al., 2023) believe that motivation is tied to the manner of designing interfaces have a direct impact on the learner. For this purpose, e-learning application designers should adapt this way of thinking as a part of their conception.

According to Jonassen et al. (1996) and Barr & John (1995), there exist four distinct attributes of learner-centered effective principles: (1) Context: Learning tasks should have a real-world application that allows learners to connect personally with what they are learning. (2) Construction: Learners should be able to link their own experiences and prior knowledge with new learning. (3) Collaboration: A problem-solving scenario allows learners to develop, test, and analyze their ideas while being exposed to other's opinions. Each learner will arrive at an individual conclusion after the collaboration is completed. (4) Conversation: Communication with and within learners is important to effective e-learning courses. Learners should spend time in conversation while planning and making sense of new learning. Evolving into a learner-centered approach for a such e-learning course doesn't require you to sacrifice the content or your rigorous expectations. Small variations could me make over time to implement a fully learner-centered e-learning course. These changes could involve four essential points: (1) Shifting your mindset: The design of learner-centered instruction requires a paradigm shift, challenging the beliefs about learning and the role the instructor plays in the process (Huba et al., 2000).

The first step to converting your e-learning courses to learner-centered experiences is to change your mindset. This new way of thinking will influence all of your instructional decisions: What content to include or not include, how to organize it, teach it and how to evaluate it. (2) Getting to know your learner: Learners are different and there are multitudes of ways to craft courses that work for a variety of individuals. Failing to pay attention to the diversity of your target community can make a big difference in the success of the course and may result in the inability to facilitate learner-centered content. This concept has same meaning to a common well-known concept called, marketing persona. This notion is aiming to fix each user's preferences and orientations, that must be taken into account during the interface design process. (3) Rethinking the learning objectives: this point deals with the fact of considering to shift perspective to create objectives aimed at describing what the learner will be able to do at the end of the course. This is an essential part of the learner-centered model. (4) Shifting the balance of power creating: E-learning courses that move on the balance of power to the learner is a good way of requiring more learner responsibility. Instead of waiting for content to be delivered learners are engaged in creating their own meaning and learning. Shifting the balance of power provides learners the opportunity to manage their own knowledge, their own place, which could reflect on their individual efforts.

Learner-centered approach focuses on the quality of learners (Barr & John, 1995). It is based on the understanding that each learner has different criteria, which should be employed to improve both training and learning. In addition, the papers (Jonassen et al., 1996) (Barr & John, 1995), explained that, learner-centered approach considers learning while doing; it supports learners in the ways of using the software and throughout the use of it. It also indicates how and why to do the learning tasks. Furthermore, these authors have defined learner-centered approach as a new challenge for human-computer interaction, with the objectif of providing support for both learning a task and doing it. In our approach, we aim to extend an existing usability evaluation interface heuristics, these common list (Granollers, 2018) has been enriched with a learner-centered strategies, to give a birth of an exhaustive e-learning interface design assessment. We considered a set of evaluation heuristics lists presented in Granoller's research (2018) this proposed heuristic list has been extended to adapt a new questionnaire. The new heuristics have been extracted and inspired by learner-oriented approach concepts. We selected only 7 well-known learner-oriented strategies references to apply our assessment. Additionally, we have applied a common method to experiment a case study which complements the usual qualitative data.

This document is structured as follows: In Section. 2, related work about some usability assessment are presented briefly. In Section. 3, we detail our contribution by giving the new assessment questionnaire that has been used to measure the level of usability of the chosen e-learning website. In Section. 4, we present the obtained results and a brief analysis of them. Finally, in Section. 5, conclusion and future work are established.

## 2. RELATED WORK

An experiment demonstrated how useful Heuristic Evaluation (HE) is and how easy it was to put our proposal into practice has been presented in Lewis's work (1995), including an application of a new questionnaire to measure the usability. Also, a case study in the e-commerce domain is presented in this paper, each proposed heuristic that is covered by the questionnaire is approximately composed of four/five items. All the items are written in an affirmative way asking the specialists about the fulfillment of a specific usability guideline related to the heuristic in reference (Paz & Granollers, 2019). J. Nielsen (1995) proposed ten usability heuristics for user interface design. A System Usability Scale (SUS) (Brooke, 1996), a method is described in the paper presenting the System Usability Scale, a reliable, low-cost usability scale that can be used for global assessments of systems usability. SUMI (Software Usability Measurement Inventory) (Kirakowski & Corbet, 1993) is considered as a solution of measuring user's perceptions of the usability of software. Another questionnaire has been presented in Lewis's research (1995), called CSUQ : Computer System Usability Questionnaire.

All these usability questionnaires such as, SUS (Brooke, 1996), SUMI (Kirakowski & Corbet, 1993), CSUQ (Lewis, 1995), ASQ (Bango et al. , 2008) are commonly recognized by the academia and industry.

Concerning the qualitative perception, work presented in Baker et al. (2018), Bonastre & Granollers (2014), Martínez et al. (2009), Quiñones & Rusu (2017), Suarez (2011) have been looking to complement them with a number to quantify how usable the evaluated interface design is. Bonastre & Granollers (2014) contribute in this work to present the standardization of recommendations by formulating them in interrogative sentences to facilitate the evaluation of e-commerce sites. Each heuristic is accompanied by examples and suggestions that facilitate their evaluation by a provided scoring system. However, the heuristics described in this paper are gathered in 6 groups according to the stages of the buying decision process described by Engel et al. (1978). Another work, which complements Nielsen's (1995) and Tognazzini's list is referenced in Tognazzini (2018). Work published by Tognazzini (2014) proposes a usability method of e-commerce websites is being analyzed with the help of automated testing tools already available in literature. Usability evaluation of e-Commerce websites has been presented in this research paper. Usability could also be measured by using automated testing tools such as : webage Analyzer, website Grader tool and Nibbler tool.

## 3. CONTRIBUTION AND DESIGN OF OUR APPROACH

In our new approach, we focused on the HE technique. It is one of the important topics in Human-Computer Interaction (HCI) when talking about usability assessment. Different research work have been discussing the effectiveness of the current HE, but it is important to improve its effectiveness. A substantial improvement is presented in this paper, consisting of: (i) a new heuristic list for evaluating e-learning interfaces, based on existing ones, that have been extended with learner-centered strategies ; (ii) a set of specific questions to be answered when analyzing every principle ; (iii) an easy rating scale for each question, and finally; (iv) a method to obtain a quantitative value, called the Usability Percentage. It gives a numeric idea about how usable the evaluated interface is. An experiment by a group of experts helped to validate the implications of the proposed solution.

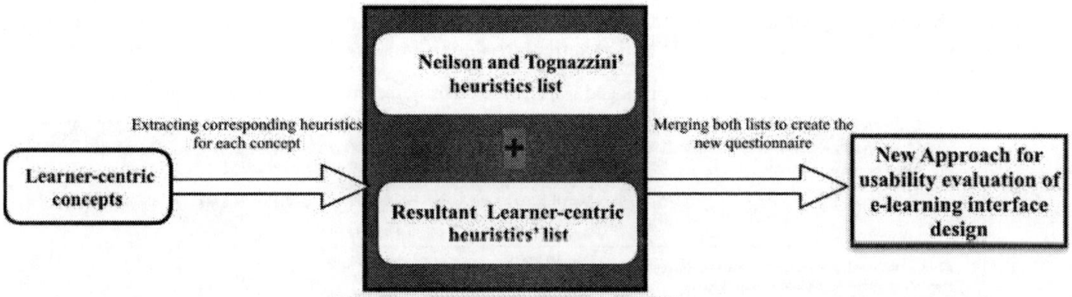


Figure 1. Methodology steps of our approach to construct the usability evaluation questionnaire

Figure 1. is illustrating the different steps to our approach to achieve an exhaustive usability evaluation questionnaire. The principale goal is to extend an existing usability evaluation interface heuristics. The common heuristics list proposed in Granollers's research (2018) has been enriched with a learner-centered strategies to give a birth of an exhaustive e-learning interface design assessment questionnaire. Therefore, a set of interrogative questions has been extracted to create an adapted assessment dedicated to e-learning interfaces, The new heuristics are inspired by learner-oriented approach concepts as shown in Figure. 1. We selected only 7 well-known learner-oriented strategies references to apply our assessment. Additionally, we have applied a common method (Granollers, 2018) to experiment a case study which complements the usual qualitative data.

Basing on the 15 heuristics (cf. Table .1 and 2) proposed by Granollers (2018), and to adapt a new usability heuristic list seven additional heuristics has been added to the general assessment method to enrich the assessment with learner-centered concepts (cf. Table .3). The heuristics proposed by Granollers (2018) contains 64 questions, grouped in a set of validated 15 heuristics. Moreover, 10 questions are resulting from learner-centered approach that are grouped in 7 heuristics. In total, the exhaustive evaluation questionnaire will contain 74 questions, grouped in 22 heuristics, that would be used to adapt the evaluation in e-learning interface. The process results a new questionnaires adapted to e-learning platforms. The main learner-centered strategies approach that have been taken into account to extend the questionnaire are defined as follow:

Heuristic (#1) Making courses responsive and accessible: Online lessons are consumed on a variety of devices and platforms. No longer are learner only accessing courses in a designated classroom or on the desktop computer, tabs, laptops and mobiles have all become a source of e-learning. This presents instructional designers with both deals and opportunities. A significant deal is ensuring accessibility and and responses in design and delivery of the course. Furthermore, e-learning can now reach a wider audience of learners because there is less dependency on the mode of delivery.

Heuristic (#2) Branching scenarios: For better efficacy and retention on the learner's end, online lessons should be customized for more personalization. For instance, is in classroom learning solutions for young children.

Branching scenarios can be created for learners with different areas of interest within one topic.

Based on responses and actions during a course, different prompts could widely rise engagement.

Heuristic (#3) Gamification: Gamification is an excellent way to make learning more fun for the learner. It creates a sense of achievement for the learner and a wide connection with the course material. Gamification also brings the power of narrative and checkpoints to a course. For instances, rewards, points, even in-course credits, create an atmosphere where the learner is getting positive reinforcement for their time invested in learning. With a concept like levels, learners can make progress and evaluate how much they have absorbed from the course.

Heuristic (#4) Breaking up the course into smaller pieces: Long lectures and tasks are not enough effective for engagement or retention. Instead, topics broken down into short and concise subheads leave a greater impression on the learner. This concept has already gained traction with microlearning. Learners absorb the material better in short bursts and have greater control over how they can proceed effectively.

Heuristic (#5) Moving beyond slide thinking: There are obstacles that come with thinking of the lesson in terms of slides and screens. For the reason that it's not interactive and ties the learner into a one-way train moving from slide to slide. There is no way for the learner to explore more by clicking and navigating on their own. Another issue is that as the lesson material and structure becomes more complicated and elaborate, the slide based approach still not a flexible and practical  to accommodate the inescapable complexity.

Table 1. Heuristic list previously proposed in [1], with their corresponding evaluation questions (Part 1)

| | Neilson and Tognazzini | |
|---|---|---|
| | **#1.- Visibility and system state** | - Does the application include a visible title page,section or site?<br>- Does the user always know where it is located?<br>- Does the user always know what the system or application is doing? - Are the links clearly defined?<br>- Can all actions be visualized directly?(No other actions are required) |
| Granoller's list | **#2.- Connection between the system and the real world, metaphor usage and human objects** | - Does information appear in a logical order for the user?<br>- Does the design of the icons correspond to everyday objects?<br>- Does every icon do the action that you expect?<br>- Does the system use phrases and concepts familiar to the user? |
| | **#3.- User control and freedom** | - Is there a link to come back to initial state or homepage?<br>- Are the functions"undo"and"re-do"implemented?<br>- Is it easy to come back to an earlier state of the application? |
| | **#4. -Consistency and standards** | - Do link labels have the same names as their destinations?<br>- Do the same actions always have the same results?<br>- Do the icons have the same meaning everywhere?<br>- Is the information displayed consistently on everypage?<br>- Are the colours of the links standard? If not, are they suitable for its use?<br>- Do navigation elements follow the standards?(Buttons,checkbox,...) |
| | **5.- Recognition rather than memory, learning and anticipation** | - Is it easy to use the system for the first time?<br>- Is it easy to locate information that has already been searched for before?<br>- Can you use the system at all times without remembering previous screens?<br>- Is all content needed for navigation or task found in the "current screen"?<br>- Is the information organized according to logic familiar to the end user? |
| | **#6 - Flexibility and efficiency of use** | - Are there keyboard shortcuts for common actions?<br>- If there are,is it clear how to use them?<br>- Is it possible to easily perform an action done earlier?<br>- Does the design adapt to the changes of screen resolution?<br>-  Is the use of accelerators visible to the normal user?<br>- Does it always keep the user busy?(without unnecessary delays) |
| | **#7- Help users recognize, diagnose and recover from errors** | -- Does it display a message before taking irreversible actions?<br>- Are errors shown in real time?<br>- Is the error message that appears easily interpretable?<br>- Is some code also used to reference the error? |

Table 2. Heuristic list previously proposed in [1], with their corresponding evaluation questions (Part 2)

Granoller's list

| | |
|---|---|
| **#8.-Preventing errors** | - Does a confirmation message appear before taking the action?<br>- Is it clear what information needs to be entered in each box on a form? -<br>- Does the search engine tolerate typos and spelling errors? |
| **#9.- Aesthetic and minimalist design** | -Is used a design without redundancy of information?<br>- Is the information short,concise and accurate?<br>- Is each item of information different from the rest and not confused? -<br>- Is the text well organized,with short sentences and quick to interpret? |
| **#10. Help and documentation** | -Is there the"help"option?<br>-If so,is it visible and easy to access?<br>-Is the help section aimed at solving problems?<br>-Is there a section of frequently asked questions(FAQ)?<br>-Is the help documentation clear,with examples? |
| **#11 -Save the state and protect the work** | - Can users continue from a previous state (where they had previously been or from another device)?<br>- Is"Autosave"implemented?<br>- Does the system have a good response to external failures? (Power cut, internet not working, ...) |
| **#12 .- Colour and readability** | - Do the fonts have an adequate size?<br>- Do the fonts use colours with sufficient contrast with the background?<br>- Do background images or patterns allow the content to be read?<br>- Does it consider people with reduced vision? |
| **#13- Autonomy** | - Does it keep the user informed of system status?<br>- Moreover,is the system status visible and updated?<br>- Can the user take their own decisions?(Personalization) |
| **#14.- Defaults** | -Does the system or device give the option to return to factory settings?<br>- If so,does it clearly indicate the consequences of the action?<br>- Is the term"Default"used? |
| **#15.-Latency reduction** | - Is the execution of heavy work transparent to the user?<br>- While running heavy tasks, is remaining time or some animation shown? |

Heuristic (#6) Integrate learner feedback to Improve And Iterate: Learner feedback needs to be a continuous process. As the Internal course reviews are important to measure accuracy, functionality, and consistency. But in a learner-centered approach, learners are the central stakeholders. And so, gathering their feedback is absolutely vital This way you can gauge whether the objectifs of the learners are satisfied by the lesson or not, what areas are effectively communicating the point and which ones can be improved etc. Understanding the learner and their environment can also inform the Designer about the inherent deals in consuming the lessons, such as technical obstacles.

Table 3. Additional proposed heuristics list (Corresponding to each learner-centered strategy)

Additionnal heuristics extension list

| Resultant heuristic (according to Learner centric concepts) | Resultant items/questions |
|---|---|
| **#1: Making Courses Responsive And Accessible** | Is the application system responsive enough?<br>Do the application items accessibles? |
| **#2: Branching Scenarios** | Is the application employing scénarios into the manner to present its lessons? |
| **#3: Gamification** | Is the application system design employing gamification concept, such : Rewards, points, even in-course credits, ?<br>Is the application design allowing an atmosphere where the learner is getting positive reinforcement for their time invested in learning? |
| **#4: Breaking Up The Course Into Smaller Pieces** | Is the courses broken down into short and concise subheads which leave a better impression on the learner? |
| **#5: Moving Beyond Slide Thinking** | Is the application design employing slides to enroll important points? |
| **#6: Integrate Learner Feedback To Improve And Iterate** | Is the application desgin employing and gathering learners feedback?<br>Is the apllication dsign inclusing the learners opinions whether the goals of the learners are satisfied by the course or not? |
| **#7: Encouraging Social Interaction** | Is the application design involve social interactions , like Learning simultaneously with others, and allowing sharing the experience from and with learners? |

Heuristic (#7) Encouraging Social Interaction: One of the most common feelings associated with learning is being part of a community. Learning simultaneously with others, and sharing the experience, is one thing that eLearning needs. This is one reason why massively organized online courses (MOOCs), like Coursera, have been so successful. With each course, there is a thread or forum where learners can help each other, discuss challenges and brainstorm ideas.

For each a learner-centered strategy involved above, we have extracted one or more than one question, corresponding to the relevant concept, the resultant list has been added to the existing questionnaire proposed by Granollers (2018). Ten questions has been proposed in our approach, grouped in 7 items/heuristics. Therefore, the full questionnaire contains 74 questions, as the already existing Granoller's list contains 64 questions grouped in 15 heuristics as shown in Table 1 and Table 2.

# 4. EXPERIMENTATION AND DISCUSSION

Regarding the process, Granollers and Bonastre (2014) determine that in each evaluation of a specific website, 3 to 5 evaluators must participate. In this experimental case study, seven (7) specialists in the field of Human-Computer Interaction were requested to participate. Once the evaluators kindly and voluntarily agreed to participate in our study, each of our evaluator received an excel template, where each evaluator has to answer to the 74 questions. It can be established that all specialists that agreed to contribute with this study, have the same degree of expertise because they all have multiple projects carried out in this area as well as published research papers. In addition, all they have technological formation in careers related to Computer Science. Many usability evaluation proposal have been established by previous common evaluation proposals, done by Granollers and Bonastre (2014), Tognazzini (2018) and Freddy (2018). In this study, to validate our proposal by applying Freddy formula usability assessment method (Kaur & Kaur, 2018). The research was intended to test this new approach in the assessment of an Algerian well-known e-learning website: Moodle: https://elearning.univ-usto.dz)

## 4.1 Evaluation Method: Inspired Freddy Formula (2018)

The experimented proposal has been inspired by the proposed work in Kaur & Kaur (2018). These assessment set was constructed considering a five-point Likert scale. Each heuristic that is covered by the questionnaire is approximately can be rated from 1 to 5, where: 1 represents "strongly disagree", 2 "disagree", 3 "neither agree nor disagree", 4 "agree" and 5 "strongly agree". The usability final score of the system is obtained with basis on the following mathematical formula **(1)** ;

$$\frac{\sum_{i=1}^{m} \frac{\sum_{j=1}^{n} s_{ij}}{n}}{m} \quad (1)$$

Where, $S_{ij}$ is the assigned score by each evaluator "i" to each item "j" of the questionnaire. n is the number of items of the questionnaire. And m is the number of evaluators. The Average scores of the evaluators by heuristic are illustrated in Table .4.

To calculate the final numerical usability value of the system, the sum of all the values is necessary (Tognazzini, 2014). However, those questions that are YES/NO would receive a value of 5 if the answer is affirmative and a value of 1 if the answer is negative. Considering that there are 22 items : 15 from Granoller's list including 64 questions, plus 7 extracted from our learner-centered approach including 10 additional questions. The maximum value obtainable will be then equal to 370 point, and the minimum value is 74 points. However, in an update of the Tognazzini's proposal (2014). Granollers establishes that when those answers that 74 points would represent a null level of usability of the system (because it is induced that all the answers are of type "Strongly disagree" or in their defect negative for the case of questions YES/NO) and 370 points would represent that the system fully covers all the aspects that are necessary to guarantee an appropriate and satisfactory level of user experience. The final score of usability evaluation, calculated from the result evaluation of each evaluator candidate of the e-learning website is shown in Table 5.

Table 4. The Average scores of the evaluators heuristic using Freddy Formula

| Heurictics | Eval 1 | Eval 2 | Eval 3 | Eval 4 | Eval 5 | Eval 6 | Eval 7 |
|---|---|---|---|---|---|---|---|
| | Neilson and Tognazzini | | | | | | |
| Heuristic 1 | 2.7 | 2.5 | 2 | 2.34 | 2.34 | 2 | 2.5 |
| Heuristic 2 | 2.34 | 2 | 2.43 | 2.43 | 2.34 | 2.5 | 3 |
| Heuristic 3 | 2.43 | 2.2 | 2.3 | 2.4 | 2 | 2.3 | 2 |
| Heuristic 4 | 2 | 2.45 | 2.45 | 3 | 2 | 3 | 3 |
| Heuristic 5 | 3 | 2.43 | 2 | 2.4 | 3 | 3 | 2.34 |
| Heuristic 6 | 2.43 | 2.4 | 2.3 | 2.4 | 2.4 | 2.3 | 2 |
| Heuristic 7 | 2.77 | 2.44 | 2.4 | 3 | 3 | 2.5 | 2.4 |
| Heuristic 8 | 3 | 3 | 2.3 | 3.4 | 3.4 | 3.4 | 3.4 |
| Heuristic 9 | 2.34 | 2.77 | 2.77 | 2.77 | 2.77 | 2.77 | 2 |
| Heuristic10 | 2.43 | 3.2 | 2.77 | 3.2 | 3 | 3.2 | 2.77 |
| Heuristic 11 | 2 | 2.34 | 2.34 | 2.34 | 2.34 | 2.34 | 2 |
| Heuristic 12 | 2.43 | 2.43 | 2.43 | 2.43 | 2.43 | 2.43 | 2.34 |
| Heuristic 13 | 2.43 | 2.34 | 2.34 | 2.34 | 2.34 | 2.34 | 2.43 |
| Heuristic 15 | 2.77 | 2.43 | 2.43 | 2.43 | 2.43 | 2.43 | 2.44 |
| | Learner-centric heuristics | | | | | | |
| Heuristic 1 | 2.34 | 2.34 | 1 | 1 | 2 | 2 | 2.34 |
| Heuristic 2 | 3 | 2 | 2 | 2 | 2 | 2 | 2 |
| Heuristic 3 | 1 | 1 | 1 | 1 | 1 | 1 | 1 |
| Heuristic 4 | 2 | 3 | 2 | 2 | 2 | 1 | 2 |
| Heuristic 5 | 3 | 2 | 1 | 2 | 2 | 2 | 2 |
| Heuristic 6 | 2 | 2.34 | 2 | 2 | 2.34 | 2 | 2.45 |
| Heuristic 7 | 2 | 2 | 1 | 2 | 1 | 2 | 2 |
| Usability purcentage | 45.82% | 45.1% | 39.32% | 44.43% | 42.84% | 43.19% | 44% |

Table 5. Percentage covered of usability (UP), average of usability evaluation of the e-learning website

| E-learning website | Final score | Score remaining | Pourcentage covered Usabiity pourcentage(UP) |
|---|---|---|---|
| Moodle (https://elearning.univ-usto.dz) | Sum of poins from 1 to 5 of 22 items / of 74 questions | Maximum point-Sum 370 points -sum | 43.52% |

## 5. CONCLUSION

Participants in learner-centered classes and programs are incorporated in the planning, execution, and evaluation process. The main question to ask is to how to create a more engaging learning environment using the learner-centered approach. When we talk about transforming your online classroom into a more engaging learning environment it can be easy to fall into the trap of implementing the latest so called trend. To avoid this, it's clear that emphasize that engaging learning is not about adding more flash words. Rather it's about shifting your mindset and thinking about what it means to create an adapted interface design that supports learner engagement.

The approach presented in this paper is born of our extensive experience, evaluating the usability of user interfaces and observing that some traditional methods need to be updated and improved. By involving learner-centered concepts and including this concept to reach an exhaustive manner to evaluate e-learning interfaces design. In future work, we aim to validate our approach by using the rest of other usability methods, and by testing a diversity of another well-known e-learning websites.

## REFERENCES


Baker, K., Greenberg, S., & Gutwin, C. (2018). Empirical development of a heuristic evaluation methodology for shared workspace groupware. Paper published in Proceedings of Conference on Computer Supported Cooperative Work, 96-105.

Bango, A., Kortum, P. T., & Miller, J. T. (2008). An empirical evaluation of the system usability scale. International Journal of Human–Computer Interaction, 24 (6), 574-594

Barr, R. B., & Tagg, J. (1995). From teaching to learning, A new paradigm for undergraduate education. Change, 27, 18-25.

Bonastre, L., & Granollers, T. (2014). A Set of Heuristics for User Experience Evaluation in E- commerce websites. Paper published in Proceedings of the 7th International Conference on Advances in Computer-Human Interactions (ACHI), 27-34, Barcelona, Spain.

Brooke, J. (1996). SUS: A quick and dirty usability scale. Usability Evaluation in Industry, Jordan, 189-193.

Elliot, S., Shari, L. J., Jonathan, K., Chris, Q., James, R., Jeff, S., & Steven, J. Stratford Learning (1996). Theory in Practice: Case Studies of Learner-Centered Design.Case Studies of Learner-Centered Design. International Conference on Human Factors in Computing Systems. Published in 13 April 1996. Education, Computer Science.

Engel J. F., Blackwell, R. D., & Kollat, D. T, (1978). Consumer behavior, 3rd ed. Hinsdale: Dryden Press.

Granollers, T. (2016). Validación experimental de un conjunto heurístico para evaluaciones de UX de sitios web de comercio-e. Paper published in Proceedings of the IEEE 11th Colombian Computing Conference (CCC), 1-8.

Granollers, T. (2018). Usability Evaluation with Heuristics, Beyond Nielsen's List. Paper published in Proceedings of the 11th International Conference on Advances in Computer-Human Interactions, ACHI 2018,ISBN: 978-1-61208-616-3.

Huba, M. E., & Freed, M. E., (2000). Learner-Centered Assessment on College Campuses: Shifting the Focus from Teaching to Learning , Community College Journal of Research and Practice. http://assessment.uconn.edu/what/index.html.

Kaur, J., Kaur, P. (2018). Usability Evaluation of E-Commerce Websites. The International Journal of Scientific Engineering and Research.

Kirakowski, J., & Corbet, M. (1993). SUMI: The software usability measurement inventory. British Journal of Educational Technology, 24(3), 210-212 . DOI:10.1111/j.1467-8535.1993.tb00076.x.

Lewis, J.R. (1995). IBM computer usability satisfaction questionnaires: Psychometric evaluation and instructions for use. International Journal of Human Computer Interactions, 7(1), 57-78. http:// dx.doi.org/10.1080/10447319509526110

Marius, O. J., Sigrunn, E., Jeno., L., M. (2023). The bright and dark side of autonomy: How autonomy support and thwarting relate to student motivation and academic functioning, Front. Educ., Sec. STEM Education . Volume 8 . https://doi.org/10.3389/feduc.2023.1153647

Martínez, A.B., Juan ,A.A., Álvarez, D., & Suárez, M.C. (2009). WAB: A quantitative metric based on WAB. Paper published in Proceedings of the 9th International Conference on Web Engineering (ICWE), LNCS, Vol. 5648, 485-488, Springer, Verlag , Berlin Heidelberg.

Nielsen, J. (1995). Accessed (2019). 10 Usability Heuristics for User Interface Design, [Online link].https://www.nngroup.com/articles/ten-usability-heuristics/

Paz, F. & Granollers, T. (2019). Redesign of a Questionnaire to Assess the Usability of Websites. Paper published in Proceedings of the 2nd International Conference on Human Systems Engineering and Design: Future Trends and Applications, Munich, Ger- many,423-428.

Paz, F. (2019). Application of a new questionnaire to measure the usability: A case study in the e- commerce domain. International Journal of Advanced Trends in Computer Science and Engineering, 8 (1.4), 395-297. https://doi.org/10.30534/ijatcse/2019/6081.42019

Quiñones, D., & Rusu, C. (2017). How to develop usability heuristics: A systematic literature review. In Computer Standards and Interfaces, Vol. 53, 89-122, ISSN 0920-5489.

Suarez, M. C. (2011). SIRIUS: Sistema de Evaluación de la Usabilidad Web Orientado al Usuario y basado en la Determinación de Tareas Críticas". Doctoral Thesis, Oviedo, Spain.

Tognazzini, B. (2014). First Principles of Interaction Design., "First Principles, HCI Design, Human Computer Interaction (HCI), Principles of HCI Design, Usability Testing [Online link]. https://www.designprinciplesftw.com/collections/first-principles-of-interaction-design

Tognazzini, B. (2014). Interaction design solution for the real world. Paper published in Proceedings of First Principles, Human Computer. Interaction (HCI), Principles of HCI Design, Usability- Testing. [Online link].https://asktog.com/atc/principles-of-interaction-design/.

# THE OBSTACLES TO DIGITAL TRANSFORMATION AND THEIR IMPACT ON CYBER THREATS IN MOROCCAN HIGHER EDUCATION: A MULTI-CRITERIA APPROACH

Abdelilah Chahid, Souad Ahriz, Kamal El Guemmat and Khalifa Mansouri
*Department of Mathematics and Computer Science,*
*Higher Normal School of Technical Education of Mohammedia, Morocco*

**ABSTRACT**

The COVID-19 pandemic has accelerated the digital transformation of higher education institutions, aiming to enhance the efficiency of teaching and operational management. However, this rapid transition has also revealed an increase in cyber risks, threatening the security and sustainability of educational systems. Specialized literature indicates that obstacles to digital transformation can heighten vulnerabilities to cyberattacks. This study evaluates and ranks the impact of these obstacles on the emergence of cybersecurity threats, focusing on the specific context of higher education in Morocco. The results show that the main barriers to digital transformation that facilitate the emergence of threats are the lack of strategy, the absence of a clear vision, difficulties in integrating information technologies into higher education, and the low level of digital literacy. These conclusions enable stakeholders to allocate resources wisely and develop robust cybersecurity strategies, thereby establishing the necessary foundations to secure the digital evolution in the educational sector.

**KEYWORDS**

Digital Transformation, Cybersecurity, AHP, Barriers, Higher Education Institutions, Obstacles

## 1. INTRODUCTION

Digital transformation, resulting from social evolution, stimulates economic development and impacts all domains (Kozarkiewicz 2020). Experts analyze digital transformation in detail to grasp its effects, advantages, and disadvantages, especially for businesses that could disappear without it (Thomas M. Siebel, 2017). Multiple studies aim to optimize digital transformation, recognized as a lever for change (Yoo et al. 2012). An analysis of the literature on digital transformation reveals that diverse perspectives lead to heterogeneous definitions (Reis et al. 2018). In this climate of change, organizations and governments understand the importance of cultivating an evolving capability to implement necessary adaptations relevantly and effectively to survive and remain competitive in a dynamic market (Morrison 2003).

Digital transformation in higher education encompasses social, organizational, and technological changes, affecting teaching, infrastructure, curricula, administration, research, operations, human resources, knowledge dissemination, governance, and information management, as well as promoting this transformation (Benavides et al. 2020). During the COVID-19 pandemic, higher education institutions were forced to quickly convert to digital and revise their teaching methods to meet specific needs (Jensen 2019). Given their diverse sizes (Jensen 2019), these institutions must adopt comprehensive evolution and effectively use the numerous digital technologies at their disposal to ensure their central role in this transformation. Currently, Morocco is deploying digitization programs in universities under the Morocco Digital 2020 initiative (Ferhane and Yassine 2022). However, according to the recent Global Innovation Index 2021 (GII) report, Morocco ranks 77th globally with a score of 29.3 (Tamer and Knidiri 2023), revealing the complexity of digitizing higher education in the country amid numerous existing obstacles. Recent studies have helped leaders formulate effective strategies to overcome these barriers and increase the success probabilities of this digital evolution (Aditya, Ferdiana, and Kusumawardani 2022; Laufer et al. 2021). The persistent presence of these barriers can exacerbate vulnerabilities and increase cybersecurity risks, hindering the progress of digital transformation (Sandhu 2021; Saeed et al. 2023; Pavlova 2020; Kendall 2022).

To determine the importance level of each barrier and the strategies to adopt for cybersecurity risk management, we conducted a case study within Moroccan universities. This study collected perceptions from IT managers, industry professionals, and decision-makers on obstacles and associated strategies. It tested multi-criteria decision analysis methods, such as the Analytic Hierarchy Process (AHP), to identify and rank the obstacles to digital transformation in Moroccan universities that facilitate the emergence of cyber threats. Applying this methodology and interpreting its results provide a solid foundation for understanding the evolution of these threats and enhancing competitiveness by choosing the most appropriate preventive strategies, thereby minimizing their impact on the digital transformation of the educational sector.

## 2. THE BARRIERS AND CYBERSECURITY THREATS IN THE CONTEXT OF DIGITAL TRANSFORMATION WITHIN HIGHER EDUCATION INSTITUTIONS

Cyber threats constitute a significant obstacle to the progression of digital transformation, as evidenced by sources (Sandhu 2021; Taherdoost, Madanchian, and Ebrahimi 2021) indicating that cybercrime represents a major impediment in this area. Furthermore, the work of (Durojaye and Raji 2022) reveals that most state-sponsored cyberattacks specifically target the weaknesses of critical infrastructures. Another study (Williams, Chaturvedi, and Chakravarthy 2020) has established guidelines to counter opportunistic attackers. According to (Chitre, Dhoke, and Shriniwar 2022), developing effective automated systems to monitor and analyze internal threats presents notable difficulties. This article presents the use of the AHP methodology to determine the obstacles to digital transformation facilitating the emergence of cyber threats in Moroccan higher education, based on the evaluation of factors and sub-factors specific to Moroccan universities. This analysis was conducted within the context of Morocco's digital strategy for 2015-2030, with collective reflection sessions involving 20 educational executives, IT experts, and decision-makers.

To examine cybersecurity threats and the barriers and sub-barriers to digital transformation, a relevant literature review was conducted, with the results synthesized in Tables 1. These data, mapping the extent of primary and secondary barriers as well as the associated threats.

Table 1. The barriers and sub-barriers to digital transformation

| Category of Barriers | References | Sous Barriers |
|---|---|---|
| Strategic (S) | (Marks and AL-Ali 2022),(AL-Ali and Marks 2022),(Mohamed Hashim, Tlemsani, and Duncan Matthews 2022), (Abdelilah et al., 2024a) | Deficiency in strategic planning (S1) |
| | (Marks and AL-Ali 2022),(AL-Ali and Marks 2022) ,(Alenezi 2021) | Absence of a defined vision (S2) |
| | (Khan, Hasan, and Clement 2012),(Watty, McKay, and Ngo 2016),(Sinclair and Aho 2018),(Abdelilah et al., 2024b) | Governmental vision, planning, and policies (S3) |
| | (Reid 2014) ,(Aditya, Ferdiana, and Kusumawardani 2020),(Aditya, Ferdiana, and Kusumawardani 2021a) | Inadequate implementation planning, time constraints, competing priorities (S4) |
| Technologica l(T) | (Marks and AL-Ali 2022),(AL-Ali and Marks 2022) ,(Jhones and Larramendi 2019), | Challenges in integrating IT into higher education (T1) |
| | (Marks and AL-Ali 2022),(AL-Ali and Marks 2022),(Benavides et al. 2020), (Kaminskyi, Yereshko, and Kyrychenko 2018) | IT security risks (T2) |
| | (Marks and AL-Ali 2022),(AL-Ali and Marks 2022),(Reid 2014),(Jensen 2019) | Inadequate IT infrastructure (T3) |
| | (Benavides et al. 2020),(Sanchez 2020),(Alenezi 2021),(Aditya, Ferdiana, and Kusumawardani 2021b) | Unsuitable IT infrastructure and support services (T4) |

| | | |
|---|---|---|
| Skills and Human Resources) (SHR) | (Marks and AL-Ali 2022),(AL-Ali and Marks 2022), (Reid 2014) | Lack of pedagogical expertise and experience (SHR1) |
| | (Reid 2014),(Jensen 2019),(Aditya, Ferdiana, and Kusumawardani 2020),(Jhones and Larramendi 2019),(Kerroum et al. 2020) | low level of digital literacy(SHR2) |
| | (Aditya, Ferdiana, and Kusumawardani 2021b),(Aditya, Ferdiana, and Kusumawardani 2021a),(Alenezi 2021),(Kalolo 2019) | Lack of human resources (SHR3) |
| | (Marks et al. 2021),(Rafiq et al. 2021),(Kopp, Gröblinger, and Adams 2019) | Leadership skills and behavior (SHR4 |
| Organizational and cultural barriers (OC) | (Gkrimpizi and Peristeras 2022),(Lašáková, Bajzíková, and Dedze 2017),(Alzahrani et al. 2021),(Marshall 2018) | Lack of coordination between departments (OC1) |
| | (Marks and AL-Ali 2022),(AL-Ali and Marks 2022),(Stüber 2018),(Thoring, Rudolph, and Vogl 2017) | Human resistance to change (OC2) |
| | (Marks and AL-Ali 2022),(AL-Ali and Marks 2022),(Stüber 2018) | The shortfall in innovation (OC3) |
| Environmental (E) | (Marks and AL-Ali 2022),(AL-Ali and Marks 2022),(Chipembele and Bwalya 2016),(Sanchez 2020) | Economic climate (E1) |
| | (Marks and AL-Ali 2022),(AL-Ali and Marks 2022),(Benavides et al. 2020),(Vicente et al. 2020),(Marshall 2018) | Budgetary limitation (E2) |

# 3. AHP

Thomas Saaty developed AHP as a decision-making method in the 1970s [40], it is a systematic decision making method which includes both qualitative and quantitative techniques. It is useful for obtaining single assessment value which is based on different indicators or criteria. It simplifies the process of decision making by subdividing a complex problem into a series of structured steps where each element in the hierarchy of criteria is supposed to be independent from others. The analytic network process is used when there is interdependence among criteria. AHP builds a hierarchy of decision items using comparisons between each pair of items expressed as a matrix. Paired comparisons produce weighting scores that measure how much importance items and criteria have with each other. Decision maker examines two alternatives by considering one criteria and indicates a preference. The standard numeric scale used for AHP is 1-9 scale which lies between "equal importance" to "extreme importance", the value 9 indicates that one factor is extremely less important than the other, while value 1 indicates equal importance. At each level of the criteria hierarchy we obtain an n*n square matrix, where n is the number of elements of the level. AHP allows building consensus among decision makers, each member can compare their judgments to those of the other members and it gives them better understanding of the impact of their priorities. AHP decomposes the decision into the following steps (Saaty 2008):

- **Step 1)** Define the problem and state the goal or objective.
- **Step 2)** Define the criteria or factors that influence the goal. Structure these factors into levels and sublevels.
- **Step 3)** Use paired comparisons of each factor with respect to each other that forms a comparison matrix with calculated weights, ranked eigenvalues, and consistency measures.

# 4. RESULT

**Step 1)** The objective is (to identify the obstacles to digital transformation facilitating the emergence of cyber threats in Moroccan higher education). The first phase of this study involved identifying the factors and sub-factors, including five categories of barriers and 17 specific sub-barriers to digital transformation, based on a detailed literature review (referenced in Tables 1). The framework of the analytic network model used in this research is explicitly illustrated in Figure 1.

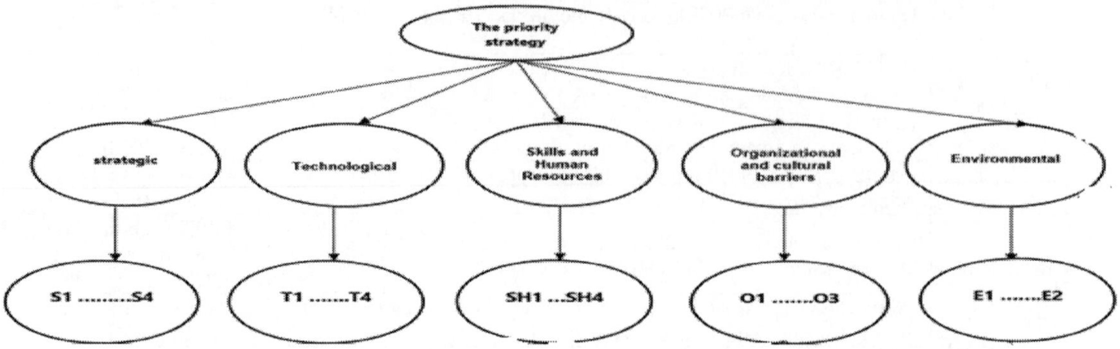

Figure 1. The hierarchical structure for the selection of the obstacles to digital transformation

**Step 2)** Based on the rankings provided by the team of experts, the importance of each factor (category of barriers) in the model is determined. While their internal interdependence was not considered, only their importance relative to the objective set at level 1 was evaluated. The importance levels of each factor are shown in Table 3, where it is noted that the strategic factor is predominant, with an importance of 49%, according to the expert evaluation.

Table 2. Pairwise comparison of barrieres groups without interdependences between them

| | S | T | SHR | OC | E | |
|---|---|---|---|---|---|---|
| S | 1 | 5 | 4 | 6 | 3 | 0,495 |
| T | 1/5 | 1 | 4 | 4 | 2 | 0, 224 |
| SHR | 1/4 | 1/4 | 1 | 4 | 2 | 0,128 |
| OC | 1/6 | 1/4 | 1/4 | 1 | 1/3 | 0,106 |
| E | 1/3 | 1/2 | 1/2 | 3 | 1 | 0 ,045 |

CR=0 ,11

In this step, the local importance of the sub-barriers is determined by the team of experts, while the ranks of the comparative pairs of the sub-barriers are provided in Tables 3 to 7.

Table 3. Pairwise comparison matrix for S

| | S1 | S2 | S3 | S4 | |
|---|---|---|---|---|---|
| S1 | 1 | 3 | 5 | 4 | 0.546 |
| S2 | 1/3 | 1 | 3 | 2 | 0.232 |
| S3 | 1/5 | 1/3 | 1 | 1/2 | 0.083 |
| S4 | 1/4 | 1/2 | 2 | 1 | 0.137 |

CR=0,017

Table 5. Pairwise comparison matrix for SHR

| | SHR1 | SHR2 | SHR3 | SHR4 | |
|---|---|---|---|---|---|
| SHR1 | 1 | 1/3 | 2 | 3 | 0,232 |
| SHR2 | 3 | 1 | 4 | 5 | 0 ,546 |
| SHR3 | 1/2 | 1/4 | 1 | 2 | 0,137 |
| SHR4 | 1/3 | 1/5 | 1/2 | 1 | 0,083 |

CR=0,019

Table 4. Pairwise comparison matrix for T

| | T1 | T2 | T3 | T4 | |
|---|---|---|---|---|---|
| T1 | 1 | 2 | 4 | 3 | 0,484 |
| T2 | 1/2 | 1 | 3 | 2 | 0,260 |
| T3 | 1/4 | 1/3 | 1 | 1/2 | 0,096 |
| T4 | 1/3 | 1/2 | 2 | 1 | 0,159 |

CR=0,015

Table 6. Pairwise comparison matrix for O

| | O1 | O2 | O3 | |
|---|---|---|---|---|
| O1 | 1 | 1/3 | 1/2 | 0,163 |
| O2 | 3 | 1 | 2 | 0,539 |
| O3 | 2 | 1/2 | 1 | 0,296 |

CR=0.00

Table 7. Pairwise comparison matrix for E

| | E1 | E2 | |
|---|---|---|---|
| E1 | 1 | 2 | 0,666 |
| E2 | 1/2 | 1 | 0,333 |

CR=0,008

**Step 3)** Table 8 presents an overall evaluation of the sub-factors of the different barriers to digital transformation in Moroccan higher education, based on their priority. The barrier factors are ranked by order of importance, and each sub-factor is analyzed in terms of individual priority and overall priority. Here are the results:

Table 8. Overall priority of the barriers sub-factors

| BARRIERES factors | Priority of the factors | BARRIERES sub-factors | Priority of the sub-factors | Overall priority of the sub-factors | rank |
|---|---|---|---|---|---|
| Strategic (S) | 0,495 | Lack of strategy (S1) | 0.546 | 0,270 | 1 |
| | | Lack of clear vision (S2) | 0.232 | 0,115 | 2 |
| | | Government vision, plan, and policy (S3) | 0.083 | 0,041 | 8 |
| | | Lack of implementation action plan, lack of time, and other priorities (S4) | 0.137 | 0,068 | 5 |
| Technological(T) | 0,224 | Difficulties embedding IT into higher education (T1) | 0,484 | 0,108 | 3 |
| | | IT security risks (T2) | 0,260 | 0,058 | 6 |
| | | Lack of adequate IT infrastructure (T3) | 0,096 | 0,022 | 13 |
| | | Unsuitable IIT infrastructure and support services (T4) | 0,159 | 0,036 | 9 |
| Skills and Human Resources (SHR) | 0,128 | Lack of pedagogical skills and experience (SHR1) | 0,232 | 0,030 | 12 |
| | | The low level of the digital literacy (SHR2) | 0 ,546 | 0,070 | 4 |
| | | Lack of human resources (SHR3) | 0,137 | 0,018 | 14 |
| | | Leadership skills and behavior (SHR4) | 0,083 | 0,011 | 17 |
| Organizational and cultural barriers (O) | 0,106 | Lack of coordination between departments (O1) | 0,163 | 0,017 | 15 |
| | | Human resistance to change (O2) | 0,539 | 0,057 | 7 |
| | | Lack of innovation(O3) | 0,296 | 0,031 | 10 |
| Environmental(E) | 0,045 | Economic environment (E1) | 0,666 | 0,030 | 11 |
| | | Budgetary constraints(E2) | 0,333 | 0,015 | 16 |

## 5. DISCUSSION

The results show that strategic, technological, skills and human resources, organizational and cultural, and environmental barriers significantly contribute to the emergence of cybersecurity threats. Strategic barriers are the most significant with a priority of 0.495, notably the lack of strategy (S1) with an overall priority of 0.270, the absence of a clear vision (S2) at 0.115, and poor implementation planning (S4) at 0.068. These elements indicate that the lack of planning and clear strategic vision significantly increases the risks of cyberattacks.

Technological barriers follow with a priority of 0.224. Difficulties in integrating information technologies (T1) have an overall priority of 0.108, while IT security risks (T2) have a priority of 0.058. Inadequate technological infrastructure and challenges in technological integration increase vulnerabilities to cyber threats.

Barriers related to skills and human resources have a priority of 0.128. Low digital literacy (SHR2) with a priority of 0.070 and the lack of human resources (SHR3) at 0.018 are critical factors. A weak digital culture and a lack of qualified personnel increase the risks of human errors and security breaches.

Organizational and cultural barriers have a priority of 0.106. Resistance to change (O2) with a priority of 0.057 indicates that organizational adaptation difficulties can also facilitate cyber threats.

Finally, environmental barriers, although the least prioritized with a value of 0.045, include budgetary constraints (E2) with a priority of 0.015, which limit the ability to invest in adequate cybersecurity measures. Our article adds unique value by focusing specifically on the Moroccan context, using a multi-criteria approach to identify and rank obstacles to digital transformation in relation to cybersecurity threats. Unlike general studies by (Gkrimpizi and Peristeras 2022b; Rima Aditya, Ferdiana, and Suning Kusumawardani 2021), this study offers strategic recommendations tailored to the local context through empirical data collection from Moroccan officials and professionals. The use of the Analytic Hierarchy Process (AHP) allows for quantifying the importance of barriers and proposing specific, applicable strategies to improve cybersecurity in Moroccan institutions, thus making a significant and pragmatic contribution to the existing literature.

## 6. CONCLUSION

The acceleration of digital transformation in higher education institutions due to the COVID-19 pandemic has highlighted increased vulnerabilities to cyber threats. This study revealed that the main obstacles to this transformation are digital culture, formulating adequate strategies, and technological infrastructure. These barriers significantly increase cyber risks, thereby threatening the security and sustainability of educational systems.

Applying the AHP methodology to evaluate and rank these obstacles in the context of Moroccan higher education has identified the most critical barriers. The results show that insufficient digital culture, lack of clear strategies, and inadequate technological infrastructures are the primary sources of vulnerability to cyberattacks.

These conclusions provide stakeholders with a solid foundation to wisely allocate resources and develop robust cybersecurity strategies. By aligning these strategies with Morocco's digital vision for 2015-2030, educational institutions can not only strengthen their digital security but also support a sustainable and competitive digital transformation. It is imperative to continue monitoring and evaluating these obstacles to minimize their impact and effectively secure the educational sector against cyber threats.

## REFERENCES

Abdelilah, C., Ahriz, S., El Guemmat, K. & Mansouri, K. (2024). Building a Specialized IT Governance Strategy for Higher Education: A Strategic Model. Journal of Computer Science, 20(7), 768-782. https://doi.org/10.3844/jcssp.2024.768.782

Abdelilah, C., Ahriz, S., El Guemmat, K., & Mansouri, K. (2024b). Implementation of suitable information technology governance frameworks for Moroccan higher education institutions. International Journal of Electrical and Computer Engineering (IJECE), 14(3), 3116-3126

Aditya, B. R., Ferdiana, R., & Kusumawardani, S. S. (2021a). Barriers to Digital Transformation in Higher Education : An Interpretive Structural Modeling Approach. International Journal of Innovation and Technology Management, 18(05), 2150024. https://doi.org/10.1142/S0219877021500243

Aditya, B. R., Ferdiana, R., & Kusumawardani, S. S. (2021b). Categories for barriers to digital transformation in higher education : An analysis based on literature. International Journal of Information and Education Technology, 11(12), 658-664.

Aditya, B. R., Ferdiana, R., & Kusumawardani, S. S. (2022a). A barrier diagnostic framework in process of digital transformation in higher education institutions. Journal of Applied Research in Higher Education, 14(2), 749-761.

Aditya, B. R., Ferdiana, R., & Kusumawardani, S. S. (2022b). Identifying and prioritizing barriers to digital transformation in higher education : A case study in Indonesia. International Journal of Innovation Science, 14(3/4), 445-460.

Aghasafari, H., Karbasi, A., Mohammadi, H., & Calisti, R. (2020). Determination of the best strategies for development of organic farming : A SWOT–Fuzzy Analytic Network Process approach. Journal of Cleaner Production, 277, 124039.

AL-Ali, M., & Marks, A. (2022). A digital maturity model for the education enterprise. Perspectives: Policy and Practice in Higher Education, 26(2), 47-58. https://doi.org/10.1080/13603108.2021.1978578

Alenezi, M. (2021). Deep dive into digital transformation in higher education institutions. Education Sciences, 11(12), 770.

Alzahrani, B., Bahaitham, H., Andejany, M., & Elshennawy, A. (2021). How ready is higher education for quality 4.0 transformation according to the LNS research framework? Sustainability, 13(9), 5169.

Benavides, L. M. C., Tamayo Arias, J. A., Arango Serna, M. D., Branch Bedoya, J. W., & Burgos, D. (2020). Digital transformation in higher education institutions : A systematic literature review. Sensors, 20(11), 3291.

Chipembele, M., & Bwalya, K. J. (2016). Assessing e-readiness of the Copperbelt University, Zambia : Case study. The International Journal of Information and Learning Technology, 33(5), 315-332.

Chitre, M. D., Dhoke, M. A., & Shriniwar, M. A. (2022). A Review : Insider Attack in New Normal. https://www.academia.edu/download/94528597/A_20Review_20Insider_20Attack_20in_20New_20Normal.pdf

Donnelly, R. (2017). Blended problem-based learning in higher education : The intersection of social learning and technology. Psychosociological Issues in Human Resource Management, 5(2), 25-50.

Ferhane, D., & Yassine, L. (2022). La transformation numérique de l'université marocaine à l'épreuve de la covid 19 : Transition vers un modèle universitaire agile. IJTM International Journal of Trade and Management, 1(1), 55-69.

Gkrimpizi, T., & Peristeras, V. (2022). Barriers to digital transformation in higher education institutions. 15th International Conference on Theory and Practice of Electronic Governance, 154-160. https://doi.org/10.1145/3560107.3560135

Jensen, T. (2019). Higher education in the digital era. The current state of transformation around the world. International Association of Universities, 2019, 28-42.

Jhones, A. R., & Larramendi, J. V. (2019). Informatizing the Universities : Reflections on One Cuban Experience. Library Trends, 67(4), 655-667.

Kalolo, J. F. (2019). Digital revolution and its impact on education systems in developing countries. Education and Information Technologies, 24(1), 345-358. https://doi.org/10.1007/s10639-018-9778-3

Kaminskyi, O. Y., Yereshko, Y. O., & Kyrychenko, S. O. (2018). Digital transformation of university education in Ukraine : Trajectories of development in the conditions of new technological and economic order. Інформаційні технології і засоби навчання, 64,№ 2, 128-137.

Kendall, C. L. (2022). The Openness of Higher Education and Implications on Cybersecurity [PhD Thesis, Utica University]. https://search.proquest.com/openview/628bf9ae362721c1afea428ac3f51732/1?pq-origsite=gscholar&cbl=18750&diss=y

Kerroum, K., Khiat, A., Bahnasse, A., & Aoula, E.-S. (2020). The proposal of an agile model for the digital transformation of the University Hassan II of Casablanca 4.0. Procedia computer science, 175, 403-410.

Khan, M. S. H., Hasan, M., & Clement, C. K. (2012). Barriers to the introduction of ICT into education in developing countries : The example of Bangladesh. International Journal of instruction, 5(2). https://dergipark.org.tr/en/download/article-file/59739

Kopp, M., Gröblinger, O., & Adams, S. (2019). Five common assumptions that prevent digital transformation at higher education institutions. Inted2019 Proceedings, 1448-1457. https://library.iated.org/download/KOPP2019FIV

Kozarkiewicz, A. (2020). General and Specific : The Impact of Digital Transformation on Project Processes and Management Methods. Foundations of Management, 12(1), 237-248. https://doi.org/10.2478/fman-2020-0018

Lašáková, A., Bajzíková, L., & Dedze, I. (2017). Barriers and drivers of innovation in higher education : Case study-based evidence across ten European universities. International Journal of Educational Development, 55, 69-79.

Laufer, M., Leiser, A., Deacon, B., Perrin De Brichambaut, P., Fecher, B., Kobsda, C., & Hesse, F. (2021). Digital higher education : A divider or bridge builder? Leadership perspectives on edtech in a COVID-19 reality. International Journal of Educational Technology in Higher Education, 18(1), 51. https://doi.org/10.1186/s41239-021-00287-6

Marks, A., & AL-Ali, M. (2022). Digital Transformation in Higher Education : A Framework for Maturity Assessment. In M. Alaali (Éd.), COVID-19 Challenges to University Information Technology Governance (p. 61-81). Springer International Publishing. https://doi.org/10.1007/978-3-031-13351-0_3

Marks, A., AL-Ali, M., Atassi, R., Elkishk, A. A., & Rezgui, Y. (2021). Digital Transformation in Higher Education : Maturity and Challenges Post COVID-19. In Á. Rocha, C. Ferrás, P. C. López-López, & T. Guarda (Éds.), Information Technology and Systems (Vol. 1330, p. 53-70). Springer International Publishing. https://doi.org/10.1007/978-3-030-68285-9_6

Marshall, S. J. (2018). Shaping the University of the Future. Springer Singapore. https://doi.org/10.1007/978-981-10-7620-6

Mohamed Hashim, M. A., Tlemsani, I., & Duncan Matthews, R. (2022). A sustainable University : Digital Transformation and Beyond. Education and Information Technologies, 27(7), 8961-8996. https://doi.org/10.1007/s10639-022-10968-y

Morrison, J. L. (2003). Organizational change for corporate sustainability-a guide for leaders and change agents of the future. Journal of Education for Business, 79(2), 124.

Pavlova, E. (2020). Enhancing the organisational culture related to cyber security during the university digital transformation. Information & Security, 46(3), 239-249.

Peng, S., Zhou, D., & Xie, B. (2022). Analysis of LNG storage tank safety : A comprehensive model approach with ANP and normal cloud. Applied Sciences, 12(23), 11941.

Rafiq, M., Batool, S. H., Ali, A. F., & Ullah, M. (2021). University libraries response to COVID-19 pandemic : A developing country perspective. The Journal of Academic Librarianship, 47(1), 102280.

Reid, P. (2014). Categories for barriers to adoption of instructional technologies. Education and Information Technologies, 19(2), 383-407. https://doi.org/10.1007/s10639-012-9222-z

Reis, J., Amorim, M., Melão, N., & Matos, P. (2018). Digital Transformation : A Literature Review and Guidelines for Future Research. In Á. Rocha, H. Adeli, L. P. Reis, & S. Costanzo (Éds.), Trends and Advances in Information Systems and Technologies (Vol. 745, p. 411-421). Springer International Publishing. https://doi.org/10.1007/978-3-319-77703-0_41

Rima Aditya, B., Ferdiana, R., & Suning Kusumawardani, S. (2021). Digital Transformation in Higher Education : A Barrier Framework. 2021 3rd International Conference on Modern Educational Technology, 100-106. https://doi.org/10.1145/3468978.3468995

Saaty, T. L. (1982). The Analytic Hierarchy Process : A New Approach to Deal with Fuzziness in Architecture. Architectural Science Review, 25(3), 64-69. https://doi.org/10.1080/00038628.1982.9696499

Saeed, S., Altamimi, S. A., Alkayyal, N. A., Alshehri, E., & Alabbad, D. A. (2023). Digital Transformation and Cybersecurity Challenges for Businesses Resilience : Issues and Recommendations. Sensors, 23(15), 6666.

Sanchez, M. A. (2020). University e-readiness for the digital transformation : The case of Universidad Nacional del Sur. Revista Gestão & Tecnologia, 20(2), 75-97.

Sandhu, K. (2021). Advancing Cybersecurity for Digital Transformation : Opportunities and Challenges. Handbook of Research on Advancing Cybersecurity for Digital Transformation, 1-17.

Shafiee, M. (2022). Wind Energy Development Site Selection Using an Integrated Fuzzy ANP-TOPSIS Decision Model. Energies, 15(12), 4289.

Shahabi, R. S., Basiri, M. H., Kahag, M. R., & Zonouzi, S. A. (2014). An ANP–SWOT approach for interdependency analysis and prioritizing the Iran′ s steel scrap industry strategies. Resources policy, 42, 18-26.

Sharma, V., Gidwani, B. D., Sharma, V., & Meena, M. L. (2019). Implementation model for cellular manufacturing system using AHP and ANP approach. Benchmarking: An International Journal, 26(5), 1605-1630.

Sinclair, J., & Aho, A.-M. (2018). Experts on super innovators : Understanding staff adoption of learning management systems. Higher Education Research & Development, 37(1), 158-172. https://doi.org/10.1080/07294360.2017.1342609

Stüber, J. (2018). Barriers of digital technologies in higher education : A teachers' perspective from a Swedish university. https://www.diva-portal.org/smash/record.jsf?pid=diva2:1201871

Taherdoost, H., Madanchian, M., & Ebrahimi, M. (2021). Advancement of Cybersecurity and Information Security Awareness to Facilitate Digital Transformation : Opportunities and Challenges. Handbook of Research on Advancing Cybersecurity for Digital Transformation, 99-117.

Tamer, H., & Knidiri, Z. (2023). University 4.0 : Digital Transformation of Higher Education Evolution and Stakes in Morocco. American Journal of Smart Technology and Solutions, 2(1), 20-28.

T. M. Siebel, « Why digital transformation is now on the CEO's shoulders », *McKinsey Quarterly*, vol. 4, n° 3, p. 1-7, 2017.

Thoring, A., Rudolph, D., & Vogl, R. (2017). Digitalization of higher education from a student's point of view. European Journal of Higher Education IT, 1. https://www.eunis.org/download/2017/EUNIS_2017_paper_47.pdf

Vicente, P. N., Lucas, M., Carlos, V., & Bem-Haja, P. (2020). Higher education in a material world : Constraints to digital innovation in Portuguese universities and polytechnic institutes. Education and Information Technologies, 25(6), 5815-5833. https://doi.org/10.1007/s10639-020-10258-5

Watty, K., McKay, J., & Ngo, L. (2016). Innovators or inhibitors? Accounting faculty resistance to new educational technologies in higher education. Journal of Accounting Education, 36, 1-15.

Williams, C. M., Chaturvedi, R., & Chakravarthy, K. (2020). Cybersecurity risks in a pandemic. Journal of medical Internet research, 22(9), e23692.

Xu, J., Li, L., & Ren, M. (2022). A hybrid ANP method for evaluation of government data sustainability. Sustainability, 14(2), 884.

Yoo, Y., Boland, R. J., Lyytinen, K., & Majchrzak, A. (2012). Organizing for Innovation in the Digitized World. Organization Science, 23(5), 1398-1408. https://doi.org/10.1287/orsc.1120.0771

# USING AN ESCAPE ROOM GAME TO EXPLORE THE RELATIONSHIPS BETWEEN GAMEPLAY SELF-EFFICACY, EXPERIENTIAL VALUE AND LEARNING PROGRESS IN A PERFORMING ARTS COURSE

Jon-Chao Hong[1], Yi-An Chien[2] and Chih-Hsuan Ho[3]
[1]*Department of Industrial Education, National Taiwan Normal University*
*Taipei, Taiwan*
[2]*Continuing Education Master's Program of Creativity Development, National Taiwan Normal University*
*Taipei, Taiwan*
[3]*Department of Education and Learning Technology, National Tsing Hua University*
*Hsinchu, Taiwan*

**ABSTRACT**

To assess the effectiveness of applying the digital game-based approach, this study explored the correlation between GSE for using an Escaping Room App, named Brain Challenge, epistemic value of game experience, hedonic value of game experience, and learning progress in a junior high school performing arts course. A total of 186 valid data were collected after Brain Challenge practices for structural equation modeling. The results showed that GSE was positively related to hedonic value and epistemic value, and could also positively predict learning progress in the performing arts course. The results imply that performing arts teachers can implement learning content in an escape room to motivate students with high levels of GSE to learn performing arts.

**KEYWORDS**

Performing Arts, Gameplay Self-Efficacy, Epistemic Value, Hedonic Value, Escape Room

## 1. INTRODUCTION

There is a significant disparity in educational outcomes between urban and rural schools. Some learners in rural areas are considered disadvantaged compared to their urban counterparts (Ge & Wang, 2019) and often exhibit inferior academic performance (Gehrke et al., 2019). According to Fernandez-Rio et al. (2020), GBL promotes meaningful learning for learners across different educational stages (primary and secondary schools), socio-economic backgrounds (low, medium, high), and educational settings (urban and rural schools). For example, in Taiwan it was found that GBL could better promote rural students' intrinsic motivation to learn science than using video games (Liao et al., 2019). Video games combine visual processing technologies with advanced human–computer interaction and have changed the learning practice in education (Dong, 2023), while the instructional innovation of escape room games creates meaningful positive effects on learners' learning experiences (Fotaris & Mastoras, 2019; Taraldsen et al., 2022). However, escape room games have not been implemented for rural students to learn performing arts; thus, this study targeted rural junior high school students who were beginners in performing arts to play an escape room game to learn about participation frameworks and spatial arrangements.

From a cognitive and emotional perspective (Mayer, 2019), game designers aim to attract players to have immersive experiences, which engage players to perceive value in their own gaming worlds. For example, an escape room can stimulate medical, nursing and pharmacy students' value of the experience (Lepp et al., 2023). This is especially evident when students engage in escape room games, as their gaming ability and confidence can facilitate their epistemic curiosity to acquire various types of knowledge (Lepp et al., 2023). To date, students' personal self-efficacy and perceived experiential value in performing arts have not been extensively studied; thus, the present study examined the correlates between game self-efficacy, experiential value and learning progress when rural students used an escape room game to learn performing arts.

# 2. METHOD

## 2.1 Research Model

The Trait Activation Theory (TAT; Tett et al., 2013) suggests that traits are activated in response to situational cues, with varying relevance of situations to specific traits. This theory focuses on how personal attributes, such as personality factors and abilities, interact with stable characteristics to influence behavior (Tett et al., 2013). We applied TAT to our theoretical model, proposing that individual differences in gameplay self-efficacy moderate the relationship between perceived experiential value and learning progress (see Figure 1).

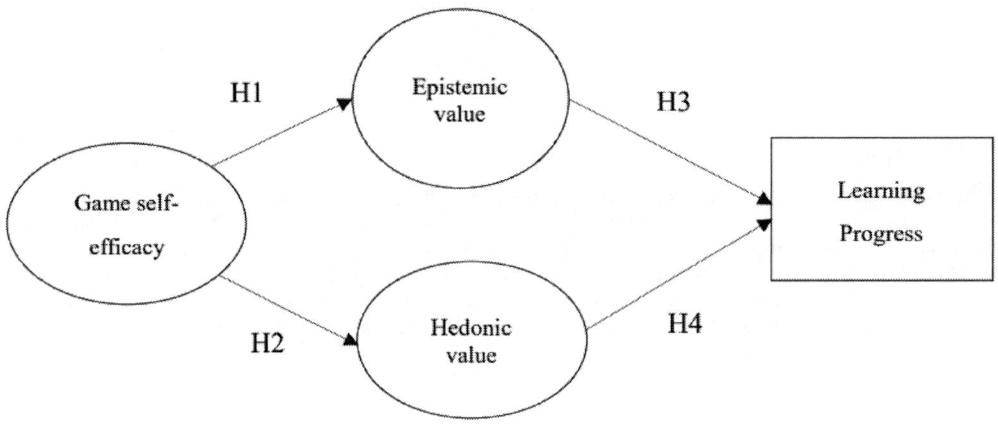


Figure 1. Research Model

## 2.2 Hypotheses

Vidergor (2021) suggested that when integrating digital escape room games into instruction, gameplay self-efficacy can enhance positive emotional responses. In this study, when participants were working in the game context, how their experiential value was driven by their gameplay self-efficacy was hypothesized as follows:

H1: Gameplay self-efficacy is positively related to epistemic value.

H2: Gameplay self-efficacy is positively related to hedonic value.

Epistemic value influences individuals' ability to recognize errors and find information to correct, and improves their game performance (Muis et al., 2018). If learning content can be successfully gamified, learners can develop a curiosity for seeking knowledge, actively acquire new information and knowledge, and significantly improve their learning performance (Hong et al., 2021). How epistemic value affects students' learning progress in performing arts cognitive learning with Brain Challenge was hypothesized as follows:

H3: Epistemic value is positively related to learning progress.

When the focus is on enhancing performance, people who are successful in a competitive situation will show higher enjoyment (Sepehr & Head, 2018). GBL can provide a motivational scheme, playing in the game activity with the enjoyment that lead players to have better game progress (Tavares, 2022). By fostering hedonic value in the learning process, educators can optimize learners' progress and outcomes (An et al., 2021). How hedonic value affects students' learning progress in performing arts with Brain Challenge was hypothesized as follows:

H4: Hedonic value is positively related to learning progress.

## 2.3 The Escape Room - "Brain Challenge"

The research tool in this study, "Brain Challenge," was designed by the Digital GBL Lab of National Taiwan Normal University. It is a digital escape room game. There are two levels and one type of room which involve a matching game that includes a mastery knowledge learning stage and a knowledge management stage. Learners must provide correct answers within the 3-minute limit at each stage to obtain the key to proceed to the next level. Content included the scope of performing arts, the golden triangle that constitutes a performance, physical level, stage and curtain type, and stage location terminology related to performing arts (see Figure 2). Secondly, the sixth lessons are game experience courses by playing room escape games, particularly in matching tasks of performing arts. When playing the game, students could escape from the virtual escape room in a limited amount of time, obtain the virtual room key for the next level, unlock the next level, and ultimately succeed in escaping.

Figure 2. Matching Questions of Brain Challenge

## 2.4 Data Collection and Participants

This study utilized a single-group quasi-experiment to investigate the research hypotheses, purposive sampling was conducted, involving seventh to ninth-grade students from a rural area in northern Taiwan during the spring semester of 2022. Participants completed the questionnaire using paper and pencil, yielding 208 responses from three music classes. After eliminating invalid responses, data from 186 students were included in the statistical analysis, with 87 (46.8%) males and 99 (53.2%) females.

## 2.5 Questionnaire

The questionnaire items were adapted from previous research and were translated into Chinese to meet the requirements of this study. The items were reviewed by three experts to ensure their content validity. Moreover, six students were invited to read the questionnaire to ensure its face validity. The questionnaire was based on a Likert 5-point scale, and the responses were ranked from 1 to 5, with 1 for *strongly disagree*, 2 for *disagree*, 3 for *average*, 4 for *agree*, and 5 for *strongly agree*. Because this was a confirmatory study, the reliability and validity of the questionnaire were re-tested after data collection.

# 3. RESULT

## 3.1 Item Analysis

To test the internal validity of the items, Amos 20 was used to examine the factor loadings (FL) of each item. If the FL value was over 0.5, the item was deleted, as suggested by Shrestha (2021). To test the external validity of items, an independent sample *t* test was performed by comparing the top 27% and the bottom 27% of each item. Green and Salkind (2004) suggested that a *t*-value greater than 3.0 ($p < .001$) is considered as having a significant level of item validity. The *t* values for each item were acceptable and are summarized in Table 1Accordingly, items were deleted, reducing GSE from six to five items, epistemic value from five to four, and hedonic value also from six to five.

Table 1. Item Analysis

| | $\chi^2/df$ | RMSEA | GFI | AGFI | FL | *t*-value |
|---|---|---|---|---|---|---|
| Threshold | < 5 | < .10 | ≥ .80 | ≥ .80 | > .50 | $\|t\| > 3$ |
| Gameplay self-efficacy | 1.95 | .07 | .98 | .94 | .80-.86 | 17.80 ~ 22.41 |
| Epistemic value | 1.43 | .05 | .99 | .96 | .79-.82 | 14.86 ~ 19.14 |
| Hedonic value | 1.52 | .05 | .98 | .95 | .76-85 | 12.62 ~ 20.72 |

## 3.2 Construct Reliability and Validity Analysis

Taber (2018) indicated the purpose of using Cronbach's α as an internal consistency indicator, and suggested that Cronbach's α should be over .70. In addition, Taber (2018) posited that the value of composite reliability (CR) for external consistency should be over .70. The statistical results are shown in Table 2, and indicate that the data of each construct met the thresholds. If a higher convergent validity is required, factor loading (FL) and the average variance extracted (AVE) should be greater than .50 (Hair & Alamer, 2022). Table 2 shows that the results of FL were between .80 and .83, and those of AVE were between .64 and .69, indicating that convergent validity of this questionnaire was acceptable.

Table 2. Reliability and Validity Analysis

| Construct | *M* | *SD* | Cronbach's α | CR | AVE | FL |
|---|---|---|---|---|---|---|
| Gameplay self-efficacy | 3.61 | 1.07 | .92 | .92 | .69 | .83 |
| Epistemic value | 3.58 | 1.05 | .88 | .88 | .64 | .80 |
| Hedonic value | 3.82 | .99 | .90 | .90 | .65 | .80 |

## 3.3 Learning Progress

This study analyzed the progress of students who played the escape room game—Brain Challenge. The results show that the total average of the first trial was $M = 66.20$, $SD = 22.19$, while the total average of the third test was $M = 74.36$, $SD =$ is 20.30. According to the paired sample *t* test, the *t* value was -5.22*** ($p < .001$), reaching a significant level. The result shows that after becoming familiar with Brain Challenge, the students' learning performance significantly improved. The detailed information is shown in Table 3.

Table 3. Performance in the First and Third Trials

| Performance | M | SD | *t value* |
|---|---|---|---|
| First trial | 66.20 | 22.19 | -5.22*** |
| Third trial | 74.36 | 20.30 | |

$p^* < .05$, $p^{**} < .01$, $p^{***} < .001$

## 3.4 Model Fit Analysis

According to Kline's (2023) suggestion, acceptable values of the absolute fit indices are: $\chi^2/df < 0.10$, root mean square error of approximation (RMSEA) < .08, goodness-of-fit index (GFI) > .08, and adjusted goodness-of-fit index (AGFI) > .08, and for the incremental fit indices they are: normed fit index (NFI) > .80, Tucker-Lewis index/ non-normed fit index (TLI or NNFI) > .90, comparative fit index (CFI) > .90, comparative fit index (IFI) > .90, and relative fit index (RFI) > .80. Parsimony adjusted indices threshold values are: parsimony-normed fit index (PNFI) > .50 and parsimony goodness-of-fit Index (PGFI) > .05. The results of the model fit indices in this study are all acceptable (see Table 4).

Table 4. Model Fit Analysis

| Index | Threshold | Result |
|---|---|---|
| Absolute fit indices | | |
| $\chi^2/df$ | <5 | 1.93 |
| RMSEA | <.10 | .07 |
| GFI | ≥.80 | .83 |
| AGFI | ≥.80 | .80 |
| Incremental fit indices | | |
| NFI | ≥.80 | .87 |
| TLI / NNFI | ≥.90 | .92 |
| CFI | ≥.90 | .93 |
| IFI | ≥.90 | .93 |
| RFI | ≥.80 | .85 |
| Parsimony adjusted indices | | |
| PNFI | ≥.50 | .78 |
| PGFI | ≥.50 | .67 |

## 3.5 Path Analysis

Path analysis was verified on the four research hypotheses in this study by using the software AMOS 20.0. The model validation results are shown in Figure 3 and indicate that game self-efficacy was positively related to epistemic value ($\beta = .84$, $t = 10.58$***, $p < .001$); game self-efficacy was positively related to hedonic value ($\beta = .85$, $t = 11.15$***, $p < .001$); epistemic value was positively related to learning progress ($\beta = .34$, $t = 2.63$**, $p < .01$); and hedonic value was positively related to learning progress ($\beta = .43$, $t = 3.35$***, $p < .001$). Figure 3 shows that the explanative power of game self-efficacy on epistemic value was 50% ($R^2 = .50$) with a large effect size ($f^2 = 1.00$); the explanative power of game self-efficacy on hedonic value was 46% ($R^2 = .46$) with a large effect size ($f^2 = .85$); and the explanative power of epistemic value and hedonic value on learning progress was 26% ($R^2 = .26$) with a medium effect size ($f^2 = .35$).

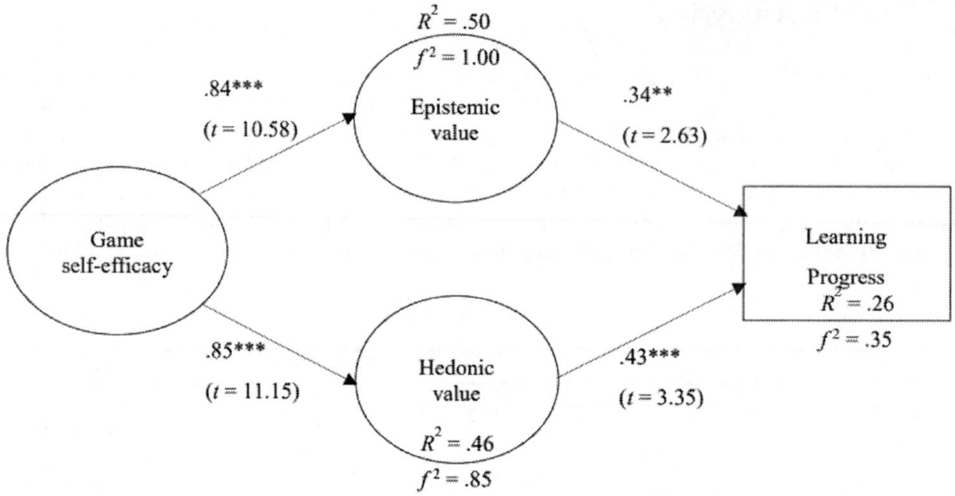


*$p < .05$, **$p < .01$, ***$p < .001$

Figure 3. The Verification of the Research Model

# 4. DISCUSSION

The discussion revolves around the application and validation of Trait Activation Theory (TAT) in the context of exploring the relationship between individual differences in gameplay self-efficacy, perceived experiential value, and learning progress. The hypotheses derived from TAT are examined and supported through the study's findings.

Higher gameplay self-efficacy correlates with a greater desire for knowledge acquisition, aligning with previous research indicating that self-efficacy influences attention, effort, and time investment in tasks (Golparvar & Khafi, 2021). This supports the hypothesis that higher self-efficacy leads to increased epistemic value (H1). Gameplay Self-Efficacy and Hedonic Value: Research demonstrates a strong correlation between gameplay self-efficacy and hedonic value in gaming contexts, where higher self-efficacy enhances enjoyment and pleasure, thereby strengthening engagement and motivation (Sharma et al., 2022). This supports the hypothesis that higher self-efficacy leads to increased hedonic value (H2).

Epistemic Value and Learning Performance: Studies show that integrating epistemic value into digital games improves learning outcomes, indicating a positive correlation between epistemic value and learning performance (Chen & Stoddard, 2020). This supports the hypothesis that higher epistemic value in gaming leads to improved learning performance (H3). Hedonic Value and Learning Performance: Positive experiences in learning, characterized by higher hedonic value, contribute to better learning performance by regulating anxiety and negative emotions (Liu & Wang, 2021). This aligns with the hypothesis that higher hedonic value in gaming leads to improved learning performance (H4).

# 5. CONCLUSION

The use of digital technologies is exerting a powerful influence on the practice of cognitive learning. The present study used “Brain Challenge” as a tool for junior high school students to reinforce their cognitive learning in performing arts and explored their emotional factors which were driven by their game self-efficacy in using Brain Challenge, and affected their epistemic value and hedonic value to trigger better learning performance.

## 5.1 Implications

Digital GBL is designed to promote learning and play value for learners, and if well designed, serious games can support effective learning (Berki & Tarjányi, 2022; Rosenthal & Ratan, 2022). From the perspective of escape room GBL, this study changed the traditional performing arts teaching, and adapted gameplay as evaluation pedagogy by incorporating an educational digital room escape game into the after-teaching evaluation. In the context of escape room games, learners in rural areas can benefit to learn performing arts knowledge, and aesthetic knowledge. Thus, it is suggested that performing arts teachers can use Brain Challenge to evaluate their students' learning effect. Moreover, in line with the results of this study, the escape room game design mechanisms and game modules that allow teachers to implement different contents based on the question-making mechanisms. The game module can significantly impact on learners' performance, so it can be used for rural teachers to motivate students to learn domain-based knowledge.

## 5.2 Limitations and Future Study

In terms of the collection of research data, the samples of this study were junior high school students in rural areas of Hsinchu County. The geographical environment, living background, resource allocation, students' physical and mental characteristics and other factors of students in rural areas are all very different from urban areas, so the sample collection is limited in this study. For example, the research lasted about 2 months, including classroom teaching, playing Brain Challenge and questionnaire filling. However, there were a few students who asked for leave, so the sample size was not as large as expected. Therefore, the period of experiments can be lengthened based on the environmental conditions in each rural school.

## REFERENCES


An, Z., Wang, C., Li, S., Gan, Z., & Li, H. (2021). Technology-assisted self-regulated English language learning: Associations with English language self-efficacy, English hedonic value, and learning outcomes. *Frontiers in Psychology*, *11*, 558466. https://doi.org/10.3389/fpsyg.2020.558466

Chen, J. A., & Stoddard, J. D. (2020). A virtual internship to prepare high school students for civic and political action. *Educational Technology Research and Development, 68*, 3449-3470. https://doi.org/10.1007/s11423-020-09847-5

Dong, Y. (2023). Research on deep learning video game simulation algorithm based on western wind music. *Entertainment Computing, 46*, 100566. https://doi.org/10.1016/j.entcom.2023.100566

Fernandez-Rio, J., de las Heras, E., González, T., Trillo, V., & Palomares, J. (2020). Gamification and physical education. Viability and preliminary views from students and teachers. *Physical Education and Sport Pedagogy*, *25*(5), 509-524. https://doi.org/10.1080/17408989.2020.1743253

Fotaris, P., & Mastoras, T. (2019). Escape rooms for learning: A systematic review. *Proceedings of the European Conference on Games Based Learning*, 235-243.

Ge, T., & Wang, L. (2019). Multidimensional child poverty, social relationships and academic achievement of children in poor rural areas of China. *Children and Youth Services Review*, *103*, 209-217. https://doi.org/10.1016/j.childyouth.2019.06.007

Gehrke, E., Lenel, F., & Schupp, C. (2022). COVID-19 crisis in Cambodia: A dataset containing linked survey and administrative data of ninth-graders in rural areas. *Data in Brief*, *43*, 108476. https://doi.org/10.1016/j.dib.2022.108476

Golparvar, S. E., & Khafi, A. (2021). The role of L2 writing self-efficacy in integrated writing strategy use and performance. *Assessing Writing*, *47*, 100504. https://doi.org/10.1016/j.asw.2020.100504

Green, S. B., & Salkind, N. J. (2010). *Using SPSS for Windows and Macintosh: Analyzing and understanding data*. Prentice Hall Press. https://doi.org/10.1198/tas.2005.s139

Hair, J., & Alamer, A. (2022). Partial least squares structural equation modeling (PLS-SEM) in second language and education research: ECECCF using an applied example. *Research Methods in Applied Linguistics, 1*(3), 100027. https://doi.org/10.1016/j.rmal.2022.100027

Hong, J. C., Hwang, M. Y., Tai, K. H., & Lin, P. H. (2021). The effects of intrinsic cognitive load and gameplay interest on flow experience reflecting performance progress in a Chinese remote association game. *Computer Assisted Language Learning*, *34*(3), 358-378. https://doi.org/10.1080/09588221.2019.1614068

Hsu, C. L., & Chen, M. C. (2018). How gamification marketing activities motivate desirable consumer behaviors: Focusing on the role of brand love. *Computers in Human Behavior*, *88*, 121-133. https://doi.org/10.1016/j.chb.2018.06.037

Kline, R. B. (2023). *Principles and practice of structural equation modeling*. Guilford publications. https://doi.org/10.25336/csp29418

Lepp, G. A., Fierke, K. K., Friedrich, C., & Sick, B. (2023). How intention/reflection fosters student learning in an interprofessional experiential escape room activity. *Journal of Interprofessional Education & Practice, 30*, 100589. https://doi.org/10.1016/j.xjep.2022.100589

Liao, C. W., Chen, C. H., & Shih, S. J. (2019). The interactivity of video and collaboration for learning achievement, intrinsic motivation, cognitive load, and behavior patterns in a digital game-based learning environment. *Computers & Education, 133,* 43-55. https://doi.org/10.1016/j.compedu.2019.01.013

Liu, E., & Wang, J. (2021). Examining the relationship between grit and foreign language performance: Hedonic value and anxiety as mediators. *Frontiers in Psychology*, *12*, 666892. https://doi.org/10.3389/fpsyg.2021.666892

Mayer, R. E. (2019). Computer games in education. *Annual Review of Psychology, 70,* 531-549.

Mondada, L. (2016). Challenges of multimodality: Language and the body in social interaction. *Journal of SocioLinguistics, 20*(3), 336-366. https://doi.org/10.1111/ josl.1_12177.

Muis, K. R., Chevrier, M., & Singh, C. A. (2018). The role of epistemic emotions in personal epistemology and self-regulated learning. *Educational Psychologist*, *53*(3), 165-184. https://doi.org/10.1080/00461520.2017.1421465

Sepehr, S., & Head, M. (2018). Understanding the role of competition in video gameplay satisfaction. *Information & Management, 55,* 407-421. http://dx.doi.org/10.1016/j.im.2017.09.007

Sharma, T. G., Hamari, J., Kesharwani, A., & Tak, P. (2022). Understanding continuance intention to play online games: roles of self-expressiveness, self-congruity, self-efficacy, and perceived risk. *Behaviour & Information Technology*, *41*(2), 348-364. https://doi.org/10.1080/0144929X.2020.1811770

Shrestha, N. (2021). Factor analysis as a tool for survey analysis. *American Journal of Applied Mathematics and Statistics, 9*(1), 4-11. https://dx.doi.org/10.12691/ajams-9-1-2

Taber, K. S. (2018). The use of Cronbach's alpha when developing and reporting research instruments in science education. *Research in Science Education*, *48*(6), 1273-1296. https://doi.org/10.1007/s11165-016-9602-2

Taraldsen, L. H., Haara, F. O., Lysne, M. S., Jensen, P. R., & Jenssen, E. S. (2022). A review on use of escape rooms in education–touching the void. *Education Inquiry*, *13*(2), 169-184. https://doi.org/10.1080/20004508.2020.1860284

Tavares, N. (2022). The use and impact of game-based learning on the learning experience and knowledge retention of nursing undergraduate students: A systematic literature review. *Nurse Education Today, 117*, 105484. https://doi.org/10.1016/j.nedt.2022.105484

Tett, R. P., Simonet, D. V., Walser, B., & Brown, C. (2013). Trait activation theory. In N. Christiansen, & R. Tett (Eds.), *Handbook of personality at work* (pp. 71-100). Routledge. https://doi.org/10.4324/9780203526910.ch5

Vidergor, H. E. (2021). Effects of digital escape room on gameful experience, collaboration, and motivation of elementary school students. *Computers & Education*, *166*, 104156. https://doi.org/10.1016/j.compedu.2021.104156

# WELCOME: TURN ON YOUR CAMERA, YOU HAVE A LEAD ROLE

Eduardo Hamuy[1], Ricardo Abarca[1], Rolando Cori[2] and José Garrote[3]
[1]*Department of Education, University of Chile, Chile*
[2]*Department of Music, University of Chile, Chile*
[3]*Department of Anthropology, Alberto Hurtado University, Chile*

**ABSTRACT**

The reluctance of students to turn on their cameras during synchronous online classes is a significant impediment that must be addressed in the Post-COVID-19 era, when the legitimacy and effectiveness of online education is no longer given by the emergency. Cameras play a pivotal role in conveying visual information and fostering effective communication within the educational domain, thereby enriching the overall learning experience. This study underscores the imperative to formulate a comprehensive strategy tailored to instill the practice of turning on cameras and microphones among students. The research extends across multiple semesters, focusing on a course that aims at developing competencies in interdisciplinary collaborative work. The methodology involves the delineation and application of a nuanced strategy designed to cultivate a habitual activation of cameras and microphones. Preliminary findings indicate a discernible enhancement in students' active utilization of cameras, thereby positing this engagement as an invaluable asset not to be relegated to optional status in the videoconferencing milieu. This paper concludes by offering practical recommendations aimed at fostering the habitual activation of cameras, with a view to optimizing the online learning environment.

**KEYWORDS**

Social Presence, Videoconferencing, Online Learning, Camera Usage, Higher Education

## 1. INTRODUCTION

The COVID-19 pandemic encouraged the use of e-learning technologies in higher education with an explosive increase in the number of online courses and online students. The hasty and involuntary shift from face-to-face to online learning in many educational institutions was a challenging transition for students, teachers, and organizations (Terblanche et al., 2023). Some problems were technical, related to Internet access or the appropriate devices, while others were related to suitable teaching and learning methodologies. The shift to online learning during the pandemic shifted the focus from teacher-centered learning and face-to-face interaction in classrooms to a prominence of a technology-based pedagogical approach, in multiple dimensions. This calls for broadening the understanding of how students learn online and their engagement in a digital environment, so that a more effective learning experience can be designed.

During the period of greatest intensity of social distancing, there was the educational phenomenon that some refer to as Emergency Remote Teaching (ERT). ERT and e-learning are two distinct approaches to delivering education, especially in response to unforeseen circumstances such as the COVID-19 pandemic (Hodges et al., 2020). While e-learning is a complex and planned approach to delivering instruction using digital tools and resources in a structured way, ERT is a temporary and reactive measure taken during crises to ensure the continuity of education.

This unintended adoption had a significant impact on social interaction during courses (Mittal et al., 2022). What is known in the framework of the Community of Inquiry as *social presence* in online education (Anderson & Dron, 2011; Garrison et al., 2010) has been widely studied as a mediating factor in collaborative learning and fostering a sense of community (Garrison et al., 2010; Rapanta et al., 2020; Zhang et al., 2023). Social presence predicts both the expectations placed on a task (and the value assigned to it), which in turn affects cognitive and behavioral engagement, respectively (Edwards & Taasoobshirazi, 2022; Tirado Morueta et al., 2016). Specifically, students who feel socially connected in online learning environments are more likely to develop positive expectations about their academic success and value their learning, leading to greater cognitive and behavioral engagement.

While both professors and university students in many cases initially faced technical glitches, leading to a slowdown in learning, these technical challenges furthermore affected the quality of social interaction and communication within the virtual environment. The nature of online student engagement can be seen as a challenge for university faculty. Many teachers' lack of experience in online education led to difficulties in maintaining effective communication and interaction with students in the online environment, including providing timely feedback, facilitating student discussions, and fostering a sense of community (Rapanta et al., 2020).

The use of technologies in higher education requires appropriate assessment of teaching methods and adoption of effective learning strategies to meet the changing needs of students and provide for effective learning in online modalities, either synchronous or asynchronous, not just in emergency contexts.

In the COVID-19 crisis, emphasis was on immediate solutions that would allow instruction to be delivered remotely and transiently, without extensive planning or preparation. In this period, some strategies were proposed to address this situation, in general these recommendations aimed to consider health security, connectivity conditions, continuity of teaching, curricular prioritization and students' socio-emotional context at that moment (Castelli & Sarvary, 2020). In many cases, an attempt was made to adapt what was familiar to emergency conditions, for example, videoconferencing became a substitute for synchronous and face-to-face classes.

The use of Zoom (as the flagship software of this modality) in distance classes allows a set of features for a group class, among which the following stand out: audio and visual communication of participants with or without a camera; chat communication; breakout rooms for group interactions; screen sharing for presentations and application windows; annotation and whiteboard tools for interactive activities; opinion polling tools; hand raising and instant reactions; recording and playback for review of the session (Trust & Goodman, 2023). Nonetheless, Zoom also has features that make users face their own visibility, such as the "self-view" feature that shows users their own video stream. This feature can arouse a sense of shyness known as "mirror anxiety" due to one's visibility during video conferencing sessions (Trust & Goodman, 2023). In numerous instances, additional factors prompt the preference to keep video/audio disabled. Both students and instructors may opt to mute their videos and microphones to mitigate excessive background noise and potential signal degradation. Additionally, privacy concerns play a significant role, as users may wish to avoid being inadvertently exposed on camera.

For these and other reasons, this modality of digital and synchronous face-to-face learning used to counteract social distancing is not equivalent to the traditional in-person classroom. According to Li et al. (2022), some of the main challenges students and teachers faced when it came to camera use in Zoom classes during the COVID-19 pandemic were decreased motivation, technical issues, privacy concerns and inadequate communication with teachers.

There were professors who pointed out that classes held through videoconferencing were poor in interaction, almost unidirectional, since many students made a habit of keeping their cameras off. The phenomenon of cameras turned off has been analyzed in some studies (Castelli & Sarvary, 2020; Li et al., 2022; Trust & Goodman, 2023; Wu et al.; 2022) and the causes of this trend could be very diverse: technical conditions, privacy, social insecurity, students doing multiple activities simultaneously with classes, household and family chores, social activities, distraction and entertainment, boredom, or stress.

Cameras provide important visual information and necessary elements to communication in an educational environment. The lack of nonverbal cues in online classes can affect teachers' ability to gauge students' understanding and adjust their teachings accordingly. Facial expressions, body language, and eye contact can provide valuable information both to teachers about student engagement, understanding, and emotional state (Li et al., 2022) and to students as well, in interaction with peers and teachers. An absence of visual communication can lead to a less effective learning experience, lower academic performance and affect the teaching process that teachers carry out.

If we consider that social presence refers to the degree to which students feel connected to their peers and teachers in an online environment (Garrison et al., 2010; Rapanta et al., 2020), then keeping the cameras on can foster a sense of community and belonging, improve student motivation and engagement, and improve the quality of communication and interactions. Fostering social presence, while using video conferencing, facilitates students' interaction with their peers and teachers and provides opportunities for collaborative learning.

Therefore, it is important to find ways to encourage students to turn on their cameras during synchronous online classes to facilitate better engagement and communication between students and teachers, because the lack of face-to-face interaction can impair the student-teacher connection that could promote students' motivation for learning and sense of belonging (Castelli & Sarvary, 2020).

Several of the studies have proposed strategies and made recommendations to encourage students to engage with their cameras. Castelli and Sarvary (2020) suggest that teachers should create an inclusive and equitable environment without imposing it. To that end, they can make one-on-one contact with students who don't turn on their cameras to understand their concerns and offer support without pressuring them. They advise against requiring students to turn on their cameras — instead — offering alternative ways to participate in class. Alternatively they suggest: explicitly encourage the use of the camera for student acceptance, explaining the benefits of using the camera such as sharing videos in the classroom, improving communication and building interpersonal relationships thanks to greater social presence; provide rules and guidelines on when it is appropriate to have cameras on or off, offer solutions to minimize distractions during class, and establish what content can and cannot be shared; designing active learning activities to keep students engaged can reduce the temptation to turn off their cameras; and collect feedback from students to understand their challenges and tailor solutions to meet their needs.

Other authors (Li et al., 2022; Wu et al., 2022) complement these strategies by pointing out that greater technological knowledge can enhance a more effective, engaging, interactive, and satisfying remote learning experience for their students. Some ways are as follows: optimize camera and microphone technology; train students in the functions and affordances of videoconferencing; provide support and resources to help students overcome the increased self-awareness and anxiety related to camera use and thereby reduce anonymity by encouraging students to engage with video, audio, and chat features; encourage interaction through chat features and raising hands to facilitate communication and engagement among students; provide opportunities for profile customization and control over personal learning environment which can contribute to a more positive remote learning experience.

The literature review identifies the need to address an area for improvement in the use of videoconferencing, which promotes students' motivation for learning and sense of belonging to a learning community. The reluctance of students to turn on their cameras during synchronous online classes is a challenge that must be addressed in the Post-COVID-19 era, when the legitimacy and effectiveness of online education is no longer given by the emergency. Furthermore, it is necessary to consider that the decision of students to opt for greater anonymity, with the cameras off, may be explained not only by a social anxiety factor, but also by social conformity:

Social learning theory suggests that learning is a social process, and thus highly influenced by our surroundings. This social environment also applies to the virtual world, as students keep an eye on their classmates in order to adjust self-behaviors during class meetings. Social conformity theory suggests a change of behavior due to group influence, resulting in congruence between individuals and group. In an online class, if many of their classmates have video cameras on, students are likely to conform and turn on their own video regardless of their original intentions. (Wu et al., 2022, p.3)

Our objective is to explore elements of a pedagogical strategy that favor greater social presence of students and their use of cameras during online learning.

## 2. METHODOLOGY

The object of study is an optional course from the transversal skills program offered to undergraduate students from all careers (social sciences, engineering, law, natural sciences, humanities, arts, architecture, communications, etc.) of a large public university in Chile. Its main curricular objective is to develop competencies in interdisciplinary collaborative work. The course has been taught for four consecutive semesters and the fifth version is underway. The study presented now, emerged after the fourth version as a form of exploratory *ex post facto* nonexperimental research in an effort of analysis and assessment of progress in teaching practice. When the course was taught for the first time, there was no intention to make it an object of study in itself, so there was no initial plan to collect data, hence it has been carried out with partial information, and of diverse nature: some quantitative measurements (percentage of time with cameras on in online classes of versions 1 and 4) and many qualitative assessments (evaluations of the 4 versions of the courses through participatory workshops with students).

This study forms an integral component of a broader research initiative aimed at cultivating a pedagogical approach that fosters outreach and collaboration between undergraduate students and external communities. To facilitate this endeavor, the research adopts the Context, Input, Process, and Product (CIPP) evaluation model developed by Stufflebeam and Shinkfield in 1985. This model, renowned for its systematic and improvement-focused methodology, is particularly suited for assessing curriculum and the efficacy of educational innovations within the realm of education (Dizon, 2022).

The present investigation operates within the Context Evaluation phase of the CIPP model, which endeavors to identify initial strengths and weaknesses, contextualize the operational environment, and diagnose any challenges that could impede the program's effectiveness and participants' well-being (Stufflebeam & Shinkfield, 1985). Hence, from this perspective it is of paramount importance the assessment of students' engagement, interaction, and communication within class sessions, recognizing these as essential components for fostering interdisciplinary collaborative learning.

Following the identification of areas necessitating improvement, modifications to the teaching methodology are explored iteratively across four course iterations (Versions). At the culmination of each iteration, a collaborative qualitative evaluation with students is conducted to assess the course's development, including strengths and weaknesses of content, methodology, and overall learning experiences. Subsequently, proposals for enhancement are formulated and implemented in subsequent iterations.

## 3. RESULTS

The cyclical process of continuous improvement that was conducted across four course iterations is detailed in Table 1.

Table 1. Progressive improvements in teaching methods

| Course Version | Improvement tested | Objective | Qualitative assessment of iteration |
|---|---|---|---|
| Version 1. Online sync. mode | (Baseline teaching methodology: theoretical and practical course; videoconferencing; lecture classes and collaborative work groups; analysis and solution proposals to complex problems from interdisciplinary perspectives) | ( Baseline objective: Develop competencies in interdisciplinary collaborative work) | Need to reduce anonymity; strengthen sense of community and engagemen cameras turned off; new understanding of camera of phenomenon. |
| Version 2. Face-to-face mode | Face-to-face mode | Increase knowledge of peers and engagement in collaborative learning; receive students' feedback. | Need to solve obstacle of distance between campuses; need for further understanding of camera off phenomenon. |
| Version 3. Online sync. mode | Back to videoconferencing; no session recordings | Improve attendance; Reduce *mirror anxiety* and social anxiety; receive students' feedback. | Need to strengthen sense of community and engagement; cameras turned off; need to encourage students to engage with video, audio, and chat features. |
| Version 4. Online sync. mode | *Entrance hall* technique and preparation room; challenging project with expert guests; more active group learning; weekly reinforcement; random groups; no session recordings until the last sessions and with class consensus | Enhance more effective, engaging, interactive, and satisfying remote learning experience; reduce anonymity by encouraging students to engage with video, audio, and chat features; install and reinforce a habit of engagement with camera use; receive students' feedback. | Increased use of cameras; higher engagement. |

The course was first offered at the beginning of 2022, when there were still health care protocols that encouraged online courses, but always in a synchronous mode. At first, a third of the students enter the virtual environment with their cameras on, but after a few sessions all the cameras were turned off and only occasionally turned on when students had a presentation, participate in breakout rooms or are particularly motivated by a specific activity. Sometimes they didn't have their microphones enabled ("I don't have a microphone", "my microphone is broken", "I'm in the library and I can't turn on the microphone") and sometimes they connected online from very unfavorable places or conditions (while traveling in a bus or car or from a courtyard of their campus). On multiple occasions there is no clear feedback regarding the motivation or understanding that students have about the topics or activities. Figure 1. accounts for an average of students with cameras turned on during each session that was recorded. This calculation is based on the LMS attendance login in relation to the total number of students with their cameras on (total attendance/students cameras turned on) in three moments of the class recording (an initial moment during the first 10-15 minutes of recording after everyone has entered to the videoconference; an intermediate moment in the middle of the recording; and a final moment about 5 minutes before students start login out).

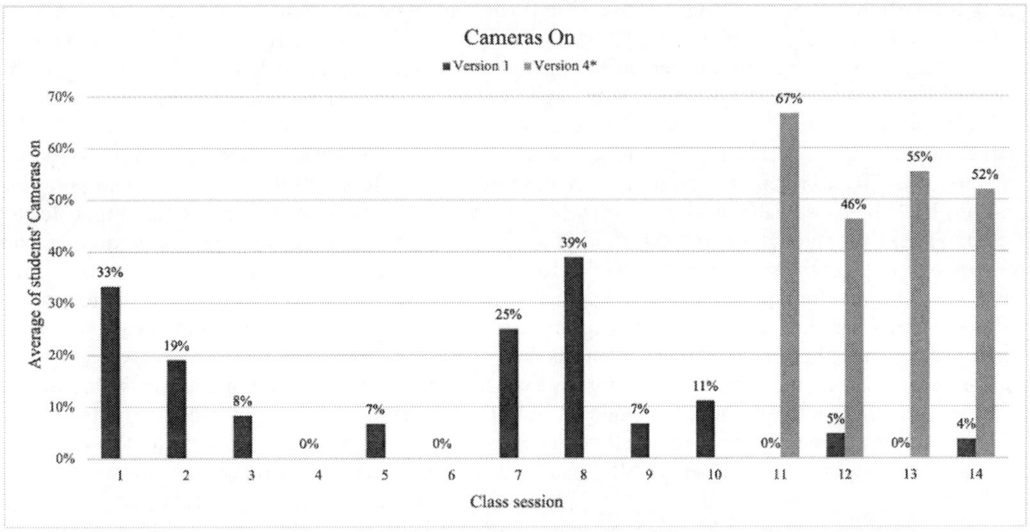


Figure 1. Average percent of cameras turned on in Version 1 and Version 4. * Version 4 only has data from the last class sessions, because the predominant criterion of that version was not to make video recordings, to help reduce student emotional stress

Across the course, students are encouraged to tackle certain collaborative tasks by means of group projects. In the first two versions, one of these tasks is to analyze the phenomenon of students turning off cameras from a systemic framework and propose innovations that may improve online communication. The result of their work allowed, on the one hand, to identify part of the academic and public discussion in the media about the phenomenon of cameras being turned off. On the other hand, very relevantly, the students present their own perceptions about this phenomenon and get involved in possible solutions. A joint evaluation with the students at the end of the first version establishes the need to reinforce their knowledge and engagement with their classmates to make collaborative work more effective in a future version. This feedback convinces us of the convenience of carrying out the second version in a face-to-face format.

However, this choice proves to be detrimental for attendance of many students and results in several dropouts along the semester because students come from various university campuses distributed around the city, some very distant from each other and cannot meet schedules or participate in all activities. For the following versions, online modality is adopted again.

Anticipating that the tendency of students to participate with the camera off would continue to be present, innovations that address social anxiety and provide meaning to interactions that require cameras turned on are introduced in the third version: recording the sessions is ruled out to reduce self-consciousness; greater emphasis is placed on group work activities in breakout rooms and less on lectures and theoretical content to leverage more intimacy and interaction in smaller groups. Because there is no recording of sessions during the third version, an average of cameras on cannot be provided. Even though there is no recording of the sessions that allows quantitative measurement, certainly the tendency towards anonymity and the lack of visual cues for communication has not been corrected. This is foreseen as a threat to the achievement of the project's objectives, so in the fourth version, we see the need to develop a comprehensive strategy, we call *Entrance Hall*, aimed to get students into the habit of turning on their cameras and microphones, and having all the technical conditions that would allow them to interact with instructors and peers in an online modality from the outset of the course.

We believe that students should make a responsible decision regarding their willingness to meet the requirements necessary to achieve the objectives of the course from the beginning. That is why, as a first step, we add the following sentence to the published course syllabus: "The online modality requires that students have adequate technical conditions for videoconferencing with complete interaction (audio, visual and written)". This informs them and leaves them with the responsibility of deciding if the methodology of the course is suitable for them. We consider essential to establish the appropriate technical conditions from the first session, and then reinforce them in the following sessions, until we install this practice as a habit and that group behavior becomes a positive motivation to favor learning in the sense that "if others turn on their cameras and communicate freely and actively... I'm going to do it too." So that the students will not be surprised by our requirements, 5 hours before the start of the opening class they receive a link to a survey in which they are asked if they have a camera and microphone and instructed that they should enter the virtual classroom with their camera on and they will receive a welcome from the teaching team. For this instance we apply an Entrance Hall dynamic (Fig, 2) were they are first given access to a Zoom session through a "waiting room" and are received individually (or in groups of maximum three students to speed up everyone's incorporation into the session) by a member of the teaching team who presents briefly himself or herself, communicates a personal welcome (asks the student's name and field of study) and conveys the meaning of activating the camera and microphone in a course were the objective is to develop interdisciplinary collaborative work. If they do not have any of these technical elements, they are informed that they must provide them for the following classes. After the welcome, they are individually referred to a common "preparation room" in which playful group dynamics are carried out with all participants as they arrive. The preparation room serves as an icebreaker and as a primer for Zoom and *Mural* (a collaborative whiteboard app used in conjunction with Zoom), this fosters a better community integration of the students and provides basic technical knowledge of the tools. Once the primer activity has been completed by all students, and briefly commented, the breakout room is closed, and the main class begins. This Entrance Hall dynamic is repeated during the first two sessions.

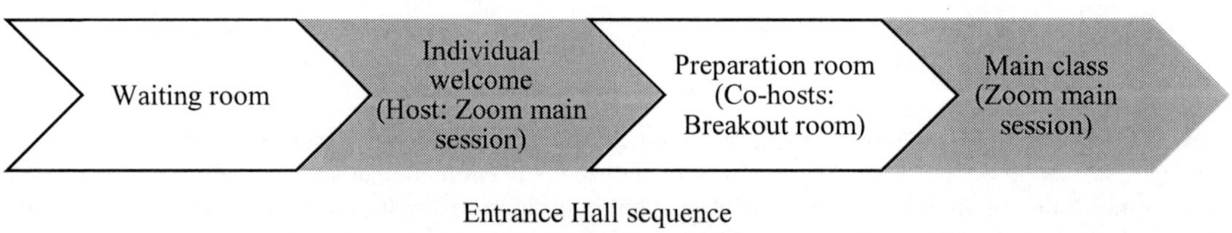


Figure 2. Sequence in Entrance Hall dynamic

For the next 14 class sessions, students are occasionally reminded of the need to turn on the camera or asked to discuss their technical difficulties in an encouraging but not punitive manner. The planning of each class always includes activities in small breakout rooms, group presentations, or the presence of external guests (other academics or practitioners from external communities) that stimulate and provide meaning to the interaction with camera and microphone turned on. Towards the end of the semester (last four sessions of the fourth version) it is collectively agreed that recording sessions would be convenient, as an archive of valuable information for the development of the final project. These 4 recordings allow a partial comparison with version 1 (Fig. 1.), which, although it does not indicate 100% of the students' cameras on, does show a clear improvement, from an average of 2.25% in the last 4 sessions of the Version 1 to 55% in Version 4.

From a qualitative perspective, students demonstrated a greater degree of commitment to collaborative work and more frequent interactions during the sessions throughout the fourth version. Improvement was reflected in the quality of the proposals in the final projects and the feedback they provided about satisfaction with their learning experience at the end of the semester.

## 4. CONCLUSION

This work constitutes an ongoing endeavor, delineating preliminary quantitative and qualitative outcomes integral to the evolution of a broader pedagogical strategy. Central to our argument is the notion that the Entrance Hall dynamic, as illustrated in the steps outlined in Figure 2, can significantly imprint students' initial interactions with the course, particularly during the introductory sessions. This, coupled with a series of provisions and subsequent actions throughout the semester, posits a cohesive strategy for online course design anchored in videoconferencing, fostering a palpable sense of social presence.

Among the recommended follow-up actions are the proactive notification of students about the mandatory use of cameras and microphones at the course's outset, the establishment of individualized contact with each student early in the course, cultivation of an understanding regarding the significance of activating cameras, and responsiveness to potential technical challenges. Additionally, the strategy emphasizes ongoing reinforcement of these principles through individual and group communications, incorporating active and collaborative learning activities. Clear guidelines for camera use and online participation are provided to underpin effective and successful learning. Furthermore, a feedback loop is instituted to comprehend students' challenges and tailor solutions accordingly.

While our recommendations align with existing literature and are validated through a trial-and-error process across four implemented versions, a crucial departure is evident. Unlike certain alternative approaches (Castelli & Sarvary, 2020; Trust & Goodman, 2023), we advocate that students' active use of cameras transcends optional status. We contend that in the post-COVID-19 era e-learning landscape, where students consciously enroll in courses leveraging videoconferencing for communication, treating camera usage as optional becomes incongruous. As educators, tasked with ensuring the quality of the learning experience, our choice of videoconferencing as a teaching resource, intertwined with collaborative methodologies, necessitates framing camera activation as a requisite for student involvement. The pedagogical strategy, therefore, integrates elements aimed at instilling the habit of turning on cameras.

The outcomes and insights presented derive from a specific case study conducted in an exploratory manner within a defined context. Recognizing the inherent complexity of educational phenomena, we acknowledge the multifaceted nature of influencing factors, encompassing intrinsic and extrinsic course-related elements, student and teacher characteristics, and dynamics of motivation, content, and methods, hence, limiting generalization. Future iterations of this course and subsequent studies can leverage these recommendations, employing refined data collection and analysis techniques to evaluate their efficacy more reliably.

## ACKNOWLEDGEMENT

This study has received partial funding from the FIDOP 2023 – University of Chile grant, dedicated to advancing research in undergraduate teaching.

## REFERENCES

Anderson, T. and Dron, J. (2011). Three generations of distance education pedagogy. *The International Review of Research in Open and Distributed Learning*, 12(3), p.80. doi:https://doi.org/10.19173/irrodl.v12i3.890.

Castelli, F. R., & Sarvary, M. A. (2021). Why students do not turn on their video cameras during online classes and an equitable and inclusive plan to encourage them to do so. *Ecology and Evolution*, 11(8), 3565–3576. https://doi.org/10.1002/ece3.7123

Dizon, A.G. (2022). Historical development of CIPP as a curriculum evaluation model. *History of Education*, pp. 1–20. doi:https://doi.org/10.1080/0046760x.2022.2098390.

Edwards, O.V. and Taasoobshirazi, G. (2022). Social presence and teacher involvement: The link with expectancy, task value, and engagement. *The Internet and Higher Education*, 55, p.100869. doi:https://doi.org/10.1016/j.iheduc.2022.100869.

Garrison, D.R., Anderson, T. and Archer, W. (2010). The first decade of the community of inquiry framework: A retrospective. *The Internet and Higher Education*, 13(1-2), pp.5–9. doi:https://doi.org/10.1016/j.iheduc.2009.10.003.

Hodges, C., Moore, S., Lockee, B., Trust, T. and Bond, A. (2020). The difference between emergency remote teaching and online learning. [online] *Educause Review*. Available at: https://er.educause.edu/articles/2020/3/the-difference-between-emergency-remote-teaching-and-online-learning.

Li, N., Romera Rodriguez, G., Xu, Y., Bhatt, P., Nguyen, H.A., Serpi, A., Tsai, C. and Carroll, J.M. (2022). Picturing One's Self: Camera Use in Zoom Classes during the COVID-19 Pandemic. *Proceedings of the Ninth ACM Conference on Learning @ Scale*. doi:https://doi.org/10.1145/3491140.3528284.

Mittal, A., Mantri, A., Tandon, U. and Dwivedi, Y.K. (2021). A unified perspective on the adoption of online teaching in higher education during the COVID-19 pandemic. *Information Discovery and Delivery*. doi:https://doi.org/10.1108/idd-09-2020-0114.

Rapanta, C., Botturi, L., Goodyear, P., Guàrdia, L. and Koole, M. (2020). Online University Teaching During and After the Covid-19 Crisis: Refocusing Teacher Presence and Learning Activity. *Postdigital Science and Education*, 2(1). doi:https://doi.org/10.1007/s42438-020-00155-y.

Stufflebeam, D. L. and Shinkfield, A. J. (1985) *Systematic Evaluation*. Dordrecht: Springer Netherlands. doi: 10.1007/978-94-009-5656-8.

Terblanche, W. *et al.* (2023) "Acceptance of e-learning applications by accounting students in an online learning environment at residential universities," *South African Journal of Accounting Research*, 37(1), pp. 35–61. doi: 10.1080/10291954.2022.2101328.

Tirado Morueta, R. *et al.* (2016) "Exploring social and cognitive presences in communities of inquiry to perform higher cognitive tasks," *The Internet and Higher Education*. Elsevier Inc., 31, pp. 122–131. doi: 10.1016/j.iheduc.2016.07.004.

Trust, T. and Goodman, L. (2023) "Cameras Optional? Examining Student Camera Use from a Learner-Centered Perspective," *TechTrends*. Springer. doi: 10.1007/s11528-023-00855-9.

Wu, Y., Sun, Y. and Sundar, S. S. (2022) "What Do You Get from Turning on Your Video? Effects of Videoconferencing Affordances on Remote Class Experience During COVID-19," *Proceedings of the ACM on Human-Computer Interaction*, 6(CSCW2), pp. 1–21. doi: 10.1145/3555773.

Zhang, S. *et al.* (2023) "Teammate Familiarity in Distributed Computer-Supported Collaborative Learning: The Mediating Role of Social Presence," *The International Review of Research in Open and Distributed Learning*, 24(4), pp. 233–251. doi: 10.19173/irrodl.v24i4.7332.

# PLAY2ESCAPE - A DIGITAL ESCAPE ROOM IN MATH CLASSES FOR UNDERGRADUATE TOURISM STUDENTS - CASE STUDY

Sónia Pais[1,2], Ana Elisa Sousa[2] and Maria Sofia Lopes[2]
[1]*CIDMA - Center for Research and Development in Mathematics and Applications, University of Aveiro, Portugal*
[2]*CITUR - Centre for Tourism Research, Development and Innovation, Polytechnic of Leiria, Portugal*

**ABSTRACT**

The younger generation, particularly those born in the late 20th and early 21st centuries known as "Generation Z," are deeply integrated with technologies in their daily live. Developing a better understanding of how such technologies influence learning is important to pedagogical approaches designed to co-opt technological applications in novel strategies for teaching and self-learning. Educational technology offers numerous advantages, including: enhancing students' learning experiences, increasing motivation and ultimately, improving student performance. Recognising these advantages, an educational resource was developed, where students explore their problem-solving skill in digital escape room.
This paper presents a case study of first-year undergraduate students from a Portuguese higher education institution using a math-related digital escape room intended to increase students' interest and motivation. The aim of the study is to assess how the escape rooms contributes to student motivation, interest and the consolidation of learning.
Assessment of students' perception was based on a quantitative survey and on the results of an evaluation test. Our findings suggest that Escape Room represent an effective educational approach that increases student interest and helps consolidate learning.

**KEYWORDS**

Higher Education, Educational Escape Rooms, Mathematics Education, Game-Based Learning, Active Learning, Pedagogical Innovation

## 1. INTRODUCTION

In recent decades, technology has become a fundamental part of western society, significantly changing the way we communicate, socialize and work. These changes are also reflected in an educational context. Education is facing a paradigm shift and it is crucial to integrate school strategies with the new social realities (Roldão, 2000, p. 3). Technologies are part of the young people's lives, especially those who were born at the end of the 20th and beginning of the 21st century, often referred to as "Generation Z". There is, therefore, an imperative to reflect on their needs, the way they learn, the role of technology and the potential to incorporate technology-based pedagogical tools in their education.

Those who are nowadays students of Higher Education were born in the time of the internet explosion and the origins and development of new ways of "on-line" interacting. The widespread use of cell phones and smartphones, the proliferation of digital content and the emergence of the social networks are part of generation Z's childhood. Many authors (Dana Fusco, 2012, 2018; Jennifer Skuza, 2020; Vanda, Pais and Hall, 2021; Billett et al, 2022) have posited that generation Z learn best with educational strategies based on challenges, with methodological applications that incorporate technology and with predominantly visual content.

One of the benefits of incorporating technology in educational contexts is that it can increase the motivation and engagement of students in the learning process (Ryan and Deci, 2017; Esteves et al, 2018; Pais and Hall, 2021; Vanda, Pais and Hall, 2021; Kukulska-Hulme et al, 2022; Pais, Sousa and Pires, 2023). According to several authors (Entwistle and Ramsden, 2015; Pais and Hall, 2021; Maharjan et al., 2022; Pais, Sousa and Pires, 2023), motivation plays a crucial role in any learning process, impacting not only the ability to acquire knowledge but also the quality of learning itself. Motivation significantly influences engagement and, subsequently, success in the learning achievements.

As a mathematics teacher, the first author of this article is confronted year after year with the unmotivated students, a general lack of interest mathematical analytical approaches and a subsequent increased failure rate of students in mathematics related subjects.

With the aim of seeking strategies to foster interest in Mathematics, motive student enthusiasm and make the learning process itself more engaging, the author developed a math-related digital escape room activity for first-year students in a tourism and hotel management undergraduate degree. The aim of the study was to assess how (digital) escape rooms can contribute towards student motivation, interest and the consolidation of their learning.

This study represents a classroom application experience. After working on the escape room activity, students were invited to participate in a survey questionnaire to share their perceptions about the use of the digital escape room activity. In order to check whether the ER activity had contributed to improving learning outcomes, a few days later the students also had to take an assessment test on the syllabus contents covered in ER.

# 2. ESCAPE ROOMS AND LEARNING

## 2.1 Escape Rooms

Escape rooms (ERs) are a relatively new type of entertainment that aim to provide an unforgettable experience. According to Nicholson they can be defined as "live-action team-based games where players accomplish tasks while participating in multiple procedures, in one or more spaces, with a specific goal, in a limited period of time" (2015, p.1). ERs first appeared in Japan in 2007, but have grown in popularity worldwide, with rooms available across most continents including Asia, Europe and America (Makri, Vlachopoulos and Martina, 2021). ERs provide players with an immersive narrative experience, creating a meaningful and memorable escape scenario. Players are transported into a narrative circle where they are faced with intriguing challenges and missions. Whether it is helping Alice escape from Wonderland, hunting ghosts in a haunted mansion, retrieving important notes from a locked school locker before an upcoming exam, or discovering the formula to cure the world from a pandemic, these narrative-driven scenarios evoke emotional investment and curiosity among players. "Although escape rooms are adventure activities, while playing, players must use multiple analytical skills, thinking skills, problem-solving, cooperation, and communication skills, which takes the whole activity to a higher cognitive level" (Huraj, Hrmo and Sejutová Hudáková, 2022, p. 2). As they navigate through the rooms and solve the presented challenges, the game atmosphere is set, fostering engagement and a sense of adventure.

## 2.2 Educational Escape Rooms

In recent years Escape Rooms have become a popular recreational activity which has also drawn the attention of teachers and educators. "An educational escape room (EER) is an innovative pedagogical concept wherein participants, typically students, engage in a purposefully designed immersive environment filled with puzzles, challenges, and riddles that are linked to unambiguous learning objectives" (Fotaris, Mastoras and Lameras, 2023, p. 180).

Educational Escape Rooms have been shown to be a fun and entertaining activity if brought to the classroom. At the same time, they have been proven to be a solid activity where, while trying to break out of the room, students can develop soft skills as well as subject-matter related competencies (Morrell, Eukel and Santurri, 2020; Sowell 2021). A systematic review carried out by Fotaris and Mastoras (2019) highlights their positive impact on student motivation and soft skills development such as teamwork, creativity, decision-making, leadership, communication, and critical thinking, and emphasizes the enjoyable experience that immerses students as active participants in the learning environment. These benefits have contributed to educators feeling increasingly motivated to incorporate Escape Room games into various educational contexts.

# 3. MATERIALS AND METHODS

To understand how the particular methodology of this Escape Room space was perceived by students, data were collected using a survey questionnaire developed by the researchers. The questionnaire included a list of statements and respondents were asked to rate their level of agreement using a 5-point agreement (Likert) scale. The statements were about the ER activity, such as contribution for learning, difficulty level, and the entertainment component of the activity. A preliminary version of this instrument was analysed by three higher education teachers, tested with three individuals from the same group, and modified based on the suggestions and the difficulties detected during its completion. The questionnaire was distributed online at the end of the activity.

In order to find out whether the escape room activity helped to improve learning outcomes, an evaluation test on the contents covered in the activity was conducted. Results were analysed and compared with the results obtained by a similar group of students from the previous school year that did not participate in the ER experience.

Statistical analyses were performed using MS Excel. Descriptive statistics (frequencies, mean, standard deviation, quartiles, maximum and minimum) were computed to summarize and compare results. A Mann-Whitney test was performed using IBM SPSS Statistics to compare the evaluation results of the two groups.

## 3.1 Participants

This case study was conducted at a higher education institution in the central region of Portugal. The participants were undergraduate students, majoring in tourism and hotel management, enrolled in the Mathematics Applied to Management course during the first semester of the academic year 2023-2024. The case study included a total of 30 students; 28 students completed the questionnaire.

It is important to highlight that the class was heterogeneous in terms of previous mathematical knowledge. Some students had studied maths for 9 years, while others had studied maths for 12 years. Among those who answered the questionnaire, 17 (63%) were female, and 10 (37%) were male. The age of the participants ranged from 18 to 38 years (average 20.8).

# 4. THE DESIGN OF THE DIGITAL EDUCATIONAL ESCAPE ROOM (DEER)

To design the DEER, the author took in consideration some aspects thought to be relevant in the creation of educational escape rooms (Nicholson, 2018; Huraj, Hrmo and Sejutová Hudáková, 2022). Although physical ERs are more immersive, they are expensive and difficult to implement, Therefore, we decided to implement a digital ER. The ER had 4 levels and the difficulty increases with each level. Each respective level included 3 problems. A linear concept of the escape room was implemented, i.e., students could only move to the next problem after successfully completing the previous one.

The activity lasted 1h15mins and was implemented in teams of 2 students, since working in pairs would require students to communicate their ideas, listen to other opinions, discuss strategies, justify critiques and thereby contribute towards a positive and comprehensive learning process. Furthermore, working in pairs is frequently employed in the classes of this course and students have previously communicated that they enjoy it. The learning aims were to practice some mathematical contents, specifically on primitives and integrals, which are usually considered a difficult topic in this course.

The DEER was developed using Genially – a platform for the creation of interactive and animated contents. This platform has a free version; it is easy to use and requires no previous knowledge in web-design or programming.

In order to increase students' engagement with the activity, the DEER development also aspired to the following principles:

- To create an attractive and challenging narrative:

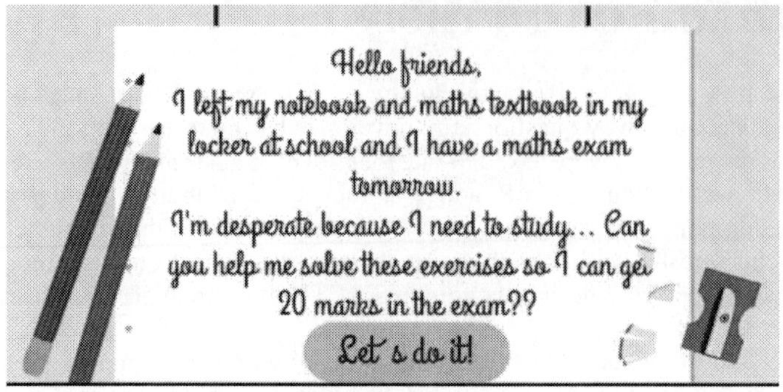


Figure 1. The ER narrative

- To provide clues: clues were related to the contents explored in the ER;

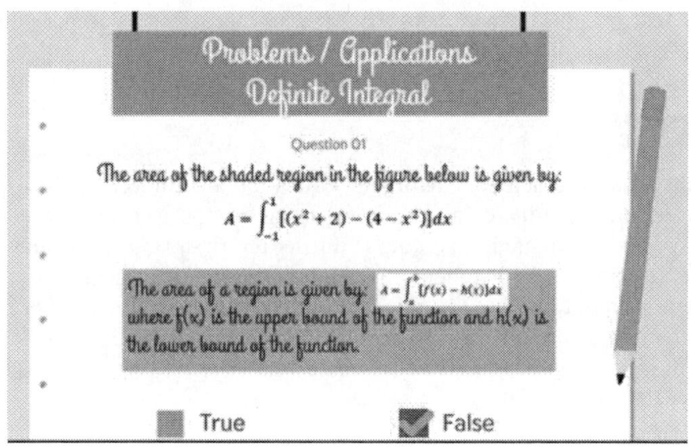


Figure 2. Example of a clue

- To create digital locks: to move to the next level, students had to successfully complete all the tasks on the previous one. So, to unlock the digital locks, the participants had to correctly solve the challenges.

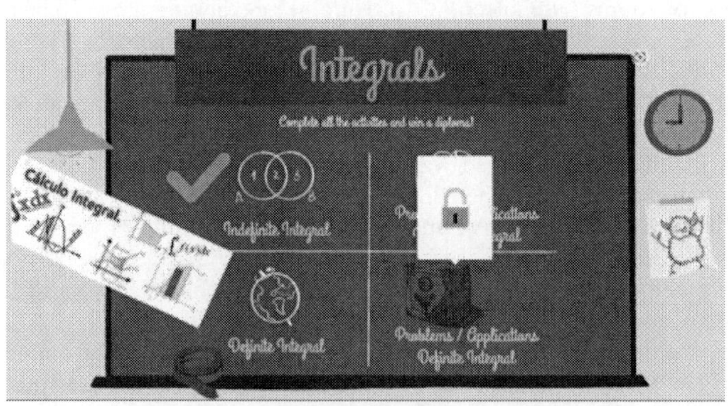


Figure 3. One of the digital locks used in the ER

Prior to carrying out the activity, the Escape Room was tested by a group of 4 students and 2 teachers aged 18-48 years.

## 5. ANALYSIS OF THE QUESTIONNAIRE

Respondents rated globally the ER activity as very good (54%) and good (43%).

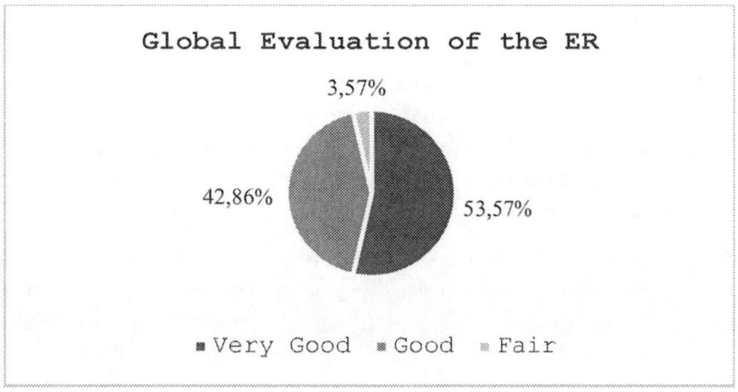


Figure 4. Rate your general evaluation of the ER activity

The majority of students reported that the activity had contributed towards their learning process, "allowing the consolidation of the syllabus contents" (36% totally agreed, 46% agreed) and considered that they "learnt more from the ER activity than in a conventional class dedicated to consolidation of syllabus contents" (32% totally agreed, 21% agreed). Most participants also "recommended the use of ER as pedagogical methodology" (36% totally agreed, 39% agreed).

Participants thought the activity "was well organized" (57% totally agreed, 39% agreed), but some students found "the ER was difficult" (18% totally agreed, 25% agreed) and 39% considered that the time available was not adequate. The authors believe that the high percentage of students who found the activity difficult and had insufficient time to complete the challenges is due to the fact that the contents covered in the activity are often considered difficult by tourism students. Despite this, it should be noted that the overwhelming majority of students did not find the activity difficult and considered the timeframe adequate. Only one team failed to finish the activity.

Students were also asked about the entertainment dimension of this activity. They revealed they "had fun" (57% totally agreed, 32% agreed) and "enjoyed it more than a conventional class for exercises resolution" (64% totally agreed, 14% agreed).

In spite being an enjoyable activity, students did not consider that it was just a distraction (72%). They considered that it "contributed towards a more positive vision of Mathematics" (29% totally agreed, 64% agreed), making the "learning process more interactive" (29% totally agreed, 61% agreed) and more challenging, interesting and stimulating (36% totally agreed, 39% agreed). Table 1 shows the detailed responses to these questions in a heatmap with the observed frequencies as well as the respective mean and standard deviation values.

Table 1. Heatmap with frequencies of responses to questions about the use of an ER activity in the classroom; mean and standard deviation
(1 – total disagreement; 2- partial disagreement; 3 – neutral; 4 – partial agreement; 5 – total agreement)

| Statement: The use of an ER activity in math classes... | 1 | 2 | 3 | 4 | 5 | mean | s.d. |
|---|---|---|---|---|---|---|---|
| contributes to a more positive attitude towards mathematics | 1 | 0 | 1 | 18 | 8 | 4.14 | 0.62 |
| makes mathematics more boring and demotivating | 14 | 10 | 1 | 3 | 0 | 1.75 | 0.90 |
| makes the learning process more interactive | 1 | 1 | 1 | 17 | 8 | 4.07 | 0.78 |
| serves only to entertain for a while | 10 | 11 | 3 | 4 | 0 | 2.04 | 1.03 |
| provides a less rigid method of learning, making it more challenging, interesting, and stimulating | 1 | 2 | 4 | 11 | 10 | 3.96 | 1.11 |

Students recommend that their colleagues participate in the ER activity (82%) and wished that this kind of activities would also be developed in other courses (90%).

Although the activity was on the last topic of the syllabus and was carried out towards the end of the semester, it was clear from direct observation that the students were more enthusiastic than usual about this subject and that, for many, their attitude changed after carrying out the ER activity. This personal observation is supported by the results obtained in the evaluation.

With regard to the results obtained in the evaluation test on the contents worked on in the ER activity and compared with those obtained in the previous academic year, there was a clear increase in the number of students who obtained a positive grade (Table 2). It was also noted that, in general, students obtained higher marks this academic year. A Mann-Whitney U test was performed to compare the results in the evaluation test in the two groups. There was a significant difference in the results ($z = -1{,}984$; $p = 0{,}047$).

Table 2. Results obtained in the evaluation test (on the contents worked on in the ER activity)

| academic year 22-23 | academic year 23-24 |
|---|---|
| n=28 | n=22 |
| percentage of positive scores: 57,14% (assessment test) | percentage of positive scores: 68,18% (assessment test) |
| highest grade: 19,86 | highest grade: 20,00 |
| lowest grade: 3,00 | lowest grade: 6,00 |
| $\bar{x}=10{,}40$<br>$Q_1=6{,}18$<br>$Q_2=9{,}86$<br>$Q_3=14{,}50$ | $\bar{x}=13{,}30$<br>$Q_1=8{,}50$<br>$Q_2=11{,}86$<br>$Q_3=19{,}50$ |

## 6. CONCLUSION

In this study a digital ER was developed and applied in an undergraduate course in Mathematics. The number of students participating in the activity was small, which is the main limitation of this study. However, the results are emphatically positive with students perceiving the Digital Escape Room (DER) activity as a valuable tool that had a positive effect on their motivation and interest and as an otherwise entertaining activity. These findings align with evidence from other researches, such as Babazadeh and Frigerio (2021), Huang et al. (2020), and Papadakis and Stavrakis (2020). Moreover, students considered that the ER activity contributed towards their learning process and helped shift their vision of Mathematics topics in general. As a result of these opinions, students expressed that it should be adopted as a pedagogical tool in other courses.

It also appeared to promote student success, with an observed improvement in learning outcomes, corroborating results from similar studies (Nicholson, 2018; Huang et al, 2020; Sowell, 2021, Pais et al, 2023).

The results of this research suggest that a DER activity is an efficient strategy of motivating and fostering interest in Mathematics and making the learning process more engaging and, therefore, more efficient and ultimately successful.

Although this study was developed in a Mathematics course, the focus is on the use of an educational resource, combining formal contents with technology and entertainment. These findings are, in essence, generally applicable and could be applied by teachers and educators wishing to explore alternative pedagogical approaches.

For future research, it would be insightful to apply this activity to a larger recipient group and in different educational courses. It would also be interesting to analyse the relationship between students' perceptions about this kind of activity and their attitude towards other characteristics, such as the application of interactive technologies, compared with their views about more conventional educational approaches.

## ACKNOWLEDGEMENT

This work was supported by CIDMA and CITUR and is funded by the Fundação para a Ciência e a Tecnologia, I.P. (FCT, Funder ID = 50110000187) under Grants https://doi.org/10.54499/UIDB/04106/2020, https://doi.org/10.54499/UIDP/04106/2020 and https://doi.org/10.54499/UIDB/04470/2020.


## REFERENCES


Billett, S., Dymock, D., Hodge, S., Choy, S., Le, A.H., 2022. Shaping Young People's Decision-Making About Post-School Pathways: Institutional and Personal Factors. In: Billett, S., Stalder, B.E., Aarkrog, V., Choy, S., Hodge, S., Le, A.H. (eds) *The Standing of Vocational Education and the Occupations It Serves. Professional and Practice-based Learning*, vol 32. Springer, Cham. https://doi.org/10.1007/978-3-030-96237-1_6

Entwistle, N. & Ramsden, P., 2015. *Understanding student learning* (Routledge revivals), Routledge. doi: 10.4324/9781315718637

Esteves, M., Pereira, A., Veiga, N., Vasco, R., Veiga, A., 2018. The Use of New Learning Technologies in Higher Education Classroom: A Case Study. In: Auer M., Guralnick D., Simonics I. (eds) Teaching and Learning in a Digital World. ICL 2017. *Advances in Intelligent Systems and Computing*, vol 715. Springer, Cham, 2018.

Fotaris, P., & Mastoras, T., 2019. Escape Rooms for Learning: A Systematic Review. In L. Elbaek, G. Majgaard, A. Valente, & S. Khalid (Eds.), *Proceedings of the 13th International Conference on Game Based Learning*, ECGBL 2019, pp. 235-243. (Proceedings of the European Conference on Games-based Learning; Vol. 2019-October). Academic Conferences and Publishing International Limited. https://doi.org/10.34190/GBL.19.179

Fotaris, P., Mastoras, T., Lameras, P., 2023. Designing Educational Escape Rooms With Generative AI: A Framework and ChatGPT Prompt Engineering Guide in *Proceedings of ECGBL* 2021 15th European Conference on Game-Based Learning, 180–189

Fusco, D., 2012. On becoming an academic profession. In: D. Fusco (Ed.), Advancing youth work: Current trends, critical questions, pp. 111-126. *Routledge.* https://doi.org/10.4324/9780203829769

Fusco, D., 2018. Some conceptions of youth and youthwork in the United States. In P. Alldred, F. Cullen, K. Edwards, & D. Fusco (Eds.), *Handbook of youth work practice* (pp. 44-58). Sage.

Huang, S.-Y., Kuo, Y.-H. and Chen, H.-C. (2020) "Applying digital escape rooms infused with science teaching in elementary school: Learning performance, learning motivation, and problem-solving ability", Thinking Skills and Creativity, Vol. 37, 100681 (online journal).

Huraj, L.; Hrmo, R.; Sejutová Hudáková, M., 2022. The Impact of a Digital Escape Room focused on HTML and Computer Networks on Vocational High School Students. *Educ. Sci.*, 12, 682. https://doi.org/10.3390/educsci12100 682

Kukulska-Hulme, A., Bossu, C., Charitonos, K., Coughlan, T., Ferguson, R., FitzGerald, E., Gaved, M., Guitert, M., Herodotou, C., Maina, M., Prieto-Blázquez, J., Rienties, B., Sangrà, A., Sargent, J., Scanlon, E., Whitelock, D., 2022. Innovating Pedagogy 2022: Open University Innovation Report 10. *Milton Keynes*: The Open University.

Maharjan, M., Dahal, N., & Pant, B.P., 2022. ICTs into mathematical instructions for meaningful teaching and learning. *Adv Mobile Learn Educ Res*, 2(2): 341-350. https://doi.org/10.25082/AMLER.2022.02.004

Makri, A., Vlachopoulos, D., Martina, R., 2021. Digital Escape Rooms as Innovative Pedagogical Tools in Education: A SystematicLiterature Review. *Sustainability* 2021,13, 4587. https://doi.org/10.3390/su13084587.

Morrell, B., Eukel, H. and Santurri, L., 2020, Soft skills and implications for future professional practice: Qualitative findings of a nursing education escape room, *Nurse Education Today*, Vol 93, 104462.

Nicholson, S., 2015. Peeking behind the locked door: A survey of escape room facilities. Retrieved from http://scottnicholson.com/pubs/erfacwhite.pdf

Nicholson, S., 2018. Creating Engaging Escape Rooms for the Classroom. *Childhood Education*, 94(1), 44-49.

Pais, S. & Hall, A., 2021. Enhancing Tourism Students' Motivation in Mathematics Classes With Kahoot! - A Case Study, in *Proceedings of ECGBL 2021* 15th European Conference on Game-Based Learning, 591–597.

Pais, S; Sousa, A. E. and Pires, A., 2023. Using an Escape Room Activity to Enhance the Motivation of Undergraduate Life Science Students in Mathematics Classes: A Case Study in *Proceedings of ECGBL 2023* 17th European Conference on Game-Based Learning, 490–496.

Roldão, M. C., 2000. *Currículo e Gestão das Aprendizagens: As Palavras e as Práticas.* Aveiro: Universidade de Aveiro.

Ryan, R. M., & Deci, E. L., 2017. *Self-Determination Theory: Basic Psychological Needs in Motivation, Development and Wellness.* New York: Guilford Press. https://doi.org/10.1521/978.14625/28806

Skuza, J. A., 2020. The Experience of Learning: Early Adolescents in Organized Youth Programs. *Journal of Human Sciences and Extension*, 8(3), 3. https://doi.org/10.54718/SAAC8798

Sowell, M., 2021. Incorporating Curriculum Content into Educational Escape Games for Middle School Students, The Clearing House: A Journal of Educational Strategies, *Issues and Ideas,* 94:2, 47-52, DOI: 10.1080/00098655.2020.1857324

V. Santos, S. Pais, and A. Hall., 2021. Mathematics Classes for Tourism Undergraduate Students and Pre-service Teachers with Active Methodologies using Technologies, *International Journal of Technology in Mathematics Education*, Vol 28, No 3, pp. 203-212, 2021.

# FORMATION OF BACHELOR'S SKILLS IN THE FIELD OF MACHINE LEARNING AND INTELLIGENT DATA ANALYSIS

Marina Lapenok[1], Sergei Patrushev[2] and Lidiya Shestakova[3]
*[1]Ural State Pedagogical University*
*Yekaterinburg, Russia*
*[2]Smart business technologies d.o.o.*
*Belgrade, Serbia*
*[3]Perm State National Research University*
*Solikamsk, Russia*

**ABSTRACT**

The article presents the content of the training courses “Workshop in Python Programming” and “Fundamentals of Artificial Intelligence”, included in the module of academic disciplines “Data Analysis. Machine learning. Artificial intelligence". The objectives of the module are to develop students’ knowledge of basic data analysis algorithms, develop skills in visualization and interpretation of data to solve applied problems using object-oriented programming technologies and artificial neural network technologies. As a result of the students’ project work, a number of neural network predictive systems were developed that allow solving pedagogical problems. The effectiveness of the developed teaching methodology was confirmed by the results of a pedagogical experiment.

**KEYWORDS**

Neural Network, Predictive Systems, Machine Learning, Training Courses

## 1. INTRODUCTION

One of the in-demand skills of the future is a high level of proficiency in artificial intelligence (AI) tools. V. Sadovnichy (2021) analyzes the impact of the human factor on labor productivity in the era of the digital economy and concludes that the main driving force of the digital economy will be the symbiosis of “human plus artificial intelligence” (p. 57), which works everywhere and effectively under the guidance of humans.

Already now, in the context of the explosive growth of artificial intelligence technologies, there is a shortage of specialists capable of developing and researching in the field of artificial intelligence, as well as teachers capable of training such specialists. Various countries are taking steps to attract AI talent as they face intense competition. Digitalization of education and the use of artificial intelligence is not only about equipping classrooms with computers, interactive whiteboards, tablets, high-speed Internet, chat bots, etc. AI-powered educational systems are used to analyze classroom dynamics and student engagement. New technologies require special training of specialists who can skillfully work with educational AI technologies. Higher education programs are already available that teach how to set problems that meet the needs of a high-tech digital economy and solve them using AI tools. According to M. Saikina:

Training in AI-related competencies is in great demand today. The lack of specialists in this field is a recognized problem that will not be solved in the coming years. The presence of competencies in the field of working with AI provides competitive advantages in the labor market for a wide range of specialties, which makes it advisable to include them in programs in various areas of training. (p. 75)

According to a study by the Higher School of Economics, “school education in the coming years may face large-scale digitalization in connection with new AI technologies and the construction of a global educational digital infrastructure” (Mayer, 2023, para 2). The emergence of AI and cloud data storage can make some educational formats obsolete, which can ultimately lead to a significant reduction in the role of the school as a social institution. L. Bosova (2018) notes that “AI is becoming today a new teaching tool that should be mastered and used in work. This is what we must prioritize in preparing the next generation of teachers." (p. 2) A. Stepanova (2022) argues that a modern teacher must develop technical competencies and enrich their portfolio with the skills necessary for confident navigation in the areas of Data Science, data analysis, and the use of AI. M. S. Rahman and H. Reza (2021) point out the need to develop skills in analyzing structured, semi-structured and unstructured data among all members of modern society, especially highlighting specialists in the social and educational fields, since consumers often use social networks to obtain various information and reviews for the purpose of making decisions regarding education. The large amount of data generated on various social media platforms in the form of text, images, audio, video, GIFs, blogs, etc. should be used to obtain accurate and in-depth recommendations and hence make the right decision.

O. Andreeva (2019) suggests assessing students' abilities to analyze the results of an experiment through expert assessment of the products of students' activities, such as cases, research projects, and coursework. As N. Danilova (2021) notes, the formation of analytical skills is included in the Federal State Educational Standards for Higher Education in pedagogical areas of training.

# 2. THE TRAINING COURSE MODULE "DATA ANALYSIS. MACHINE LEARNING. ARTIFICIAL INTELLIGENCE"

Intelligent modeling is currently mainly implemented on the basis of the development and use of artificial intelligence networks - multilayer neural networks that make it possible to simulate the linear and nonlinear behavior of complex systems in fuzzy conditions. Modeling is based on a system analysis of essential (in the context of the study) elements and connections between them, and the synthesis of a network of information transformation units in the form of oriented graphs.

Typical pedagogical tasks (in the context of developing a smart network model) are learning, classification, forecasting, and decision making. In pedagogy, neural networks are developed to implement adaptive learning systems (reinforcement (Ausin et al., 2019) / individual (Caro et al., 2015); categorization (learners / pedagogical problems); development of individual recommendations (nutrition / physical activity / class schedule / rhythm of training load, taking into account preferences and psycho-physiological characteristics of health); predicting the success of activities (educational-cognitive / project); diagnosing educational achievements of students.

## 2.1 Contents of the Training Course "Workshop on Python Programming"

In some Ural universities (Ural State Pedagogical University, Solikamsk branch of Perm State National Research University, Ural Federal University) the training course module "Data Analysis. Machine learning. Artificial Intelligence" is being implemented, the objectives of which are to form an understanding of the capabilities and limitations of the main methods of processing and analyzing big data; developing the skills to visualize and interpret data to solve applied problems using object-oriented programming technologies, data analysis methods and artificial neural network technologies.

This module is implemented in the academic disciplines "Workshop in Python Programming" and "Fundamentals of Artificial Intelligence" for bachelors in the areas of study "Applied Mathematics and Computer Science" and "Pedagogical Education".

Target outcomes of students mastering the discipline "Workshop in Python Programming" are: knowledge of the theoretical foundations of data analysis methods; basic data analysis algorithms in Python; types of problems solved using such methods; modern machine learning technologies, as well as the ability to present and interpret the results of data analysis to solve applied problems.

Based on the goals and expected learning outcomes, the content of the laboratory work in the discipline "Workshop on programming in Python" is highlighted (presented in Figure 1).

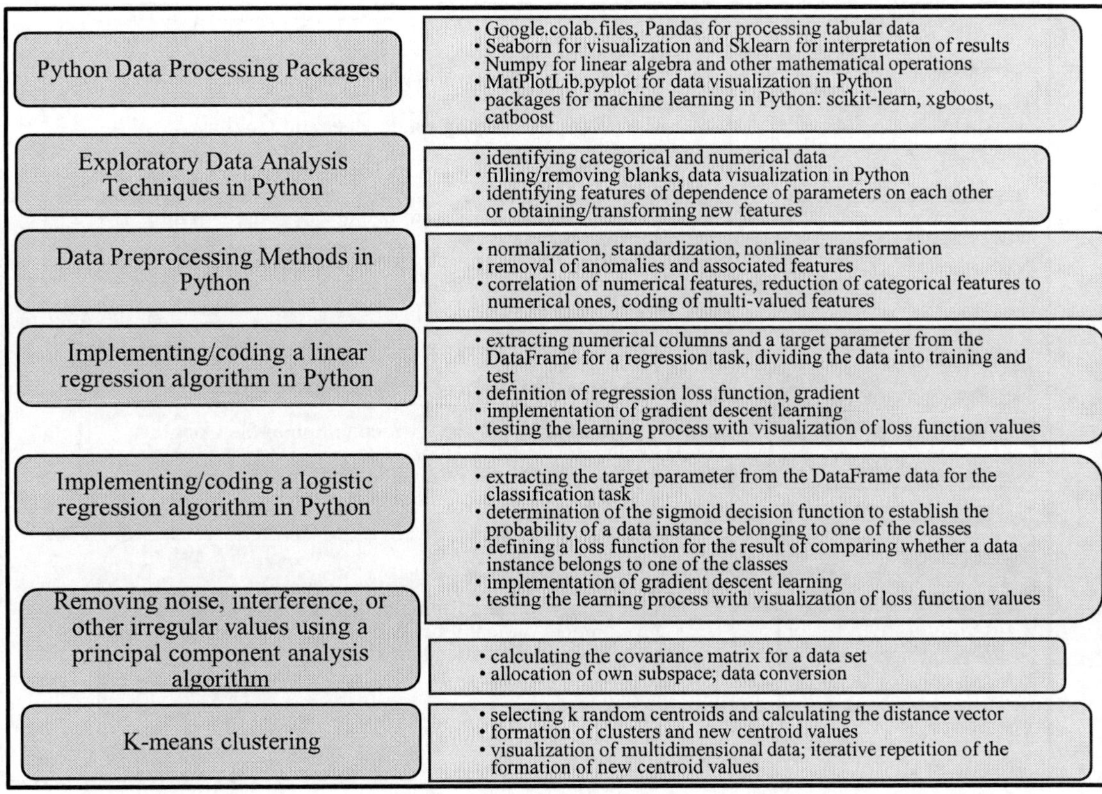


Figure 1. Contents of laboratory work in the discipline “Workshop on programming in Python”

## 2.2 Contents of the Academic Discipline “Fundamentals of Artificial Intelligence”

When selecting the content of the academic discipline “Fundamentals of Artificial Intelligence”, presented in Figure 2, the following goals were pursued: the formation of a conceptual apparatus in the field of neural network technologies; developing skills in the field of neural network mathematical modeling of subject areas in accordance with the applied problems being solved; accumulation of experience in implementing the stages of data mining (formalization of the problem; generation of training/test examples; initial design of the structure of the neural network; training of the neural network; testing and optimization of the neural network; research of the subject area).

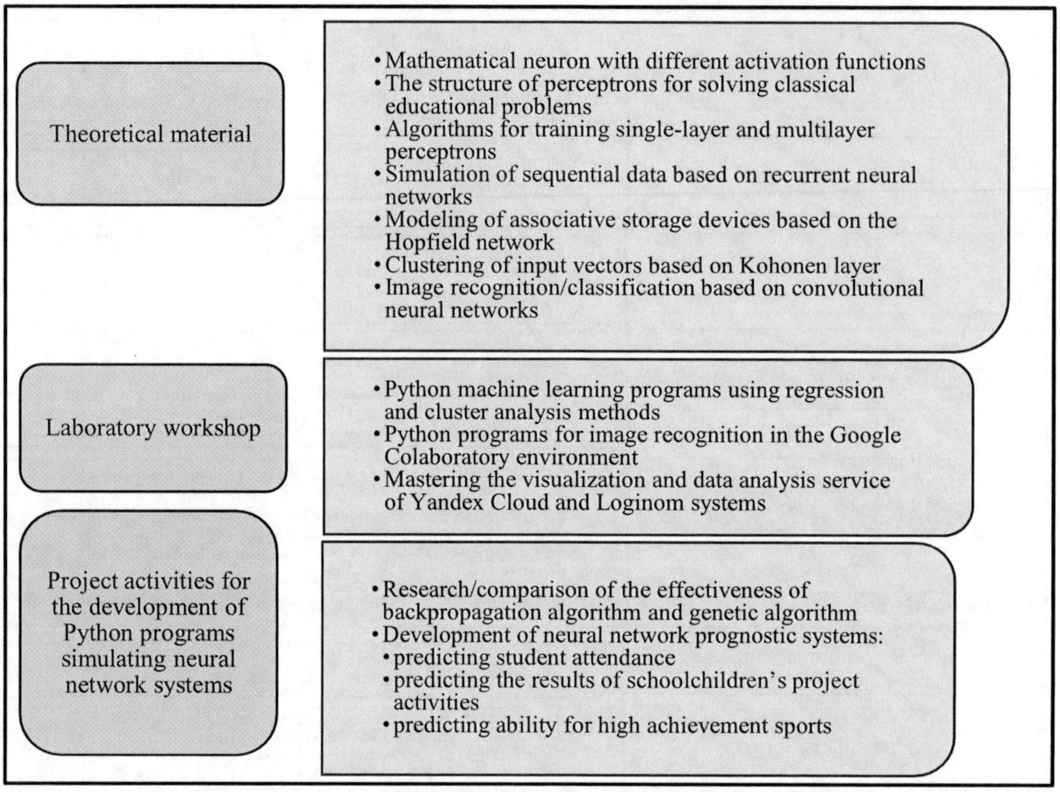


Figure 2. Contents of the discipline "Fundamentals of Artificial Intelligence"

Students' project activities are aimed at developing their experience in creating computer programs for solving research and applied problems. An example of a research project is a task to study and comparatively evaluate the effectiveness of two training algorithms for multilayer perceptron: the backpropagation algorithm and the genetic algorithm.

During the implementation of this project, students create programs in Python that implement the specified algorithms, run the developed programs on several different datasets, taken, for example, from the site https://www.kaggle.com/datasets, then record and compare the results of both programs (number iterations required to achieve an acceptable value of the root mean square error of the loss function).

Examples of application projects:

— forecasting the results of schoolchildren's project activities (and developing recommendations for selecting a promising project task);

— predicting student attendance at classes based on their personal qualities, goals and class schedule (and developing recommendations for changing the schedule in order to improve class attendance) (Lapenok et al., 2019);

— forecasting (and developing recommendations for the development) of a person's abilities for a certain type of activity (business, high-achievement sports, research, etc.) based on a model of human activity (Lozinskaya et al., 2021).

Application project №1. Students begin the implementation of an applied project to create a neural network system for predicting the results of schoolchildren's project activities by generating an input vector X and an output vector Y.

Features of project assignments that are "promising" for schoolchildren must be invented. The term "promising" project task means that the topic of the task will be of interest to the schoolchild, that the schoolchild has a complex of knowledge and skills sufficient to complete the project task, that the schoolchild's personal characteristics (dedication, perseverance, ability for analytical work, etc.) will allow him to cope with the expected volume of analytical and/or experimental work.

When choosing a "promising" task, you can base it on the identified interests and activity of the schoolchild (for example, whether the schoolchild is involved in clubs as part of additional education or electives, whether he participated in competitions and how successfully); on the wishes of the schoolchild in relation to the scientific field within which he intends to carry out the project; on the results of educational activities for previous periods of study; at the level of diligence (for example, in relation to completing homework) and discipline of the schoolchild (for example, whether he missed classes without a good reason, whether he received reprimands for bad behavior). You can also take into account the schoolchild's sociability, type of temperament and other data that determine personality characteristics and, therefore, affect the likelihood of systematic, purposeful work (collective or individual) on a project task. Thus, several parameters accumulate that characterize the schoolchild's personality and the history of his educational activities, which define the vector X.

Possible results of schoolchildren's project activities should be encoded in the output vector Y, taking into account the components of a comprehensive assessment (for example, whether the goal of the research has been achieved; whether the research objectives have been specified and solved; whether an analysis of literature sources has been carried out within the framework of the topic of the project assignment; whether the results are presented in the form of an explanatory note; whether the oral report was successful in the process of defending the project; what is the share of the schoolchild's personal contribution to solving the problem in the case of a collective project, etc.). In the case of using a dichotomous scale of 0 and 1, the output vector of the neural network Y will consist of many ones (corresponding to the implemented components of the complex assessment) and one or more zeros (if the corresponding component of the complex assessment is not implemented).

Then students perform search work at school to generate training examples for the neural network. Each training example includes a pair of vectors (X and D), where X is an input vector, and D is a vector of the desired responses of the neural network (Fig. 3). Next, students develop programs in Python for training and testing the neural network, as well as for a convenient user interface.

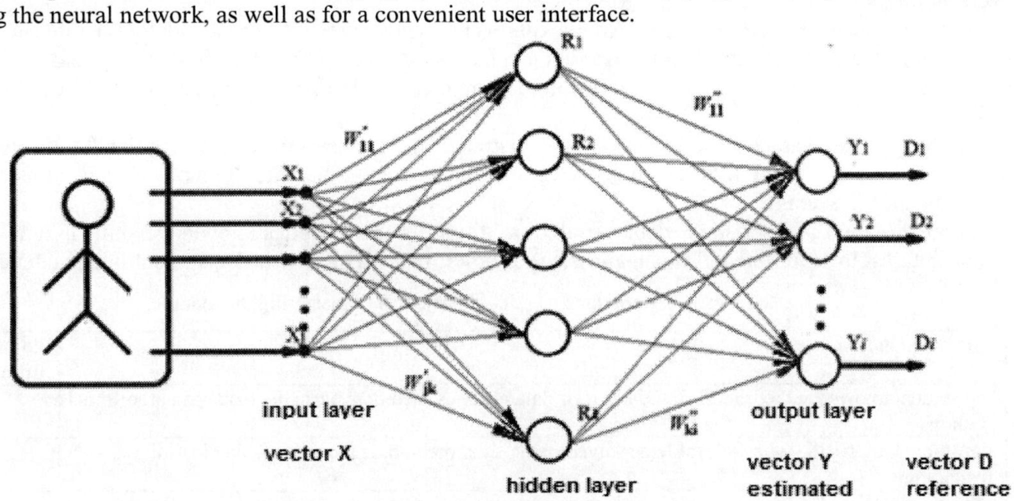


Figure 3. The scheme of the neural network

Application project №2. Students begin to implement an applied project to create a neural network system for predicting student attendance by forming an input vector X and an output vector Y. Personal qualities and goals are the parameters of the X vector used to calculate the neural network. These characteristics can be identified using a survey:

— organization (the ability to withstand weaknesses);
— psychotype (extrovert, introvert, ambivert);
— temperament (choleric, sanguine, phlegmatic, melancholic);
— responsibility (or irresponsibility) of the student in relation to the fulfillment of public assignments;
— the prevalence of the desire to acquire new knowledge in the learning process over the desire to acquire a document confirming the presence of this knowledge as a result of studying (or vice versa);

— the student's assessment of the importance of his own time in terms of the feasibility of time spent on transport depending on the expected duration of classes (for example, only one class in the class schedule or the presence of "windows");

— concern about one's own health in terms of intention to attend (or not attend) training sessions;

— area of subject interests (programming, computer games, theoretical foundations of computer science);

— focus on combining studies with professional activities (or not), etc.

The output vector Y should encode the results of the surveyed students' attendance at training sessions (for example, in specialized disciplines), taken from the dean's group journals. Empirical data obtained as a result of a survey and from the dean's office data are reduced to a dichotomous "yes"-"no" scale and are used to design and train a neural network (Figure 3).

## 2.3 Pedagogical Experimental Work

At the Solikamsk branch of the Perm National Research University, 30 students specializing in "Applied Mathematics and Computer Science" studied academic disciplines and took part in pedagogical experimental work in 2023–2024.

Stages of the pedagogical experiment: ascertaining (identification of the initial level of formation of the selected characteristics); formative (organization of training students in the module "Data Analysis. Machine Learning. Artificial Intelligence"); control (identification of the final level of formation of the selected characteristics). The disciplines "Workshop on programming in Python" and "Fundamentals of artificial intelligence" were mastered by students sequentially.

To diagnose learning outcomes, students' knowledge and skills were identified as criteria for assessing the success of mastering the module (Table 1), and a rating range of 0–3 points was used. The scoring indicators were as follows:

0 points - characteristic (i.e., knowledge/skills) is not demonstrated or practically not demonstrated;

1 point - the characteristic is demonstrated at a mandatory level, while the student may make mistakes and inaccuracies; the student mainly acts according to a model, an algorithm, the student's level of independence is low;

2 points - the characteristic is mainly demonstrated; the student confidently applies knowledge/skills in familiar situations; there may be minor errors or inaccuracies that the student corrects himself; student independence is average;

3 points - the characteristic is demonstrated in full; the student applies knowledge/skills in new situations; shows interest in acquiring additional knowledge, experiments with new tasks; student independence is high.

Table 1. Criteria for assessing the success of mastering the module

| Characteristic | Points, from 0 to 3 |
|---|---|
| the student knows the theoretical foundations of data analysis methods; basic data analysis algorithms Python | |
| the student knows the types of problems solved using such methods; modern machine learning technologies | |
| the student knows how to use data analysis methods and basic data analysis algorithms in Python to solve professional problems | |
| the student is able to present and interpret the results of data analysis in Python to solve professional problems | |
| the student knows basic terminology and basic theoretical material in the field of artificial intelligence | |
| the student knows how to perform and the expected results of the stages of data mining (statement/formalization of the problem; generation of training/test examples; initial design of the structure of the neural network; training of the neural network; verification and optimization of the neural network; research of the subject area) | |
| the student is able to solve professional problems using artificial intelligence systems | |

To summarize the results, the points received by the student were summed up (the maximum possible score is 21 points). Students at the ascertaining and control stages were distributed according to levels: unsatisfactory (less than 7 points scored); satisfactory (from 7 to 10 points scored); good (scored from 11 to 17 points); high (from 18 to 21 points scored). The results of the ascertaining and control stages of assessing the success of mastering the module "Data Analysis. Machine learning. Artificial Intelligence" are presented in Table 2 and Figure 4.

Table 2. Results of the ascertaining and control stages of assessing the success of mastering the module

| Level | Ascertaining stage | | Control stage | |
|---|---|---|---|---|
| | Number, persons | Fraction, % | Number, persons | Fraction, % |
| Unsatisfactory | 6 | 20 | 0 | 0 |
| Satisfactory | 20 | 67 | 9 | 30 |
| Good | 3 | 10 | 14 | 47 |
| High | 1 | 3 | 7 | 23 |
| Sum | 30 | | 30 | |

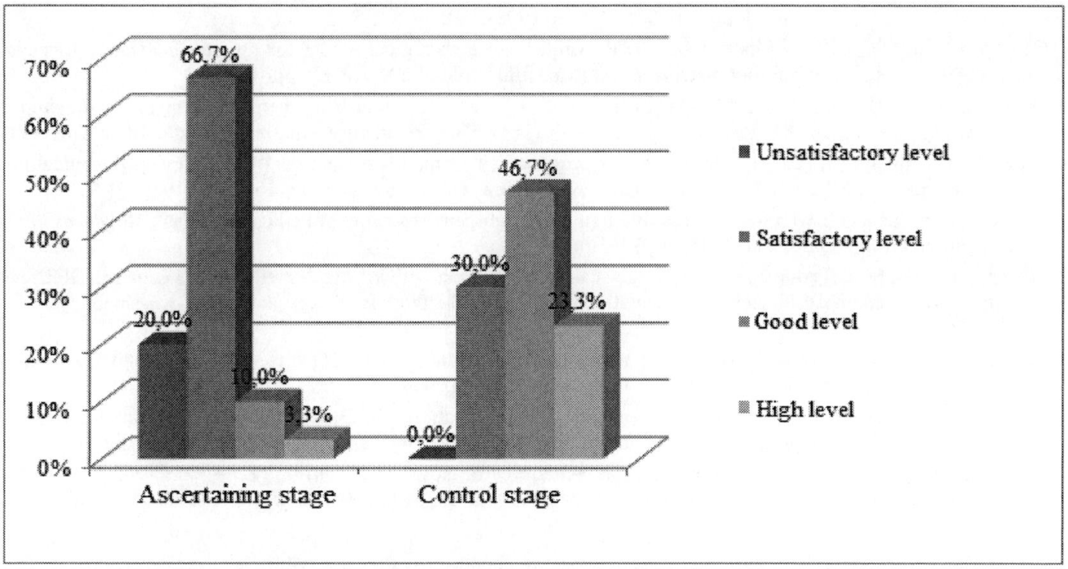


Figure 4. Distribution of students by level of module mastery

Based on the Chi-square test, the difference between the indicators of the ascertaining and control stages is statistically significant. The Chi-square value is 21.790 (Chi-square critical 11.345 at p=0.01 significance level). The relationship between factor and performance characteristics is statistically significant at a significance level of $p<0.01$. Significance level $p<0.001$. Results of mastering the module "Data Analysis. Machine learning. Artificial Intelligence" prove the effectiveness of the developed teaching methodology.

# 3. CONCLUSION

Methodological educational materials for module disciplines have been developed, including video lectures and presentations for them, laboratory work (including those based on GoogleColab notebooks), datasets, sets of tests, questions on theory and practical tasks for exams and tests.

The effectiveness of the developed teaching methodology for the module "Data Analysis. Machine learning. Artificial Intelligence" was confirmed by the results of a pedagogical experiment.

As a result of the students' project work, a number of neural network predictive systems were developed that allow solving pedagogical problems. You can change some of the input parameters and observe how the values of the output vector Y change. In this way, the laws of the subject area can be identified and the problems of predicting the future properties of the modeled object and its optimization can be solved.

## REFERENCES


Akaev, A., Sadovnichy, V. (2021). The human factor as a determinant of labor productivity in the era of the digital economy. *Studies on Russian Economic Developments*. No. 1, pp. 56-58. Moscow, Russia: International academic publishing company "Science/Interperiodica".

Andreeva, O. (2019). Comprehensive diagnostics of the components of research competence among students of pedagogical areas of training. *The Education and Science Journal.* Vol. 21, No 1., pp. 37-58. Yekaterinburg, Russia: Russian State Vocational Pedagogical University Press.

Ausin, M.S., Azizsoltani, H., Barnes, T., Leveraging, Chi M. (2019). Deep Reinforcement Learning for Pedagogical Policy Induction in an Intelligent Tutoring System. *Proceedings of the 12th International Conference on Educational Data Mining*, Montreal, Canada, July 2-5, 2019, pp. 168 – 177.

Bosova, L., Samylkina, N. (2018). Modern computer science: from robotics to artificial intelligence. *Computer science at school*, 1(8), pp. 2-5. https://doi.org/10.32517/2221-1993-2018-17-8-2-5

Caro, M.F., Josyula, D.P., Jiménez, J.A. (2015). Multi-level pedagogical model for the personalization of pedagogical strategies in intelligent tutoring systems. *Dyna* (Medellin, Colombia), Vol. 82. pp. 185-193.

Danilova, N. (2021). Improving the analytical activity of students - future teachers. *Bulletin of the Leningrad State University named after A.S. Pushkin.*, Vol. 2, pp. 461-471. Saint Petersburg, Russia: Leningrad State University Press.

Lapenok, M., Rozhina, I. (2019). Neural network prognostic systems for research and solution of pedagogical problems. *Proceedings of ISERD 174 International Conference*, New York, USA, October 16 -17, 2019, pp. 13-17.

Lozinskaya, A., Lapenok, M. (2021). Cognitive Issues in Intelligent Modeling of Pedagogical Task. *Advances in Inteligent Systems and Computing.*, Vol. 1346, pp. 355-360.

Mayer A. (2023). School education awaits large-scale digitalization with artificial intelligence. *Vedomosti*. Retrieved from https://www.vedomosti.ru/society/articles/2023/09/18/995660-shkolnoe-obrazovanie-zhdet-masshtabnaya-didzhitalizatsiya.

Rahman, Md. S., Reza, H. (2021). Big data analytics in social media: a triple T (Types, Techniques and Taxonomy) study. *Advances in Intelligent Systems and Computing*, Vol. 1346, pp. 479-487.

Saikina, M. (2024). Recommendations for educational organizations. In Bayramkulova L., Kazaryan K. (Eds.), *The influence of artificial intelligence on education*. Pp. 74-75. Moscow, Russia: ANO "Digital Economy".

Stepanova, A. (2022). AI in education: what can technology do now? *Sber University. EduTech.* No. 4, Retrieved from https://sberuniversity.ru/upload/iblock/09f/85v0n3to7fvy3awqz3p1lboeq0sk464r/EduTech_49_ web.pdf.

# STUDENT-CENTERED PERSONALIZATION OF INDIVIDUAL EDUCATION THROUGH REUSABLE AND AUTONOMOUS LEARNING UNITS - THE SPIRAL MODEL

Sorana Cimpan[1], Francesco Floris[2], Marina Marchisio Conte[2] and Sergio Rabellino[2]
[1]*University of Savoie Mont-Blanc, France*
[2]*University of Turin, Italy*

**ABSTRACT**

The COVID-19 pandemic acted as a catalyst, accelerating the transition to e-learning worldwide. Educators and students swiftly transitioned to online teaching and learning platforms as educational institutions closed their doors to curb the spread of the virus. A key trend in Education 4.0 is personalized learning, ensuring that the pace, content, and delivery methods are adapted to maximize individual growth and achievement, thereby increasing the effectiveness of student-centered learning environments. Personalization at scale is only possible through technology. Intelligent e-learning systems enable personalized learning by dynamically tracking and managing the learning needs of all students. These systems provide a platform that delivers a wide range of engaging learning content, resources, and opportunities, accessible anytime, anywhere, far beyond the confines of a traditional classroom. However, in parallel with technology, teachers and their role in the learning design process play an essential role in effective personalization. In this article, we therefore present the model we have developed for online course design that seeks to converge two models (that of adaptive learning platforms and that of competency-based learning). The model fits into the design phase of a classic ADDIE educational design cycle and proposes a strategy for the macro-design of the course. This model also aims to standardize the metadata available for learning units, making finding and sharing learning materials easier. This uniformity is crucial for applying comparison algorithms to personalize the learning experience for students. The model is currently undergoing testing and validation.

**KEYWORDS**

Online Course Design, Personalized Learning, SPIRAL Model, Student-Centered Learning

## 1. INTRODUCTION

The COVID-19 outbreak in early 2020 has impacted education at an unprecedented level, and almost all educational institutions worldwide have had to completely move to online education (Banihashem et al., 2023). The COVID-19 pandemic acted as a catalyst, accelerating the transition to e-learning across the globe. With educational institutions forced to close their doors to curb the spread of the virus, educators and students quickly adapted to online platforms for teaching and learning. This shift highlighted the importance of digital literacy and revealed the potential of e-learning to provide flexible and accessible education. As a result, many institutions and individuals have invested in technology and digital infrastructure to enhance the e-learning experience, paving the way for more widespread adoption of online education even beyond the pandemic era. The University of Turin has considerable experience in the creation of online courses. The Orient@mente initiative was launched in 2015 with the objective of assisting students in the transition between secondary school and university through a platform that enables them to interactively explore the study paths proposed by the University (Barana et al., 2017a, Barana et al, 2017b). In 2018, Start@Unito was established as a selection of university modules in a broad range of topics, administered through open online courses that are freely available. These courses could also be used to facilitate the transition between secondary and tertiary education and to enhance success in higher education (Marchisio, Operti et al., 2019, Marchisio et al., 2020). During the pandemic, numerous practices emerged to ensure continuity of education in student-centered learning environments (Otto et al., 2024). An analysis of the cycle of visioning, planning, and implementing learner-centered ecosystems in teaching-learning can be found in Patey's work (Patey,

2023). One of the nine crucial trends identified by Patey in Education 4.0 is personalized learning: "*Education 4.0 emphasizes learner-centered learning through `flipped' or `blended' learning modes, in which students can choose the content and courses they want based on their needs, to help them explore their skills and abilities*" (Patey, 2023). The design of a course is fundamental to effectively personalize learning. A well-structured course framework allows for the incorporation of diverse learning styles, paces, and preferences, ensuring that each student can engage with the material in a way that suits them best. By thoughtfully integrating adaptive learning technologies, varied instructional methods, and flexible assessments, course design can address individual needs and promote a deeper, more meaningful understanding of the subject matter. This tailored approach not only enhances student engagement and motivation but also maximizes educational outcomes by recognizing and nurturing the unique potential of every learner. To make this approach possible, it is essential to use Learning Analytics techniques to monitor, evaluate and eventually modify the learning path of individual students (Floris et al., 2020, Marchisio et al., 2019; Fissore et al., 2023). The role of learning analytics is also very important in blended learning, and it is necessary to inform and raise awareness among teachers on this topic (Bournaveas et al. 2022). This article proposes a design model for online courses and their content. The model aims to provide a way for collecting and analyzing the data required for effective personalization of learning. It also aims to provide teachers with a common design methodology, enabling them to share information, collaborate on course implementation, and highlight any design problems. In the next section, the concepts of student-centered learning, personalized learning, and an overview of recommendation models and systems will be presented with their connection to the instructional design practices. To understand teachers' perceptions of learning design, we conducted an international survey, the results of which are presented in the section "Teachers' perspectives on learning design". This analysis is based on data collected through 3 questions in a questionnaire administered to participants of the online course "INVITE training module," which was developed within the context of the INVITE project (https://invite-erasmus.eu/). The SPIRAL model will thereafter be introduced, without going into too much detail, focusing on the main entities and main aspects that its application can have, both for students and teachers. A testing and validation phase of the model is currently in progress. The concluding section, on the other hand, will give space for future developments of this work.

## 2. STUDENT-CENTERED AND PERSONALIZATION OF LEARNING

Student-centered learning is an educational approach that places students' needs, interests, and learning styles at the core of the learning process. It emphasizes active learning, where students engage in activities, discussions, and projects that foster more profound understanding and critical thinking (Wright, 2011; Overby, 2011). This approach is closely connected to personalized learning, which tailors educational experiences to individual student's strengths, weaknesses, and preferences (Patey, 2023). By focusing on each student's unique path, personalized learning ensures that the pace, content, and delivery methods are adapted to maximize individual growth and achievement, thereby enhancing the overall effectiveness of student-centered learning environments. Beese gave another definition of personalized learning: "*Educational personalization is best conceived, in a broad sense, as what occurs in any process that uses information from or about a student to generate plans or educational decisions for that student*"(Beese, 2019). One theory that describes the design characteristics of personalized learning is that of Walkington and Bernacki. This theory focuses on how a learning environment can be modified to benefit students' cognitive, motivational, and affective processes that influence their learning. In this theory, three dimensions relevant to learning theories are identified, based on which personalized learning strategies can vary. The first dimension is the "variable degrees of depth", which measures how much the daily life experiences of students affect the design of teaching activities. The second dimension, called "different grain sizes", is the granulometry of the personalization intervention, which can be at the level of the individual student, of small groups (with one or more common characteristics) or of larger groups based on more general parameters. The third dimension is "ownership", that is the degree in which students are given control and choice of learning situations (Walkington et al., 2014). Wolf (2010) stated that technology also enhances personalized learning: "*Personalization cannot take place at scale without technology. Personalized learning is enabled by smart e-learning systems, which help to dynamically track and manage the learning needs of all students and provide a platform to access myriad engaging learning content, resources, and learning opportunities needed to meet each student's needs everywhere at any time, but which are not all available within the four walls of the traditional classroom.*" In their review work, Otto S. and his colleagues (Otto et al., 2024) identify three technological models to support personalized learning: *Open Digital Badges* – offer personalized learning by allowing students to choose badges relevant to their competence level, prior knowledge, and progress; *Competency-Based Learning* - enables learners to progress by mastering competencies aligned with their prior knowledge, skills, and abilities (an example of application is reported in Pasterk et al, 2022); *Adaptive Learning Platforms* - create personalized learning pathways by analyzing students' responses to lessons and assessments, thereby providing tailored resources and learning materials; this way, class time can be used for discussions and extending learned material.

In particular, for the last model, the related instructional design theory refers to learning objects. The term "learning object" is often used interchangeably with "learning material", "learning resource" or "educational resource". The Learning Technology Standards Committee uses the term learning object to describe small instructional components: "*Learning Objects are defined here as any entity, digital or non-digital, which can be used, reused, or referenced during technology-supported learning. Technology-supported learning includes computer-based training systems, interactive learning environments, intelligent computer-aided instruction systems, distance learning systems, and collaborative learning environments. Examples of Learning Objects include multimedia content, learning objectives, instructional software and software tools, and persons, organizations, or events referenced during technology-supported learning.*" (LTSC, 2007). David Wiley (Wiley, 2002) defines a learning object (LO) as any digital resource that can be reused to support learning, emphasizing its digital nature. He also created a taxonomy of LO types and listed characteristics to distinguish them, with the aim of differentiating types for instructional design. Wiley employs the metaphor of famous compostable bricks game to illustrate LOs. He posits that small instructional pieces can be assembled into larger structures (e.g., castles) and reused in new configurations (e.g., spaceships). However, unlike the bricks, not all learning objects (LOs) can be easily combined. Another metaphor Wiley employs is that of the atom, representing LOs as small units that can be combined to form larger entities. This metaphor highlights granularity and combination properties yet leaves unresolved issues. Granularity pertains to the minimum size of an LO, which is likened to subatomic particles forming atoms. Establishing this minimum size is, however, necessary. Furthermore, the combination issue remains to be clarified, as Wiley's taxonomy needs to address how LOs can be effectively combined explicitly. However, in line with what Bruschi and Perissinotto stated in their work (Bruschi, Perissinotto, 2003), there may be problems using learning objects to design online courses. One of the most complicated points is that there is a certain autonomy in terms of the structure and the amount of knowledge that each learning object must contain. In this way, however, the creation of customized pathways, even in an automated way, although possible given the technological progress achieved, would result in internally very diverse and uneven pathways. Furthermore, another possible problem is the amount of work the teacher requires to design a single learning object, considering all the metadata that would need to be defined. It is not surprising that in the literature of the last few years, there is a large amount of work on teachers' perceptions of online teaching and the use of technology (An et al., 2021; Hassan, 2020). The various studies reveal that while teachers recognize the potential value of technology and online/blended learning, they feel they need more time to make a complete transition and prefer traditional teaching. One of the points on which many teachers agree is that creating e-content takes more time and effort than classroom teaching (Hassan, 2020). On the other hand, the literature should focus on teachers' perceptions of learning design strategies and their implementation and effectiveness. A survey was therefore conducted in order to gain insight into teachers' perceptions of design practices and to gain an overview of which aspects they consider most relevant in design practice. The results of this survey will be presented in the next section.

## 3. TEACHER'S PERSPECTIVES ON LEARNING DESIGN

The INVITE (Developing Competencies and Innovative Designs for International Virtual and Blended Modalities) project aims to enhance teaching and learning competencies for designing and implementing new virtual and blended modes of international collaboration within European Higher Education Institutions (https://invite-erasmus.eu/). Despite the rapid response of many institutions to the educational challenges posed by the COVID-19 pandemic, there is now a collective understanding among EU Member States of the need to develop further and integrate virtual and blended learning modalities post-pandemic. In alignment with the European Commission's Digital Education Action Plan 2021-2027, the INVITE project seeks to support educators in acquiring digital skills to improve the quality of digital practices in education. The project also emphasizes the development of new Erasmus+ Blended Intensive Programs, in line with the Green Agenda's support for internationalization efforts. The INVITE project addresses various challenges associated with digitalization in higher education, aiming to create a robust educational foundation for the future. Focusing on student-centered approaches and active learning, the project aims to promote a more inclusive and effective digital education environment in Europe (Barana & Marchisio, 2024).

One of the results of this project is a new training module for higher education teachers and staff to promote the activation of virtual and blended international learning activities. One of the first steps during the training program is an initial questionnaire (in two versions, one for teachers and one for staff), in which three questions are related to the course design. We included these questions (and some in the final questionnaire) to explore teachers' perceptions of learning design. The training program started in the second half of April. Registration was free and open to all; over 1000 users signed up for the course. The registration form asked the user to indicate which of the two categories the user feels most similar to teacher or staff member. In order to choose the correct training path, the following definitions were present in the module: Teachers (or educators) are people interested in developing digital international teaching and learning activities; the staff members want to create international bridges among institutions and support teachers and/or students in developing or implementing digital international activities. We decided to analyze the three questions on learning design, and so far, we have the answers from 124 teachers and 47 staff members. The first question was "*Do you have experience in course design?*". There is little difference between teachers and staff, as only 55% of teachers have experience in course design and 36% of staff members. Then we asked teachers and staff members who answered *yes* to the following question: "*On a scale from 1 to 5, where 1 represents minimal planning, and 5 represents comprehensive planning, please indicate how thoroughly you typically design the following aspects of an online course*", and the results are reported in Figure 1 and Figure 2. The results show that teachers who claim to have experience in course design focus mainly on the course structure and learning outcomes, while not thoroughly structuring the prerequisites. The same attitude is reflected in the behavior of the staff. This second result was entirely predictable, as the creation (and therefore planning) of individual materials and evaluation is usually more associated with the role of the teacher. What is surprising, however, is a certain disregard for the role of prerequisites.

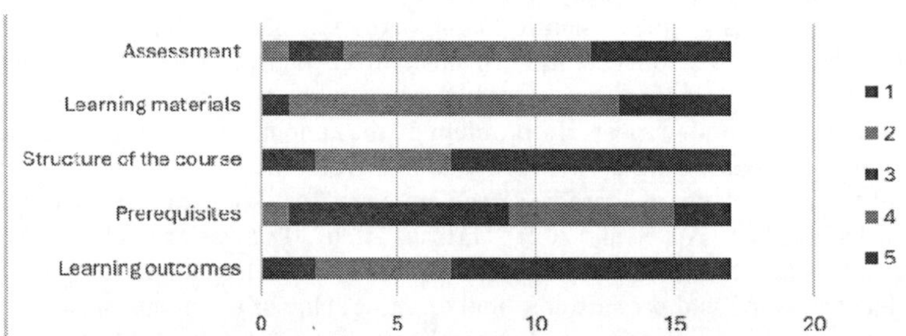


Figure 1. Answers from the staff members to the question n.2

The lack of focus on prerequisites stems from the fact that students' pathways are usually placed in a broader context, and therefore, many teachers somehow think that they come naturally. However, for effective adaptation, it is essential to correctly and in-depth outline the aspects related to the prerequisites. In the final question, which was also asked of those participants who answered *yes* to the first question, we asked what tools and/or software they use for course design. The survey shows that almost 50% of people in each group do not use specific tools and/or software. The responses vary significantly from those who say they use them, and most do not indicate fundamental course planning/design tools. For example, some mention Moodle (about 25% in both groups) or h5p, but these are tools for creating and using content, not designing it. In fact, only one person included in the group of staff members stated that he or she uses a specific tool for learning design (the ABC learning design model). The fact that the instructional design is clearly perceived as an important aspect, combined with the almost inexistent use of tools to support this design highlights the need and opportunity for proposing effective support for instructional design. This is one of the main goals of the SPIRAL model we propose in the following section.

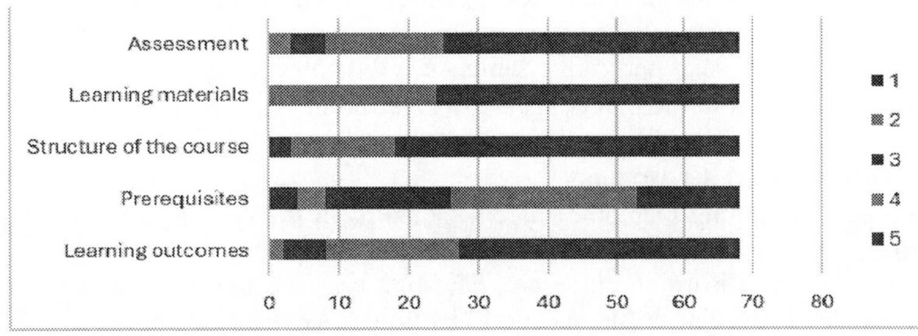


Figure 2. Answers from the staff members to the question n.2

# 4. SPIRAL MODEL

In an attempt to integrate the two models discussed in section 2 (competency-based learning and adaptive learning platforms), we developed a new model, the SPIRAL model, for designing digital learning courses based on a new type of object: the learning unit. SPIRAL stands for ***S**tudent-centered **P**ersonalization of **I**ndividual education through **R**eusable and **A**utonomous **L**earning* units. The principal objective of the model is to identify and formalize all entities and their relationships in order to facilitate personalized learning for students. Furthermore, it seeks to provide teachers with both a methodology and a tool to design online courses and teaching units, thereby simplifying their work and also helping to identify potential problems in course design. The design phase can be divided into macro-design and micro-design, with one following the other. The macro-design phase defines course outcomes, target audience, content, and learning units. During the micro-design phase, designers create learning activities, assessment methods, and instructional materials for every designed learning unit. The model proposed here focuses on the macro-design phase. The standardization of metadata associated with teaching units aims to to facilitate the search and sharing of learning units. Furthermore, this uniformity is essential for the application of comparative algorithms that can automatically propose personalized learning experiences for students. The "SPI" part of the acronym refers to student-centered learning, which places the student at the heart of the learning process. This approach allows students to decide both what materials to learn and how to learn it. These decisions involve two levels of content customization: a general level and a specific level. At the general level, recommendation systems suggest courses to students based on their aptitudes, preferences, and past experiences, which is a significant area of research. Once a topic has been chosen, there are various possible learning paths at the specific level. Each path, composed of a sequence of learning units, is designed to achieve the same learning outcomes through different means. This model proposes a structure for these units that allows for comparison and highlights differences, enabling students to choose their preferred path and retain control over their learning process. The second part of the acronym highlights the key characteristics of these learning units. Each learning unit is connected to a specific single learning outcome, defining the learning pathway by a set of outcomes. Students can choose the units they wish to use to achieve these outcomes. The R part emphasizes that learning units should be reusable, meaning they can be applied in different contexts. The A part indicates that the units should be autonomous and self-contained, allowing them to function independently of their context. This modular design approach ensures flexibility and adaptability. In the next section we will focus on the entities and the idea behind the model, analyzing the advantages we can derive from it for both teachers and students.

## 4.1 Learning Outcomes and Learning Units

The first entity that we define is *Learning Outcomes (LO)*. We started with the definition of learning outcomes given in the "European Quality Framework 2017" (European Commission, 2008): "[. . .] statements of what an individual should know, understand and/or be able to do at the end of a learning process, defined in terms of knowledge, skills and responsibility and autonomy", but we added some descriptors. Basically, in our model, when we talk about learning outcome, we mean an object represented by:

- The sentence that represents the learning outcome. This sentence will consist of "Verb + Object". In our case, the verbs are selected from the verbs associated with a particular level of Bloom's Taxonomy. The second part of the sentence is the object of the verb.
- Topics. Compared to the object previously declared, the topic defines a more general topic that can be inserted from a hierarchical topic structure.
- The level of Bloom's Taxonomy
- The type of Learning Outcome (parent or child) in the context in which the Learning Outcome is defined.

There are two types of learning outcomes. The first type are *macro-LOs*, which refer to the level of the whole online course. They answer the question "*What will learners be able to do when they have completed the course?*". The second type are *micro-LOs*, which refer to the level of the learning units in the learning pathway and answer the question "*What will learners be able to do when they have completed this unit?* ". The main reason for this distinction is that macro-LOs can only be assessed indirectly because they are too broad. However, we can construct a learning path made up of micro-LOs that will enable them to be acquired. To emphasize this difference, we have chosen to use the parent-child relationship of the many-to-many type. A clear implication of this classification is that it is inherently contextual.

The fundamental concept of our model is the *Learning Unit* (LU), whose definition is derived from Wiley's definition of learning object (Wiley, 2002), but has many points in common with the definition of "Autonomous Learning Object" given in (Bruschi, 2003). A learning unit is a collection of three types of digital learning materials: resources, activities, and summative assessment. When a learner completes a learning unit, he/she acquires one and only one learning outcome. In order to be sure that a learner has acquired a learning outcome, a summative assessment is a mandatory component of a learning unit.

Starting from this definition, we have identified all the descriptors necessary for each learning unit to define a learning pathway and, simultaneously, to make them comparable: Title - a short title of the Learning Unit; Description - a short description of the Learning Unit; Prerequisite (0+) - prerequisites are learning outcomes that define the basic knowledge required to access the LU. It can be 0 or more; Outgoing Learning Outcome (1) - we use Bloom's taxonomy to indicate the Learning Outcomes: ECTS - this is the time allotted for study in order to complete the learning unit; Scale of difficulty - the teacher can indicate here the estimated the difficulty for this LU. We use the term *Learning Path (LP)* to denote a structured and sequenced progression of Learning Units designed to guide individuals through the process of acquiring a Learning Outcome. This sequence of LUs is constructed starting from the LU to which the LO is associated and, recursively, adding all the learning paths necessary to acquire the prerequisites of that learning unit. We can therefore extend the relationships and characterizations introduced for LO and LU also to learning paths, such as a measure of path difficulty, a measure of interactivity or even an expected completion time. The formalization introduced allows us to calculate a measure of similarity between Learning Outcomes which can be exploited for the generation of alternative learning paths, with a dual meaning. On the one hand, in fact, if two Learning Outcomes are similar, we can say that their respective learning paths are equivalent, in the sense that following one or the other makes no difference in terms of learning outcome acquired. On the other hand, since the prerequisites are also defined as Learning Outcomes but seen as "input", the similarity can be exploited to create a sequence between learning units. In this way it is not necessary to define a specific path a priori for each student but to show each student all the possible learning paths to acquire an appropriate learning outcome, with all the information to make an informed choice about which path to follow. In this way, going back to the 'ownership' dimension of (Walkington et al., 2014), the control of the Learning Units to be followed is entirely in the hands of the individual student. Furthermore, all the information collected by the system can also be used to identify possible problems in the design phase. For example, when we think about macro and micro learning outcomes, an important consideration is that the level of a macro learning outcome must be higher than or equal to that of its child LOs when referring to Bloom's Taxonomy. An LO can only be of a lower level if all its children are of different themes. In this case, it is worth considering how and to what extent knowledge of different topics at an advanced level can provide skills and knowledge in another topic. These discrepancies can be analyzed automatically to identify possible gaps in the course design. Another crucial aspect to consider is that the formalization introduced allows for the sharing of information, which, in a community of practice of teachers/educators, can be leveraged to utilize the learning units even outside the context in which they are defined. This also facilitates their sharing. To achieve this, it is essential to prioritize autonomy, ensuring that the units are self-contained and contain all the necessary contents, activities, and evaluation tools to function and integrate seamlessly in another context.

## 5. CONCLUSION AND FUTURE DEVELOPMENTS

Implementing a student-centered approach in higher education has the potential to yield significant and far-reaching consequences. This paradigm shift fosters a more engaging and interactive learning environment, encouraging students to participate actively in their education. Individual needs and preferences are better met by introducing personalized learning pathways, thereby promoting equity and inclusion. To realize customized courses at individual units within a curriculum and across entire programs, we have developed a SPIRAL model that defines the entities involved and their relationships. Our model revolves around the learning unit, a structured collection of digital learning materials, including resources, activities, and summative assessments. Each learning unit leads to a specific learning outcome, aligning with the European Quality Framework 2017. The model utilizes descriptors such as title, description, prerequisites, learning outcomes, estimated learning time, and difficulty defining and comparing learning units. We can create personalized learning paths by establishing relationships between these units, such as prerequisite connections. Instead of prescribing a specific path for each student, our approach presents multiple pathways, allowing students to choose their preferred route to achieving the desired learning outcomes. The model's data collection capabilities also help identify potential gaps and discrepancies in course design, ensuring a cohesive and comprehensive learning experience. In conclusion, this model aims to provide a flexible and adaptable design methodology for online courses that meet the needs of different learning contexts, like BIPs, COILs, and MOOCs. The model is currently undergoing an extensive phase of testing and validation to ensure its accuracy, reliability, and performance before it is deployed for use. Building on the SPIRAL model, we are developing an online editor to collect, organize, and utilize educational data. This editor will be integrated with existing learning management systems (LMSs), creating a unified platform where teachers can create, manage, and adapt the content in real time based on individual student needs. Leveraging advanced data analytics, the editor will monitor student progress, engagement, and performance, providing valuable insights that enable educators to refine instructional strategies and materials, thus better supporting each student's journey.

## ACKNOWLEDGEMENT

The research activity described in this paper has been carried out within the Erasmus+ K.A.2 N. KA220-HED-2021-006 “Developing Competences and Innovative Designs for International Virtual and Blended – INVITE” and within the GNSAGA group of INDAM – “Istituto Nazionale di Alta Matematica F. Severi”.

## REFERENCES

An, Y., Kaplan-Rakowski, R., Yang, J., Conan, J., Kinard, W. and Daughrity, L., (2021). Examining K-12 teachers' feelings, experiences and perspectives regarding online teaching during the early stage of the COVID-19 pandemic. *In Educational technology research and development*, 69, pp. 2589-2613.

Barana A. and Marchisio Conte M., (2024). Eight Sustainable Practices for Digital Activity Development: Drivers and Barriers in International Higher Education Collaboration. In *Computer*, vol. 57, no. 03, pp. 117-122.

Barana, A., Bogino, A., Fioravera, M., Marchisio, M., & Rabellino, S., (2017b). Open Platform of self-paced MOOCs for the continual improvement of Academic Guidance and Knowledge Strengthening in Tertiary Education. In *Journal of e-Learning and Knowledge Society*, *13*(3), pp. 109-119.

Barana, A., Marchisio, M., Bogino, A., Operti, L., Fioravera, M., Rabellino, S. and Floris, F., (2017a). Self-paced approach in synergistic model for supporting and testing students. *Proceedings of 41st Annual Computer Software and Applications Conference (COMPSAC)*, Vol. 1, pp. 407-412.

Beese, E. B., (2019). A process perspective on research and design issues in educational personalization. *In Theory and Research in Education*, v. 17(3), pp. 253–279.

Bournaveas, V., Gueye, M., Cimpan, S. and Exposito, E., (2022). Penser l'hybridation avec l'appui des Learning Analytics: l'approche HyPE 13 pour la formation des enseignants. *In Questions de pédagogies dans l'Enseignement Supérieur*, pp 1-19.

Bruschi, B. and Perissinotto, A. (2003). *Come creare corsi on line*. Carocci, 2003.

da Silva, F. L., Slodkowski, B. K., da Silva, K. K. A. and Cazella, S. C., (2023). A systematic literature review on educational recommender systems for teaching and learning: research trends, limitations and opportunities. *In Education and Information Technologies*, 28(3), pp. 3289-3328.

European Commission, (2017). Council recommendation on the European qualifications framework for lifelong learning and repealing the recommendation of the European Parliament and of the Council of 23 April 2008 on the establishment of the European qualifications framework for lifelong learning. Official Journal of the EU.

Fissore, C., Floris, F., Marchisio, M. and Rabellino, S., (2023). Learning analytics to monitor and predict student learning processes in problem solving activities during an online training. *Proceedings of 47th Annual Computers, Software and Applications Conference (COMPSAC)*, pp. 481-489.

Floris, F., Marchisio, M., Sacchet, M., and Rabellino, S., (2020). Learning Analytics to Evaluate the Effectiveness of Higher Education Student Failure Prevention. In *International Association for Development of the Information Society*, pp. 265-272.

Hassan M. M., Tabasum M. and Mirza, W. H., (2020). A critical review by teachers on the online teaching-learning during the COVID-19. *In International Journal of Education and Management Engineering*, pp 17–27.

Kazem Banihashem, S., Noroozi, O., den Brok, P., Biemans, H. J.A. and Taghizadeh Kerman, N., (2023). Modeling teachers' and students' attitudes, emotions and perceptions in blended education: Towards post-pandemic education. *In The International Journal of Management Education*, Volume 21, Issue 2, pp. 1472-8117.

LTSC, Learning Technology Standards Committee. The learning object metadata standard, (2007), available at https://www.ieeeltsc.org/working-groups/wg12LOM/lomDescription/

Marchisio, M. and Sacchet, M., (2020). Analysis Items to Assess the Quality of Open Online Courses for Higher Education. In *International Association for Development of the Information Society,* pp. 63-70.

Marchisio, M., Operti, L., Rabellino, S, and Sacchet, M., (2019). Start@unito: Open online courses for improving access and for enhancing success in higher education. *In Proceedings of the 11th International Conference on Computer Supported Education*, pp. 639-646.

Marchisio, M., Rabellino, S., Roman, F., Sacchet, M. and Salusso, D., (2019). Boosting up data collection and analysis to learning analytics in open online contexts: an assessment methodology. *In Journal of E-learning and Knowledge Society*, 15(3), pp. 49-59.

Otto, S., Bertel, L. B., Lyngdorf, N. E. R., Markman, A. O., Andersen, T. and Ryberg, T., (2024). Emerging digital practices supporting student-centered learning environments in higher education: A review of literature and lessons learned from the COVID-19 pandemic. *In Education and Information Technologies*, 29(2), pp. 1673-1696.

Overby, K., (2011). *Student-centered learning.* Essai, 9(1), art. 32.

Pandey, A., (2023). E-Learning and Education 4.0: Revolution in Education of 21st Century. *In International Conference on Digital Technologies and Applications*, pp. 431-438.

Pasterk, S., Pagitz, L., Weiß, A. and Bollin, A., (2021). DigiFit4All–Conceptualisation of a Platform to Generate Personalised Open Online Courses (POOCs). *In Open Conference on Computers in Education*, pp. 247-258.

Walkington, C. and Bernacki, M., (2014). Motivating students by 'personalizing' learning around individual interests: a consideration of theory, design and implementation issues. *In Advances in motivation and achievement*, v. 18, pp. 139-176.

Wiley, D. A., (2002). *The instructional use of learning objects* (Vol. 1). Bloomington: Agency for instructional technology.

Wolf, M., (2010). Innovate to education: System [re]design for personalized learning. A report from the 2010 symposium. *In Software and Information Industry Association Washington DC.*

Wright, G. B., (2011). Student-centered learning in higher education. *In International journal of teaching and learning in higher education*, 23(1), pp 92-97.

# DIGITAL FOR WELFARE. STORIES OF "THIRD SECTOR" REALITIES FACING THE CHALLENGES OF DIGITAL TRANSFORMATION

Dalia Gallico
*Università San Raffaele Roma*
*Italy*

**ABSTRACT**

The third sector, also known as the nonprofit sector, encompasses organizations that operate independently from the government and the private sector. These organizations primarily focus on social, cultural, environmental, and community goals rather than profit. In the past three years, the topic of digital transformation has become increasingly relevant for virtually all nonprofit organizations, which have sought to address it through various initiatives. These efforts have targeted both internal dimensions—such as work processes, service development, communication, and evaluation—and external dimensions—such as service offerings, promotion, and delivery.

We specifically analyze a call for proposals under the PNRR (National Recovery and Resilience Plan) titled "Social Digital" and the related digital project proposal by Società Umanitaria (a historic nonprofit organization founded in 1893 with the mission of promoting social welfare, education, and cultural development). This proposal aims to enhance the digital skills of employees, permanent collaborators, and volunteers through an innovative training program. The primary objective is to create a digital ecosystem that improves the effectiveness of the institution's social and cultural activities.

The project stands out for its ability to combine technical digital training with a strong component of social inclusion and human skills development. Advanced technologies are employed for online learning and digital tools to optimize internal processes, with a particular focus on integrating new technologies in the fields of fashion, design, and craftsmanship.

The creation of an e-learning platform, which can subsequently be offered to other nonprofit organizations (scalability and sharing) for training and specialization, is a key feature. This interactive educational platform utilizes AI to provide a personalized learning experience, tailored to the students' knowledge level and learning pace. Student support is provided via chatbots that respond to questions, offer educational materials, assist in navigating courses, coordinate courses, and send reminders and surveys. Additionally, VR/AR is used for specialized courses, and a mobile application is available for monitoring progress and communication.

The studies, examples, and research we have conducted demonstrate that the synergy between people and technology has yielded excellent results in project development, communication, and education. These results are evident both in terms of optimizing routine processes and in achieving exceptional and beautiful outcomes.

**KEYWORDS**

Social, Digital Education, AI, No Profit, Welfare, Edutech

## 1. INTRODUCTION

The third sector, also known as the nonprofit sector, encompasses organizations that operate independently from the government and the private sector. These organizations primarily focus on social, cultural, environmental, and community goals rather than profit. They include charities, foundations, voluntary associations, social enterprises, and NGOs (non-governmental organizations), all dedicated to addressing various societal needs and contributing to the public good. For a long time, the topic of digitalization was on the fringes of discussions about the present and future of the Italian Third Sector. Many organizations in this sector were reluctant to make the necessary investments to implement digital tools and processes, as their priorities lay elsewhere. Digitalization could wait. The COVID-19 pandemic forced many Third Sector entities to embark, more or less consciously, on paths of digital transformation (and not mere digitalization).

In the past three years, the digital topic has become increasingly relevant for virtually all nonprofit organizations, whether large or small, which have therefore sought to address it. They have done so through initiatives targeting both internal dimensions—such as work processes, service development, communication, and evaluation—and external dimensions—such as service offerings, promotion, and delivery. To clarify: some have adopted hardware and software to manage their activities, some have started communicating through social media, some have rethought internal processes using new digital tools, some have transitioned products and services (also) to the virtual field, some have made digitalization the cornerstone of their new projects, and some are using it to evaluate the actual impact of their activities.

The paths of digital transformation undertaken by Third Sector organizations during the pandemic had different starting points and subsequently involved various stages reached with varying methods and speeds. In general, almost all organizations began to accelerate in the early months of the emergency for the reasons mentioned above. In the following period, behaviors differed. While some entities, especially the smaller ones, slowed their digital growth trajectory after the pandemic's initial push (sometimes stopping or even regressing), others continued to "press the accelerator" beyond the pandemic. These are the organizations we want to explore, sharing some of the most significant experiences they are realizing (Russell et al., 2010).

## 2. EDUTECH – PNRR CALL "SOCIAL DIGITAL"

According to LinkedIn's Global Talent Trends Report, by 2030, the skills required for workers will undergo a significant shift, with at least 65% of current abilities needing to be updated or replaced. In Italy, job postings that reference generative AI have increased nearly fivefold in the past two years, highlighting the rapid adoption of this technology. Applications for AI-related roles have grown by 31%, indicating a rising interest among professionals.

Generative AI can enhance creativity and innovation within companies and personalize a wide range of activities. Simultaneously, this technology opens new opportunities for professionals who interact with AI-generated content, such as reviewers, curators, moderators, and consultants. To fully leverage its potential, it is essential to develop both technical skills and soft skills (Santosuosso, 2020).

The objectives of the PNRR (National Recovery and Resilience Plan) "Social Digital" call are to support innovative projects aimed at enhancing the digital skills of employees, permanent collaborators, and volunteers in organizations operating within the social economy. The main goal is to empower the digital knowledge and skills within these entities.

The target of the call includes employees, permanent collaborators, and volunteers of organizations operating in general interest sectors of the social economy. Areas of intervention include basic and/or advanced digital training for operators to increase their digital skills and awareness and the implementation of digital solutions to improve the internal and external efficiency of the organization and the services offered. The PNRR call allocates a national fund of 15 million euros.

Criteria for participation include forming a partnership between a formal or informal network of private nonprofit entities needing digital empowerment, and the presence of a public or private nonprofit entity with experience in the digital field to construct the training pathway and/or develop the digital solution. It is also possible to include a for-profit technological provider with a budget not exceeding 25% of the requested contribution.

The digital solution must aim to improve organizational efficiency and the services offered to the community. The minimum amount of funding is 500,000 euros, and the maximum is 1 million euros, with a project duration not exceeding 15 months. Evaluation criteria include the coherence of the project proposal (15%), the structure and quality of the involved entities (20%), feasibility and organization (20%), and the quality of the project proposal (45%) (Manovich, 2020).

## 3. "SOCIETÀ UMANITARIA" FOR THE "SOCIAL DIGITAL" CALL

The Società Umanitaria Foundation in Milan is a historic nonprofit organization founded in 1893 with the mission of promoting social welfare, education, and cultural development. It focuses on providing educational opportunities, vocational training, and support for marginalized communities. The organization hosts cultural events, workshops, and courses aimed at fostering social inclusion and improving the quality of life for individuals in need. The Società Umanitaria has played a significant role in the social and cultural landscape of Milan, contributing to the city's rich tradition of philanthropy and community service.

The project proposal by Società Umanitaria for the "Social Digital" call aims to enhance the digital skills of employees, permanent collaborators, and volunteers through an innovative training program. The primary objective is to create a digital ecosystem that improves the effectiveness of the institution's social and cultural activities. The project stands out for its ability to combine technical digital training with a strong component of social inclusion and the enhancement of human skills.

Società Umanitaria operates in a diverse national and social context, with a long tradition of cultural activities and professional training. The main needs that the project intends to address are:

- The need to update the digital skills of staff and volunteers to improve operational efficiency and innovation capacity.
- The inclusion of new technologies to support training activities, especially in the fields of fashion, design, and craftsmanship.
- Combating social isolation by facilitating access to digital tools to promote communication and collaboration.
- Improving access to educational and cultural resources through digital platforms.

The objectives of the project are:

- Enhance the digital skills of at least 200 employees, permanent collaborators, and volunteers of Società Umanitaria within the next 12 months.
- Implement innovative digital platforms for managing training and cultural activities.
- Reduce the digital divide among different age groups and social categories involved in the institution's activities.
- Improve the effectiveness of internal communications and collaboration among staff through advanced digital tools.

The expected outcomes of the project, in both quantitative and qualitative terms, include:

- Certified training of at least 200 individuals in specific digital skills, with the issuance of certificates valid at the European level.
- Creation of a digital platform for managing and delivering courses, with a 30% increase in online course participation.
- A 25% increase in user participation and engagement in social and cultural activities through the use of digital tools.
- Improved operational efficiency of Società Umanitaria, with a 20% reduction in administrative management time due to digital automation.

With its long tradition and commitment to the intellectual and moral elevation of citizens, Società Umanitaria is perfectly positioned to meet the requirements of the "Social Digital" call. The proposed project not only enhances the digital skills of its staff and volunteers but also strengthens the institution's role as a leader in social inclusion and cultural promotion. Its strong local roots and the validity of ongoing projects demonstrate Società Umanitaria's ability to adapt and innovate to meet contemporary challenges (Quintarelli et al., 2020).

## 4. THE DIGITAL PROJECT PROPOSAL

Società Umanitaria has identified a significant training need among its employees, stable collaborators, and volunteers. The digital and complementary skills to be developed include:

- Basic Digital Skills: Computer use, internet navigation, cybersecurity, and management of digital documents.

- Advanced Digital Skills: Use of specific software for fashion, design, and crafts, database management, and e-learning technologies.
- Soft Skills: Effective communication, time management, teamwork, and problem-solving.
- Life Skills: Adaptability, resilience, stress management, and continuous learning ability.

These skills are fundamental for improving operational efficiency, fostering innovation, and supporting social inclusion.

The training program content includes:

Basic Digital Skills Modules: Introduction to computer and digital tools usage. Internet navigation and cybersecurity. Management of digital documents and use of office automation tools.

Advanced Digital Skills Modules: Use of specific software for fashion, design, and crafts. E-learning technologies and online educational platforms. Database management and data analysis.

Soft Skills: Effective communication and presentations. Time management and work organization techniques. Teamwork and leadership. Problem-solving and critical thinking.

Life Skills: Adaptability and change management. Resilience and stress management. Continuous learning and personal development.

Course Delivery: In-person, online, and hybrid courses. Use of an e-learning platform for course management and delivery.

Weekly modules with interactive lessons and practical workshops. Virtual classes with videoconferencing sessions and group discussions. Practical labs for the application of learned skills. Digital platform for access to educational materials and resources. Technological tools and equipment for fashion, design, and craft labs. Classrooms equipped with multimedia devices for in-person lessons. Periodic assessments to monitor student progress. Issuance of certifications valid at the European level for the acquired skills.

The methodologies and formats of the training activities are innovative and effective.

Personalized Learning: Use of artificial intelligence algorithms to adapt training content to individual student needs.

Interactive Teaching: Interactive lessons with advanced technologies such as VR/AR for specialized courses.

E-learning: Digital platform for course access and management of training activities.

Virtual Assistance: Chatbots and virtual assistants to support students and answer their questions in real-time. Labs and workshops for the practical application of learned skills.

Continuous Feedback: Progress analysis and personalized recommendations for improving learning.

Communication Campaigns: To inform and raise awareness among employees, stable collaborators, and volunteers about the importance of digital training. Recognition and incentives for active participation and completion of training programs. Continuous assistance through chatbots and virtual tutors to answer questions and support beneficiaries during the training program. Organization of interactive and collaborative activities to promote engagement and motivation of participants.

Training Content:

Digital Transformation: Program of lessons and workshops on digital transformation in companies, development of fundamental digital skills to face modern market challenges.

Application of Neural Networks: Introduction to neural networks and their use in various sectors, practical workshops with real implementation examples.

Data Science: Fundamentals of Data Science and the use of tools like Python and R for data analysis, practical projects, and analysis of real datasets.

Work on Digital Platforms: Training on the use of digital platforms for collaboration and work management (e.g., Microsoft Teams, Slack, Trello), strategies for adopting and optimizing the use of these platforms.

The proposed digital solution includes the creation of a unique and interactive e-learning platform, integrating innovative technologies to enhance the learning experience and support social inclusion.

Main Elements of the Platform:
- E-learning Platform: Develop an e-learning platform that can later be offered to other institutions and companies for training and specialization (knowledge sharing). Analyze the most popular e-learning platforms

(Moodle, Blackboard, Canvas) to identify best practices and integrate specific features to meet the needs of Società Umanitaria. Integration of specialized modules such as adaptive courses, progress analysis, and virtual assistants to enhance learning effectiveness.

- Interactive Educational Platform with AI Utilization: Create a personalized learning experience for students, adapted to their knowledge level and learning pace. Personalized Educational Programs: The platform analyzes the initial knowledge level of students and automatically adjusts the educational materials, offering individual recommendations for course completion and additional materials. Use of AI to monitor student progress, identify problem areas, and provide recommendations. Dashboard for teachers with detailed analysis of each student.

- Chatbots and Virtual Assistants: Chatbots to support students, answer questions, and provide educational materials in real-time. Assistance in navigating through courses and study materials. Virtual assistants that can coordinate courses, send reminders about exams and assignment deadlines, conduct mini-tests, and surveys to check material comprehension. The platform will integrate chatbots for real-time support, progress analysis through AI, a mobile application for monitoring and communication, and specialized courses on digital skills. The platform aims to improve the effectiveness of training activities, facilitate course access, and promote digital inclusion.

The platform is designed to be sustainable in the medium to long term, with continuous updates and dedicated technical support. It will be made available for free to other non-profit organizations after the project's end to maximize social impact.

The focus is on the impact and opportunity of generative AI in Italy on the world of work and the new professions and skills that derive from it. It also presents data on how this technology can help support the activities of third-sector professionals by unleashing creativity and productivity. Training activities and the creation of new skills to evolve professionals within the labor market and create new employment opportunities, particularly for young people.

Microsoft has always invested significantly in innovation and training in Italy, offering training programs to develop the skills required by the labor market. Therefore, Microsoft Italy is involved in the innovation and launch of the new training platform. The goal is to create an ecosystem where the required digital skills can be developed and enhanced, thus contributing to a non-profit workforce ready to face the challenges and opportunities of the generative AI era (Sartor, 2016).

## 5. THE STRUCTURE AND QUALITY OF THE ENTITIES INVOLVED IN THE PROJECT

As the lead organization, Società Umanitaria bears the responsibility of coordinating and managing all project activities. Its primary duties encompass overall coordination, operational management, monitoring and evaluation, communication and dissemination, as well as interaction with partners. These responsibilities ensure the smooth execution and success of the initiative.

The project engages various partners, each assigned specific roles to contribute to its success:

- Technology Partners: Companies specializing in digital technologies and artificial intelligence will provide advanced technical solutions such as e-learning platforms, chatbots, and progress tracking tools.
- Microsoft Italy: Providing cloud infrastructure and technological support through Microsoft Azure.
- AI Startups: Integration of AI algorithms for personalized learning and student progress monitoring.
- Educational Institutions: Collaboration with universities and specialized schools for content development and validation.
- University of Milan: Development of specific training modules and certification of acquired competencies.
- Fashion and Design Schools: Creation of specialized courses in fashion, design, and craftsmanship.
- Communication Partners: Communication and marketing agencies to promote the project and engage the community.
- Development Agency: Implementation of communication campaigns to increase project visibility.

The for-profit technology provider contributes specialized expertise for the implementation and support of digital solutions:

- Platform Development: Creation of a unique e-learning platform, integrating innovative features and specialized modules.
- Technical Support: Continuous technical assistance to ensure platform stability and security.
- Chatbot Integration: Implementation of chatbots and virtual assistants to support students and coordinate training activities.
- Use of VR/AR: Development of VR/AR content for specialized courses, such as medical simulations and technical laboratories.

The contribution of the technology provider ensures the adoption of cutting-edge solutions, enhancing the effectiveness of training activities and facilitating the adoption of innovative technologies by Società Umanitaria.

The project will involve professionals with high-level technical and operational skills to ensure the effective implementation of project activities:

- Project Manager: Overall project management, coordination of activities, and communication among teams.
- Technical Writers/Analysts: Technical documentation, writing requirements, and creation of educational materials.
- Business Analyst: Definition of business needs and process analysis.
- Technical Architect: System architecture design and coordination of technical activities.
- Educational Experts: Education sector professionals for course content development and ensuring pedagogical soundness.
- AI Specialists: Experts in artificial intelligence for algorithm integration and progress analysis tools.
- Mobile App Developers: Creation of the mobile application and integration with the main platform.
- Technical Integration Partners: Outsourced IT services company for cloud infrastructure management and support.

This combination of roles and expertise ensures the effective realization of the project, with a clear focus on technological innovation and social inclusion.

## 6. FIVE BEST PRACTICES IDENTIFIED FROM THE "SOCIAL DIGITAL" CALL

We also bring the example of those who, through the PNRR grant, have initiated, accelerated, or strengthened extensive and articulated digital transformation paths that are still ongoing. These are social organizations active in various fields and territories, united by the strategic choice to invest in digital technology to enhance their ability to generate social impact. We have selected five organizations that we believe are particularly interesting. We believe that their experiences have much to teach and offer exemplary paths that can inspire other non-profit organizations (Jaoui, 1993).

The first story is that of "Il Faro Sociale di Macerata" and, specifically, what they are doing with "Come a Casa." This is a project where the social cooperative uses digital technology to improve therapies for children with autism, enhancing not only their quality of life but also that of their families and caregivers.

The second is "Edugamer," a project through which the Turin-based cooperative "Crescere Insieme" has developed an innovative methodology to guide children in their relationship with video games. From concerns shared by many parents, a training program has emerged aimed at both children and adults to explore the educational potential of gaming.

The third story is about "GAAP," the app that the "Consorzio Spazio Aperto" in Milan is experimenting with to improve the cleanliness of some cities in the hinterland and enhance the work of those who keep them clean. An innovative project that, thanks to trash bins equipped with chips, has attracted the interest of Microsoft's headquarters in Seattle, which chose to finance its experimentation.

The fourth concerns "Mission Bambini," an organization that nationally supports projects dedicated to disadvantaged youth and children. During COVID, it extensively used digital technology, restructuring its internal organization and relationships with partners. They were asked to completely rethink how they interact with assisted minors and their families. An important effort that, however, has created fertile ground for impact evaluation, among other things.

The last example is that of "Kaleidos," a reality based in Bergamo that has been operating in the telemedicine sector for years. Here too, the pandemic has given a significant boost to activities, allowing investments in genetic mapping and mental health. All this through a unique path, where the synergy between Profit and Non-Profit plays a crucial role (Warwick, 2015).

## 7. CONCLUSION

Le Organizations must become aware that technology is an infrastructure enabling critical processes for the organization's existence and its ability to create value and generate impact. Like any other infrastructure, technology investments require constant maintenance and attention. From here, there are strategies that can make these investments more or less significant. Specifically, this means working synergistically with other organizations to build innovation platforms to distribute costs according to economies of scope and scale; considering where standard technologies can be used; and evaluating, if processes are specific, the development of customized solutions.

We find a paradigm of microservices sustainable and, in most cases, successful, where a digital ecosystem is composed, in which some technological pillars remain standard, while other operational elements of the specific context can be developed ad hoc. In doing so, the organization must adopt a design system, relying on consistently similar component libraries, already produced, which it consults each time to compose the specific digital artifact or system element. This way, the organization will always have high-performing technologies supporting business processes, without incurring the costs it would have by starting from scratch each time.

Today, we have this opportunity to seize: we can develop applications personalized enough to meet specific needs without bearing the full costs of developing all the necessary components. This is provided that each organization - and here the concept of orchestration returns - has, as an internal competence, the ability to define a design system.

In a "ubiquitous computing" paradigm, the separation between digital (devices or software) and reality (the world) is becoming increasingly subtle because computing is increasingly embedded in all our work processes and relational dynamics. This pervasive vision potentially enables unprecedented availability of information about the real impact of our actions.

For organizations, it will become a characteristic of organizational action capable of enabling data availability that did not previously exist or was more costly to produce. These elements, combined with the progress that, for example, artificial intelligence is making, constitute a combination that informs us that yes, we have more data and yes, we have more capacity to analyze them.

And it is here that digital and impact assessment meet, which is gradually becoming a priority for Non-Profit organizations: first and foremost in the ability to generate data, which would otherwise be relegated to non-analyzable or entirely unnoticed analog supports, and secondly in their analysis. The challenge here is not to converge data onto a platform - from Excel spreadsheets onwards - but in the ex-ante integration of digital into the DNA of its processes and the delivery of its services. Only in this way can native information and data be generated, born from the fact that the process has enriched itself with digital elements, not just because they were collected ex post. From this need, we are experimenting with ad hoc frameworks and technologies that not only exhaust the obvious reporting needs but also help to structure impact assessment processes, empowering with IT infrastructure the collection and processing of data essential to populate impact assessment models, which by definition can only be customized.

These tools help organizations deliver their services, thus collecting data that will then be used for impact assessment. It is a "management" layer that optimizes and simplifies work while making organizations more aware of the effects of their activities.

While we wait for digital to permeate everywhere, in all processes, and for data to be generated "natively" from it, we can therefore equip ourselves with user-friendly, fast, responsive interfaces to enable the collection of data that will populate impact assessment models and ultimately make us more aware of the effects of our actions. An exciting prospect for all civil society actors.

The studies and examples reported and the research we have conducted demonstrate that the person-technology combination has yielded amazing results in project, communication, and exhibition contexts. Both in terms of optimizing routine processes and in the case of actual results and their exceptionalism and beauty.

## REFERENCES


Jaoui, H. (1993).*Creatività per tutti. Strumenti e metodi da impiegare nel quotidiano,* Milano, Franco Angeli.

Manovich, L. (2020). *L'estetica dell'intelligenza artificiale. Modelli digitali e analitica culturale* (a cura di V. Catricalà), Milano, Luca Sossella Editore.

Quintarelli, S. et al., (2020). *Intelligenza artificiale*, Bollati Boringhieri.

Russell, S. J., Norvig, P., and Davis, E. (2010). *Artificial Intelligence: A Modern Approach,* 3rd ed., Prentice Hall Series in Artificial Intelligence (Upper Saddle River: Prentice Hall).

Santosuosso, A. (2020). *Intelligenza artificiale e diritto,* Mondadori, Italia.

Sartor, G. (2016). L'informatica giuridica e le tecnologie dell'informatica: corso d'informatica giuridica.

Warwick, K. (2015). *Intelligenza Artificiale - Le basi,* Dario Flaccovio Editore.

# TINKERCAD AS A VIRTUAL LABORATORY

József Udvaros and Norbert Forman
*Budapest Business University*
*Budapest, Hungary*

**ABSTRACT**

This study examines the effectiveness of teaching programming control structures at the high school level, comparing traditional teaching methods and the use of the TinkerCad online platform. The research was conducted in two groups, involving a total of 41 students. The results were evaluated using SPSS analysis and the Mann-Whitney test. The study concludes that in some cases the use of TinkerCad can significantly improve students' performance in understanding programming control structures, highlighting the importance and effectiveness of digital teaching tools in today's education.

**KEYWORDS**

TinkerCad, Virtual Laboratory, Teaching Methods, Education, Visualization, ICT

## 1. INTRODUCTION

In today's world, access to education and self-development is a fundamental right for every individual. This right is guaranteed by the governments of different nations through a number of laws that promote equal access to education and the opportunity for individual development (Bódi, 2023; Udvaros and Bódi, 2023).

In the 21st century, it is already a basic need for students to become familiar with the basics of algorithms and the basic concepts of programming and program development within the framework of secondary school education. This approach highlights the requirement that students acquire relevant and current skills in the age of technology. The results of the two researches presented in article (Udvaros, 2019) highlight the need for modern education systems to prepare students for the challenges of the digital world. Understanding the basics of algorithms and programming is not only necessary for computer science and engineering careers, but also fundamentally important for the development of critical thinking, problem-solving skills and digital literacy. This approach not only provides proficiency in computing, but also fosters the development of broad skills such as analytical thinking, logical problem solving and creativity. These skills are essential for the 21st century labor market and society, where technological development and the challenges of the information age play an increasingly important role.

One of the key questions of modern education is how to deliver the curriculum in the most efficient way in the classroom, where interactive methods play a prominent role. To this end, animation-simulation software has been developed for many subjects, which can be used to make lessons more colorful and attractive. These tools are particularly beneficial for students who prefer a visual learning style, as they make it easier and faster to understand the course material. Incorporating these digital tools into many teaching strategies, such as experiential learning or deep learning, allows students to learn new knowledge through experimentation and learning from mistakes (Annuš et al., 2023; Végh and Takáč, 2021).

Living in the information age, we are witnessing that the methods of teaching and learning are dynamically changing under the influence of information and communication technologies (ICT). This change is strongly influenced not only by the development of technology, but also by other global events such as pandemics. This change applies to both public and higher education.

Digital technology is present in many areas of people's daily lives, including entertainment, leisure activities, information acquisition and delivery, and communication. As a result, it becomes essential to learn the digital language, which includes skills such as solving problems through coding and computational thinking as a new type of operational paradigm (Llorens-Largo, 2015).

An important question for educational professionals and researchers is what changes are taking place in the field of learning and teaching supported by digital technology, and how these tools can be effectively used in the teaching process. They examine what new teaching methods and techniques prove to be effective in this area (Udvaros and Czakóová, 2021).

Digital laboratories play a particularly important role in this paradigm shift. These labs allow students to conduct a wide range of experiments in a virtual environment that might be too dangerous or impractical in real labs. In this way, students can learn experimental methods and scientific principles in a safe and controlled environment. TinkerCad can also be considered such a digital laboratory. TinkerCad is a versatile computer-aided design (CAD) software that individuals and organizations find particularly useful in manufacturing and marketing. The software's functions also cover the needs of the healthcare and construction industries, thus ensuring a wide range of applications. This free software allows users to easily edit and modify files and recreate existing designs. TinkerCad provides an innovative approach to design, allowing users to create their own unique creations with the many modifiable shapes and objects offered by the software. Users can combine these shapes, add various gestures, and stack blocks to create complex 3D designs essential for prototyping.

A special feature of TinkerCad is the use of traces and adjustable shapes in the design of circuits, which opens up new dimensions in the field of electronic design. Furthermore, the software offers the possibility to use code blocks. It also includes user-friendly interface presets that make it easy for users to learn and use the software effectively, making TinkerCad an ideal tool for both professionals and hobbyists in various fields.

The integration of modern technical tools and methods is crucial in modern education, especially at the high school and university level of IT and other scientific fields. This need is especially valid in the technical and IT fields, where the use of easily accessible tools adapted to diverse skill levels is beneficial. In addition to emphasizing multidisciplinary and practice-oriented education, the transparency and practice-oriented nature of the curriculum is also important. The use of Arduino and Raspberry Pi microcontrollers fits perfectly into this concept. These tools allow students to connect different sensors and directly experience the physical effects of the code they write. The application of IoT (Internet of Things) is particularly relevant in teaching programming. The use of IoT devices allows students to engage in real-time data collection, network communication, and control of physical devices. In this way, students not only learn the basics of programming, but also become able to understand and manage modern, complex systems. In the modern educational environment, the use of Arduino and Raspberry Pi microcontrollers as well as IoT technologies offers students the opportunity to transform their theoretical knowledge into practical skills, thus preparing them for the technological challenges of the future. This type of education fosters critical thinking, problem-solving skills, and interest in technological innovation (Udvaros and Takáč, 2022; Acosta et al., 2022).

The aim of the research is to present the importance of the TinkerCad web application in programming education, visualization, and as a simulation tool.

## 2. LITERATURE RESEARCH

The virtual laboratory is an advanced computer-based educational tool that allows students to interactively participate in various experimental situations or other scientific activities through a digital interface. In this environment, students engage in virtual experiments that simulate real laboratory conditions, allowing them to conduct experiments without being physically present in the lab. Such typical examples are computer simulations of various scientific procedures or phenomena, where students can perform various operations with the help of interactive software, as well as remotely controlled experiments, when students control real laboratory equipment through a computer connection in real time and perform the experiments while they are physically removed from the experimental site. These methods allow students to gain experimental knowledge and study scientific concepts without direct contact with laboratory equipment or chemicals, thus providing a safe and effective learning environment (Handayani et al., 2024; Abburi et al., 2021).

The research results of Husin and his fellow researchers highlight that TinkerCad simulation software can be an extremely useful tool for teaching and learning electronic skills, especially during the COVID-19 pandemic. According to the results, the majority of students received the software positively and successfully understood the work assigned to them, only a few of them required intensive tutoring assistance. The study notes that the motivation to learn online was negatively affected by certain external factors, such as a poor Internet connection or an inadequate learning environment. Nevertheless, the majority of students agreed that online education provides an equally satisfying experience as traditional, face-to-face learning. The study highlights that, in parallel with technological development, it is important that well-structured and planned education take place between the instructor and the student for effective communication and knowledge transfer. These results can help educators select and use the most effective online teaching tools. The majority of students responded positively to using TinkerCad and successfully developed their programming skills using the software. According to the study, students unanimously used TinkerCad as a simulation tool to develop their end-of-year projects. This suggests that new approaches to online education, such as the use of TinkerCad, can help students achieve learning goals related to IoT (Internet of Things) subjects, even despite the barriers encountered in online learning. The COVID-19 epidemic created an opportunity to discover and apply new ways of digital learning, which is an important step in the future development of education (Husin et al, 2022).

According to Jacob et al.'s research, virtual laboratories play an important role in the teaching of electronics in two different post-secondary technical courses. It is considered important to emphasize that these virtual laboratories do not replace, but rather complement, physical laboratories. Especially during pandemics, they provide a valuable opportunity for students to conduct more practical experiments. This solution is especially useful for students who find it difficult to attend face-to-face lessons, whether due to distance, health reasons or other obstacles. The virtual laboratories provide them with the opportunity to access the course material and gain practical, realistic experiences.

The article also notes that the method of blended learning (b-learning), which combines traditional and online education, has proven to be effective, especially with the support of similar digital tools. The TinkerCad platform provides the necessary components and tools to effectively manage traditional laboratory routines in a virtual environment, thereby supporting the teaching of electronics subject laboratory activities in post-secondary technical courses. These virtual labs allow for new forms of experiential learning that have generated a lot of interest among students and teachers, especially if they have a stable Internet connection. The TinkerCad platform can be effectively used by students in post-secondary engineering courses to learn and understand the subject of electronics, complementing traditional learning methods (Jacob et al., 2021).

According to the lecturers of the University of Alicante, the COVID-19 epidemic has brought about significant changes in the field of education, especially in terms of laboratory exercises, which were previously difficult to imagine being carried out remotely. This global situation has prompted educational institutions to transform traditional, physical laboratories into virtual laboratories. In this context, TinkerCad, an online circuit design platform, has served as an outstanding tool for distance learning in electronics laboratories. The TinkerCad platform was used for the data collection exercises, together with the Arduino microcontroller and a number of input and output elements, in the Electronic Instruments subject of the Telecommunications Engineering Master's program at the University of Alicante. The use of TinkerCad during the practical sessions brought positive results, offering the students even wider opportunities compared to the real laboratory components. Student satisfaction surveys confirmed that the use of TinkerCad effectively contributed to the completion of laboratory exercises in a virtual environment. According to research, few other applications or virtual environments have the capabilities and interactivity of TinkerCad. The article also points out that with the return of normal educational conditions, the possibilities offered by TinkerCad should not be neglected. Moreover, the authors recommend that TinkerCad be integrated into personal teaching, providing students with extensive opportunities to develop their skills. Combining TinkerCad with traditional, real-world components can provide additional opportunities in the learning process, so incorporating TinkerCad into classroom teaching is essential to further improve practices (Alavés-Baeza and Galiana-Merino, 2022).

The situation caused by the pandemic made it necessary to rethink educational methods and applications, especially in the field of practical activities. The use of ICT (Information and Communication Technology) has become essential so that students can continue to participate in practical activities such as teaching robotics. Robotics education is a particularly important part of STEAM (Science, Technology, Engineering, Arts, Mathematics) education and the development of computational thinking (Kang and Temkin, 2022). According to Tsolakis et al., in this new context, the TinkerCad simulation platform was used to teach robotics, allowing students to design and simulate robotics projects in an online environment. These designs were later implemented in the physical classroom with a robot system on the Arduino platform, thus combining virtual and physical learning environments. This approach made it possible for students to continue their robotics studies during distance learning, while maintaining an important element of hands-on activities. The combination of TinkerCad and the Arduino platform provided the necessary flexibility and interactivity for students to creatively develop their robotics projects, first in a digital environment and later in a real environment. This hybrid approach not only allowed robotics education to continue during the pandemic, but also opened new opportunities for learning and developing practical skills, encouraging students to develop their interest in technology and computational thinking (Tsolakis et al., 2022).

# 3. RESULTS

We use the research results of our previous article (Udvaros et al., 2022) as a basis. The basic objective of our research was to investigate the understanding and correct application of programming control structures. In order to achieve this goal, we organized a special course for the students, in the framework of which they could learn the basic control structures of programming. The educational process was conducted in two separate groups, each using different pedagogical methods. One group followed traditional teaching methods, while the other group used the TinkerCad online platform as a learning tool. The performance of both groups was examined during a subsequent assessment, during which four different types of tasks were evaluated. These tasks tested the understanding and application of the following programming concepts: conditional branches (if..else statements), counting loops (for loops), and conditional loop statements (while and do..while loops).

The group participating in the research consisted of 41 students who were taught the basics of programming according to two different methods. The first group, which we identify as the 'ORIGINAL GROUP', consisted of 21 students and classical teaching methods were applied to them. The second group, the 'CONTROL GROUP', consisted of 20 students and they learned with microcontrollers and the TinkerCAD software. In the case of both groups, the aim of the education was to learn about programming control structures and to use them correctly.

The SPSS statistical software was used to evaluate the performance of the students. In the research, we used the 'Independent Samples Test' method to compare the answers of the Original and the Control Group. During the analysis, we found that the standard deviation of the two groups is the same, which means that the performance of the two groups is comparable. Thus, we took into account the results of tests based on variance matching offered by SPSS during the evaluation. The scale of the performance evaluation ranged from 0 to 5, where 5 was the maximum and 0 the minimum possible score.

**Independent Samples Test**

| | | Levene's Test for Equality of Variances | | t-test for Equality of Means | | | | | | |
|---|---|---|---|---|---|---|---|---|---|---|
| | | | | | | | | | 95% Confidence Interval of the Difference | |
| | | F | Sig. | t | df | Sig. (2-tailed) | Mean Difference | Std. Error Difference | Lower | Upper |
| Q1 | Equal variances assumed | ,211 | ,649 | -1,592 | 39 | ,119 | -,626 | ,393 | -1,422 | ,169 |
| | Equal variances not assumed | | | -1,594 | 39,000 | ,119 | -,626 | ,393 | -1,421 | ,168 |
| Q2 | Equal variances assumed | 1,283 | ,264 | -1,097 | 39 | ,280 | -,479 | ,436 | -1,361 | ,404 |
| | Equal variances not assumed | | | -1,105 | 36,280 | ,276 | -,479 | ,433 | -1,356 | ,399 |
| Q3 | Equal variances assumed | ,445 | ,509 | -,128 | 39 | ,899 | -,062 | ,485 | -1,043 | ,919 |
| | Equal variances not assumed | | | -,128 | 38,811 | ,899 | -,062 | ,484 | -1,040 | ,917 |
| Q4 | Equal variances assumed | 1,194 | ,281 | -1,132 | 39 | ,265 | -,545 | ,482 | -1,520 | ,429 |
| | Equal variances not assumed | | | -1,138 | 37,863 | ,262 | -,545 | ,479 | -1,515 | ,425 |

Figure 1. Independent Sample Test of the responses of the two groups

Based on the data in the table (Figure 1), if the significance (Sig) value for each question exceeds 5%, this indicates that the standard deviation of the answers of the two investigated groups, i.e. the 'Original Group' and the 'Control Group', is the same for each question. This statistical result indicates that there is no significant difference between the answers of the two groups according to the examined aspects.

Furthermore, since the two-tailed significance (Sig 2-tailed) value for each question exceeds 5%, this supports the acceptance of the H0 hypothesis. The H0 hypothesis, i.e. the null hypothesis, states that there is no significant difference between the two groups. Thus, based on the statistical analysis, we can come to the conclusion that there is no significant difference between the answers of the Original and the Control Group in the understanding and application of the examined programming control structures.

During the research, there was no significant difference between the answers of the two groups, the 'Original Group' and the 'Control Group' under the condition of equality of variance, as the two-tailed significance (Sig 2-tailed) exceeded 5% in all cases. This means that classical statistical tests, which assume a normal distribution, did not show a significant difference between the two groups.

However, when the responses did not follow a normal distribution, it became necessary to continue the studies with alternative, non-parametric methods. In such cases, the Mann-Whitney test is often used, which does not require the samples to be normally distributed. The Mann-Whitney test is a non-parametric statistical test used to compare two independent samples and is particularly useful when the sample sizes are small or when the data do not meet the requirements of a normal distribution.

The Mann-Whitney test allows researchers to determine whether there is a significant difference in ranked responses between two independent groups without having to meet the conditions of a normal distribution. This method therefore provides an alternative solution in situations where the use of classical parametric tests is not appropriate.

| | Null Hypothesis | Test | Sig. | Decision |
|---|---|---|---|---|
| 1 | The distribution of Q1 is the same across categories of ROUND. | Independent-Samples Mann-Whitney U Test | ,006 | Reject the null hypothesis. |
| 2 | The distribution of Q2 is the same across categories of ROUND. | Independent-Samples Mann-Whitney U Test | ,351 | Retain the null hypothesis. |
| 3 | The distribution of Q3 is the same across categories of ROUND. | Independent-Samples Mann-Whitney U Test | 1,000 | Retain the null hypothesis. |
| 4 | The distribution of Q4 is the same across categories of ROUND. | Independent-Samples Mann-Whitney U Test | ,016 | Reject the null hypothesis. |
| 5 | The distribution of SUM is the same across categories of ROUND. | Independent-Samples Mann-Whitney U Test | ,099 | Retain the null hypothesis. |

Asymptotic significances are displayed. The significance level is ,05.

Figure 2. Hypothesis Test Summary

Based on the results obtained (Figure 2), it can be concluded that significant differences between the answers of the 'Original Group' and the 'Control Group' were discovered only for the first and fourth questions using the Mann-Whitney test. This suggests that in terms of these two questions, one group performed significantly differently compared to the other.

For the other questions, as well as for the students' total scores (SUM), no significant differences were found between the two groups at the 5% significance level. This means that the performance of the two groups was similar in the other questions and in the total score, and there were no statistically significant differences.

This result may indicate that the different teaching methods (classical method for the 'Original Group' and the use of TinkerCad for the 'Control Group') brought significant differences in students' understanding or skills only in certain areas (in the first and fourth questions). In the other areas, the different teaching methods did not bring significant differences in the performance of the students.

## 4. CONCLUSION AND DISCUSSION

In the article, we conducted a literature research on the applicability of TinkerCad as a digital laboratory, as well as on international experiences. Summarizing the results of the articles, we can say that the TinkerCad simulation software can be an extremely useful tool for teaching and learning electronic skills. We also proved the results of our literature research with our own research.

The aim of the research was to investigate the understanding and correct application of programming control structures in secondary school education. The study evaluated instruction in two groups: one using traditional methods and the other using the TinkerCad online platform. On the ThinkerCad online platform, students learned the basic programming structures with the help of a virtual LED light chain (students had to illuminate the LEDs of the LED light chain in different ways). According to the results, based on the SPSS analysis, no significant difference was found between the performance of the groups. However, using the Mann-Whitney test, significant differences were discovered for the first and fourth questions, which suggests that different teaching methods may be more effective in certain areas. The achieved results show that by using the TinkerCad simulation software in education, we can achieve similar or better results than with classical teaching methods. Therefore, it is perfectly suitable for simulating the operation of electronic components, teaching and visualizing programming control structures.

The discussion topic can focus on the effectiveness of traditional and digital teaching methods, with particular attention to the areas in which there are differences in student performance. The importance of the study lies in the fact that it highlights the role of digital tools in education, especially in the context of teaching programming.

## ACKNOWLEDGEMENT

The author(s) wish to express gratitude to all those who provided insights and feedback during the preparation of this article.

## REFERENCES

Abburi, R., Praveena, M. and Priyakanth, R. (2021). Tinkercad-a web based application for virtual labs to help learners think, create and make. *Journal of Engineering Education Transformations*, 34(0), p.535.

Acosta, R.S., Cooper, E.H. and Medina, A.H. (2022). Modeling electrical circuits with tinkercad and mathematica. *In INTED2022 Proceedings* (pp. 2343-2349). IATED.

Alavés-Baeza, V. and Galiana-Merino, J.J. (2022). Expanding The Possibilities Of The Electronics Laboratory With Tinkercad. *In INTED2022 Proceedings* (pp. 7777-7783). IATED.

Annuš, N., Paksi, D. and Csóka, M. (2023). Interactivity And Animation-Simulation Tools In The Digital Presentation Of Educational Material. *In INTED2023 Proceedings* (pp. 6975-6980). IATED.

Bódi, S. (2023). Presentation of the fundamental law of Hungary. *The Lawyer Quarterly*, 13(3).

Handayani, A.N., Nafista, S.S., Kusumawardana, A., Elmunsyah, H., Arai, K. and Asmara, R.A. (2024). Development of Interactive Virtual Laboratory Media with Tinkercad Platform on Analog Electronics Material. *In E3S Web of Conferences* (Vol. 473, p. 04008). EDP Sciences.

Husin, N.A., Ramli, M. and Alwi, D.M. (2022). Tinkercad simulation software to optimize online teaching and learning in embedded internet of things. *Journal of Counseling and Educational Technology*, 5(2), pp.49-57.

Jacob, F., Alberto, A. and Guimarães, P. (2021), October. Use of Tinkercad platform for Teaching Electronics Subject in Post-Secondary Technical Courses. *In Ninth International Conference on Technological Ecosystems for Enhancing Multiculturality (TEEM'21)* (pp. 543-547).

Kang, J. and Temkin, S. (2022). Integration of Web-based Arduino/circuits Simulator in Enhancing Future Engineering Student Projects. *In AIAA SCITECH 2022 Forum* (p. 1352).

Llorens-Largo, F. (2015). Dicen por ahí…que la nueva alfabetización pasa por la programacion. *ReVisión*, 8(2), 11-14.

Tsolakis, S., Theofanellis, T. and Voulgari, E. (2022). Introducing STEAM Through Tinkercad and Arduino. *In Handbook of Research on Integrating ICTs in STEAM Education* (pp. 239-264). IGI Global.

Udvaros, J. and Bódi, S. (2023). Division and Regulation of Drones in EU and Hungary. *International Journal of Science, Engineering and Technology.* 11(4), 1-6.

Udvaros, J. and Czakóová, K. (2021). Using teaching methods based on visualizing by TinkerCad in teaching programming. *In ICERI2021 Proceedings* (pp. 5913-5917). IATED.

Udvaros, J. and Takáč, O. (2022). Technical IT solutions in teaching. *In INTED2022 Proceedings* (pp. 4047-4052). IATED.

Udvaros, J. (2019). Teaching Object Oriented Programming By Visual Devices, *eLearning and Software for Education* 15 pp. 407-413. Paper: 7, 7 p.

Udvaros, J., Forman, N. and Dobák, D.É. (2023). Application and impact of electronic solutions in teaching programming. *In Annales Mathematicae et Informaticae* (Vol. 57, pp. 78-91).

Végh, L. and Takáč, O. (2021). Teaching and Learning Computer Programming By Creating 2D Games in Unity. *In ICERI2021 Proceedings* (pp. 5696-5700). IATED.

# LEARNING ANALYTICS VISUALISATIONS OF SCHOOL DROPOUT IN VOCATIONAL EDUCATION

Christos Samaras[1], Anna Mavroudi[2] and Vassilios S. Verykios[1]
[1]*Hellenic Open University, 263 35 Patras, Greece*
[2]*University of Oslo*
*Boks 1072 Blindern, NO-0316 Oslo, Norway*

**ABSTRACT**

Vocational Education and Training (VET) suffers from an increasing rate of student dropout in the school sector, while it is largely neglected in the learning analytics (LA) research. At the same time school dropout rates are consistently rising globally and researchers are trying to better understand this rising phenomenon. Some concentrate on its relation to chronic absenteeism from school and to the student socioeconomic background. In addition, other contextual factors can come into play to explain this complex and challenging social phenome-non. Research also shows that the use of LA could contribute to alleviating school dropout. The main purpose of this paper is to present a LA system that visualizes student dropout in the VET school context providing the rea-sons behind it. The paper describes both the backend and the frontend of the system. Regarding the former, the paper focuses on the LA system design and architecture. Regarding the latter, it presents a set of interactive LA visualisations provided by our system based on existing student data on school dropout. In particular, the paper demonstrates its LA visualisations via a case study in a large VET school in Greece based on student data gathered from four consecutive school years. The LA visualisations show students' dropouts across years in the school unit, trajectories of students' dropouts in a specific school year, and the degree of the association between some geographical area (as the area of student residence) with chronic absenteeism.

**KEYWORDS**

Vocational Education and Training, Learning Analytics, School Dropout, Chronic Absenteeism

## 1. INTRODUCTION

Vocational Education and Training (VET) equips students with skilled know-how needed to enter the modern workforce directly after high school. The goal of the VET curriculum is to provide students with the abilities they need to succeed in their fields of study and find future employment (Kotsifakos et al., 2020). The successful execution of VET programs is affecting the socioeconomic structure of society and, therefore, improving student training in vocational and technical high schools and preventing school dropout have high societal importance.

At the same time, school dropout rates are consistently rising; as a result, dropout is seen as a global issue and scholars throughout the world are trying to understand the reasons behind it. Significant dropout rates exist especially in the VET, causing academics and researchers to concentrate on the relationship between Chronic absenteeism (CA) and dropout. It seems that CA is directly connected to dropout be-cause chronically absent kids typically have lower grades and greater dropout rates than their peers who regularly attend school. Moreover, we know from previous literature that extreme CA levels are more likely to occur in the VET schools (Cerda-Navarro, Sureda-Negre and Comas-Forgas, 2017). We also know that research focusing on dropout in VET schools is scarce, especially compared to secondary and higher education (Demir and Akman Karabeyoglu, 2015)

The VET field is largely neglected in the learning analytics (LA) research as well (Gedrimiene et al., 2020), while the use of LA could contribute to alleviating school dropout, which are pressing issues in VET. Thus, it has been suggested that more empirical research is needed regarding various student groups and possible interventions towards this direction. To this end, the main purpose of this paper is to present a LA system that visualizes student dropout in the VET school context and seeks to understand the reasons behind it in this particular context. The system was developed for the Greek VET system.

## 2. LEARNING ANALYTICS IN VOCATIONAL EDUCATION

LA is defined as the “measurement, collection, analysis, and reporting of data about learners and their contexts to understand and optimize learning and the environments in which it occurs” (Siemens, 2013, p. 1382). Although the topic of LA in VET has not been thoroughly investigated, there are some notable examples, such as those presented below.

A research study was conducted in Finland with the aim to analyze which factors from students’ admission data can predict dropout in a large-size initial VET school (López-Pernas et al. 2022). The authors used statistical analysis of the admission data (gender, age, native language, employment status, and educational background) and identified two vulnerable groups: students who combined their studies with a job or job-seeking and those who started a VET program after basic education.

Another relevant article had two main research questions, namely, problems and prospects of big data analytics for improving the learning process in technical VET in Nigeria (Mustapha et al., 2021). The authors distributed a survey to 117 teachers of technical VET to collect their opinions on the topic at stake. The main conclusion about the prospects is that LA can contribute to predict learners’ school failure and to help teachers to analyse and enhance their practices. The main issues are related to data privacy as well as issues related to data collection such as poor data quality and incorrectly formatted data.

## 3. DESCRIPTION OF THE CONTEXT

Unfortunately, it is well documented in the literature that the student socioeconomic background plays an important role in early school dropout eg. parents’ unemployment, poverty rate (Lavrijsen and Nicaise, 2015). Greece is no exemption to that (Castaño et al., 2020). In addition, research concerning the Greek context pinpoints gender, age and different cultural/ethnic background than the dominant (i.e. Greek herein) as predictors of early school leaving (Samaras et al., 2022). We also know that there is a regional dimension to this phenomenon in Greece (Samaras et al., 2022); finally, teachers play an important role on it (Panagiotis, 2023).

In Greece, "MySchool" is an information system supported by the Ministry of Education which support the digitalization of school- and administrative units in primary and secondary education. The "MySchool" platform facilitates both online and face-face instruction and can be integrated with several eLearning tools. It stores information about school staff and teachers, but mostly information about students, such as grades and absences, school departments, courses, programs, specialisations that they attend and so on (Sarlis, Kotsifakos & Douligeris, 2021).

## 4. DESCRIPTION OF THE LA SYSTEM ARCHITECTURE AND DESIGN

With respect to enabling data collection from a technical point of view, we developed a data warehouse (DWH). The development of DWH was used as a means for gathering digital data from dispersed information systems into a centralized integrated storage system. The information in a DWH can also be utilized as input for application systems, such as LA dashboards, something that was considered as an advantage. We collected data from the "MySchool" application, which exports data for each year, but in order to perform a multi-year analysis, the data must be linked. In order to connect, organize, prepare and visualise the data, we stored them in the developed DWH. Figure 1 shows how the student data collected from "MySchool" were converted into LA visualizations. It also shows all intermediate processes that we designed and implemented. Data from "MySchool" were exported as an Excel file format. The ETL (Extract, Transfer, Load) program that we developed imported and processed the EXCEL data and loaded it in the DWH (namely, a Cloud SQL Server).

Next, we employed ETL technical processes responsible for obtaining, cleaning, transforming, and loading data from various operational systems that carry out daily transaction processing. Data quality assurance was achieved via the use of OnLine Analytical Processing (OLAP), which was performed over simple but neatly organized "cubes" (Najm et al., 2022). In technical terms, each of these cubes is a multi-dimensional array of data. After the OLAP employment, we created the different LA visualizations that we needed with the help of several applications such as KNIME (a free and open-source data analytics, reporting and integration platform ) -see for example Maraza-Quispe et al. (2022)- Microsoft Power BI (a commercial interactive data visualization software product with a primary focus on business intelligence, R programming language and ArcGIS (a family of client, server and online geographic information system software).

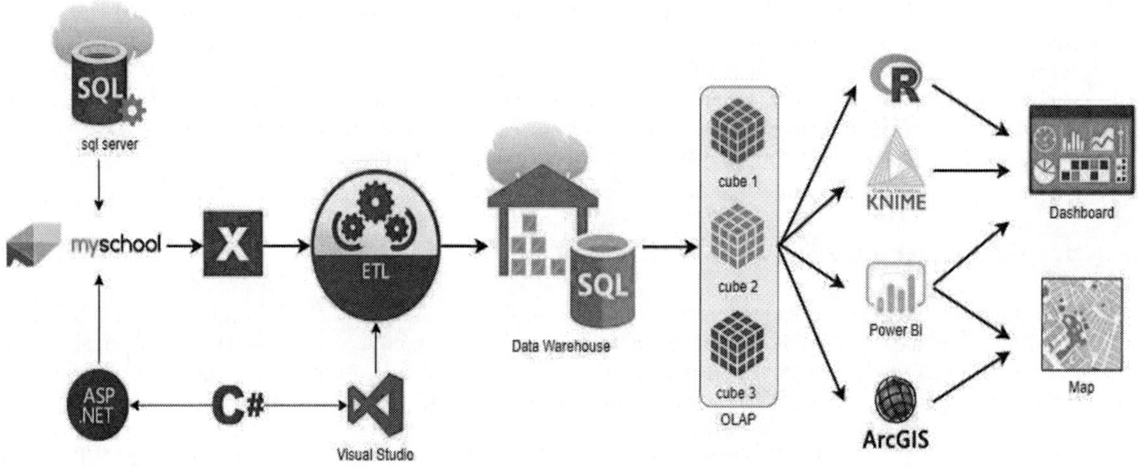


Figure 1. Workflow of the main technical processes

We designed the database of the DWH as shown in Figure 2. To capture absenteeism, we collected data (depicted in the table "Absences") on: the students' absences per hour in the daily school schedule, the total students' absences per school day and the students' absences due to illness (due to Covid or due to other reasons). We also took into account all possible characteristics that can affect students' attendance and early school leaving as indicated in the previous section, such as data on: student characteristics without including properties that reveal the student's identity (eg father job, mother job, age, gender), characteristics of the region that the student lives (e.g. house value, distance from school), teachers, school class, educational sector/domain and specialisation, and student grades.

With respect to the LA visualisations on school dropout in VET, we used the Sankey Chart visual tool. We selected this type of visual representation due to the fact that it was originally developed for visualizing the quantitative information of flows between multiple entities or processes. We also used a specific optical tool which allows users to clearly display complex, multi-stage path guides for the creation of trajectories of school dropouts. Finally, we used another optical tool to create an interactive map that helped us to visualise any regional aspect of school dropout. All these tools were included in the Microsoft Power BI gallery.

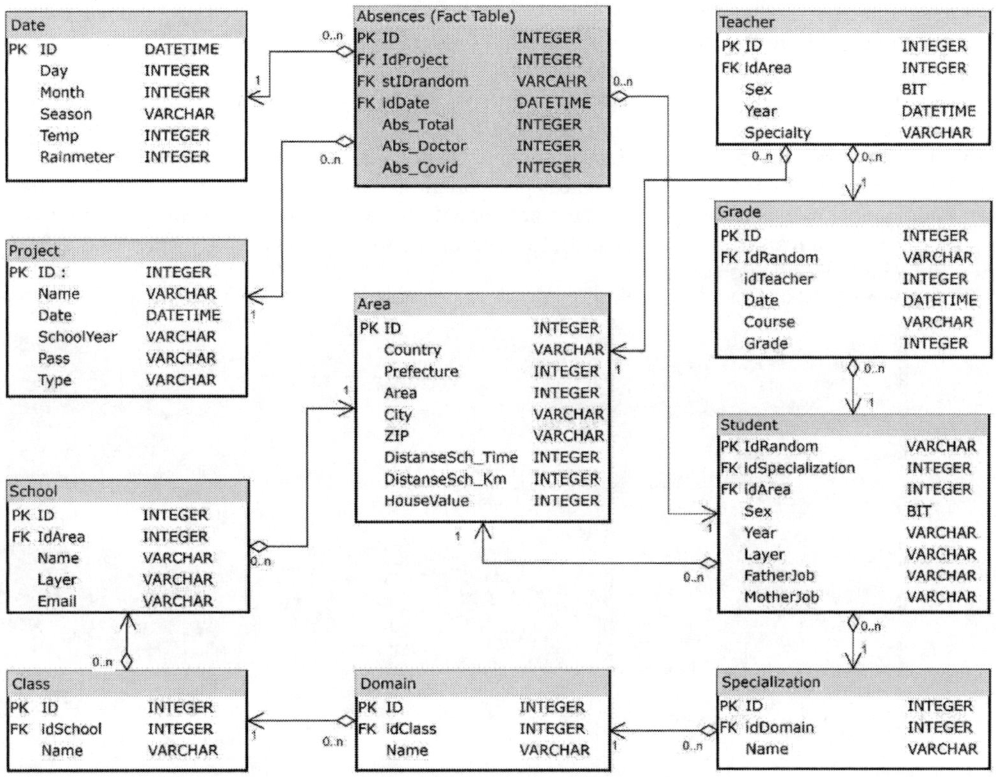


Figure 2. Cloud data warehouse (DWH)- database schema

## 5. LA VISUALISATIONS ON SCHOOL DROP OUT IN VET

We created a fully functioning prototype with interactive LA visualisations based on existing student data on school drop out in a large VET school in Greece gathered across four school years (2017- 2021). The total number of students who attended the school these years was 1150. Figure 3 depicts students' dropout in this particular school. That is, it shows the flow of all the students who have not completed their class at the particular school across four years: we can see the distribution of students across classes (A to D), domains/sectors (no domain, civil, electrical, and mechanical) and specialisations offered in the specific VET school (no specialisation, electronic and computer systems, electrical systems, and so on). From the dynamic graph of Figure 3 we can observe that the largest school dropout is happening in the first grade, a fact that was related to the registration of refugees in this specific school unit. In the period 2017-2021, a total of 82 adult refugee students enrolled in the first grade in this particular school, of which only 2 (percentage 2.4%) managed to complete the three-year study and obtain a high school diploma and a specialisation degree. We can also notice that the Domain of Construction and Geoinformatics (Civil) is problematic with respect to school dropout.

In Figure 4, we see the whole dropout trajectory for a particular school year. The path starts from the school unit name ("Root"), continues to the school class/grade level, then in the domain level, then in department level (eg "A3", "BM1", "BD" and so on). It ends by enumerating in its "leaves" all students who lost their year in a specific school year. It can be concluded from this diagram that one student left the 'COX', 'CHL' and 'BM2' classes and two students left the 'BM1' department. Also, that there seems to be a problem with the department 'BD' which was abandoned by 15 students. This particular department belongs to the domain of Construction and Geoinformatics which was detected as problematic in Figure 3. This fact must be related to the great decline of the construction industry in Greece during the last decade. We can also see that departments 'A1' and 'A2' were abandoned by 13 students each; and that department 'A3' was abandoned by 4 students. Thus, the main conclusion derived from this visualisation is that the first grade (class A) faces a rather serious problem in this VET school, something that was also detectable in Figure 3.

In the map of Figure 5, we see the places of residence of all the students of the specific VET school. It shows that this school has students from almost all areas of Thessaloniki (i.e the city where the school is located). These areas have different distances from the school and different property market value. In the dynamic map of Figure 6, we see the absences per student in each area in a visualisation that combines a bar chart and a map. On the left there are horizontal bars with the absences per students in relation to their home residence. From the map (on the main part of the figure) we are able to visually inspect and identify the areas where students were more absent.

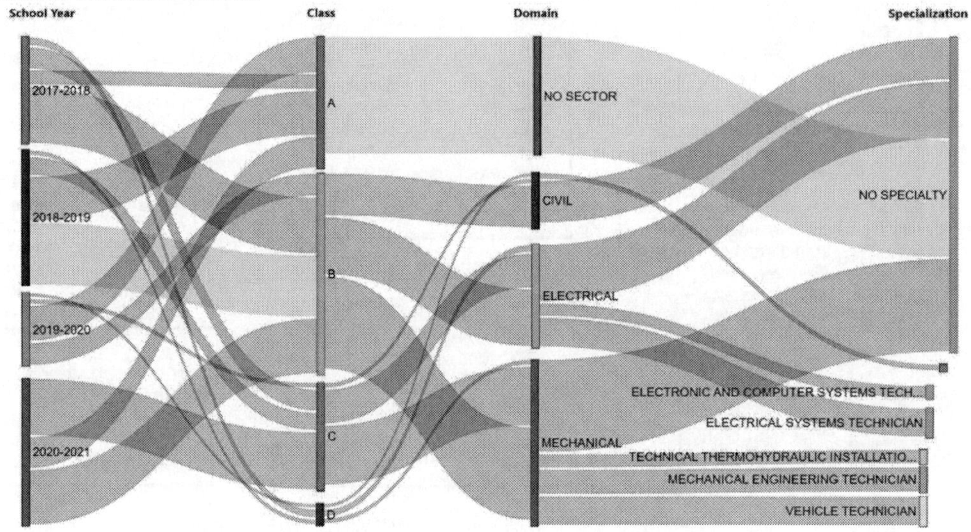


Figure 3. Students' dropouts across four years in a VET school unit

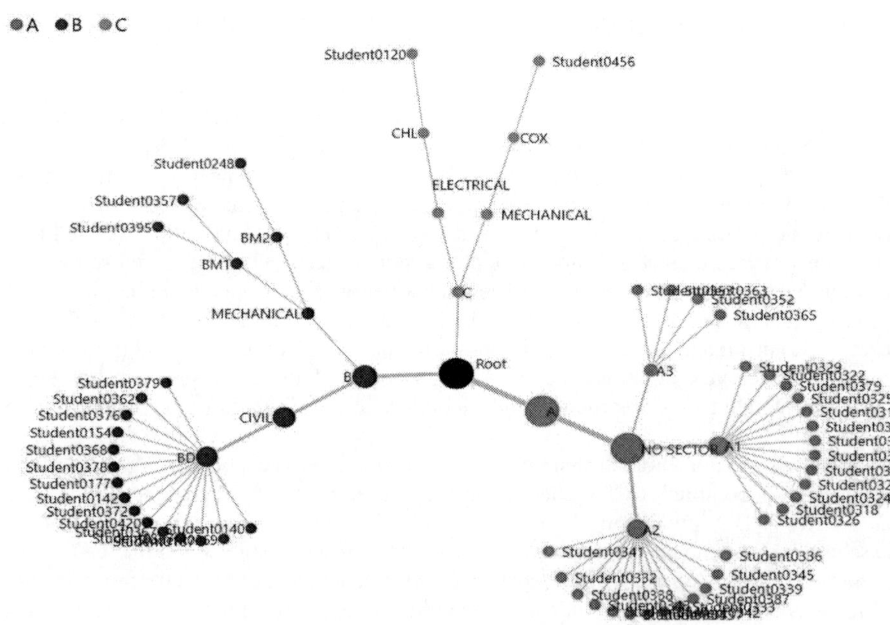


Figure 4. Trajectories of students' dropouts in a specific school year

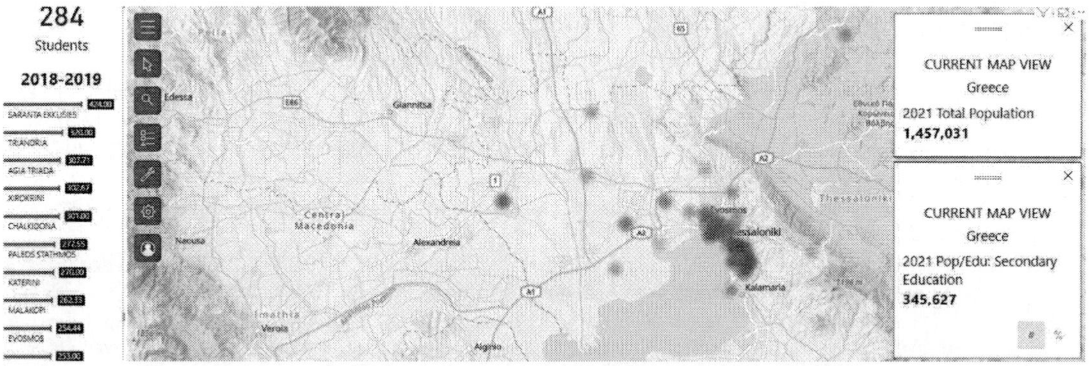


Figure 5. Spatial distribution of (adult) students' places of residence

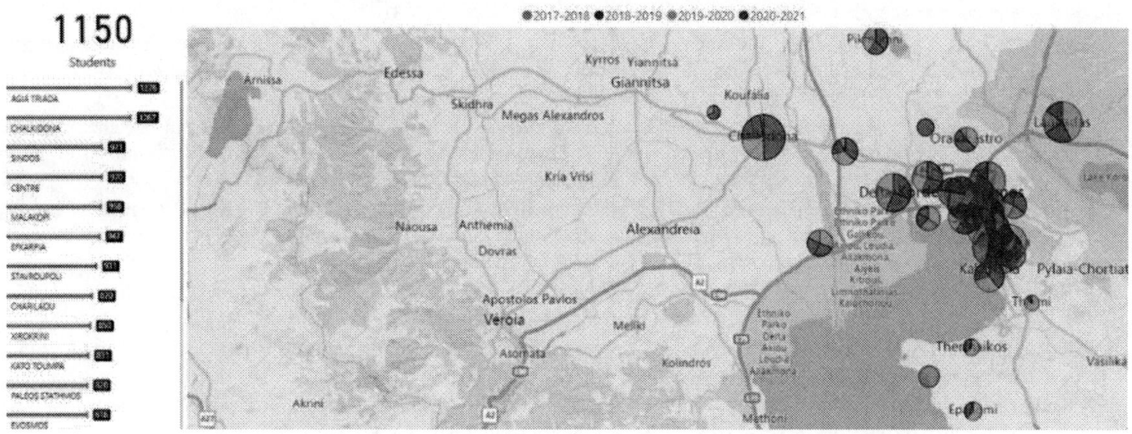


Figure 6. Spatial distribution of students' home residence that have high absenteeism in a specific school year

## 6. CONCLUSION

There is a lack of studies on how smart technologies like LA can help towards alleviating school drop out in the context of VET. The aim of the paper is to explain the design and showcase the development of a system that provides LA visualizations on the topic at stake. Also, to discuss how these visualisations can empower the professional judgment of the teachers or local educational policymakers (i.e. school principles) by helping them to identify where the problems lie in their respective school units over time (i.e. across school years) and across domains and specialisations. This insight in turn can help these stakeholders to make informed decisions on how to alleviate the problem supported by actual data. Also, the LA system can contribute in creating pragmatic strategies against dropout that match the specificities of the respective school unit. For example, in the case of the specific VET school the first grade needs to be supported better. In practical terms, that means that redirecting financial and human resources to create a better student experience for the first-grade students could have a detrimental positive effect for this school. Similar conclusions seem to hold true about students with non-Greek background (e.g. refugees), something that is not surprising taking into consideration previous research work. Plans for future work include: to create visualised correlation plots showing factors that affect dropout (absenses, students per class, property market value of students' home, home-school distance and so on); to create decision tree-based models. Also, to employ mixed methods using a combination of surveys and follow-up interviews with final users, i.e teachers from vocational school education. The focus would be to understand whether the visualizations are reasonably simple to interpret and to be used as a means of understanding the potential causes of dropout.

# REFERENCES


Cerda-Navarro, A., Sureda-Negre, J., & Comas-Forgas, R., (2017). Recommendations for confronting vocational education dropout: A literature review. *Empirical Research in Vocational Education and Training*, Vol. 9, No. 1, pp. 17.

Castaño, F.F., Segura, M.J.S., Gentile, A. and García, L.F.L., (2020). Early school leaving as a social problem in Spain, Italy and Greece. *In Social Problems in Southern Europe*, pp. 64-72.

Demir, K., & Akman Karabeyoglu, Y., (2015). Factors associated with absenteeism in high schools. *Eurasian Journal of Educational Research,* Vol. 62, pp. 37-56.

Gedrimiene, E., Silvola, A., Pursiainen, J., Rusanen, J., & Muukkonen, H., (2020). Learning Analytics in Education: Literature Review and Case Examples from Vocational Education. *Scand. J. Educ. Res.*, Vol. 64, pp. 1105–1119.

Kotsifakos, D., Adamopoulos, P., Kotsifakou, P., & Douligeris, C., (2020). Vocational education and training apprenticeship: Using teaching and learning analytics in a learning management system for improved collaboration, individual empowerment and development of apprentices. *In 2020 IEEE Global Engineering Education Conference (EDUCON)*, Porto, Portugal, 2020, pp. 1775-1782.

Lavrijsen, J., & Nicaise, I., (2015). Social inequalities in early school leaving: The role of educational institutions and the socioeconomic context. *European Education,* Vol. 47, No. 4, pp. 295-310.

López-Pernas, S., Kleimola, R., Väisänen, S., & Hirsto, L., (2022). Early detection of dropout factors in vocational education: A large-scale case study from Finland. *Proceedings of the Finnish Learning Analytics and Artificial Intelligence in Education Conference (FLAIEC22),* Joensuu, Finland, pp. 44-50.

Maraza-Quispe, B., Valderrama-Chauca, E.D., Cari-Mogrovejo, L.H., Apaza-Huanca, J.M. and Sanchez-Ilabaca, J., (2022). A predictive model implemented in knime based on learning analytics for timely decision making in virtual learning environments. *International Journal of Information and Education Technology*, Vol. 12, No. 2, pp.91-99.

Mustapha, A., Abubakar, A. K., Ahmed, H. D., & Mohammed, A., (2021). Application of Big Data Analytics for Improving Learning Process in Technical Vocational Education and Training. *In Information and Communication Technology and Applications: Third International Conference (ICTA*), Minna, Nigeria, pp. 15-25.

Najm, I., A., Dahr, J., M., Hamoud, A., K., Alasady A., S., Awadh W., A., Kamel B., M., and Humadi, A., M., (2022). OLAP Mining with Educational Data Mart to Predict Students' Performance, Informatica, Vol 46, No 5.

Panagiotis, G., (2023). 'Voices' of early school leavers in Greek educational system: A Bourdieusian approach. *Equity in Education & Society*, Vol. 0, No. 0, pp. 1-19

Samaras, C., Tsoni, R., Paxinou, E., Kotsiantis, S., & Verykios, V. S., (2022). Coping with Access Difficulties and Absenteeism through Data Visualization: A Case Study from a Rural Vocational School in Northern Greece. *Applied Sciences*, Vol.12, No. 14, pp. 6946.

Sarlis, I., Kotsifakos, D., & C. Douligeris, C., (2021). Techniques and Technologies for the Administrative Support of Distance Education in Junior High Schools in the Corona Virus Age. *In 2021 IEEE Global Engineering Education Conference (EDUCON),* Vienna, Austria, 2021, pp. 766-773.

Siemens, G., (2013). Learning Analytics: The Emergence of a Discipline. American Behavioral Scientist, Vol. 57, No. 10, pp. 1380-1400.

# DIGITAL TECHNOLOGIES AND THEIR IMPACT ON METACOGNITIVE ASPECTS IN LEARNING RECOVERY IN SCIENTIFIC AND HUMANISTIC SUBJECTS

Andrea Balbo[1], Alice Barana[2], Giulia Boetti[2], Marina Marchisio Conte[2] and Sara Omegna[2]
[1]*Department of Humanities, University of Turin, Via Sant'Ottavio 20, 10124 Torino, Italy*
[2]*Department of Molecular Biotechnology and Health Sciences, University of Turin, Piazza Nizza 44, 10153 Torino, Italy*

**ABSTRACT**

Digital technologies can support studying and learning different disciplines. For students with difficulties in their educational path, these technologies can aid in learning recovery from a metacognitive point of view. In this research study, we investigated the use of interactive materials and automatic formative assessment activities in supporting lower secondary school students in Italy (grades 6 to 8) in learning and studying scientific and humanistic subjects in relation to metacognitive aspects, such as confidence, self-esteem, motivation, and independence. The analyses were conducted within the edition 2022/2023 of the project "compiti@casa: curing educational fragility", organised by the University of Turin and by the De Agostini Foundation. Participants involved in the project were tutored by university students in weekly synchronous online meetings to facilitate their learning recovery. Tutoring activities are implemented within a Digital Learning Environment that allows the use of interactive materials and activities. Through the 555 responses to the final questionnaire submitted to the tutors, we analysed the variation of metacognitive aspects that they observed in the students they tutored, and we investigated their relationship with the use of interactive materials and automatic formative assessment activities. It turned out that their use can effectively support learning recovery across scientific and humanistic subjects in lower secondary school students from the perspective of metacognition.

**KEYWORDS**

Automatic Formative Assessment, Digital Learning Environment, Distance Learning, Interactive Materials, Learning Recovery, Metacognition

## 1. INTRODUCTION

The COVID-19 pandemic not only led to the temporary closure of many businesses and schools but also had a devastating and dramatic impact on the lives of some students. In addition to economic poverty, educational poverty also increased. Educational poverty is defined as the condition in which minors are deprived of the opportunity to train, develop, and realise their skills, talents, and aspirations (Save the Children, 2022). This is particularly the case for young people from low-income families and with a low level of education (INVALSI, 2021a; Moscoviz & Evans, 2022).

In 2020, with the aim of combating this problem, the DELTA research group of the University of Turin and the De Agostini Foundation launched the project "compiti@casa: curing educational fragility" funded by the De Agostini Foundation. The project is in line with the PNRR, which is part of the Next Generation EU programme agreed with the European Union as a result of the pandemic crisis and which has three fundamental axes: digitalisation and innovation; ecological transition; social inclusion (Ministero dell'Istruzione e del Merito, 2022). It is also in line with the DigCompEdu framework (Digital Competence Framework for Educators), developed in 2017 by the European Commission's Joint Research Centre (JRC) to support the development of digital competencies (European Commission).

The *compiti@casa* project provides full online support to lower secondary school students in several Italian cities who have learning difficulties, little autonomy in their own learning, low motivation to learn, are often socio-economically disadvantaged, and find themselves alone with their afternoon homework. The main objective is to facilitate learning recovery in scientific and humanistic subjects using digital technologies and the potential of a digital learning environment. This paper aims to investigate the impact of the use of interactive materials and automatic formative assessment activities on metacognitive aspects related to learning and studying these subjects in the students involved in the project.

## 2. THEORETICAL FRAMEWORK

The literature highlights how poor school performance, particularly in relation to disadvantaged contexts and ethnic minorities, is one of the causes of early school leaving (Szabó, 2018). It is linked to both internal factors such as self-esteem, self-confidence, motivation, attitude, cognitive style, and anxiety, and external factors such as the school environment, home environment, parental support, and the socio-economic context (Hossein-Mohand & Hossein-Mohand, 2023; Raj Acharya, 2017; Fagnani et all, 2020). Metacognition, that is a set of abilities and skills to recognise one's own learning activities, to plan, monitor, and evaluate procedures that they have applied in their learning, is a strong predictor of school success (Hrbáčková et al., 2012). Thus, acting on motivation, self-esteem, self-awareness, learning-how-to-learn, and self-assessment can support the improvement of school results and, consequently, reduce early school leaving. Many ways to promote school success have been suggested, including introducing innovative teaching practices such as formative assessment and feedback. Feedback is really useful when it is used by teachers and/or students to modify the gap between current and reference performance to ensure effective and personalised learning that promotes continuous improvement and student success (Sadler, 1989). When students have difficulty in solving a task, a step-by-step approach can be appropriate to stimulate motivation and reduce the complexity of the task, thus providing an opportunity to develop reflection and awareness of learning processes (Grange, T., & Patera, S., 2021).

Over the last few decades, several projects have been activated in Italy with the aim of reducing school failure and increase school success among lower secondary school students. Among them we remember the "Fuoriclasse" project (Ambrosini & De Simone, 2015), which for 10 years (from 2012 to 2022) provided face-to-face out-of-school tutoring for students from highly disadvantaged backgrounds and at risk of dropping out of school in different areas of Italy. The "Fuoriclasse" project aimed to act both at the cognitive level through peer tutoring actions and at the motivational, metacognitive, and relational levels through motivational laboratories and out-of-school activities. Differently from the *compiti@casa* project, the activities organised within the "Fuoriclasse" project did not include a technological component. A second example is the "Scuola dei Compiti" (which means "Homework School") project, which was activated in Turin from 2013 to 2020, funded by the Town of Turin, and organised in collaboration with the University of Turin. It involved small group afternoon courses led by university students as tutors, who helped at-risk students of lower and upper secondary school with their difficulties in different subjects, mainly Italian language, Latin language, Mathematics, Science, and Foreign Languages. The activities were face-to-face, but some of the tutors integrated the face-to-face activities with digital ones in a Digital Learning Environment (DLE). Studies we carried out in (Barana et al., 2017, Giraudo et al., 2014) show that the courses that integrated digital methodologies were more effective than the traditional ones. A DLE, as defined in (Barana & Marchisio, 2022), is a learning ecosystem in which teaching, learning, and the development of competence are fostered and supported in a face-to-face, online, blended, or hybrid setting. It involves three elements: the human component (the learning community), a technological component (a Learning Management System, specific software and tools, devices, and other systems to collect data), and the interrelations between the two (interactions, learning activities, and methodologies). Through a DLE, it is possible to identify and assess learning objectives, track the progress of the students involved, and collect data to monitor the learning process. DLEs are therefore widely used to support online educational processes, but the literature has shown how they can also implement learning in classroom-based settings (Barana & Marchisio, 2022).

## 3. SETTING AND RESEARCH METHODOLOGY

### 3.1 The *compiti@casa* Project

The *compiti@casa* project, literally *homework@home*, is organised by the University of Turin and the De Agostini Foundation, and it draws from the experience of the "Scuola dei Compiti" project. It was implemented for the first time in the 2020-2021 school year. University students act as tutors, and the programme is implemented in peripheral schools (representing critical areas) in several Italian cities. In particular, the third edition (2022-2023) involved around 290 students from 6 Italian cities (Milan, Naples, Novara, Rome, Palermo, Turin) and 105 tutors. The students are divided into very small groups of two, in order to pay close attention to the needs of each individual, and each group is assigned to two tutors, university students with expertise in science area (mainly Mathematics) and humanities area respectively (mainly Italian). Each group receives guidance and support for afternoon study in both disciplinary areas through web conferencing meetings, spending two hours a week on each area from February to May. In particular, the project is divided into four phases: (1) in November, preparation of the DLE, meetings between teachers and organisers, selection and registration of students; (2) in December and January, teacher training, selection (through public selection) of tutors and their training; (3) from February to May, tutoring activities; (4) in May, a final meeting at the University of Turin: this is an opportunity for the actors involved to meet in person.

The Moodle platform on which the DLE is based is a fundamental part of the project and is integrated with an Advanced Computing Environment (ACE) and an Automatic Assessment System (AAS). This platform has made it possible to provide tutors with interactive materials and tests with automatic assessment. In particular, interactive materials have been designed in such a way that objects can be manipulated and modified by the students themselves (Barana et al., 2019). These activities can also be adaptive, in the sense that their development depends heavily on the actions of the users. Indeed, these activities guide the learning path by introducing intermediate stages to allow the students to understand the topic in a gradual manner consistent with their needs and learning styles. Furthermore, the integration with an AAS has made it possible to enhance the activities through automatic formative assessment (Barana, Fissore & Marchisio, 2020), which supports the acquisition of knowledge, the development of disciplinary, metacognitive and problem-solving skills of students (Barana et al., 2021). In particular, the activities are designed to be repeatable with an unlimited number of attempts, finding different objects each time thanks to the algorithmic capabilities of the system.

Tutors can access materials that the University has developed on the base of the difficulties that students usually encounter in lower secondary school (in Italy "Scuola Secondaria di Primo Grado", where students are between 11 and 14 years old). Tutors can also create new resources if they feel it is appropriate, for example, because one of their students has specific needs.

To monitor the effectiveness of the tutoring action, students, tutors, and teachers involved are surveyed at the beginning and end of the project. The data has been collected through questionnaires delivered by DLE.

In this research study, we focused on the tutors' opinions regarding the students they tutored, examining the questionnaire administered to them at the end of the project in the 2022/2023 edition. Tutors were asked to fill out one questionnaire for each student, reflecting on each student's path and answering a series of questions about organizational aspects of tutoring, cognitive and metacognitive aspects found in students, and the collaboration between those involved in the project.

## 3.2 Research Method

At the end of the activities carried out within the 2022/2023 edition of the *compiti@casa* project, we collected the answers provided by science and humanities tutors to the final questionnaire concerning the students they assisted. In particular, we focused on specific items to answer the following research questions:

**RQ1.** To what extent do interactive materials contribute to learning recovery in scientific and humanistic subjects among lower secondary school students, considering their impact on metacognitive aspects?

**RQ2.** To what extent do automatic formative assessment activities contribute to learning recovery in scientific and humanistic subjects among lower secondary school students, considering their impact on metacognitive aspects?

In the final questionnaire about students, tutors indicated whether or not the students used interactive materials and automatic formative assessment activities. Tutors were also asked to reflect on metacognitive aspects related to their students' approach to learning and studying scientific and humanistic subjects. In detail, they were asked to rate these aspects on a Likert scale from 1 to 5, in which 1 is "Not at all" and 5 is "To a great extent". Two sets of items related to metacognitive aspects were considered, as shown in Table 1.

Table 1. This table reports the items taken into account to investigate the metacognitive aspects of students

| Items set 1 | Items set 2 |
|---|---|
| 1. Does the student show himself/herself to be motivated? | 1. Fostering motivation |
| 2. Does the student demonstrate disciplinary competence? | 2. Improving skills |
| 3. Does the student demonstrate learning skills? | 3. Getting students to do the tasks independently |
| 4. Does the student study independently? | 4. Teaching more effective learning strategies |
| 5. Does the student show confidence in his/her own abilities? | 5. Teaching a study method |
| 6. Does the student show self-esteem? | |
| 7. Does the student identify the real causes of his/her difficulties? | |
| 8. Was the student aware of his/her strengths? | |

Tutors were asked to provide an answer to the items of set 1 by reporting their observations on students about these aspects both at the beginning and at the end of the project. They rated the items of set 2 by reflecting on the extent to which the project was effective in improving them. For the analyses, the items of set 2, and so their ratings, were considered separately because they directly measured the effectiveness and impact of the project on educational objectives related to metacognitive aspects, whereas the ratings of items in set 1 were clustered to provide a comprehensive analysis of the students' overall metacognitive abilities and their variations within the project terms.

We conducted statistical analyses considering the responses of the science tutors and those of the humanities tutors separately to account for potential differences in disciplinary perspectives. We used the statistical software SPSS 29 (Statistical Package for Social Science) to conduct the statistical analyses. We first considered the use of interactive materials, computing the average rating of the metacognitive aspects of items set 1 at the end of the project and the average variation of it in relation to the beginning by considering two different groups of students, i.e., those who used the interactive materials and those who did not use them during the tutoring activities. We then performed an ANOVA test to analyse the variance between the two groups and determine if there were any statistically significant differences between them in relation to the metacognitive aspects of set 1. We then computed the average ratings of the metacognitive aspects of items of set 2 considering the two different groups of students, and we performed an ANOVA test by considering each item separately. For all the ANOVA tests performed we considered the use of the interactive materials as the independent variable. We then carried out the same analyses considering the use of automatic formative assessment activities as the independent variable.

# 4. RESULTS AND DISCUSSION

## 4.1 The Use of Interactive Materials

To address the RQ1, we conducted the analyses investigating the metacognitive aspects shown in Table 1 in relation to the use of the interactive materials made available in the DLE to use them to support students' recovery in scientific and humanistic subjects. A total of 555 questionnaires were filled out by tutors. Specifically, each tutor filled out a questionnaire per each student he/she tutored: 279 questionnaires were filled out by science tutors (out of 282 students who tutored science) and 276 by humanistic ones (out of 280 students who tutored humanities). We analysed the answers of science and humanities tutors separately. To verify the reliability of the data analysed, the Cronbach's alpha was computed, which was 0.95 and 0.96, respectively.

As far as the science tutors' responses are concerned, they reported that 191 students did not use interactive materials in the science area, and 88 did. In the humanities area, tutors reported that 185 students did not use interactive materials in this area and 91 did. The use of interactive materials was not mandatory: these were made available in the DLE and tutors could decide whether or not to propose their use depending on the needs of the students they tutored. However, it is interesting to consider whether their use impacted students' metacognitive aspects.

Table 2 shows the mean of the rating of the items of set 1 clustered by considering the two different groups of students, i.e., those who used the interactive materials and those who did not use them during the tutoring activities. Specifically, the first column reports the mean of the rating at the end of the project. We remind that the rating was made on a Likert scale from 1 to 5, in which 1 is "Not at all" and 5 is "To a great extent". Considering the items of set 1 clustered, the maximum value of the rating can be 40. The second column of Table 2 contains the mean of the rating variation in relation to the beginning. The variation was calculated with the difference between the tutors' observations of students' metacognitive aspects at the end of the project and the beginning.

Table 2. This table shows the average rating for items of set 1 related to the use of interactive materials

| | Use of interactive materials | Rating mean at the end of the project of items set 1 | Rating mean variation of items set 1 |
|---|---|---|---|
| **Science** | No | 27.7 | 4.34 |
| | Yes | 30.25 | 6.12 |
| **Humanities** | No | 27.22 | 5.58 |
| | Yes | 30.08 | 6.69 |

As shown in Table 2, students who used interactive materials had, on average, higher ratings than those who did not use them for what concerns skills and metacognitive aspects such as motivation, competence, learning skills, autonomous study, self-confidence, self-esteem, capacity to identify difficulties, and awareness of improving their ability in studying and learning the subjects in the two areas. The average variation from the beginning of the project in these aspects is, in all cases, an increase, but as can be seen in the second column of Table 2, it was stronger for students who used the interactive materials.

We performed ANOVA tests to investigate further whether the use of interactive materials can support the learning recovery of lower secondary school students in scientific and humanistic subjects. The results are reported in Table 3. The ANOVA tests were carried out considering both the tutors' ratings based on their observations at the end of the project ('Final' rows) and the variation from the beginning ('Variation' rows).

Table 3. This table contains the results of the ANOVA tests, considering the items of set 1 related to the use of interactive materials

| | Rating | F-value | p-value |
|---|---|---|---|
| **Science** | Final | 9.89 | 0.002 |
| | Variation | 9.84 | 0.002 |
| **Humanities** | Final | 12.01 | <0.001 |
| | Variation | 4.06 | 0.045 |

As shown in Table 3, the F-values are statistically significant in all cases since all the p-values are $< 0.05$.

Considering the items of set 2, we compared the means of tutors' ratings. In both science and humanities areas, the ratings for these items were slightly higher on average in students who used interactive materials (scientific +0.27; humanistic +0.53). We then performed ANOVA tests to investigate the impact of using interactive materials. The results are in Table 4.

Table 4. This table shows the results of the ANOVA tests, considering the items of set 2 related to the use of interactive materials

| | Item | F-value | p-value |
|---|---|---|---|
| **Science** | Fostering motivation | 5.76 | 0.017 |
| | Improving skills | 6.50 | 0.011 |
| | Getting students to do the tasks independently | 1.29 | 0.258 |
| | Teaching more effective learning strategies | 4.07 | 0.045 |
| | Teaching a study method | 7.32 | 0.007 |
| **Humanities** | Fostering motivation | 19.13 | <0.001 |
| | Improving skills | 15.19 | <0.001 |
| | Getting students to do the tasks independently | 10.52 | 0.001 |
| | Teaching more effective learning strategies | 26.02 | <0.001 |
| | Teaching a study method | 36.33 | <0.001 |

In this case, all the F-values except one are significant. This suggests that, unlike other aspects, the use of interactive materials did not significantly impact students' ability to complete exercises independently within the scientific discipline. It should be considered that the students involved in the project were attending lower secondary school, and solving exercises independently is a skill that is developed and improved over time.

These ANOVA test results suggest overall that there are substantial differences between students who used the interactive materials and those who did not, both in the metacognitive aspects investigated with items of sets 1 and 2, and in both the science and humanities areas.

## 4.2 The Use of Automatic Formative Assessment Activities

Considering the items related to automatic formative assessment to address the RQ2, science tutors reported that 209 students did not use automatic formative assessment activities in the scientific area and 70 did. In the humanities area, tutors reported that 229 students did not use automatic formative assessment activities and 47 did. We can observe that automatic formative assessment activities were used less than the interactive materials, but also in this case, their use was not mandatory.

Table 5 reports the mean of the rating of the items of set 1 clustered by considering the two different groups of students, i.e., those who used the automatic formative assessment activities and those who did not use them during the tutoring activities. Specifically, the first column contains the mean of the rating at the end of the project, and the second column contains the mean of the rating variation in relation to the beginning.

Table 5. This table shows the average rating for items of set 1 related to the use of automatic formative assessment activities

| | **Use of automatic formative assessmen activities** | **Rating mean at the end of the project of items set 1** | **Rating mean variation of items set 1** |
|---|---|---|---|
| **Science** | No | 28.07 | 4.46 |
| | Yes | 29.89 | 6.23 |
| **Humanities** | No | 27.74 | 5.70 |
| | Yes | 30.21 | 7.15 |

As we see in Table 5, the use of automatic assessment helped students to improve the metacognitive aspects investigated with items of set 1. Indeed, in both areas, students who used these tests had a slightly greater improvement in metacognitive terms than those who did not use them. To further explore this relationship, we performed ANOVA tests, which revealed significant differences between the two groups in terms of their metacognitive growth. This suggests that integrating automatic formative assessment activities can effectively enhance metacognitive skills in middle school students recovering in both scientific and humanistic subjects. The results of ANOVA tests performed considering the use of automatic formative assessment activities as the independent variable are reported in Table 6. They were carried out considering both the tutors' ratings based on their observations at the end of the project ('Final' rows) and the variation from the beginning ('Variation' rows).

Table 6. This table contains the results of the ANOVA tests, considering the items of set 1 related to the use of automatic formative assessment activities

| | **Rating** | **F-value** | **p-value** |
|---|---|---|---|
| **Science** | Final | 4.39 | 0.037 |
| | Variation | 8.38 | 0.004 |
| **Humanities** | Final | 5.62 | 0.018 |
| | Variation | 4.42 | 0.036 |

We then compared the means of tutors' ratings by performing ANOVA tests to investigate the impact of using automatic formative assessment activities on the metacognitive aspects of items set 2. In both science and humanities areas, the ratings for these items were slightly higher on average in students involved in automatic formative assessment activities (scientific +0.26; humanistic +0.35). The results of the ANOVA tests are shown in Table 7.

Table 7. This table shows the results of the ANOVA tests, considering the items of set 2 related to the use of automatic formative assessment activities

| | Item | F-value | p-value |
|---|---|---|---|
| | Fostering motivation | 2.72 | 0.100 |
| | Improving skills | 5.92 | 0.016 |
| **Science** | Getting students to do the tasks independently | 2.68 | 0.103 |
| | Teaching more effective learning strategies | 2.67 | 0.103 |
| | Teaching a study method | 6.12 | 0.014 |
| | Fostering motivation | 4.79 | 0.03 |
| | Improving skills | 9.50 | 0.002 |
| **Humanities** | Getting students to do the tasks independently | 5.72 | 0.017 |
| | Teaching more effective learning strategies | 6.81 | 0.01 |
| | Teaching a study method | 2.23 | 0.136 |

As reported in Table 7, in the scientific field, some F-values are not statistically significant as the related p-values are greater than 0.05. While improving skills and teaching a study method are significant, other aspects, such as fostering motivation and promoting independent exercise completion, do not show significant changes. As stated above, for lower secondary school students, the ability to complete tasks independently is a skill that is only just in the early phases of its development. Improving their own skills in this area is important for students, because this competence will gradually lead them to assume greater independence in their learning tasks. In humanities, all measured items of set 2 show highly significant improvements, except for teaching a study method, suggesting that the automatic formative assessment activities are particularly effective in this area for students aged between 11 and 14.

## 5. CONCLUSION

This research study investigated how digital technologies can aid in the academic improvement of lower secondary school students with learning difficulties who were involved in the *compiti@casa* project in the 2022/2023 edition. Specifically, we analysed the contribution of interactive materials and automatic formative assessment activities in learning and studying scientific and humanistic disciplines, considering their impact on metacognitive aspects.

If we consider the analyses carried out to answer the research questions and the obtained results, we can conclude that using interactive materials (RQ1) and automatic formative assessment activities (RQ2) can effectively support learning recovery across scientific and humanistic subjects in lower secondary school students under the point of view of metacognitive aspects. Integrating interactive materials and automatic formative assessment activities into educational practices could thus enhance student engagement, foster deeper understanding, and improve metacognitive aspects. Since metacognitive skills support the development of disciplinary competence and promote school success (Hrbáčková et al., 2012), we can assume that the digital methodologies adopted in the project have great potential to help at-risk students cope with their difficulties. As these resources were used to a limited extent by the tutors within the project, their use should be enhanced by making tutors understand their importance and the role they can play in supporting student learning. This can be done during the initial training of tutors, before the start of tutoring activities with students. Further research may analyse whether using this type of interactive activities implemented with digital technologies also has an effective impact on classroom learning by considering teachers' responses to the final questionnaire.

## ACKNOWLEDGEMENT

We thank Fondazione *De Agostini*, IGT, Fondazione *Alberto e Franca Riva* and Fondazione *Comunità Novarese*, which made it possible to repeat the project in the 2023/2024 school year. We would like to thank the headmasters and teachers of the schools involved who collaborated to develop the project.

We acknowledge financial support under the National Recovery and Resilience Plan (NRRP), D.M. 118/2023 by the Italian Ministry of University and Research (MUR), funded by the European Union – NextGenerationEU.

## REFERENCES

Ambrosini, M. T., & De Simone, G., (2015). Fuoriclasse: Un modello di successo per il contrasto alla dispersione scolastica. Fondazione Giovanni Agnelli - *Save The Children.* https://www.savethechildren.it/cosa-facciamo/pubblicazioni/fuoriclasse-un-modello-di-successo-il-contrasto-alla-dispersione

Barana, A., Brancaccio, A., Conte, A., Fissore, C., Floris, F., Marchisio, M., & Pardini, C., (2019). The Role of an Advanced Computing Environment in Teaching and Learning Mathematics through Problem Posing and Solving. *Proceedings of the 15th International Scientific Conference eLearning and Software for Education*, Vol. 2, pp. 11–18. https://doi.org/10.12753/2066-026X-19-070

Barana, A., Fioravera, M., Marchisio, M., & Rabellino, S., (2017). Adaptive Teaching Supported by ICTs to Reduce the School Failure in the Project "Scuola Dei Compiti". *Proceedings of 2017 IEEE 41st Annual Computer Software and Applications Conference (COMPSAC)*, pp. 432–437. https://doi.org/10.1109/COMPSAC.2017.44

Barana, A., Fissore, C., & Marchisio, M., (2020). From Standardized Assessment to Automatic Formative Assessment for Adaptive Teaching. *Proceedings of the 12th International Conference on Computer Supported Education*, pp. 285–296. https://doi.org/10.5220/0009577302850296

Barana, A., Casasso, F., Fissore, C., Marchisio, M., & Roman, F., (2021). Mathematics Education in Lower Secondary School: Four Open Online Courses to Support Teaching and Learning. *In Proceedings of the 18th International Conference on Cognition and Exploratory Learning in Digital Age CELDA2021*, pp. 95-102. https://doi.org/10.33965/celda2021_202108L012

Barana, A., & Marchisio, M., (2022). A Model for the Analysis of the Interactions in a Digital Learning Environment During Mathematical Activities. *In B. Csapó & J. Uhomoibhi (Eds.), Computer Supported Education*, Vol. 1624, pp. 429–448. Springer International Publishing. https://doi.org/10.1007/978-3-031-14756-2_21

European Commission. Digital Competence Framework for Educators (DigCompEdu) [online]. *Eu Science Hub..* Disponibile da: https://joint-research-centre.ec.europa.eu/digcompedu_en

Fagnani, Giuditta, and Scanu, (2020). Emergenza scuola: I bisogni ignorati dei nostri figli nella crisi sanitaria. *Il Leone Verde Edizioni*, Italia

Giraudo, M. T., Marchisio, M. & Pardini, C. (2014). Tutoring with new technologies to reduce the school failure and promote learning of mathematics in secondary school. *Mondo Digitale*, Vol. 13, Issue 51, pp. 834–843.

Grange, T., & Patera, S., (2021). La valutazione formativa per sostenere lo sviluppo della dimensione profonda dell'agire competente. Un caso di studio. *Education Sciences and Society*, Vol. 2, pp. 47 - 81

Hrbáčková, K., Hladík, J., Vávrová, S., (2012). The Relationship Between Locus of Control, Metacognition, and Academic Success. Procedia - Social and Behavioral Sciences 69, pp. 1805–1811. https://doi.org/10.1016/j.sbspro.2012.12.130

Hossein-Mohand, H., & Hossein-Mohand, H., (2023). Influence of motivation on the perception of mathematics by secondary school students. *Frontiers in Psychology*, Vol. 13:1111600, pp. 1-16. https://doi.org/10.3389/fpsyg.2022.1111600

INVALSI., (2021). Le diseguaglianze che non si vedono senza dati per tutti [online]. *INVALSIopen.* https://www.invalsiopen.it/leggere-diseguaglianze-dati/

Ministero dell'Istruzione e del Merito, (2022). Azioni di prevenzione e contrasto della dispersione scolastica [online]. *Futura PNRR.* https://pnrr.istruzione.it/wp-content/uploads/2022/07/m_pi.AOOGABMI.REGISTRO-UFFICIALEU.0109799.30-12-2022.pdf

Morabito C., (2022). Povertà educativa: necessario un cambio di passo nelle politiche di contrasto. *Save The Children Italia.* https://s3-www.savethechildren.it/public/files/Poverta_educativa.pdf

Moscoviz, L., & Evans, D. K., (2022). Learning Loss and Student Dropouts during the COVID-19 Pandemic: A Review of the Evidence Two Years after Schools Shut Down. *CGD Working Paper 609. Washington, DC: Center for Global Development.* https://www.cgdev.org/publication/learning-loss-and-student-dropouts-during-covid-19-pandemic-review-evidence-two-years

Raj Acharya, B., (2017). Factors Affecting Difficulties in Learning Mathematics by Mathematics Learners. *International Journal of Elementary Education*, Vol. 6, No. 2, pp 8. https://doi.org/10.11648/j.ijeedu.20170602.11

Sadler, D. R., (1989). Formative assessment and the design of instructional systems. *Instructional Science*, Vol. 18, No. 2, pp. 119–144.

Szabó, C. M., (2018). Causes of Early School Leaving in Secondary Education. *Journal of Applied Technical and Educational Sciences*, Vol. 8, No. 4, pp. 54–76. https://doi.org/10.24368/JATES.V8I4.65

# TAILORING EDUCATION TO DATABASE TRENDS: CHALLENGES IN CURRICULUM DESIGN FOR ADVANCED DATABASE TOPICS

Rosa Maria Reis, Bertil P. Marques and Ricardo Almeida
*Games Interaction and Learning Technologies - GILT- ISEP*
*Rua Dr. Bernardino de Almeida,431 – Porto, Portugal*

**ABSTRACT**

The study aims at designing a syllabus for the curricula unit, advanced database topics, that belongs of curricular plan of master's degree, computer engineering, specializing in data sciences. The syllabus was designed based on learning outcomes approach, EUR-ACE guidelines and ABET accreditation agency. The paper presents the path taken to and explains all the process carried out to obtain a syllabus that allows a software development professional obtains the skills in database require job market. The proposed syllabus has been implemented, for the first time during this semester, and at the end not only allowed us to identify some skills gaps but also conclude that some contents to need adjusting in next semester. Our main objective is known whether we should teach only NoSQL databases, failing to teach relational databases or whether we must have a hybrid system, where both subjects should be taught. If so, what type of content should we teach so that students acquire skills more targeted to the needs of the job market.

**KEYWORDS**

Learning Outcomes, Active and Cooperative Learning, Assessment, Quality, Euro-Ace, ABET

## 1. INTRODUCTION

In recent years we have been faced with an increase in the volume of data. This data is largely unstructured or semi-structured, which causes storage problems to arise when we use relational databases. To solve this problem, new types of databases have appeared on the market, namely NoSQL. At the education level, database programs have been adopting a curriculum that includes relational databases and NoSQL. However, at our institution, the database curricular unit of the computer engineering course is focused on teaching basic concepts of relational databases. As we know, the database area plays a fundamental role in data transformation and as educators we must evolve to meet the needs and demands of the industry. Thus, with the new structuring of the computer engineering master's degree, a discipline of advanced database topics emerges with the aim of resolving gaps found in terms of optimizing queries, indexes, transactions, distributed databases, as well as NoSQL databases. Knowing these limitations, we decided to incorporate the concepts mentioned above into the curriculum because we understand that this knowledge is essential for a student to face the tasks of a data engineer. When defining the curriculum, we had doubts about whether it was really expected to teach only addressing issues related to NoSQL databases or whether we should have a hybrid system, where both subjects, relational and NoSQL databases, should be taught. If so, what type of content should we teach so that students acquire skills more targeted to the needs of the job market. As it was the first time that the subject would be taught, we decided to address both subjects and structure the syllabus according to the profiles of the students from the different specializations of master's in computer engineering. This structuring is described in this article, reflecting the first study carried out to build a curriculum that ensures a broad base in key aspects of databases and allow our students the flexibility required to learning to resolve problems database both in the educational environment and the workplace.

## 2. STRUCTURING THE STUDY

The main objective of this study was to develop a study plan for the discipline of Advanced Database Topics for students studying the master's degree in informatics engineering, data sciences specialization. This study plan was developed under the Learning Outcomes (LO) approach, EUR-ACE guidelines and ABET accreditation agency. Based on these premise's, the present study was carried out through three steps: Preparation Work, Design of Syllabus and Results.

The first step consisted of carrying out a systematic mapping of the literature of programs and master's degrees that had a unit curricular of database in their curriculum and would allow to comparison to be made between the taught contents. This study required the need to define keywords, such as, database, inclusion criteria, exclusion criteria and finally research questions. The second step was based on the study carried out in step 1, allowed us to outline the curriculum of the subject that meets problem-centered design, a methodology widely used in courses at our institution, increasing the relevance of the curriculum and helping students to obtain transferable skills for the world of work. Finally, in the last step was defined a strategy to improve program weaknesses and the topics that was not worked on effectively. This was possible with the collection of data obtained through student opinions, the work and difficulties presented during the teaching process.

## 3. PREPARATION WORK

The exponential growth of data has led to database performance problems. One of the options to improve this issue is to adjust the databases using some concepts, namely, indexing and query optimization. These concepts are important in applications that make use of database management systems. However, we are seeing an increase in large datasets of different formats and the emergence of distributed computing (Chen et al, 2014), which has allowed the emergence of new types of databases, so-called NoSQL databases. These types of databases have been used by large companies, such as Google, Amazon, Facebook, etc. They have helped in data sciences, because their resources offer professionals in this area greater flexibility in scaling data and allow them to work with semi-structured and unstructured data where their storage and retrieval is easier, as they do not follow a schema or specific format. Although data science professionals are interested in learning how to use NoSQL databases to integrate them into their applications, they also need, in our opinion, to have knowledge of database management systems in general. This allows professionals to acquire a more complete knowledge of the main platforms and programming languages, database management and analysis in multiple ways. According to Bala (2003), the database skills described in figure 1 are "essential for data scientists because they provide the foundations for effective data management, analysis and interpretation".

The outline of this paper is as follows. Section 1 will describe the main steps that led to the structuring of the curriculum. Section 2 will outline a systematic mapping of the literature of programs and master's degrees that have a unit curricular of database in their curriculum plan. Section 3 will describe the structuring of the curricular unit of advanced database topics to be aligned with the different profiles of computer engineering master's students. The objectives, structure, assessment requirements, teaching and learning approaches used to improve student performance are described. Section 4. will present the results obtained during applying this syllabus and finally section 6 will draw some conclusions.

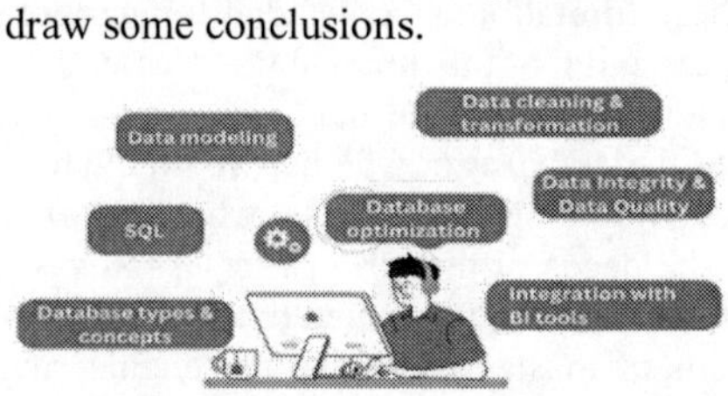


Figure 1. Key database skills (Bala, 2003)

To design the syllabus for the curricular unit, we proceeded initially to survey existing programs in the various universities and polytechnics in our country, with the aim of identifying which subjects are covered in databases and at the same time verifying whether the gaps are filled existing in the job market in terms of these skills. Three general search strings were defined in this survey, namely:

I. Portuguese: ((“Mestrados”) OR (“Engenharia Informática) OR (“Sistemas de Gestão de Base de Dados”));
II. Portuguese: ((“Mestrados”) OR (“Engenharia Informática) OR (“Sistemas de Gestão de Base de Dados”) OR (“Institutos Politécnicos”));
III. Portuguese: ((“Mestrados”) OR (“Engenharia Dados) OR (“Sistemas de Gestão de Base de Dados”));

Following this procedure, institutions that have master’s degrees in informatics engineering and data science were found. From this set, programs from institutions that did not have any curricular unit related to databases were rejected. After this, the syllabus of each subject was analyzed in detail (see table 1).

Table 1. Contents of different curricular units

| Topics | Relation Databases | | | | | NoSQL Databases | | | | Data Formats | | | | | | | | |
|---|---|---|---|---|---|---|---|---|---|---|---|---|---|---|---|---|---|---|
| UC | Relational Algebra | Otimization Query | Performance /Tuning | Concurrence and Transactions | Administration and Recuperation | Data Modelation | Indexing | Storage and recuperation of data | Languages | Semi Strutured | Xml Databases | Stream data | Distributed Databases | Knowledge Databases | Spatial Databases | New Databases types | Object-Oriented | Web Semantics |
| Advanced Databases | | x | x | x | | x | | x | | | | | | | | | | |
| NoSQL Databases | | | | | | x | x | x | x | | | | | | | | | |
| Data Modelation | | | | | | x | | | x | | | | | | | x | | x |
| Database Tecnologies | x | x | x | | x | | | x | x | x | x | | | | | | x | |
| Advanced Databases Topics | x | | | | | | | | | | | | | x | x | x | x | |

This analysis allowed verify the existence of subjects that are covered in other curricular units of the computer engineering master's plan, namely, data warehouse, database administration, text mining and information retrieval. So, we chose to incorporate content into the program that met the competencies defined by Bala (2003) and that also covered some subjects that are not covered in our institution due to time restrictions in the school calendar. It is essential that students are exposed to concepts beyond the basic topics taught in a semester in the Database discipline (Sadalage and Flowler,2012). These themes were framed into four main areas with the aim of having a link between the contents taught in the advanced database topics unit and the curricular units mentioned above. Each of these areas are subdivided into subthemes.

After identifying the topics that will be taught throughout the semester, we proceed to develop a program for the subject based on learning outcomes (LO) that goes beyond complying with current regulations and the evaluation criteria of national or international accreditation agencies and program certification.

## 4. DESIGN OF SYLLABUS

When designing a curriculum, aspects such as learning outcomes, teaching, and learning methodologies, training profiles, assessment mechanisms, actions and the organization of educational content must be carefully considered to help learning. We must not forget that learning is a complex process, which creates or changes behavior so that individuals have knowledge and skills to solve problems and adapt to real-life situations.

According to Merriam et al. “Learning is a process that brings together cognitive, emotional, and environmental influences and experiences for acquiring, enhancing, or making changes in one’s knowledge, skills, values, and worldviews” (Merriam et al.,2008).

For the purposes of the academic design, we based on the following assumptions:

1 - to follow the ACM curricular guidelines, which provide guidelines for computer science programs, including Database. These guidelines aim to help institutions develop comprehensive and up-to-date curricula that reflect market needs and prepare students for the world of work in computing (Lunt et al.,2008);

2 – using Bloom's taxonomy to help us ensure that learning objectives have of agree with student needs, allowing the teachers to create more cohesive and student-centered curricula. In fact, they contribute to improving the learning process, offering adaptability and personalization (Lunt et al.,2008);

3- the Learning be student-centered to promote acquire and improve cognitive and emotional experiences of student becoming more active, creative, participatory being and motivated to learn (Merriam et al.,2008);

4- to use the theory of constructivism which sees learning as an active process where students construct new ideas through social interactions based on their current and past experiences (Kumar and Refaei,2013);

5- to split of contents according to Davis (Davis,2009), ensuring student motivation during classes. The students " succeed at the beginning of the term and when they can relate the new material to something they already know" (Davis,2009).

6- to choose teaching materials e teaching methods to establish links between different components of the curriculum, because students learn differently.

An overview of the curriculum development process is shown in figure 2. The goal is to allow teachers to use the curriculum as a guide.

Figure 2. Steps Process of Curriculum Development

## 4.1 Course Objectives and Learning Outcomes

As one of our premise's is student-centered learning, we suggest defining learning objectives and learning outcomes simultaneously, given that "learning outcomes describe what the student is expected to know, understand and be able to do after completing of the learning process". (European Commission Education and Culture DG, 2009). As an answer to this statement, we can consider the process of backward design, which is a method of creating an educational curriculum, defining goals before choosing teaching methods and forms of assessment. Therefore, if we consider that learning outcomes have this purpose, then teachers can describe to students what is expected of them, plan the teaching strategy, evaluate, learn, and make changes to the curriculum. We improve students' learning because they can monitor their learning progress, know where they are and know in advance how they will be assessed (McCallum & Milner, 2021).

Therefore, we begin by defining the objectives and learning outcomes for our curricular unit that are in accordance with the eight learning areas established for engineering courses accredited by European Accredited Engineer (EUR-ACE) (see table 2).

Table 2. Outcomes according with the EUR-ACE criteria

| EUR-ACE Criteria | Learning Outcomes |
|---|---|
| Knowledge and understanding | CO1. Explain problems related to data management.<br>CO2. Compare and describe the latest developments in the data management field.<br>CO4. Interpret and apply database storage structures; improved performance. |
| Engineering Analysis | CO3. Understand the data modeling techniques – NoSQL and selecting appropriate solutions considering the functional requirements of the systems to be developed. |
| Engineering Design | CO5. Deploy and operate most suitable data management systems regarding domain and requirements; |
| Engineering Practice | CO2. Compare and describe the latest developments in the data management field.<br>CO4. Interpret and apply database storage structures and access techniques to improved performance.<br>CO5. Deploy and operate most suitable data management systems regarding domain and requirements. |
| Making Judgments | CO1. Explain problems related to data management.<br>CO3. Understand the data modeling techniques-NoSQL and selecting appropriate solutions considering the type of information and the functional requirements CO2. Compare and describe the latest developments in the data management field. |
| Communication and Team-working | CO4. Interpret and apply database storage structures and access techniques to improved performance.<br>CO5. Deploy and operate most suitable data management systems regarding domain and requirements. |
| Lifelong Learning (#LL) | CO1. Explain problems related to data management.<br>CO3. Understand the data modeling techniques (NoSQL)<br>CO5. Deploy and operate most suitable data management systems regarding domain and requirements. |

We were careful to use an active verb with the aim of facilitating learning and articulating actions that students will be able to carry out. At the same time, it helps us apply Bloom's Taxonomy to ensure that learning objectives are aligned with students' needs and that the curriculum promotes the development of cognitive skills at different levels, resulting in a comprehensive and meaningful learning experience. This alignment allows teachers to create more cohesive and student-centered curricula. In fact, they contribute to improving the learning process, offering adaptability and personalization (Lameras and Arnab, 2022).

## 4.2 Teaching Strategies

The curriculum plan of the master's degree in informatics engineering, specialization of data sciences, has different types of classes for each curricular unit, namely theoretical, theoretical-practical and laboratory classes. Given this, several student-centered teaching strategies were chosen with the aim of allowing the application of what they learned through discussion and collaborative work. Several studies, notably Bain (2004), state that students learn effectively when they try to solve problems, collaborate with each other, and can correct mistakes when they receive feedback as they continue trying to solve. These strategies will be efficient if we keep in mind that the students learn differently. It is necessary to make choices regarding teaching materials, teaching methods and factors related to, for example, location, time, groups so that we can establish links between the different components of the curriculum. We have made some decisions for this subject, namely:

- we structured the contents in essential and optional subjects. This division will allow us to have some basic materials, which must be mastered by all students, some materials recommended for students looking for more knowledge and finally optional material to be mastered by students with special interests and skills. However, "in exams we should focus on basic topics. Others should be included in class sessions, supplementary material, group work and readings"(Davis, 2009);
- we applied problem-based learning (PBL), promoting individual and collective study of the subjects covered in the curricular unit. Thus, learning becomes a process that allows the student to acquire and improve cognitive and emotional experiences, becoming an active, creative, participatory being and motivated to learn (Merriam et al., 2007);
- we select some subjects related with the real world that will be presented by external invitees whenever possible. According Tzenios (2022) "Engaging students with real-life complexities helps them develop critical thinking, a useful skill in their professional careers. Students learn to innovate and find smart solutions to everyday problems".

## 4.3 Teaching Methodologies and Learning Outcomes

To achieve the defined learning objectives, there must be coherence between teaching/learning methodologies and learning outcomes because according to Shuell (1986) "If students are to learn desired outcomes in a reasonably effective manner, then the teacher's fundamental task is to get students to engage in learning activities that are likely to result in their achieving those outcomes... It is helpful to remember that what the student does is actually more important in determining what is learned than what the teacher does." (Shuell, 1986: 429).

To have this coherence, we cannot only define the contents, we must also define the level of understanding that we want students to reach, forcing students to get involved in learning activities. We want students to reach the objectives of course, forcing them to get involved in learning activities. We understand as fundamental the use of a teaching-learning methodology that promotes work, students' reflection on the content to be learned and the ability to apply them in practical terms to real-life situations. In our case, the teaching and learning methodologies are designed to help the students develop their skills as future professional in informatics engineering and the coherence is reflected in:

1. Lectures - Exposing contents and case studies in lectures classes, promotes CO1, CO2 and C03 objectives. Exposing and discussing examples promote CO4 and CO5 objectives;
2. Practical classes will complement some contents of the theoretical classes through the practical resolution of exercises and/or the (partial) development of projects that will allow establishing the bridge to the laboratory classes, thus promoting objective C04, and consequently, C01, C03 objectives;

3. Laboratory classes will be for supervised database development using a problem-based learning approach, promoting C03, CO4 objectives. Also, the students perform tasks or develop small supervised in group. Group work will be given priority to promote discussion between students (Peer Learning, Cooperative Learning). The execution of tasks and development of small projects promotes students' autonomy and teamwork in the practical and integrated application of the various skills acquired, thus promoting the CO4 and CO5 objectives.

## 4.4 Evaluation Procedures

Assessment is a continuous and guided process that allows us to assess whether students can achieve the planned results and at the same time allows us to identify which areas need to be improved.

When this assessment is based on learning outcomes and it's impacts, then the assessment requires a planning, a selection or design of methods and tools appropriate to the LO level. This requires a new attitude from the teacher, as "he must provide precise information about what the student learned, under what conditions he understood it and at what level of performance he acquired such learning."(Mendoza et al., 2021). Assessment is an integral part of the curriculum design for any course. It helps us obtain information to identify the strengths and weaknesses of the learning process, improve and/or adjust teaching activities, meeting the requirements of the training program quality accreditation.

According ABET criterion 4 - Continuous improved - which states:

"The program must regularly use appropriate, documented processes for assessing and evaluating the extent to which the student outcomes are being achieved. The results of these evaluations must be systematically utilized as input for the continuous improvement of the program. Other available information may also be used to assist in the continuous improvement of the program" (ABET, 2014), authors decided to involve direct and indirect assessment measures in the evaluation of our curricular unit.

Direct assessments with the aim of measuring student performance, and they are related to the reports, projects, presentations, and exams carried out by students. In our opinion, we can more easily articulate the specifics of our program with ABET guidelines. The assessment of each project requires the presentation and face-to-face discussion of each student. Each project is done in a group, but the assessment is individual, so members of the same group may have different classifications.

Indirect assessments have the objective allow of students intervening in each other's assessment, namely student self-assessment. Students evaluate their cooperative performance using a rubric such as the one shown in figure 3. "With the help of rubrics, students can understand what level of work is considered good, very good, excellent or bad. A well-designed rubric can also help students by delivering constructive feedback about their common mistakes and by providing further information about how to improve their work" (Chowdhury, 2019).

| Criteria | Overview (purpose or Objective) of work | Organization | Mechanics (presentation between group members) | Presentation aids (powerpoint, visual aids, handouts) | Results & Discussion | Duration of Presentation |
|---|---|---|---|---|---|---|
| 1- Not Acceptable | The central idea is not clear throughout the presentation. (It's never cleary stated) | The flow flow of ideas is unclear. There is no logical sequence. It´s hard to figure out and things aren't interconnected | The conversation is unclear and prevents the audience from understanding a significant portion of the ideas. Presentation problems are frequent and annoying | Presentation aids are not used ot are misured. Any aids used interfere with the communitation of ideas. The slide has poor quality; very small letter, did not use images | The discussion is either significantly unclear or conclusions are not supported. The work seems incomplete or incorrect | Too long or too short. More than 10 minutes over the allotted time or more than 10 minutes under the time |
| 2- acceptable | By the time the student reaches the end of the presentation, the main idea is clear, but it took a lot of work for audience.... | The main idea are understandable, but there's a sense of clutter or awkwardness to the flow. There are minor problems throughout the lecture. | The conversation is just standard, without ditinction. Presentation problems occur repeatedly, but they are not excessive. The audience perceives the conversation as just "okay" | Presentation aids are not used ot are misured. Any aids used interfere with the communitation of ideas. The slide has poor quality; very small letter, did not use images | The discussion is sometimes incomplete or the rationale for conclusion is correct. However the core content and conclusions are present | about 7 minutes of alloted time |
| 3- Good | the main idea is usally clear, but presentation sometimes stays from the topics. The importance of work may have to be inferred rather than explicit | The reasoning flows well overall. Well organized and all information is presented in a logical order. | The conversation is clear, but unusually interesting. There are some presentation issues that distract the audience, but they are not major interruptions to the flow of the lecture. | Suitable presentation aid are used. Material is cleary displayed and both text and graphics are used. Some problems with level or detail or graphic style are present. | Arguments or experimental analyses or creative explanation are clearly made. The depth of the discussion cam sometimes be a bit lacking or rationale for a conclusion is incomplete. | about 5 minutes of alloted time |
| 4- Excellent | The central idea of the work is immediately clear to the audience. The general ideis clearly has merit | ideas flow seamlessly. Well organized. The information is presented in a clear and logical way, easy to follow. | The lecture is interesting and the audience would describe the speaker as he also spoke. Prrsentaiton manerisms | Presentation aids clearly enhance the presentation and have profissional feel. The leel detail is adequate. | The students' main ideais/creation or experimental ideas are clearly presented and discussed. The rationale behind the conclusions is clearly set out. | used the time actually stipulated |

Figure 3. Assessment Rubric

## 4.5 Planning and Assessment Tools

To ensure quality in the course, we develop a lesson plan and assessment, whose objective is helping students remember the topics, follow the topic in its presentation and understand the logic of the suggested activities.

This plan was developed in part based on work done by Bresciani in 2006 that states “as faculty members, we must decide on the intended learning outcomes of the curriculum and the measures that are used to assess them”.

Our plan is divided into types of classes, where in each type of class we include the subject, the activities to be carried out or evaluated, the evaluation method, strategies to check student understanding. For each activity to be evaluated, we define the type of tools, type of assessment, implementation, description, criteria, rubrics, and rating scales. From our point of view, lesson planning is a factor to consider in the design because it helps teachers in determining the success of the teaching-learning process. That is, it helps to achieve goals, objectives, prevent or avoid problems, and gives insight into the development of our daily performance. on the other hand, it allows equity, inclusion of students in the teaching-learning process and can increase student involvement, learning results and overall satisfaction.

## 5. RESULTS

The master's course in informatics engineering, specializing in data sciences, began operating this academic year. And as such, the curricular unit on advanced database topics was taught for the first time, with no history on learning outcomes. We didn't know if the defined curriculum would be ideal to achieve the defined objectives. It was during the semester that we observed that:

- the way we structure the curriculum allowed us to verify that the combination of various methods and tools used to evaluate the outcomes of the discipline to provide a comprehensive, balanced, and inclusive assessment. It helped capture different aspects and dimensions of student learning. Through them we made small adjustments to the planning carried out, which allowing to focus on the acquisition of some basic skills that are important for those who use databases.
- It is important to incorporate a variety of assessment and assessment methods throughout the course or program, as well as encouraging student involvement and choice in assessment. The direct and indirect methods chosen provided us with information about the quality of learning.
- Collecting data and using feedback from participants is important to identify learning gaps. Allowed us to check the existence of different lacks, namely, knowledge, skills, and performance. This is due to the diversity of student profiles existing in the classroom. The students who entered the master's degree in informatics engineering come from several universities and polytechnic institutes that have a database curriculum that is quite different from what is taught at our institution. SO, one of the strategies adopted was to let students solve the exercises and projects based on the programming languages and relational databases they already knew. However, in the next academic year the prerequisite will be Python language, as it is covered in other curricular units of the master’s degree and at the database level it is still a point to be decided.

## 6. CONCLUSION

This work allowed us to develop a flexible curriculum in which it was possible to adjust content and learning activities based on what students were interested in knowing throughout the semester. This helped us plan and teach classes with students' interests in mind, allowing us to incorporate information that helped distinguish advantages and disadvantages of using relational databases and NoSQL databases. Furthermore, it enabled us to observe that by incorporating practical experiences, inviting people from companies to address new concepts, namely, vault data, incorporating real-world case studies, research, and development group work, we can help students with critical thinking and to develop skills aimed at the needs of the job market.

Designing the curriculum based on learning outcomes, applying ABET and EUR-ACE guidelines helps instructors to articulate its components and to focus on creating a learning environment based on student centered learning, following strategies that allow you to obtain information about the path and how a student achieves their learning. There is no single form of curriculum design, not only a single route implementation of LO. It depends on the organizational culture, structure, intentions, philosophy, and axiological and teleological framework. (Mendoza et al, 2022).

We are aware that although these results are positive, we still have a long way to go, because the skills acquired do not guarantee that they are the most suitable for the needs of software development at the database level. Therefore, as future work, we will be in contact with some companies in our region, with the aim of obtaining answers to some questions that we think are relevant, such as, how do organizations currently understand and use NoSQL and relational databases? and what type of database skills, whether relational or NoSQL, should students acquire to address the existing gaps in data science?

We hope that this study will serve as a basis for discussion for together with the responses obtained by the various companies, we can develop a syllabus that can adequately capture knowledge oriented towards the best practices of the industry and our discipline, following ABET guidelines.

## REFERENCES


Bain, K. (2004). *What the best college teachers do*. Harvard University Press.

Bala, P.C (2003). https://www.kdnuggets.com/introduction-to-databases-in-data-science.

Bresciani, M. J. (2006). Outcomes-based academic and co-curricular program review: A compilation of institutional good practices. Sterling, VA: Stylus

Chen, Z. K. et al. (2014b). *The Data Allocation Strategy Based in Load in NoSQL Database*. Applied Mechanics and Materials 513–517: 1464–1469.

Chowdhury, F. (2019). Application of Rubrics in the Classroom: A Vital Tool for Improvement in Assessment Feedback and Learning. International Education Studies, 12(1), 61–68.

Davis, B, (2009). *Tools for teaching*. San Francisco: Jossey-Bass.

Kumar, R. and Refaei, B., (2013). *Designing a problem-based learning intermediate composition course. College Teaching*, 61(2), 67-73.

Lameras, P., & Arnab, S. (2022). Power to the Teachers: An Exploratory Review on Artificial Intelligence in Education. Information (Switzerland), 13(1), 1–38

Lunt, B., Ekstromet al., (2008). *Curriculum guidelines for undergraduate degree programs in information technology*. ACM.

Malmqvist J. et al, (2020). *A First Set of Optional CDIO Standards for Adoption*, Proceedings of the 16th International CDIO Conference, hosted online by Chalmers University of Technology, Gothenburg, Sweden, June 8–11, 2020.

McCallum, S.; Milner, M.M. (2020). The effectiveness of formative assessment: Student views and staff reflections. Assess. Eval. High. Educ. 2020, 46, 1–16.

Mendoza, W., Ramírez, G. M., González, C., & Moreira, F. (2022). Assessment of curriculum design by learning outcomes (LO). *Education Sciences*, *12*(8), 541.

Merriam, S. B. et al, (2007). Learning in adulthood: A comprehensive guide (3rd ed.). San Francisco: Jossey Bass.

Sadalage, P. J. and Fowler, M. (2012). *NoSQL distilled: a brief guide to the emerging world of polyglot persistence*. Pearson Education.

Shuell, T. J. (1986). Cognitive conceptions of learning. Review of Educational Research, 56, 411-436. http://dx.doi.org/10.3102/00346543056004411

Stonebraker M. (2010). *SQL databases v. NoSQL databases Commun. ACM*, 53 (4), pp. 10-11.

Tzenios, Nikolaos. (2022). LEARNER-CENTERED TEACHING. International Journal of Current Research in Science Engineering & Technology. 04. 916-919. 10.56726/IRJMETS32262.

# RATING UNIVERSITIES IN HONG KONG, MACAO AND SINGAPORE BY GENERATIVE ARTIFICIAL INTELLIGENCE (AI): ARE VARIOUS AI MODELS CONSISTENT?

Victor K. Y. Chan
*Macao Polytechnic University*
*Rua de Luís Gonzaga Gomes, Macao*

**ABSTRACT**

This article seeks to study the consistency, among other ancillary comparisons, between popular generative artificial intelligence (AI) models in rating various dimensions of 26 universities in Hong Kong, Macao, and Singapore. Having experimented with six models, only PaLM and Llama ended up being amenable to analysis where the duo were individually requested to award rating scores to the five dimensions (1) Teaching, (2) Research, (3) Citations, (4) International Outlook, and (5) Industry Income of the universities. For each of the two models, the minimum, the maximum, the range, and the standard deviation of the rating scores for each of the five dimensions were computed across all the universities. The rating score difference for each of the five dimensions between the two models was calculated for each university. The mean of the absolute values, the minimum, the maximum, the range, and the standard deviation of the differences for each dimension between the two models were calculated across all universities. A paired sample $t$-test was then applied to each dimension for the rating score differences between the two models over all the universities. Finally, a correlation coefficient of the rating scores was computed for each dimension between the two models across all the universities. Among other collateral findings, the two models' ratings were found almost flawlessly consistent for the five dimensions with the correlation coefficients ranging from .851 to .908 ($p$ all at 0.000). Consistency implies at least the likely trustworthiness of both PaLM's and Llama's ratings.

**KEYWORDS**

Generative Artificial Intelligence, AI, Large Language Model, Universities Rating, Consistency, Convergent Validity

## 1. INTRODUCTION

Universities play a momentous role in society, especially in the advancement of knowledge, education, and innovation. Traditionally, the rating of universities has employed such manual methods as peer reviews, rankings, and surveys. While these methods have rendered invaluable insights, they are often subject to limitations like subjectivity, bias, and limited scope.

Recent advancements in generative artificial intelligence (AI) offer promising avenues for transforming the rating of universities. AI techniques, such as natural language processing (NLP) and machine learning (ML), enable the analysis of huge amounts of data, including scholarly publications, patents, citations, student feedback, and online reviews. If properly implemented, this data-driven approach may be able to provide relatively objective, impartial, and comprehensive ratings of universities across multiple dimensions.

Traditional methods of rating universities have primarily concentrated on academic reputation, research output, and teaching quality. These ratings often involve peer reviews (Musselin, 2013) by experts in the field, rankings (Conard and Conard, 2000; Anonymous, 2024a; Anonymous, 2022; Yusoff et al., 2015; Anowar et al., 2015; Huang, 2011; Anonymous, 2024b; Anonymous, 2023; Olcay and Bulu, 2017), surveys (Conard and Conard, 2000; Anonymous, 2024a; Anonymous, 2022; Bell and Brooks, 2018; Stolzenberg et al., 2019; Honeycutt et al., 2023; Alao et al., 2009), or combinations of them (Huang, 2011; Anowar et al., 2015).

While traditional rating methods have served their purpose, they are not without limitations such as subjectivity (King et al., 2018), bias (King et al., 2018; Safón and Docampo, 2020), and limited scope (Kayyali, 2023).

Generative AI techniques are nascent approaches to university rating, leveraging data analysis and pattern recognition at scale. These techniques can process and analyze mammoth amounts of information, including scholarly publications, patents, citations, online reviews, and social media contents. They make use of such techniques as Natural Language Processing (NLP) (Chowdhary, 2020) and Machine Learning (ML) (Carleo et al., 2019). In particular, the generative AI adopted in this study was Large Language Models (LLMs) (IBM, 2024), which incorporated both the above two techniques.

Notwithstanding, in the context of rating universities or otherwise, there exists no absolutely neutral "yardstick" to benchmark the trustworthiness of any particular AI model against. At best, one may be able to discern cues as to which AI model to trust, by gauging the consistency between multiple AI models. If a set of AI models happen to be somewhat consistent, the odds are that all models in the set are trustworthy to some extent, albeit in theory all being coherently fallacious can never be ruled out. This way to estimate the trustworthiness of AI models is analogous to the concept of convergent validity of an operationalized instrument to measure an abstract construct. This article aims to exam such consistency, among other ancillary comparisons, between as many popular generative AI models as possible in the rating of universities.

# 2. METHODOLOGY

## 2.1 Data and Materials

In March 2024, the present study commenced by experimenting with six very popular generative AI models, namely, Assistant (Poe, 2024), ChatGPT (de Souza et al., 2023), Claude-instant (de Souza et al., 2023), Microsoft Coplilot (Cambon et al., 2023), Google PaLM (Anil et al., 2023), and Meta Llama (Oxford Analytica, 2023) as candidates for rating universities and for comparison between such ratings. Copilot was bundled with the Microsoft Edge browser and to a lesser extent the Microsoft Windows operating system, whereas all the remaining models have been incorporated into the AI portal poe.com. The five dimensions (1) Teaching, (2) Research, (3) Citations, (4) International Outlook, and (5) Industry Income, which originated from THE's World University Rankings 2023 (Anonymous, 2022), to rate universities were adopted and were rated by the above models in this study. According to THE (Anonymous, 2022), (1) Teaching, with a weight of 30%, refers to a university's learning environment and is subdivided into the sub-dimensions (the outcomes of a) Reputation Survey, the Staff-to-Student Ratio, the Doctorate-to-Bachelor's Ratio, the Doctorates-Awarded-to-Academic-Staff Ratio, and Institutional Income with respective sub-weights of 15%, 4.5%, 2.25%, 6%, and 2.25% (Anonymous, 2022). It is worth noting that the sub-dimension Doctorate-to-Bachelor's Ratio is normalized to take account of a university's unique subject mix, reflecting that the volume of doctoral awards varies by discipline (Anonymous, 2022). Also, the sub-dimension Institutional Income is scaled against academic staff numbers and normalized for purchasing-power parity (PPP) (Anonymous, 2022). (2) Research, with a weight of 30%, covers a university's output volume, income and reputation in connection with research of the university and is operationalized by the sub-dimensions (the outcomes of a) Reputation Survey, Research Income, and Research Productivity with sub-weights of 18%, 6%, and 6% (Anonymous, 2022). It is worth noting that the sub-dimension Research income is scaled against academic staff numbers, adjusted for purchasing-power parity (PPP), and fully normalized to take account of each university's distinct subject profile, reflecting the fact that research grants in science subjects are often bigger than those awarded for the highest-quality social science, arts and humanities research. (Anonymous, 2022). Also, the sub-dimension Research Productivity is measured by counting the number of publications published in the academic journals indexed by Elsevier's Scopus database per scholar, scaled for institutional size and normalized for subject (Anonymous, 2022). (3) Citations, with a weight of 30%, are equal to the average number of times a university's published work is cited by scholars globally (Anonymous, 2022). This metric is normalized to reflect variations in citation volume between different subject areas such that institutions with high levels of research activity in subjects with traditionally high citation counts do not gain an unfair advantage (Anonymous, 2022). (4) International Outlook, with a weight of 7.5%, concerns how much a university's staff, students, and research are international and is measured by the sub-dimensions the Proportion of International Students, the Proportion of International Staff, and (the amount of) International Collaboration with sub-weights of 2.5%, 2.5%, and 2.5% (Anonymous, 2022). In particular, the sub-dimension International Collaboration is calculated as the proportion of a university's total relevant publications that have at least one international co-author and reward higher volumes and is normalized to account for a university's subject mix (Anonymous, 2022). (5) Industry Income means the knowledge transfer (to industry) and is rated by how much research income a university earns from industry (adjusted for purchase price parity or PPP), scaled against the number of academic staff employed. Despite such formal "definitions" of these five dimensions, how the AI models defined and interpreted these five dimensions, how much each model's own definitions and interpretations differed from these formal "definitions," how much the definitions and interpretations varied across the models, how the models microscopically rated these five dimensions for each university, and how the models gained access to data on each dimension, especially (5) Industry Income, (presumably during the models' training) were all beyond the scope and control of this study. As a matter of fact, such "hows' have always been and are still inside the black-box of AI and thus fall into the domain of AI's "explanability," which has always been and is still a yet-to-accomplish "Holy Grail" in AI research (Adadi and Berrada, 2018). As long as a particular model produced its ratings for these five dimensions of the universities in question, this study just took these ratings as what they were for the purpose of comparison between the models and thus the assessment of the consistency between them. In other words, this study's purpose was to macroscopically evaluate the consistency between AI models in rating universities regardless of the microscopic "hows" underlying each model's ratings. The practical implication was that as long as macroscopic consistency was ascertained, trustworthiness of all the AI models concerned was likely by analogy with the concept of convergent validity whether or not the microscopic "hows" were known and well understood.

In particular, rating of universities' five dimensions by each model was implemented by a request for rating scores submitted to each of the six models. Specifically, the request was for giving a rating score on a scale of 1 to 10 (1 being the worst and 10 the best) to each of the major universities in Hong Kong, Macao, and Singapore for each of the five dimensions (1) Teaching, (2) Research, (3) Citations, (4) International Outlook, and (5) Industry Income. The major universities were spelt out as the University of Hong Kong, the Chinese University of Hong Kong, the Hong Kong University of Science and Technology, the City University of Hong Kong, Hong Kong Polytechnic University, Lingnan University, the Education University of Hong Kong, Hong Kong Metropolitan University (or formerly the Open University of Hong Kong), Hong Kong Baptist University, the Hang Seng University of Hong Kong, the University of Macau, Macao Polytechnic University (or formerly Macao Polytechnic Institute), Macao Institute for Tourism Studies, the Macao University of Science and Technology, the City University of Macau, the University of Saint Joseph, National University of Singapore, Nanyang Technological University, Singapore Management University, the Singapore University of Social Sciences, the Singapore Institute of Technology, James Cook University, Singapore, Curtin University, Singapore, the DigiPen Institute of Technology Singapore, Duke-NUS Medical School, and ESSEC Business School, Asia-Pacific. The request accentuated deriving the scores from global users' textual comments on these five dimensions of these universities as appear all around the web.

The 26 "universities," inclusive of institutions not expressly named as universities, spelt out in the above request were the prominent tertiary institutions in Hong Kong, Macao, and Singapore and were mostly recognized by the corresponding local governments. These three cities were targeted in this study simply because they were three major international metropolises that the author was familiar with and were characterized by influence of the Chinese culture, which the author was interested in. After all, this study is extensible to universities elsewhere if preferred in the future. Please note that the request above expressly accentuates deriving the scores from global users' textual comments on these five dimensions of these universities as appear all around the web. In other words, each model presumably derived its rating scores from global users' textual comments appearing on the worldwide web instead of simply echoing any analogous scores already existing somewhere, for example, the ratings published by the major university rankings like THE, QS, ARWU, etc. Also, of the 26 universities in 2023, 10 or 38.5% were ranked by, for instance, THE and thus had their THE ratings published in the public domain while the remaining 16 or 61.5% were not so ranked and thus had no THE ratings disclosed to the public. Therefore, whereas there was still plausibility that the AI models parroted the publicly published THE ratings or their scaled versions when providing the models' own rating scores, the effects of such parroting were appreciably diluted even if not totally ruled out. Here, the term "users" referred to all individuals and entities that made use of the outputs from the universities where such outputs might include the graduates from the universities, the knowledge generated by the universities' researches, the education provided by the universities, the consultancy offered by the universities, etc., and the corresponding users were the graduates' employers, those who applied the knowledge generated by the universities, the students at the universities, those served by the universities' consultancy services, etc.

In their replies, Assistant, ChatGPT, Claude-instant, and Microsoft Coplilot all declined to provide any rating scores on various grounds and were thus eliminated from further steps of this study. In contrast, only PaLM and Llama replied with complete rating scores for all the five dimensions and all the 26 universities enumerated above. Collaterally, these two models were asked to explain their rating scores. PaLM responded with its own interpretations of the five dimensions and the way it rated in accordance with such interpretations whilst Llama shortlisted 10 universities out of the 26 and narratively expounded the rationales behind its rating scores for the 10 universities.

## 2.2 Analysis

For each of the two AI models PaLM and Llama, the minimum, the maximum, the range, and the standard deviation of the rating scores for each of the five dimensions were calculated across all the 26 universities. If there is a considerable range and standard deviation for a particular dimension, it is revealed that the corresponding model accords discrimination in rating the dimension across the universities. Take the standard deviation as an example. If the standard deviation of the rating scores for a particular dimension is sufficiently large across all the 26 universities, the rating scores for that dimension vary adequately across the universities. In other words, the model does not indiscriminately assign the same rating score to all the 26 universities.

Subsequently, the rating score difference for each of the five dimensions between the two models was computed for each of the 26 universities. The mean of the absolute values, the minimum, the maximum, the range, and the standard deviation of the differences for each dimension between the two models were computed across all the 26 universities. If the mean of the absolute values, the range, and the standard deviation for a particular dimension are sufficiently small, it is ascertained that the two models neither overrate nor underrate erratically, with respect to each other, the dimension across the universities. Take the standard deviation as an example. If the standard deviation of the rating score differences for a particular dimension between the two models is small across all the 26 universities, the rating score difference for the dimension of any university between the two models must be nearly the same regardless of which university is in question. In other words, PaLM's rating score for the dimension of any university must differ from Llama's counterpart by only an almost constant margin irrespective of which university is concerned. Equivalently, the two models must not overrate or underrate, with respect to each other, the dimension with capriciously fluctuating margins across all the 26 universities.

A paired sample *t*-test was then applied to each dimension for the rating score differences between the two models over all the 26 universities. If the *t*-test is significant for a particular dimension and the corresponding mean difference is positive (negative), it is verified that the first model systematically overrates (underrates) the dimension with respect to the second model.

Finally, a correlation coefficient of the rating scores was calculated for each of the five dimensions between the two models across all the 26 universities. If the correlation coefficient is positively high enough (for instance, over 0.6) for a particular dimension, it is attested that there is consistency between the two models in rating the dimension across the universities or equivalently the convergent validity of the two models is high in rating the dimension across the universities.

# 3. RESULTS

Table 1 lists the minimum, the maximum, the range, and the standard deviation of the rating scores for each of the five dimensions across the 26 universities as rated by each of the two models PaLM and Llama. Both PaLM and Llama rated all the five dimensions with substantial discrimination as evidenced by the considerable ranges and standard deviations. Nevertheless, Llama rated the dimension (1) Teaching with comparatively less discrimination than it rated the remaining four dimensions and than PaLM rated all the five dimensions. This is indicated by the slightly smaller range and standard deviation corresponding to Llama's rating of (1) Teaching in contrast with its ratings of the remaining four dimensions and PaLM's ratings of all the five dimensions. More specifically, Llama' rating of (1) Teaching yielded a small range of 3, which was paralleled only by Llama's rating of (5) Industry Income whilst the ranges corresponding to Llama's ratings of the remaining three dimensions and PaLM's ratings of all the five dimensions were 4. Likewise, Llama's rating of (1) Teaching rendered a standard deviation of 0.9829, which was substantially less than all the standard deviations corresponding to its ratings of all the remaining four dimensions and than those corresponding to PaLM's ratings of all the five dimensions, all these being well above 1.

Table 1. The minimum, the maximum, the range, and the standard deviation of the rating scores for each of the five dimensions across the 26 universities as rated by each of the two models PaLM and Llama

| Model (sample size *n*) | Minimum/ maximum/ range/ standard deviation | (1) Teaching | (2) Research | (3) Citations | (4) International Outlook | (5) Industry Income |
|---|---|---|---|---|---|---|
| PaLM ($n = 26$) | Minimum | 6 | 6 | 6 | 6 | 5 |
| | Maximum | 10 | 10 | 10 | 10 | 9 |
| | Range | 4 | 4 | 4 | 4 | 4 |
| | Standard derivation | 1.2310 | 1.2310 | 1.2310 | 1.2310 | 1.2310 |
| Llama ($n = 26$) | Minimum | 6 | 6 | 6 | 5 | 5 |
| | Maximum | 9 | 10 | 10 | 9 | 8 |
| | Range | 3 | 4 | 4 | 4 | 3 |
| | Standard derivation | 0.9829 | 1.4718 | 1.5216 | 1.3062 | 1.2159 |

Table 2 depicts the mean of the absolute values, the minimum, the maximum, the range, and the standard deviation of the rating score differences for each of the five dimensions between the two models across the 26 universities. The disparity between PaLM and Llama in rating each dimension was rather even across all the five dimensions. Having said that, one may opine that such disparity between PaLM and Llama in rating the dimension (1) Teaching was slightly smaller than in rating the remaining four dimensions given that the mean of the absolute values of the corresponding rating score differences for (1) Teaching stood at 0.3333 and was somewhat less than those for the remaining four dimensions, which ranged from 0.5 to 0.75. However, the range and the standard deviation of the corresponding rating score differences for (1) Teaching were not far less than those for the remaining four dimensions. Such ranges for the remaining four dimensions were either 2 or 3, and such corresponding standard deviations were in the neighborhood between 0.5 and 0.8. In other words, the two models overrated or underrated erratically, with respect to each other, (1) Teaching across the universities only marginally less than the remaining four dimensions.

Table 2. The mean of the absolute values, the minimum, the maximum, the range, and the standard deviation of the rating score differences for each dimension between the two models across the 26 universities

| Differences (sample size *n*) | Mean of the absolute values/ minimum/ maximum/ range/ standard deviation of the differences | (1) Teaching | (2) Research | (3) Citations | (4) International Outlook | (5) Industry Income |
|---|---|---|---|---|---|---|
| Llama − PaLM ($n = 26$) | Mean of the absolute values | 0.3333 | 0.5 | 0.5417 | 0.75 | 0.5833 |
| | Minimum | -1 | -2 | -2 | -2 | -2 |
| | Maximum | 1 | 1 | 1 | 0 | 0 |
| | Range | 2 | 3 | 3 | 2 | 2 |
| | Standard derivation | 0.5335 | 0.7736 | 0.8000 | 0.5870 | 0.5711 |

Table 3 depicts the paired sample *t*-test of the rating score differences for each of the five dimensions between the two models over the 26 universities. Relative to Llama, PaLM appeared to overrate the three dimensions (1) Teaching, (4) International Outlook, and (5) Industry Income respectively at the 5% ($p < 0.05$), 1% ($p < 0.01$), and 1% ($p < 0.01$) significance levels.

Table 3. The paired sample *t*-test of the rating score differences for each of the five dimensions between the two models over the 26 universities

| Differences (sample size *n*) | Dimension | Mean difference / [95% confidence interval] | *t* (*p*-value) / degrees of freedom |
|---|---|---|---|
| PaLM – Llama ($n = 26$) | (1) Teaching | .26923 / [.05375, .48471] | 2.573 (.016*) / 25 |
| | (2) Research | .03846 / [-.27400, .35093] | .254 (.802) / 25 |
| | (3) Citations | .00000 / [-.32313, .32313] | .000 (1.000) / 25 |
| | (4) International Outlook | .76923 / [.53212, 1.00634] | 6.682 (.000**) / 25 |
| | (5) Industry Income | .61538 / [.38471, .84606] | 5.494 (.000**) / 25 |

* $p < 0.05$; ** $p < 0.01$

Table 4 illustrates the correlation coefficient of the rating scores for each of the five dimensions between the two models over the 26 universities, the 95% conference interval for the correlation coefficient, and the *p*-value to test whether the coefficient differs from zero. The two models were highly, positively correlated and thus highly consistent and had high convergent validity in rating all the five dimensions as confirmed by their corresponding positive correlation coefficients and by the corresponding *p*-values being significant at the 1% significance level ($p < 0.01$). It is noteworthy that a highly positive correlation coefficient and thus high consistency for a particular dimension imply that a high rating score of a university for the dimension as awarded by one model is generally associated with a high rating score of that university for that dimension as awarded by another model and vice versa even if these two rating scores may not necessarily be the same or not necessarily be even close.

Table 4. The correlation coefficient of the rating scores for each of the five dimensions between the two models over the 26 universities

| Dimension | Correlation coefficient / [95% confidence interval] | *p*-value |
|---|---|---|
| (1) Teaching | .908 / [.8031, .9583] | .000** |
| (2) Research | .851/ [.6916, .9313] | .000** |
| (3) Citations | .852 / [.6935, .9318] | .000** |
| (4) International Outlook | .895 / [.777, .9522] | .000** |
| (5) Industry Income | .891 / [.7691, .9504] | .000** |

** $p < 0.01$

# 4. CONCLUSION AND DISCUSSION

Both AI models PaLM and Llama rated all the five dimensions with substantial discrimination. In other words, both PaLM and Llama did not muddle through the job by rating the five dimensions constantly (or nearly constantly) and indiscriminately across the 26 universities in Hong Kong, Macao, and Singapore, and did perform the rating "seriously," "conscientiously," and "responsibly." Also, the two models overrated or underrated erratically, with respect to each other, all the five dimensions across the 26 universities to some extent, and did so to (1) Teaching only marginally less than to the remaining four dimensions.

To further analyze such overrating and underrating by the two models, this study verified that with respect to Llama, PaLM appeared to overrate the three dimensions (1) Teaching, (4) International Outlook, and (5) Industry Income. Overrating or underrating a particular dimension by one model with respect to another model is no testament to consistency or inconsistency between the two models' ratings for the dimension over the universities. Consistency or inconsistency depends on whether such overrating or underrating is erratic across the universities. To further pinpoint consistency (or inconsistency), the study uncovered that the two models were highly consistent and had high convergent validity in rating all the five dimensions across the 26 universities. The practical implication is that both models are likely to be trustworthy to an extent in rating all the five dimensions of the universities although the possibility of both models' ratings not being truthful about the universities can never be eradicated.

This study itself is not without its critics. First, whereas six generative AI models Assistant, ChatGPT, Claude-instant, Microsoft Coplilot, Google PaLM, and Meta Llama were experimented with in this study, only the last two yielded rating scores for analysis. Against the backdrop of profuse AI models available globally, this study could have been better off scrutinizing the consistency between and thus the convergent validity of a larger number of models so as to reinforce confidence in the trustworthiness of AI models in general in rating the dimensions of universities. Second, these two models were trained on data up to some cutoff dates, so even the rating scores generated by them as of March 2024 for this study could not catch up with the 26 universities' most current statuses. Therefore, in consideration of these two points, broadening the scope of AI models, especially, those having been trained on data with most recent cutoff dates, will be a priority in future researches. Third, it is true that the 26 universities of this study varied drastically in their sizes, academic disciplines, and reputational prominence, so the volume of global users' textual comments on each of them might also differ dramatically, probably undermining the fairness in the rating of each of them by the two models. Whilst this disparity was rather inevitable from the author's standpoint, this study chose to "absorb" such disparity and accept the ratings in the form of rating scores as what they were when estimating the consistency between and thus the convergent validity of these two models' rating scores given to each dimension of the 26 universities. In view of the high consistency and convergent validity found, both the two models would likely in theory be trustworthy to an extent irrespective of such disparity's existence. In fact, the author envisions no effective solution to this disparity issue in future researches. Fourth, of the 26 universities in this study, 38.5% of them were ranked by, for example, the major university ranking THE, and the corresponding ratings by THE were published. It may be conceivable that the two models might disproportionately rely on such published THE ratings in deriving the models' own rating scores, precluding the two models from generating their own ratings equitably and genuinely from global users' textual comments at large. This undue influence was muffled by the remaining 61.5% of the 26 universities that were not so ranked. In future researches, more fine-grained methodologies differentiating consistency between and convergent validity of AI models' ratings for ranked universities from those for unranked universities may be advisable.

In light of the high consistency between and thus convergent validity of the two AI models PaLM and Llama as attested in this study, these two models and very likely generative AI models in general are admittedly poised to be a major means of universities rating and probably most other types of opinion evaluations whether in academia or industry. If appropriately tuned and optimized, such AI model-based ratings might even be less subjective and less biased and might cover a more compendious scope than traditional methods of rating universities.

## REFERENCES


Adadi, A. and Berrada, M., (2018). Peeking Inside the Black-Box: A Survey on Explainable Artificial Intelligence (XAI). *In IEEE Access 6*, 52138-52160. https://doi.org/10.1109/ACCESS.2018.2870052

Alao, A. A. et al., (2009). Employer Satisfaction Survey of the University of Botswana Graduates. *In International Journal of Contemporary Issues in Education and Psychology*, Vol. 1, No. 3, pp. 17-34. https://core.ac.uk/download/pdf/12356682.pdf

Anil, R. et al., (2023). PaLM 2 Technical Report. https://doi.org/10.48550/arXiv.2305.10403

Anonymous, (2022). *World University Rankings 2023: Methodology*. THE - Times Higher Education. https://www.timeshighereducation.com/world-university-rankings/world-university-rankings-2023-methodology

Anonymous, (2023). *Academic Ranking of World Universities*. ShanghaiRanking Consultancy, Shanghai, China. https://www.shanghairanking.com/rankings/arwu/2023

Anonymous, (2024a). *Academic Reputation*. QS Quacquarelli Symonds Limited. https://support.qs.com/hc/en-gb/articles/4405952675346-Academic-Reputation#:~:text=The%20Academic%20Reputation%20Index%20is,the%20greatest%20interest%20and%20scrutiny

Anonymous, (2024b). *QS World University Rankings*. QS Quacquarelli Symonds Limited. https://www.topuniversities.com/qs-world-university-rankings

Anowar, F. et al., (2015). A Critical Review on World University Ranking in Terms of Top Four Ranking Systems. In: Elleithy, K. and Sobh, T. (eds.) New Trends in Networking, Computing, E-learning, Systems Sciences, and Engineering. *In Lecture Notes in Electrical Engineering*, Vol. 312, pp. 559–566. Springer, Cham, Switzerland. https://doi.org/10.1007/978-3-319-06764-3_72

Bell, A. R. and Brooks, C., (2018). What Makes Students Satisfied? A Discussion and Analysis of the UK's National Student Survey. *In Journal of Further and Higher Education*, Vol. 42, No. 8, pp. 1118-1142. http://dx.doi.org/10.1080/0309877X.2017.1349886

Cambon, A. et al., (2023). Early LLM-based Tools for Enterprise Information Workers Likely Provide Meaningful Boosts to Productivity. https://www.microsoft.com/en-us/research/uploads/prod/2023/12/AI-and-Productivity-Report-First-Edition.pdf

Carleo, G. et al., (2019). Machine Learning and the Physical Sciences. *In Reviews of Modern Physics*, Vol. 91. https://journals.aps.org/rmp/abstract/10.1103/RevModPhys.91.045002

Chowdhary, K. R., (2020). Natural Language Processing. *In Fundamentals of Artificial Intelligence*, pp. 603–649. Springer, New Delhi, India. https://doi.org/10.1007/978-81-322-3972-7_19

Conard, M. J. and Conard, M. A., (2000). An Analysis of Academic Reputation as Perceived by Consumers of Higher Education. *In Journal of Marketing for Higher Education*, Vol. 9, No. 4, pp. 69-80. https://doi.org/10.1300/J050v09n04_05

de Souza, C. et al., (2023): Are the New AIs Smart Enough to Steal Your Job? IQ Scores for ChatGPT, Microsoft Bing, Google Bard and Quora Poe. https://papers.ssrn.com/sol3/papers.cfm?abstract_id=4412505

Honeycutt, N. et al., (2023). The Academic Mind in 2022: What Faculty Think About Free Expression and Academic Freedom on Campus. *In Foundation for Individual Rights and Expression*. https://doi.org/10.31219/osf.io/surq2

Huang, M.-H., (2011). A Comparison of Three Major Academic Rankings for World Universities: From a Research Evaluation Perspective. *In Journal of Library and Information Studies*, Vol. 9, No. 1, pp. 1-25. https://jlis.lis.ntu.edu.tw/files/journal/j32-1.pdf

IBM, (2024). Homepage, https://www.ibm.com/topics/large-language-models. Last accessed April 21, 2024.

Kayyali, M., (2023). Critiques and Limitations of University Rankings. *In International Journal of Governance in Higher Education*, Vol. 2, No. 1, pp. 1–25.

King, E. B. et al., (2018). Systematic Subjectivity: How Subtle Biases Infect the Scholarship Review Process. *In Journal of Management*, Vol. 44, No. 3, pp. 843-853. https://journals.sagepub.com/doi/full/10.1177/0149206317743553

Musselin, C., (2013). How Peer Review Empowers the Academic Profession and University Managers: Changes in Relationships between the State, Universities and the Professoriate. *In Research Policy*, Vol. 42, No. 5, pp. 1165-1173. https://www.sciencedirect.com/science/article/abs/pii/S0048733313000309

Olcay, G. A. and Bulu, M., (2017). Is Measuring the Knowledge Creation of Universities Possible? A Review of University Rankings. *In Technological Forecasting and Social Change*, Vol. 123, pp. 153-160. https://www.sciencedirect.com/science/article/abs/pii/S004016251630021X

Oxford Analytica, (2023). Meta LLaMa Leak Raises Risk of AI-Linked Harms. *In Expert Briefings*. Emerald, Leeds, England, UK. https://doi.org/10.1108/OXAN-ES276597

Poe, 2024. Homepage, https://poe.com/chats. Last accessed March 7, 2024.

Safón, V. and Docampo, D., (2020). Analyzing the Impact of Reputational Bias on Global University Rankings Based on Objective Research Performance Data: The Case of the Shanghai Ranking (ARWU). *In Scientometrics*, Vol. 125, pp. 2199–2227. https://doi.org/10.1007/s11192-020-03722-z

Stolzenberg, E. B. et al., (2019). *Undergraduate Teaching Faculty: The HERI Faculty Survey 2016–2017*. University of California, Los Angeles, USA. https://vtechworks.lib.vt.edu/server/api/core/bitstreams/2a6b09c5-7ac8-4690-9648-e26e288f7553/content

Yusoff, M. et al., (2015): Dimensions Driving Business Student Satisfaction in Higher Education. *In Quality Assurance in Education*, Vol. 23, No. 1. https://www.emerald.com/insight/content/doi/10.1108/QAE-08-2013-0035/full/html

# DIGITAL EDUCATION IN THE TEACHING AND LEARNING OF EXPONENTIAL AND LOGARITHMIC FUNCTIONS

Alberto Codeço, Pedro Miguel Faria and Isabel Araújo
*Instituto Politécnico de Viana do Castelo, Viana do Castelo, Portugal*

**ABSTRACT**

Exponential and logarithmic functions have always been part of high-school math's *curricula* in Portugal. They are subjects in which students experience many difficulties, in understanding and applying their properties. Several authors argue that there is a need to improve the way these functions are taught, shifting the focus from simply performing calculation procedures to understanding them. It is recognized in the literature that one of the areas of knowledge that can benefit from the use of Information and Communication Technologies is mathematics. In this context, digital educational resources (DER) have been developed, taking advantage of the functionalities of the GeoGebra dynamic geometry software relating to the study of functions. Therefore, the aim of this research is to understand how students overcome their difficulties by working collaboratively in small groups and carrying out activities with exploration tasks, supported by the use of GeoGebra, in the learning of exponential and logarithmic functions. In this article we present a study carried out in a twelfth grade class at a school in the north of Portugal. The study favored the use of a qualitative, interpretive methodology. To this end, the following data collection techniques were used: enquiry, observation and document analysis. Data analysis points out to a consistent evolution in the students' performance since the beginning of this research.

**KEYWORDS**

Exponential Function, Logarithmic Function, GeoGebra, Digital Educational Resources, Technologies

## 1. INTRODUCTION

We live in an era where transformations are increasingly evident, especially regarding climate change, infrastructure, and technological advances that contribute to scientific development in general. Education is not immune to these changes. The new Information and Communication Technologies (ICT) have transformed society itself, the way we live, communicate, and consequently, how we teach, fostering new forms of teaching and learning that complement traditional teaching. According to Unser (2017), "the use of technology in the classroom is a positive influence for students, in their academic experience and in their success" (p. 47). Although available for all areas of knowledge, mathematics was the first course to use them in education, particularly from the late 1980s (Cunha et al., 2010). Mathematics programs have long encouraged and stimulated the thoughtful use of new technologies to provide new opportunities for students to improve their learning. Given the unfortunate academic failure that continues to be associated with this course (Masola & Allevato, 2019; Serrão et al., 2021), it is acknowledged in the literature that one of the areas of knowledge that can benefit from the use of quality digital educational resources (RED) is mathematics (Araújo, 2014; Ponte & Canavarro, 1997). (Despacho n.º 702/2023, 2023), issued by the Office of the Minister of Education of the Portuguese Republic, approves the (New) Essential Learning (AE) of the curriculum component/subject of Mathematics in high-school. In its preamble, it is assumed that the development of mathematically competent individuals is a fundamental purpose of the Mathematics curriculum. One of the key ideas of these new AE is the systematic use of technology, integrating it as a level for understanding and problem-solving. In high-school, exponential and logarithmic functions are key mathematical concepts. They are also concepts in which students often have difficulties assimilating. Due to this, many researchers argue for the need to improve how these functions are taught, shifting the focus from procedural calculation to understanding (Sawalha, 2018). Therefore, researchers set out to investigate, integrated into the content of functions, the subthemes of the logarithmic and exponential functions, where traditionally students perform poorly. Aware of students' motivation for using technologies in the classroom and the importance of studying functions and the difficulties most of them present in this domain, research was conducted to identify potential benefits of the dynamic geometry environment GeoGebra in teaching these functions. In this article, the use of GeoGebra by students in a 12th-grade class is presented in the didactic sequence related to exponential and logarithmic functions that are part of the Mathematics A curriculum in high-school, describing this journey and providing reflective comments on it. And thus, understanding how students in a 12th-grade class, working collaboratively in small groups and carrying out activities with exploration tasks supported by the use of GeoGebra, overcome their difficulties in learning this type of functions.

## 2. THEORETICAL FRAMEWORK

Digital technologies have increasingly asserted themselves, especially in the last two decades, across all professional fields. Regarding this, Rapp (2017) states that "the dependence on these technologies is ever-growing and today it is impossible to imagine the future without them" (p. 1). Meirinhos and Osório (2014), warn that, although they favor the production and circulation of information, they also contribute to making the knowledge lifecycle increasingly shorter and, consequently, lead to a faster obsolescence of professional qualifications. In the field of education, the importance of ICT has grown exponentially in the last two decades, and teachers have increasingly used them in their educational institutions. It is one of the areas of human knowledge that has developed the most, standing out both for the depth of the knowledge involved and for its applicability (Cunha et al., 2010). For Coutinho and Lisbôa (2011), the challenge imposed on schools by this new society is immense. Schools must be "able to develop in students the skills to participate and interact in a highly competitive, globally interconnected world, which values being flexible, creative, capable of finding innovative solutions to tomorrow's problems" (p. 5). Thus, it is necessary to develop new skills, including digital ones, that empower teachers, while also educating students with this profile. In Portugal, there are measures aimed at promoting the implementation of technologies in classrooms, such as the Digital Transition Action Plan (PTD), dated April 21, 2020 (Resolution of the Council of Ministers No. 30/2020 of April 21). Numerous studies conducted over the past years have proven that the use of technologies, both by teachers and students, has improved the overall academic performance of the latter (Unser, 2017). However, the use of ICT in the teaching-learning process does not only bring advantages (NCTM, 2008; Pereira, 2019; Santos & Moraes, 2009; Unser, 2017). In terms of advantages, we can clearly highlight the promotion of collaborative work, increased multiculturalism, greater control and autonomy on the part of the student, increased motivation, greater creativity, greater flexibility in the space and time of learning, and greater interaction between teacher-student and student-educational resources (Rapp, 2017; Ribeiro & Vairinhos, 2020; Stošić, 2015; Trindade & Bulegon, 2017). Difficulties in learning Mathematics are evident in children from the beginning of their academic journey, and despite all the efforts of educational agents, a significant percentage of failure persists. The use of technology in mathematics influences how it is taught and learned, as students can now explore and solve more complex problems (NCTM, 2008, p. 28). Trindade and Bulegon (2017) argue that "for the learning of mathematical content, it is understood that the use of technology, specifically the computer, is a strong trend in the teaching of Mathematics and opens up space to facilitate the knowledge of the discipline in question, providing valuable work and significant and quality learning" (p. 143). The (good) use of computers and new technologies can be an indispensable aid in the teaching of mathematical contents. According to Ponte et al. (1998), technologies have a close connection with Mathematics. Dynamic geometry environments (DGE) have an increasingly prominent role in education, especially in the Mathematics discipline and, particularly, in the field of geometry. According to Ponte et al. (2003), a DGE, besides allowing the construction and manipulation of geometric objects, facilitates the exploration of conjectures and the investigation of conjectures that precede formal knowledge (p. 83). By quickly verifying their ideas or conjectures visually, students engage in the exploration and discovery of concepts through clicking and dragging a mouse. DGEs are essential in the exploration and validation of properties, proofs, reflection, and interpretation of relationships (Bairral & Barreira, 2017; Hanna & Sidoli, 2007, cited by Brunheira & Ponte, 2016; Trindade & Bulegon, 2017). GeoGebra (https://www.geogebra.org/), as a DGE, introduced in 2001, freely accessible and available on different platforms, is currently the most well-known and globally used software in this domain. It initially started being developed for high-school, but has evolved to be used in other educational cycles, from preschool to university education (Brunheira & Ponte, 2016). GeoGebra, allows for the construction of geometric figures using points, lines, line segments, polygons, conics, etc., and currently also allows for the insertion of functions and dynamically changing them. When working with GeoGebra, students are generally successful in transforming function graphs, both in algebraic and graphical representation, as its interface includes both ways of representing functions (Abar & Almeida, 2018). According to Abar and Almeida (2018), "the extent and capability of GeoGebra to integrate new mathematical functionalities place this software at the forefront of educational research and, no less important, in the research of new mathematical results". Considering these functionalities, it is intended to explore the use of this software in an educational context regarding the theme of logarithmic and exponential functions.

# 3. METHODS AND TECHNIQUES

In order to identify potential benefits of the GeoGebra Dynamic Geometry Environment (DGE) in teaching exponential and logarithmic functions, a study was conducted with the primary objective of interpreting how the GeoGebra dynamic geometry software can contribute to the teaching of exponential and logarithmic functions, and what impact the use of technology has on student learning.

## 3.1 Methodological Options and Target Audience

Considering the objective of this research, a study was chosen that involved the use of the application by students in a real context. Since the aim is to study an educational phenomenon in the environment where it occurs, a methodology predominantly qualitative in nature was selected because, according to Bogdan and Biklen (1994) and Creswell (2003), qualitative research involves obtaining descriptive data, obtained through direct contact of the researcher with the phenomenon in its natural context. Furthermore, interpretive research was chosen as it values understanding and explanation, aiming to develop and deepen knowledge of a phenomenon in a context (Amado, 2014; Bogdan & Biklen, 1994). This research was conducted in high-school education during the academic year 2022/2023 at Caldas das Taipas High School. It was carried out in a class consisting of 23 students and covered the domain of functions, specifically focusing on the content of exponential and logarithmic functions taught in the final year of high-school education.

## 3.2 Data Analysis: Techniques, Tools and Procedures

Data collection took place in a real context, during Mathematics A classes, where students engaged in exploration and investigation tasks related to exponential and logarithmic functions proposed, by the teacher/researcher. Data collection was conducted by the researcher, acting as a participant observer. Considering the qualitative nature of the methodology adopted in the research, data collection was primarily descriptive, aiming to obtain a detailed characterization of the situation under study. To achieve this, it was decided to diversify data collection methods, as having evidence from various sources allows for data triangulation, ensuring greater credibility, validity, reliability, and scientific rigor. Therefore, in this study, 3 main techniques were chosen – observation, inquiry, and document analysis – which correspond to specific data collection instruments – questionnaires, interviews, field notes, observation grids, analysis of physical and digital documents, student productions, and tests. The data presentation is primarily descriptive, relying on evidence from student responses to various instruments and diary records. Given the collected information, both qualitative and quantitative data analysis was chosen, guided by the nature of this information, as data analysis involves examining, categorizing, comparing, and testing, i.e., recombining both quantitative and qualitative evidence to address research questions (Yin, 2003). It is noteworthy that the responses to the tasks completed by students in pairs, resulting from the use of DER throughout the fieldwork, were subjected to interpretative analysis to understand the impact of ICT use on student learning, in the study of these transcendental functions.

## 3.3 Study Implementation

In the preparatory phase of the study implementation, the research study was presented to the class, and ethical imperatives were established. Subsequently, two tests were administered to the students: an initial diagnostic test (pre-test) on the eve of the empirical study implementation, in the last week of January 2023, and a second test (post-test) conducted at the end of March to assess the skills acquired by the participants throughout the study. Both tests had an equal number of items related to the study of function properties, using two main variants: interpretation of the graphical representation of a function on one hand, and the analytical study of the functions on the other. They also had the same number of construction items (with restricted and extended response) and similar levels of difficulty. They focused on the following contents: tabular representation of functions, domain and codomain, monotony, parity, inverse function, injectivity, intersection of the function graph with the coordinate axes, asymptotes to the function graph, and sketching of the graph. During the period between the two tests, considering that current curricula should support learning through diversified tasks, contexts, and resources (Silva et al., 2023), and that tasks are organizing elements of the learner's activity and value the active role of the student in learning (Ponte, 2014), 12 tasks were implemented in the classroom using the chosen digital educational resource (DER) - GeoGebra. The tasks were solved through pair work, considering that "dyads [pairs] in which there is some kind of interaction and with the same initial level are the ones that most favor the co-construction of solutions" (Ponte et al., 1998, p. 194). In this research, the proposed tasks formalized for pairs of students using GeoGebra aimed to diversify their nature, focusing mainly on exercises and exploration tasks. Among the range of tasks presented, there was a concern to vary their degree of difficulty, their degree of openness (more open or more closed), or the time required for their resolution. In eight of these tasks, the researchers constructed GeoGebra applications that guided the completion of these tasks. The tasks were applied to the students in two phases: in the first phase, tasks involving exponential functions were performed, and in the second phase, tasks on logarithmic functions were explored, following the annual planning of the discipline, defined in the disciplinary section.

At the end of this research, a questionnaire regarding the contribution of using the GeoGebra software in learning topics of exponential and logarithmic functions was administered to the participants. Conducted using the Google Forms application, the questionnaire mainly consisted of closed questions, with two open questions about the specific contribution of GeoGebra to the teaching of exponential and logarithmic functions and the appreciation of the study of exponential and logarithmic functions using ICT. The questionnaire also aimed to triangulate the data collected in the tests and the tasks developed by the students during the research.

# 4. RESULTS AND DISCUSSION

In this section, the results obtained from the various data collection instruments supporting this research are analyzed: tests, tasks performed by pairs, and a questionnaire.

## 4.1 Tests

The tests performed with the class at two different time points, aiming to assess the evolution of students' performance regarding exponential and logarithmic functions, were completed by all 23 students in the class, with only two students not taking both tests. The minimum scores remained at 3.7 points, while the maximum scores, corresponding to the same student (A16), increased from 15.4 to 19.9 points. The average of the participants' scores in the two tests is 10.8 points in the pre-test and 14.5 points in the post-test, indicating an increase of 3.7 points. From the analysis of the results obtained in the tests taken by the students, it is observed that among the 21 students who took both tests, only 2 obtained a lower score in the second test compared to the score obtained in the pre-test. This indicates that the majority of students improved their performance after the study conducted on exponential and logarithmic functions using RED. Analyzing the students' responses to the tests, it is noted that the students were able to solve, either fully or partially, tasks or exercises for which they were unable to perform at the beginning of the study, despite the structure being similar, and the post-test including a greater number of questions on exponentials and logarithms. For a better understanding of students' learning, a more detailed analysis was conducted regarding gains and losses. Concerning progress in students' learning, gross gains or losses (Table 1) were calculated for the participants who took both tests.

Table 1. Gross and relative gains and losses at the assessment moments

| | | | Gains | | Losses | |
|---|---|---|---|---|---|---|
| Student | Pre-test | Post-test | Gross | Relative (%) | Gross | Relative (%) |
| A2 | 7,6 | 8,3 | 0,7 | 5,6% | --- | --- |
| A3 | 10,7 | 17,6 | 6,9 | 74,2% | --- | --- |
| A4 | 8,7 | 11,3 | 2,6 | 23,0% | --- | --- |
| A5 | 12,9 | 12,7 | --- | --- | 0,2 | −1,6% |
| A6 | 12,8 | 16,7 | 3,9 | 54,2% | --- | --- |
| A7 | 10,6 | 10,7 | 0,1 | 1,1% | --- | --- |
| A8 | 5,5 | 14,5 | 9,0 | 62,1% | --- | --- |
| A9 | 12,9 | 19,0 | 6,1 | 85,9% | --- | --- |
| A11 | 11,4 | 17,5 | 6,1 | 70,9% | --- | --- |
| A12 | 9,4 | 7,7 | --- | --- | 1,7 | −18,1% |
| A13 | 9,5 | 12,0 | 2,5 | 23,8% | --- | --- |
| A14 | 9,2 | 10,4 | 1,2 | 11,1% | --- | --- |
| A15 | 9,8 | 17,7 | 7,9 | 77,5% | --- | --- |
| A16 | 15,4 | 19,9 | 4,5 | 97,8% | --- | --- |
| A17 | 3,7 | 3,7 | --- | --- | --- | --- |
| A18 | 10,5 | 15,2 | 4,7 | 49,5% | --- | --- |
| A19 | 15,1 | 18,4 | 3,3 | 67,3% | --- | --- |
| A20 | 12,0 | 17,7 | 5,7 | 71,3% | --- | --- |
| A21 | 9,3 | 14,2 | 4,9 | 45,8% | --- | --- |
| A22 | 14,4 | 19,4 | 5,0 | 89,3% | --- | --- |
| A23 | 16,1 | 19,6 | 3,5 | 89,7% | --- | --- |

Analyzing Table 1, it is observed that only 2 students experienced losses, and one student obtained the same score in both instances described. Eighteen students improved their scores from one test to the other. The average gross gain was 4.37 points, while the average gross loss was -0.95 points. Regarding the averages of relative gains from the pre-test to the post-test, a value of 55.6% was obtained. The average of relative losses (-9.9%) was calculated based on only 2 values, thus lacking significant statistical meaning, due to the small sample size. Therefore, It can be deduced that the class has made progress in learning exponential and logarithmic functions.

## 4.2 Tasks

The 12 tasks were applied in the regular classroom of the class, totaling 6 lessons, in the months of February and March 2023. In the first task, which addressed the number e, the responses given by the students were very satisfactory, with only a few poorly made numerical approximations. The objectives proposed for this task were achieved, and it was executed without difficulty. In task 2, the performance of the pairs was good, but some common flaws were evident: in the perception that the function $f(x) = 1^x$ is not an exponential function, in establishing the correspondence between the function's graph and its analytical expression, and in mathematical communication. In task 3, regarding the algebraic properties of the exponential function, some pairs showed difficulties in understanding negative exponent powers. In the remaining items, the participants, after understanding how the application worked, achieved the established purposes. Task 4 consisted of graphically solving exponential equations and inequalities using the used RED, and it was executed in about 40 minutes. After correcting the pairs' responses, it was observed that more gaps were detected in the graphical resolution of inequalities. Solving equations was unproblematic. Another difficulty encountered was the lack of number sense in some groups, as the software presents approximate values instead of exact values due to inherent limitations. For example, some students did not understand 1.33333 as the fraction four-thirds or 0.999999 as the representation of one. There were also significant errors in mathematical symbolism in operations with conditions and operations on sets, resulting from a deficient mobilization of previous knowledge. Regarding this task, it can be concluded that the class graphically solved exponential equations accurately, with more inaccuracies in the graphical resolution of inequalities. In task 5, regarding the limit of the function $(e^x - 1)/x$, as x approaches 0, the performance of the pairs was very good, as expected, given the simplicity of the exercise. During the resolution of the sixth task, which involved considering the properties of the logarithmic function, some difficulties were encountered regarding the study of the function's monotonicity, transformations to the function's graph, and the identification of asymptotes. The flaws identified in the items related to the transformation of the function's graph can be explained by a lack of prerequisites and a lack of connection with previously taught content. In task 7, concerning the algebraic properties of the logarithmic function, it was found that almost all participants were able to correctly infer the indicated properties. Two conclusions were drawn from the execution of the task: one) the need to reinforce the algebraic properties of logarithms in class with practical exercises; two) the properties to be applied in solving these exercises regarding the operational properties of powers were assimilated. The eighth task was hastily completed. The decision to introduce this task at the end of the class was ambitious and proved to be incorrect. Nevertheless, some pairs managed to complete the task with relative success. As a result, upon classifying the responses of the pairs, it was found that the results obtained were not satisfactory. Task 9 had a smooth development, given that the procedures were similar to those of task 4, which consisted of students graphically solving equations and inequalities with logarithms. It was confirmed during its execution that the process was already consolidated. Compared to the previous one, the tenth task was more complex. The main difficulty encountered by the pairs in completing the activity was the syntax of the commands. When opening the software, the language may be predefined as Portuguese (Portugal) or Portuguese (Brazil). Since there were different presets in the room, this situation caused confusion. For example, depending on the language, commands in the application's Algebra menu could be labeled as Interseção or Intersetar, EixoX or EixoOx, LimiteSuperior or LimiteàDireita. In reality, since GeoGebra distinguishes between Portuguese from Portugal and Portuguese from Brazil as languages, the syntax of some commands led to this confusion. Nevertheless, upon analyzing the responses provided, it was considered that they were very satisfactory, except for those of two pairs. The number of wrong answers was minimal, with some typos related to mathematical symbols and writing inaccuracies. Task 11, addressing the transformations of the exponential function's graph, consisted of five items. The first item involved (re)checking the characteristics of the initial function $f(x) = e^x$ and posed no difficulty to the work groups. In the subsequent items, the degree of complexity gradually increased, with the composition of an increasing number of geometric transformations according to the stated functions: $g(x) = e^x - 1$, $h(x) = -e^x + 3$ e $i(x) = 2e^{x+2} + 3$. The groups' performance was good, consistent with the guidelines provided by the NCTM (2008), which advocates for the investigation of functions and observation of changes in their graphs as the parameter values vary, enabling a better understanding of transformations and alterations to their coordinates. The students carried out the activity with the commitment and concentration required by the task's complexity. Regarding the study of the function concerning its domain, codomain, coordinates of intersection points of its graph with the coordinate axes, injectivity, monotonicity, sign, asymptotes, and extremes, the performance of the pairs was very satisfactory. However, in the domain of mathematical communication, gaps were evident in all pairs. Despite efforts to improve mathematical communication, the use of rigorous mathematical language is still not a consolidated topic. The last task, on geometric transformations of the graph of the natural logarithm function, was not completed by five groups during the lesson. The evaluation of this activity (task 12) was conditioned by the fact that it was carried out in the same lesson as the previous task (task 11), in which the students took longer than expected, somewhat jeopardizing this last task. What's more, given that the tasks were very similar, the students showed little commitment and collaboration between the members of the pairs diminished.

In view of what has been said in this subsection, it can be concluded that the work done by the pairs in solving the tasks was quite satisfactory, further proving that the dyads that interact are the ones that show the best performance, as advocated by Ponte et al. (1998).

## 4.3 Questionnaire

At the end of the fieldwork, 19 students anonymously responded to the opinion questionnaire regarding the use of RED, in this case, GeoGebra, in learning exponential and logarithmic functions. The majority of students (84.2%) enjoyed using GeoGebra, both in class and at home. According to the positively responding students, GeoGebra is "a good program", "quite interactive", and "a better way to learn mathematics". Other justifications noted include: "it is a very useful tool in solving exercises involving functions", "it allows visualizing the graphical representations of functions, which facilitate their study, such as intersections or zeros. It is a practical application and easy to use", "it is a very practical tool, and we can observe graphs while understanding the material", and "I like using GeoGebra because it helps students understand the content". Regarding negative responses (15.8%), it was mentioned: "I don't find it an easy application to use, and often some functions do not work as expected. In addition to various problems with the lack of internet during classes that prevent the use of GeoGebra" and "I don't like the method because I believe I don't have a better understanding of the subject". Regarding assessing how useful the participants considered the tool, on a 5-point Likert scale, with anchors verbalized as 1 and 5, respectively, "Not useful at all" and "Very useful", it is noted that, after the effective use of the tool throughout the empirical study, the majority of participants recognized it as useful or very useful (89%). No student indicated it as not useful at all. Regarding the tasks developed in class with the RED used, when questioned about the influence they had on learning these functions, 95% of respondents agreed that these tasks helped them understand them and helped improve their ability to work in GeoGebra; 74% agreed that these tasks encouraged their interest in studying these functions, while only 42% recognized the requirement of mathematical prerequisites for their understanding. There is absolute agreement that the tasks proposed by the teacher provided a good introduction to the topics taught, while 95% recognized that GeoGebra allowed confirming results obtained analytically in solving exercises with paper and pencil. The reverse path, i.e., confirming analytically the solution obtained using the software, is only done by 58% of the respondents. 58% of the respondents disagreed that they had difficulty in solving the tasks requested. Finally, assessed on a scale of 1 to 10, the contribution of GeoGebra to clarify the difficulties experienced in learning exponential and logarithmic functions, it is noted that only 2 students (11%) indicated a value below 5 (value 4). For almost all respondents, it positively contributed to addressing their gaps. The majority of students indicated grades of 8, 9 or 10, which allows us to conclude that they recognize that GeoGebra contributed significantly to help them to overcome difficulties in learning these functions. When asked about the specific contribution of GeoGebra to the teaching of exponential and logarithmic functions, respondents highlight the ease with which it draws function graphs and manipulates transformations to the function's graph, as emphasized by authors like Rosa (2018). In their assessment of studying logarithmic and exponential functions using ICT, it is observed that the majority of participants enjoyed studying them. When compared to other functions already taught, it was found that one-third of the respondents consider them more difficult to understand, which is in line with some studies (Kenney & Kastberg, 2013; Sawalha, 2018).

## 5. CONCLUSION

This article provides a brief contextualization of the contribution of the GeoGebra Dynamic Geometry Environment (DGE) to the learning of exponential and logarithmic functions, aiming to identify the underlying students' perspective. It can be considered that GeoGebra contributed to the integration of algebraic and graphical representations of exponential and logarithmic functions in a natural and intuitive way. Particularly, it is concluded that students became more aware of what happens to the graph of a function when it undergoes a transformation (or composition of various transformations) in its analytical expression. Regarding other characteristics, such as points related to extremes and intersection with coordinate axes, domain, codomain, monotony, and asymptotes, it is also concluded that they were able to establish a correct relationship between the characteristics of the original graph of the functions and those of its transformed form. The analyzed data also suggest that participants valued the assistance of tasks in understanding the algebraic properties of functions and corresponding characteristics associated with their graphs. The use of technology in learning exponential and logarithmic functions had an impact on increasing students' interest in this content, especially because it promoted more active learning, which was evident in the participants' involvement in task completion, as well as in the results obtained in the post-test. The conclusion to be drawn is that it is essential to approach with naturalness a greater use of digital educational resources in the classroom. It is hoped that this work will spur new research in mathematics education, particularly in high-school and at a time when new curriculum guidelines are set to take effect. In the future, active learning integrating technology will be increasingly sought after, constituting enriching learning practices. Upon hearing the students' opinions, they expressed a liking for using tasks using GeoGebra, showing interest in applying it to other content. Therefore, given their appetite to perform them, it would be interesting to diversify the thematic areas in which a study of this type could be applied, in the first two years of high-school.

# REFERENCES


Abar, C., & Almeida, M. (2018). GeoGebra como organizador de recursos tecnológicos para o ensino e aprendizagem da matemática em uma formação de professores. *Ensino da Matemática em Debate*, *4*(2), 136-144.

Amado, J. (2014). *Manual de investigação qualitativa em educação* (2.ª ed.). Imprensa da Universidade de Coimbra.

Araújo, I. (2014). *Aprendizagem matemática no ensino superior: a influência da plataforma M@T-educar com sucesso* [Tese de doutoramento, Universidade de Aveiro]. Repositório Institucional da Universidade de Aveiro.

Bairral, M. A., & Barreira, J. C. F. (2017). Algumas particularidades de ambientes de geometria dinâmica na educação geométrica. Revista do Instituto GeoGebra Internacional de São Paulo, 6(2), 46-64.

Bogdan, R., & Biklen, S. (1994). *Investigação qualitativa em educação: uma introdução à teoria e aos métodos*. Porto Ed.

Brunheira, L., & Ponte, J. P. (2016, Novembro, 19-20). *Realizar construções geométricas com o GeoGebra: O contributo do AGD para a estruturação geométrica*. Encontro de Investigação em Educação Matemática, Universidade de Évora.

Coutinho, C. P., & Lisbôa, E. (2011). Sociedade da informação, do conhecimento e da aprendizagem: Desafios para educação no século XXI. *Revista de Educação*, *17*(1), 5-22.

Creswell, J. W. (2003). *Research design: Qualitative, quantitative, and mixed methods approaches* (2$^{nd}$ ed.). Sage Pub.

Cunha, B., Duarte, E., & Martins, J. (2010). *A matemática com as TIC no processo de ensino-aprendizagem: construção de uma unidade didática* [Pós-Graduação em TIC em Contextos de Aprendizagem]. Repositório Científico da Escola Superior de Educação de Paula Frassinetti.

Despacho normativo n.º 702/2023 do Ministério da Educação (2023). Aprendizagens essenciais de Matemática do ensino secundário. Diário da República n.º 10, Série II de 13-01-2023.

Kenney, R., & Kastberg, S. (2013). Links in learning logarithms. *Australian Senior Mathe-matics Journal*, *27*(1), 12-20.

Masola, W., & Allevato, N. (2019, Janeiro). Dificuldades de aprendizagem matemática: algumas reflexões. *Educação Matemática Debate*, *3*(7), 52-67.

Meirinhos, M., & Osório, A. J. (2014). A colaboração em ambientes virtuais: aprender e formar no século XXI. Associação ArcaComum.

National Council of Teachers of Mathematics (NCTM) (2008). *Princípios e Normas para a Matemática Escolar* (2.ª ed.). Associação de Professores de Matemática.

Pereira, G. M. (2019). O impacto das TIC: no desenvolvimento e aprendizagem das crianças e no papel da escola [Dissertação de mestrado, Faculdade de Letras da Universidade do Porto]. Repositório Aberto Universidade do Porto.

Ponte, J. P. (2014). Tarefas no ensino e na aprendizagem da Matemática. In J. P. Ponte (Org.), *Práticas profissionais dos professores de Matemática* (pp. 13-27). Instituto de Educação da Universidade de Lisboa.

Ponte, J. P., Brocardo, J., & Oliveira, H. (2003). *Investigações matemáticas na sala de aula*. Autêntica Editora.

Ponte, J. P., & Canavarro, A. P. (1997). *Matemática e novas tecnologias*. Universidade Aberta.

Ponte, J. P., Matos, J. M., & Abrantes, P. (1998). *Investigação em educação matemática: implicações curriculares*. Instituto de Inovação Educacional.

Rapp, M. L. (2017). *Integração das TIC nos processos de ensino e aprendizagem pelos professores do 1.o e 2.o Ciclo de uma escola portuguesa*. Faculdade de Ciências e Tecnologias da Universidade Nova de Lisboa.

Ribeiro, S., & Vairinhos, M. (2020). Realidade Aumentada no Ensino de Matemática.

Rosa, M. (2018). *Transformações geométricas dos gráficos de funções: um estudo no 10.º ano, com recurso à tecnologia* [Dissertação de mestrado, Instituto de Educação da Universidade de Lisboa]. Repositório da Universidade de Lisboa.

Santos, C. A., & Moraes, D. R. (2009). *Tecnologia educacional no contexto escolar: contradições, desafios e possibilidades*.

Sawalha, Y. (2018). *The effects of teaching exponential functions using authentic problem solving on students' achievement and attitude*. [Tese de Doutoramento, Wayne State University]. Wayne State University Dissertations.

Serrão, A., Simões, P., & Pires, R. (2021). *Estudo de Aferição Amostral do Ensino Básico 2021, Volume I – Resultados nacionais*. Instituto de Avaliação Educativa, IP.

Silva, J. C. (Coord.), Rodrigues, A., Domingos, A., Albuquerque, C., Cruchinho, C., Martins, H., Almiro, J., Gabriel, L., Martins, M. E., Santos, M. T., Filipe, N., Correia, P., Espadeiro, R. G., & Carreira, S. (2023). *Aprendizagens Essenciais de Matemática no Ensino Secundário*. ME-DGE.

Stošić, L. (2015). The importance of educational technology in teaching. *International Journal of Cognitive Research in Science, Engineering and Education*, 3(*1*), 111-114.

Trindade, L. P., & Bulegon, A. M. (2017). GeoGebra: recurso tecnológico no ensino da matemática. In L. Tarouco, & C. Abreu (Orgs.), *Mídias na Educação: a pedagogia e a tecnologia subjacentes* (pp. 139-159). Editora Evangraf.

Unser, C. E. (2017). A Study on The Positives and Negatives of Using Technology In The Classroom. *Undergraduate Honors College Theses 2016-18*.

# DETECTING BLENDED LEARNING, COLLABORATIVE LEARNING AND INTERNATIONALISATION IN MILITARY EDUCATION: A SYSTEMATIC LITERATURE REVIEW

Alice Barana[1], Giulia Boetti[1], Antonino Cambria[1,2], Marina Marchisio Conte[1] and Enrico Spinello[3]
*[1]Department of Molecular Biotechnology and Health Sciences, University of Turin, Piazza Nizza 44, 10153 Torino, Italy*
*[2]High Defence Study Centre, Piazza della Rovere, 83, 00165 Roma, Italy*
*[3]Education and Training Command and School of Applied Military Studies, Via Arsenale 22, 10121 Torino, Italy*

**ABSTRACT**

Since the 1990s, the topic of Military Education has begun to receive greater attention from researchers following the transition of the armed forces into a configuration as a professional entity. This transition required specific training that adhered to national standards and, over time, included both European and international ones. Moreover, interest has also been directed towards teaching methodologies to improve the officer education process, considering technological evolution and openness to the foreign as essential factors within this process. Therefore, alongside the theme of military education, thematic areas that seek to improve and innovate military education have begun to coalesce. These areas include Blended learning, Collaborative learning, and Internationalisation. The purpose of the paper is to understand the specific research state of the art in the world of reference literature on the topic. This research, through a systematic literature review, aims to investigate the combination of military education with forms of blended learning, collaborative learning, and internationalisation. The paper intends to be a point of reference for future research and carry the reins of innovation in the field of military education.

**KEYWORDS**

Blended Learning, Collaborative Learning, Internationalisation, Literature Review, Military Education

## 1. INTRODUCTION

In a constantly evolving world where technology makes giant strides and becomes increasingly more efficient, the learning and teaching ecosystem is influenced by a wide range of factors. These factors are the reflection of all those situations that require change and adaptation to the future, caused by events such as the COVID-19 pandemic and the outbreak of minor conflicts around the planet.

Especially if we delve into the analysis of a particular environment, such as that of military education in a VUCA (Volatility, Uncertainty, Complexity, and Ambiguity) world, we can observe that the need for innovative teaching and digital methodologies is increasingly becoming urgent and predominant (Cambria et al., 2023a; and Ratiu, et al., 2023). We are on the threshold of a period in which states must limit their budgets dedicated to the armed forces, where training has a cost both in terms of purely economic capital and in terms of trained human capital, where time becomes more a bargaining chip and removing an officer from the field to dedicate him to training becomes difficult (Spinello et al., 2019, Spinello et al., 2020). Therefore, a major issue is understanding what is the trade-off that needs to be implemented in order to connect people even in an asynchronous way almost in any place and at any time. For this reason, in the field of military education, teaching methodologies based on concepts of blended learning (BL), collaborative learning (CL), and internationalisation (I) brings a breath of fresh air that meets the needs of armed forces throughout the world.

This vision is shared at the national level but also in the Europe a landscape; step by step, all states have begun to study and conceive training models, which initially lead to a standardization of training and education elements, and finally converged on necessity for a process of harmonization. In the panorama of Education and Research in Europe, the number of various projects and theoretical frameworks related to the field of Defence and Security and Military Education has increased exponentially.

The purpose of this paper is to carry out an in-depth literature review to understand the existing framework/materials in the literature on the topic of military education and related to blended learning, internationalisation, and collaborative learning. At the SUISS (Interdepartmental University School in Strategic Sciences) of the University of Turin (Italy), innovative approaches based on blended learning, collaborative learning, especially between civilians and military personnel, and internationalisation are being experimented with. The research group wants to track other studies or experiences dealing with these elements, with the aim to introduce in the academic research landscape the seeds of investigative purpose, as well as the outcomes of innovative research results, in the field of military education.

The following section reports the theoretical framework on the topics, consolidating the most important terms of the research and defining and describing them for subsequent use in the practical research process. The methodology section poses the research question and explains the literature review process step by step, trying not to overlook possible methodological gaps and eventually explicitly setting out the research question. In the fourth part, the results section, a matrix was created as a useful way to analyse data and create graphs that are later discussed in the fifth part. Lastly, conclusions are drawn, answering the research question and highlighting possible future research programmes using this paper as a stepping stone for the next research.

# 2. THEORETICAL FRAMEWORK

## 2.1 Military Education

The professionalism of the commanding personnel starts from the traditionally classic modernist view on education, which promotes an environment of learning that embraces uniformity and enhances scenario-based pre-planned drills as ways of conducting military operations. However, an alternative perspective, one that embraces difference rather than uniformity as a means of developing military units and their soldiers, can help in finding new efficient ways of learning and teaching in the field of military education (Sookermany, 2017).

Many factors and variables must be considered in the system. Most European Union Member States (EUMS) and their officers' basic education and training institutions are engaged in enhancing exchanges between themselves or with civilian counterparts. Therefore, in an environment where there is a need for cooperation to solve problems, Education and Training in the field of Security and Defence aims to prepare military personnel (officers) and highly specialised civilians able to cooperate in national and international environments with colleagues of other nations. Thus, efforts for harmonising the systems to make them "compatible" for exchanges and efforts for facilitating mobility is needed (Paile-Calvo, 2016).

General Dempsey identifies the development of critical thinking as the key ingredient to the future of Military Education (ME). Indeed, he reported that in order to express the majesty of mission and command, the joint education shall instil the cognitive capability to understand, receive, and clearly express intent, to take decisive initiative within intent, accept prudent risk, and build trust within the force (Murray, 2016). In accordance with the General, the ME in a professional environment is devoted to develop the "habits of mind essential to our profession based principally on critical thinking." Taking into consideration the General words (Murray, 2016) in more detail, making explicit reference to the education system, it is possible to say that: "*Joint and service doctrine, education, and training are keys to achieving the habit of mission command. [...] schools must teach it, and must train individually and collectively to it. the education of our officer corps—joint and service—must begin at the start of service to instil the cognitive capability to understand, to receive and express intent, to take decisive initiative within intent, and to trust.*"

Military officers' basic education, indeed, is generally made up of two aspects: "academic education", which will also be referred to as "education," and "vocational training." The academic aspect derives from the definition of an academic path defined as any curriculum leading to an undergraduate or post-graduate degree equivalent to a degree conferred by the civilian higher education system. The academic topics are most commonly research subjects (Paile, 2011).

Therefore, from this point of view, the research process with ME will designate those aspects of learning and education which shape the path of military officers or civilian officers who are attending a bachelor's or master's degree in the field of defence and security in a nearby Military Institution. The education and training courses require a solid and complete theoretical multidisciplinary preparation and participation in practical activities, obtained through specific internships or specific modules. Furthermore, in Europe a widespread concept focuses on the aim of institution to enable them to learn how to work in collaboration with other colleagues coming from other Member States, and to develop a European culture for safety and Defence in a constantly changing international scenario (Marchisio et al., 2017).

Looking at the Italian situation, since 1998 the IT-Army Education and Training Command and School of Military Applied Studies (COMFOR) in Turin and the University of Turin (UniTO) at the SUISS have developed a Bachelor's Degree and a Master's Degree in Strategic Sciences for Young Military Officers and Civilian Students in order to prepare professionals able to operate and take part in operations and missions in a complex national and international context. Since the beginning of 2000, the Italian Armed Forces have been experimenting with new solutions for training purposes, and the e-learning system has been offering a new opportunity for innovative methodologies. In the last decade, these solutions have been developed in an international environment with the creation of an international common module (Marchisio et al., 2022a; Marchisio et al., 2022b; Marchisio et al., 2023). Other common modules took place in the following years and evolved into Blended Intensive Programmes (BIPs).

In 2012 the IT Army Staff invited the Education and Training Command to join the EMILYO program in accordance with the rules established by the Implementation Group and the first Erasmus program was launched (2013-2014) (Marchisio and Spinello, 2021). The first Common Module on Common Security and Defence Policy (CSDP) was organised in December 2013. The first BIPs were launched in 2023 (Cambria et al., 2023b). In 2024, BIPs were developed and improved. These programmes were prepared and executed to merge the paramount elements of blended learning, collaborative learning, and internationalisation within the condensed environment of Military Education.

## 2.2 Blended Learning

In general, taking into consideration the handbook of blended learning for high-quality and inclusive primary and secondary education, promoted and published by the European Commission (EU, 2021), blended learning in formal education happens when a school, educator, or learner takes more than one approach to the learning process: blending school sites and distance learning environments; blending different learning tools that can be digital (including online) and non-digital. This new approach to the learning and teaching is adopted not only in the primary and secondary education but also in the post-secondary education.

Through blended learning in multiple contexts, students experience the possibility to be independent and autonomous in their learning and study at their own pace. Blended learning classes include the integration between the face-to-face session and the online learning phase, which is assumed to involve students in active learning through diverse learning approaches (Cambria et al., 2023c; Namyssova et al., 2019). In these phases, there is always active peer communication, information processing, checking to understand, and knowledge organisation. Teachers must also pay attention to the motivation and engagement of the students as part of their professional development because their roles as instructors and mediators in the interaction with peers make them important elements of these new methodologies.

In line with this description and definition, blended learning is a flexible approach that can support a project or course of study whilst not requiring teachers and learners to be in the same physical space at all times. It can help improve the inclusiveness of education, particularly due to its flexibility, if resources and school organization allow. A blended approach may also be described along a spectrum of less-to-more time spent learning at a distance compared to on the school site (EU, 2021). In the Italian context, some BIPs were delivered at the University of Turin and the Army Education and Training Command and School of Applied Military Studies. In the academic years 2022-23 and 2023-24 BIPs were created in different fields: Common Security and Defence Policy; Biosafety and Bioterrorism; Problem Solving and Critical Thinking; Law of Armed Conflicts. During these BIPs, a digital learning environment was used to deliver theoretical lessons and create a digital environment where participants could meet and put the knowledge gained for practical activities into practice.

## 2.3 Collaborative Learning

Collaborative learning is a powerful classroom tool that can be used in different contexts, from classroom games and project work to team-building activities. Working collaboratively helps students form more caring, supportive, and committed relationships with one another. It also gives them the chance to practice their oral communication and leadership skills, as well as develop critical thinking, problem-solving, and creativity. A comprehensive definition was provided by Laal and Laal (2012), who defined Collaborative Learning as follows: "*CL is an educational approach to teaching and learning that involves groups of learners working together to solve a problem, complete a task, or create a product. In the CL environment, the learners are challenged both socially and emotionally.*"

More or less during the same time, in their work, Wilson and Wilson (2019) after Johnson, et al. (2007), advocated a theory of social interdependence, where three ways in which students can learn in a context of reciprocal relationships were proposed: they can work together cooperatively to achieve their learning goals; they can compete against each other; or they can work individually.

Collaboration is an important indicator in the development of officer and civilian actors' professional paths. Their roles will be complementary and useful for both parties one day. Therefore, there is the need to experiment in safe places, namely laboratories, where students start to interact and share their insights into the process. In this way it will be possible to leave the standard references and produce outstanding outcomes. Education cooperation and the knowledge exchange foster a wide range of aspects in the learning process: (Cambria et. al, 2023b) to increase global competitiveness; to help tackle global challenges; to transform higher education through capabilities and capacity building; to nurture equality.

## 2.4 Internationalisation

In his work, Hans de Wit (2015) promoted a new agenda for internationalisation for the future, by extending the definition of Knight (2008), defining internationalisation as follows: "*The intentional process of integrating an international, intercultural or global dimension into the purpose, functions and delivery of post-secondary education, in order to enhance the quality of education and research for all students and staff and to make a meaningful contribution to society*" (De Wit et al., 2015).

As far as the first configuration of the above-mentioned concept is concerned, among all the strategies provided, there was the pillar that seeks to measure the increasing number of exchanges, the mutual recognition, the use of Erasmus mobility, the opening of national educational opportunities to young European officers, and the learning and teaching about Europe and its defence, such as the creation of common modules as well as the promotion of the learning of several foreign languages (EMILYO, 2015; Gell, 2015).

The mobility of young military students, their teachers, and their trainers, exchanging knowledge and know-how within the Union, pave the way for greater integration and coherence of our future actions in favour of peace and security (Gell, 2017). Internationalisation is an answer to the current context and scenarios that EUMS and strategic alliances such as NATO are trying to give to their contributors. International training is adopted not only in terms of mobility - both outgoing and incoming - but also to create more international curricula. The aim is to develop officers' ability to work in any international operating theatre with colleagues from all over the world, to fit into different cultural contexts, and to create collaborative networks Marchisio and Spinello 2021; Cambria et al. 2023b).

## 3. METHODOLOGY

In order to produce this paper a timeline dedicated for the systematic review was put in action. The first phase was to develop a research question. After that there was the phase to develop the protocol, which was meant to be use for the codification of the work. Afterwards, a full search was carried out, and a title/abstract screening brings to the end. Once the matrix was coordinated a full text retrieval and screening was conducted. The outcome was a critical appraisal, data extraction, and synthesis process that summed up in the writing and submission of the paper for publication. The research question that this paper aims to answer is:

**RQ: what is the state of the art in the literature on the use of innovative digital methodologies for officer training based on blended learning, collaborative learning, and internationalisation?**

To answer the RQ, a systematic literature review was adopted. In this process 5 steps were identified and adopted in consideration of the PRISMA guidelines, which consist of a four-phase flow diagram. The flow diagram describes the identification, screening, eligibility and inclusion criteria of the reports that fall under the scope of a review. The five steps were Framing questions for a review, identifying relevant work, assessing the quality of studies, summarizing the evidence, and interpreting the findings.

The platform identified was limited to Scopus since it is the reference platform of the international academic world based on bibliometric indicators. In the first step, keywords were identified for the research, and then, how the keywords should be arranged during the research was defined. The keywords chosen were Military Education (ME), Blended Learning (BL), Internationalisation (I), and Collaborative Learning (CL). The first research was done only with the keyword "military education" in the "Abstract, Title, and Keywords" research field of the Scopus database. Then, the search was repeated, adding all the other keywords (BL, CL, and I) to ME, linking them through the operator "AND", and then removing one of the keywords and using the same query structure in the research field. Once the results were obtained for all pairs of queries that could be formed, another keyword was excluded, thus leaving a simple pair. To filter only the results relevant to this study, some topics were excluded from the research (for example, medicine, nursery, biology, chemistry, religion). Once the result was obtained, to identify relevant work, the abstracts of the returned articles were read carefully, and the items that were not relevant to the topic of Defence and Security or related academic areas and study fields were excluded. The remaining papers were searched for on the net, read, and analysed. All the collected documents were consolidated into an Excel matrix and analysed to assess the quality of the studies. Other papers included in the bibliography of those chosen for their relevance were included in the Excel matrix and analysed. In the matrix, "yes" or "no" items were included to indicate which of the keywords (ME, BL, I, CL) were deepened by the paper. The evidence was then summarized and interpreted in light of the theoretical framework.

## 4. RESULTS

From the results obtained using only ME as a keyword, it is possible to say that the landscape of publications related to the world of ME is quite broad, with a bullish trend after the 1990s. We can assume that the period corresponds to when the army became a profession rather than an obligation. There was a need for innovation and research to cope with these new realities and understand the military world and its education for creating professional military personnel. We found more than 8000 articles, subsequently reduced to 5972 with the exclusion of the areas of non-interest. The research was then repeated, adding all the other keywords linked with "AND" to ME. For the categories of (ME+BL+I+CL), (ME+BL+I) and (ME+BL+CL), no documents matching the keywords were found. For (ME+I+CL), the results are 2. As far as ME+BL is concerned, following the exclusion operation, 24 results were found, including those published by our research group. For ME+I, 30 results were considered after the exclusion operation. Lastly, for ME+CL, with the following exclusions, it reached 153, which were exported and analysed for relevance. The result found were more than 200. Over these 200 papers, a selection reading the titles, the abstracts, and the scientific sector of reference, only 30 were chosen for consideration in the research process. The reason for this cut is based on the relevance to the topic of the research that was depicted in the papers' abstract. The selection of relevant papers was followed by the process of research on the web to find papers and go through the analysis process. The results were merged and the duplicates removed, as well as other papers that were deemed not relevant for the purpose of this literature review were eliminated by the research; 4 of them were not available on the web and thus they were excluded. In the end, 26 papers were added to the matrix where other papers were present because they were regarded as touching the topic but without any reference in Scopus. The matrix of 54 papers in total, adequately dealt with the topic of ME combined with BL, I, and CL were analysed. From there we managed to create a graph, Figure 1 shows how many papers in the field of military education dealt with BL, CL and I. The result of the intersection of the three areas of study with the overlap of the theme of military education led to the finding of 8 articles. Then, by excluding one of the keywords, we managed to fill the intersections of only two study areas, always overlapping ME, finding 13 for ME+BL+CL, 6 for ME+CL+I, and 7 for ME+BL+I. In turn, the same reasoning was used to find those papers that always dealt with ME as a background and only one of the aspects, thus finding the following results: 13 ME+I, 5 for ME+CL, and 2 for ME+I. Figure 1 shows the result of the analysis. After reading all the relevant papers, we selected the most interesting examples which show innovative solutions in military education, including one or more of the considered aspects. They are listed in Table 1.

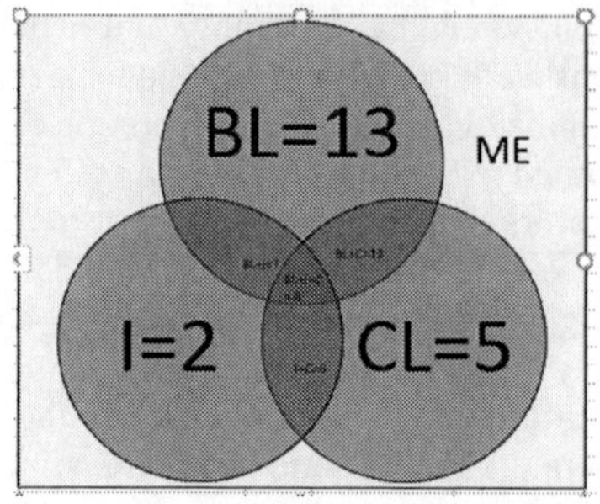


Figure 1. Combined results in their intersections

Table 1. Summary of the results

| Keywords included | Number of items | Selected papers |
|---|---|---|
| ME+BL+CL+I | 8 | P1. DIGICODE experience at SUISS (Cambria et al. 2023b)<br>P2. Artic Simulation (Last, D., et al., 2021) |
| ME+BL+CL | 13 | P3. Experience at the Montenegro Academy (Bauk,S., and Radlinger, R., 2013) |
| ME+CL+I | 6 | P4. Experience of Intermobility and cooperation in the Brazilian academy (Guimarães-Honorato H. 2020) |
| ME+BL+I | 7 | P5. English learning in order to save cost and operative people (Almeida et al., 2020) |
| ME+BL | 13 | P6. New approach and infrastructure for delivering various forms of e-learning in the Slovenian Army (Scanvikar D., 2014) |
| ME+CL | 5 | P7. Professional development of Chinese soldiers (Liu, et. al, 2023) |
| ME+I | 2 | P8. Officer education policy development in the context of the changing European security and defence identity (Melnikovas A.,2019) |

## 5. DISCUSSION AND CONCLUSIONS

One first element of analysis that emerges from the results is that the studies including innovative learning approaches for ME such as BL, CL and I are few in consideration with the total number of studies dealing with ME, and therefore, there is a need for research in these areas. We suppose that there are probably many experiences, but few are analysed at a research level, thus highlighting a gap in the research.

As far as a temporal analysis of the results is concerned, from the 54 results, it emerges that the range of years in which papers related to the topics were published goes from 2003 to 2024. The mode of the most popular publication year is 2019. Moreover, 18 papers were published after 2019 and 27 before. Restricting the analysis to the 30 results found in Scopus, the mode resulted in the year 2023. Of these, just 5 were published before 2017. The highest point of publication of the research is between 2017 and 2023. Indeed, the mode, in general, such as the median is the year 2019. Regarding a locational analysis of the results, papers reported experiences from all continents. The majority were published in the United States and in the EU.

We selected two papers from the intersection of all the keywords, that is, experiences of ME where BL, CL and I are considered. (Last et al., 2021) discusses the artic simulation, a laboratory of exploratory learning combining collaboration, internationalisation, and blended learning. Students must attend the first part at their respective university, where they learn the structure and theoretical details that should be put into practice in the second part. The second paper aims to present the pilot project and its effectiveness. (Cambria et al. 2023b) reports the results of an investigation aimed at understanding the perceptions, strengths, weaknesses, opportunities, and problems of blended international modules based on collaborative learning. The other selected papers report experiences depicted in the military academies of other countries. One common element is the expressed need for an international experience to learn languages and soft and cross-cultural skills that students will need in their professional lives. Considering the time required, the best time for nurturing them is during education, adding elements of BL, CL, and I to the traditional learning experience.

What emerges from the analysis is that the combination of Blended Learning, Collaborative Learning, and Internationalization in military education is an innovative approach that addresses the current needs for training and professionalism in the sector of Defence and Security. It is crucial to evaluate how innovative training techniques can help develop the mindset, availability, and adaptability of officers for the missions they must undertake. These elements are encoded in a set of challenges that should be addressed when planning and considering new military education pathways, such as the technological and infrastructure challenges, which can be faced with the use of digital environments; challenges of international management and coordination, which lead to the creation of mixed working groups with various expertise and composition; challenges also for specialised teachers training in innovative teaching techniques, considering the constants of security and privacy in international environments. From the analysis of the papers, it is possible to state that the three components, in addition to military education, are the countermeasures for the problems that nowadays, the sector of Defence and Security are coping with. They could help make Military education efficient, effective and proactive against the problems of lack of time and lack of budgets. If there is a need to provide transversal competencies and soft, cultural, and critical thinking skills, the solution is to merge sectors in a transversal way.

It is important to involve military students in blended learning, collaborative learning and internationalisation activities with a view to the future creation of highly professional personnel. The type of personnel that know how to interact and are comfortable with national and international colleagues, manage their operational career with that training in any place and at any time, but above all, satisfy the needs of the armed forces with cost reduction and specialised training of personnel. These three ingredients are an ideal combination for an innovative education methodology that matches the needs of the armed forces.

These new inputs must be experimented with, collected, elaborated and divulgated among the scientific communities because this field of research is still young, fertile and ready to be deepened. The items found in this research can be used as a good stepping stone for future research about edginess methodologies in the field of military education. The European frameworks (EU, 2016; Gell 2017) say that it is important to include blended, collaborative, and internationalisation experiences, but we have not found a huge number of examples, so with this work, we want to set a starting point and we have highlighted a new potential field of innovation and exponential growth. The work, in a long-term view, is to stress the importance of developing the components to become pillars in the world of military education. The future directions that the research group wants to follow are to expand the pool of items by searching on specialised journals and Google Scholars as well as Scopus, in order to reach those experiences that perhaps meet lower standards at the research level but that can may a starting point for what it is supposed to experiment.

## ACKNOWLEDGEMENT

We acknowledge financial support under the National Recovery and Resilience Plan (NRRP), D.M. 118/2023 by the Italian Ministry of University and Research (MUR), funded by the European Union – NextGenerationEU.

## REFERENCES

Almeida V. et al., (2020). Online Platform to Teach Aviation English at a Military School in Salinas. *In Developments and Advances in Defense and Security: Proceedings of MICRADS 2020.* Springer Singapore, pp. 237-249.

Bauk, S., and Radlinger, R., (2013). Concerning Web-based e-learning at a Maritime Higher Education Institution: Case Study. *Transactions on Maritime Science*, Vol. 2, No. 02, pp. 115-122.

Cambria, A., et al., (2023a). Educational Innovation In Hybrid Warfare, *Proceedings of 16th International Conference Of Education, Research And Innovation (iCERI2023)*, pp. 9433-9440. https://dx.doi.org/10.21125/iceri.2023.2432

Cambria, A., et al., (2023b). Blended Intensive Programs For Fostering Collaboration And Knowledge Exchange In Security And Defence Education. *Proceedings of 16th International Conference Of Education, Research And Innovation (iCERI2023)*, pp. 9441-9448. https://dx.doi.org/10.21125/iceri.2023.2433

Cambria, A., et al., (2023c). Higher Education training activities in international environments for developing skills of civil-military cooperation. *Proceedings of the 20th International Conference On Cognition And Exploratory Learning In The Digital Age (CELDA2023)*, pp. 402-406.

De Wit, H., et al., (2015). Internationalisation of higher education: A study for the European parliament. *Brussels: European Parliament Policy Department B: Structural and Cohesion Policies*

EU European Commission, (2016). *Learning Opportunities and Qualifications in Europe – Information about courses, work-based learning and qualifications.* Brussels. Available from: https://ec.europa.eu/ploteus/content/descriptors-page

EU European Commission, Directorate-General for Education, Youth, Sport and Culture, (2021).*Blended learning for high quality and inclusive primary and secondary education: handbook.* Publications Office of the European Union. https://data.europa.eu/doi/10.2766/237842.

Gell, H., (2015). Increase of Students' personal Development by Internationalisation. *Habilitation Process at the University of Defence Brno*, Faculty of Military Leadership Czech Republic, Brno.

Gell, H., (2017). Increase of Military Leadership Skills and Competences of future Leaders by Internationalisation, Scientific *Journal "Economics & Management",* Brno, February

Guimarães-Honorato H., (2020). Temporary migrants and students in the brazilian military institution of higher education: Experiences of human mobility. *Estudios de Derecho,* Vol. 77, No. 169, pp. 185-212.

Johnson, D. W., et al., (2007). The state of cooperative learning in postsecondary and professional settings. *Educational Psychology Review*, Vol. 19, No. 1, pp. 15–29.

Knight, J. (2008). Higher education in turmoil: The changing world of internationalisation. *Sense Publishers,* Vol. 13.

Laal, M. and Laal, M., (2012). Collaborative Learning: What is it? *SciVerse ScienceDirect Procedia*, Social and Behavioral Sciences, Vol. 31, pp. 491-495. https://doi.org/10.1016/j.sbspro.2011.12.092.

Last, D., et al., (2021). Collaborative Multinational Military Education: Arctic Simulation. *Scandinavian Journal of Military Studies*, Vol. 4, No. 1, pp.1-12.

Liu, Y., et al., (2023). On "Officer Professional Military Education Policy" and its enlightenment. *SHS Web of Conferences.* EDP Sciences, Vol. 179.

Marchisio, M., et al., (2017). Advanced e-learning for IT-Army officers through Virtual Learning Environments. *Journal of e-Learning and Knowledge Society (Je-LKS), The Italian e-Learning Association Journal, Peer Reviewed Papers – EMIMITALIA Conference 2016*, Vol. 13, No. 3, pp. 59-70. https://doi.org/10.20368/1971-8829/1382

Marchisio, M., et al., (2022a). Teachers' Digital Competences Before And During The Covid-19 Pandemic For The Improvement Of Security And Defence Higher Education. *Proceedings of the International Conference on E-Learning 2022*, pp. 68–75.

Marchisio, M., et al., (2022b). Teachers' Perception of Higher Education in a Transition Scenario. *Proceedings - 2022 IEEE 46th Annual Computers, Software, and Applications Conference, COMPSAC 2022*, pp. 139–144. https://doi.org/10.1109/COMPSAC54236.2022.00028

Marchisio, M., et al., (2023). EU Security And Defence Students' Perception And Use Of Digital Competencies In Higher Education. *17th International Conference on E-Learning and Digital Learning 2023, ELDL 2023-Held at the 17th Multi-Conference on Computer Science and Information Systems, MCCSIS 2023*, pp. 35–42.

Marchisio, M., Spinello, E., (2021). Internationalisation For Enhancing The European Security And Defence Higher Education. *Proceedings of International Conference e-Learning 2021*, pp. 99-106.

Melnikovas, A., (2019). Officer education policy development in the context of the changing european security and defence identity. *Šiuolaikinės visuomenės ugdymo veiksniai*, Vol. 4. Bulgaria.

Murray, N., (2016). The Role of Professional Military Education in Mission Command, National Defence Universtiy, *Joint Force Quartely* 72, pp.13-15.

Namyssova, G., et al., (2019). Challenges and benefits of blended learning in higher education, *International Journal of Technology in Education*, Vol. II, No. 1, pp. 22-31.

Paile-Calvo, S., (2016). From European Mobility To Military Interoperability Exchanging Young Officers, Knowledge And Know-How. *European Security and Defence College.*

Paile, S., (2011). Europe for the Future Officers, Officers for the Future Europe Compendium of the European Military Officers Basic Education. Polish Ministry of National Defence.

Ratiu, A. et al., (2023). Military actions characteristics in Hybrid Operational Environments. *International Conference knowledge -Based Organization*, Vol. XXIX, No 1, Sciendo, https://doi.org/10.2478/kbo-2023-0013.

Scanvikar D., (2014). E-learning in the slovenian armed forces - from its first step to its wider use. *Security and Defence.*

Spinello, E., et al., (2019). A Full Spectrum Lifelong e-Learning Project for the Army. *Proceedings of 15th International Scientific Conference eLearning and Software for Education (eLSE2019).* Vol. 1, pp. 152-158.

Spinello, E., et al., (2020). The E-learning project for institutional learning. *Proceedings of 16th International Scientific Conference eLearning and Software for Education (eLSE2020).* National Defence University-Carol I, pp. 174-181.

Sookermany, A.M., (2017). Military Education Reconsidered: A Postmodern Update. *Journal of Philosophy of Education*, Vol. 51, Issue 1, pp. 310–330. https://doi.org/10.1111/1467-9752.12224.

Wilson, K.F., and Wilson, K., (2019). Collaborate to graduate: what works and why? *Higher Education Research & Development*, Vol. 38, No. 7, pp. 1504-1518. https://doi.org/10.1080/07294360.2019.1660311.

# STUDENTS' PERCEPTION OF INTERNATIONAL COLLABORATION AND CULTURAL UNDERSTANDING IN DIGITALLY DRIVEN EDUCATION

Marina Marchisio Conte[1], Fabio Roman[1], Matteo Sacchet[1], Enrico Spinello[2], Anna Molnár[3], Shay Attias[4], Paul Tudorache[5], Jaroslav Kompan[6] and Andrew Dolan[7]
[1]*Department of Molecular Biotechnology and Health Sciences, University of Turin, Via Nizza 52, 10126 Torino, Italy*
[2]*Education and Training Command and School of Applied Military Studies, Via Arsenale 22, 10121 Torino, Italy*
[3]*Department of International Relations and Diplomacy, University of Public Service, Ludovika tér 2, 1083 Budapest, Hungary*
[4]*Department of Political Science and School of Communication, Bar-Ilan University, 5290002 Ramat Gan, Israel*
[5]*Nicolae Bălcescu Land Forces Academy, Strada Revoluției 3-5, 550170 Sibiu, Romania*
[6]*General Milan Rastislav Štefánik Armed Forces Academy, Demänová 393, 031 01 Liptovský Mikuláš, Slovakia*
[7]*The Centre for the Study of New Security Challenges, 20 Morningside Gardens, EH10 5LE Edinburgh, United Kingdom*

**ABSTRACT**

One of the fields in which internationalization plays a pivotal role in today's world is education, with internationally developed study programs enrolled by students from various countries. While such kinds of collaborations allow for fostering unity especially when digital learning accompanies them, we also have to take into account how some entities may be more interested in spreading dissension than supporting collaboration. Actions such as these often fall under the category of "hybrid warfare", which includes all the efforts aimed at behaving with hostility, without relying uniquely on military strength; a significant portion of these makes use of digital instruments. In this research, we examine an event part of a project specifically devoted to educating in hybrid warfare. In particular, we focused on how it was able to internationally improve collaboration and mutual understanding, with the aim of providing general guidance to different teaching and learning initiatives where e-learning plays a role. We conducted a survey among military and civilian students participating in the event, detecting appreciable effects on interaction, respect, observation, and understanding. Participants deemed these aspects as important both for setting the stage for sector-specific collaboration and their personal experience. This research is part of the activities undertaken in the European project HYBRID, whose goal is promoting awareness of all the practices related to hybrid warfare and their interpretation, also by using digital didactics and tools, including the not less important purpose of being able to defend from them.

**KEYWORDS**

Blended Learning, Digital Education, Hybrid Warfare, Online Learning, Student Training

## 1. INTRODUCTION

Internationalizing education is becoming increasingly important in a globalized world: it is nowadays quite common to develop study programs based upon international reference frameworks, and a relevant number of students choose to perform at least part of their paths abroad, possibly intending to work in a different country. However, we are still far from an ideal concept of a united humanity that has put aside divisions and contentions: on the contrary, the ongoing geopolitical situation dramatically shows that we still have a long way to go in this regard. On the one hand, international collaboration should be further enhanced as a means to counteract disunity, with e-learning playing the role of one of the most prominent strategies (Mihalova, 2006; Marchisio and Spinello, 2021). Nevertheless, on the other hand, we cannot negate the existence of bad actors that have an interest in hindering collaborative efforts, also through subtle actions aimed at dividing the counterpart while keeping a low profile. In order to face these challenges, as part of broader action realized in several Erasmus+ Key Action 2 Strategic Partnerships, the European Union funded the Hybrid Warfare (or simply HYBRID) project. This initiative, whose full name is "Interdisciplinary Education and

Training on Hybrid Warfare", aims to promote awareness of these practices and how to interpret conflictual situations and their evolutions as multifaceted phenomena, with the additional scope of effectively defending from these threats. Several nations, with their relative civilian and military institutions, are involved: Hungary (University of Public Service), Italy (University of Turin; Army Education and Training Command and School of Applied Military Studies), Romania ("Nicolae Balcescu" Land Forces Academy), Slovakia (Armed Forces Academy of General Milan Rastislav Stefanik), Israel (Bar-Ilan University), and the United Kingdom (The Centre for the Study of New Security Challenges). Note that the latter two nations are not part of the EU: their involvement highlights how the topic is of global interest and cannot be reduced to a few actors. These institutions already performed research on hybrid warfare, in various facets (Hrnčiar, 2017; Ratiu and Tudorache, 2021; Tudorache et al., 2023). In this paper, we made the most of an event that was part of the project, by considering how initiatives including novel forms of teaching and learning can foster collaboration and understanding at an international level. Such cooperation has to become a reality not only regarding the specific features concerning the reasons why the stakeholders collaborate, but also from the cultural point of view, which is equally important to guarantee a common understanding and proficient joint effort. To detect this, we administered a survey to teachers and students taking part in the event, assessing several aspects comprising (but not limited to) interaction, respect, observation, and understanding; they were polled both quantitatively, employing scales and ratings, and qualitatively, through open-ended questions. The respondents were well distributed among nationality, gender, and civilian or military affiliation, thus giving a transversal relevance to the opinions that emerged. The paper is organized as follows: Section 2 presents the theoretical framework in which we conducted this research, while Section 3 describes in detail the Hybrid Warfare project. Section 4 discusses the research question and methodology, with Section 5 presenting the results and the relative discussion. Finally, Section 6 provides some concluding remarks.

## 2. THEORETICAL FRAMEWORK

International cooperation in education has been a trending topic in recent times, since the demands to internationalize curricula are constantly increasing, acting as a catalyst towards the globalization of the perspectives relative to the various disciplines and its positive effects on the students, (de Wit, 2015). Universities are investing with valuable efforts in strategies involving internationalization, putting them at the center of their actions along with other contemporary challenges such as inclusiveness and sustainability (Caniglia et al., 2018). This implies solutions from both the technological and human points of view; regarding the former, a widely recognized solution is the use of Open Educational Resources, which can be accessed under licenses granting more rights to the users than traditionally copyrighted materials (Wiley, 2016). They allow learners to enhance collaboration and knowledge sharing, while concurrently saving time and resources (Nascimbeni et al., 2021). Concerning the human aspect, it is important to devise pathways allowing to support and activate reflection on intercultural issues, since this aspect does not necessarily emerge naturally while putting together learners from different backgrounds (Bruhn, 2017). A solution different authors and institutions took into account considers the blended paradigm (Marchisio et al. 2020; Marchisio et al. 2022), by fine-tuning it in such a way that the resulting learning experience would enhance students' intercultural competence (Lacaste et al., 2022; Vromans et al., 2023). This allows students to combine the best from online and face-to-face learning, thus overcoming the limitations in social interactions implied by a purely online modality, as depicted during the COVID-19 crisis (Galluzzi et al., 2021). Indeed, socially interacting between international peers, is recognized as having important positive effects on the learning environment, also from the emotional point of view (Huang and Lajoie, 2023), and this extends even outside formal situations (Szabó et al., 2020), including the recreational use of digital tools (Sleeman et al., 2020). The acquisition of a new cultural setting, which is a process generally occurring when a student is immersed into a new culture, should take place as well when other cultures are brought to them, in the form of an *acculturation* without a prevailing culture (Sam and Berry, 2010).

## 3. THE HYBRID WARFARE PROJECT

We describe now the project within which our experimentation took place. The main premise motivating its development is that various security challenges have nowadays the potential to undermine the security and the values that characterize our societies, requiring us to be ready to address these challenges with the appropriate means. These include of course technical tools, but also an indispensable human factor, for example in terms of raising people's awareness of these kinds of threats and how to counteract them, since these aspects are often underestimated. The concept of *hybrid warfare*, first defined around twenty years ago, refers to all the practices in which hostile action between nations or international entities is taken through not only military effort but also employing other means, such as cyberattacks on critical network infrastructure or malicious use of artificial intelligence. This can result for instance in the dissemination of fake news that can influence public opinion and destabilize communities and societies to the advantage of the attacker. Such operations usually try to exploit disunities within organizations - which are particularly common in the geopolitical setting of the 21st century since the world has become multipolar - without the presence of defined hegemonic actors in the international balance of power. In light of all this, this project aims to present a conceptual framework that can help key national stakeholders, including the government and civic society, better understand current and potential future conflicts. To achieve this goal, an academic and professional curriculum is necessary to improve situational and contextual awareness, especially regarding the potential consequences of these conflicts. The project aligns with the clear requirement of security studies institutes to gain a better understanding of the complexities associated with hybrid warfare (HW), and to initiate a consolidated familiarization with a refined appreciation of the various risks associated with HW. This cannot be separated from fostering a culture of mutual appreciation in order to promote a wider understanding and dissemination of knowledge and to support the development of common responses to HW. Indeed, to have a proficient collaboration, particularly during periods of high tension and crisis, it is pivotal to address several issues jointly, ranging from practical facts such as mechanics of force or policy posture, to more theoretical aspects like definitions and lexicon. In this spirit, the project aims to provide lecture material for civilian, police, and military higher education institutions. This allows for addressing a significant number of issues associated with the policy and operations of most forms of HW. In terms of learning methods, it is central to highlight that this objective is mainly reached by developing a Hybrid Warfare Reference Curriculum (HWRC), forming the basis for either the partial or the entire re-design and update of courses within the curriculum of military, police, and civilian students of higher education institutions. Furthermore, a series of online and blended modules based on the HWRC and proper methodologies constitutes the Hybrid Warfare Massive Open Online Course (HWMOOC). The combination of HWRC, methodology, and HWMOOC allows on the one hand for blended learning methods for accredited university classes, and on the other hand for mass online learning, thus reaching a much wider audience.

## 4. RESEARCH QUESTION AND METHODOLOGY

Inspired by the opportunities the Hybrid Warfare Project enables and based on the positive feedback obtained during the multiplier event (ME) conducted in Sibiu on the proposed conceptual model (Tudorache et al., 2023), we were interested in providing an answer to the following research question (RQ): *how is it possible to improve international collaboration and cultural understanding through teaching and learning initiatives that include e-learning and digital learning?* Aiming at this scope, we considered a Learning, Teaching, and Training Activity (LTTA) taking place in the Slovak city of Liptovský Mikuláš from 29 January to 2 February 2024. On the one hand, the event was the occasion to beta test the course design, including the HWMOOC and some simulation exercises, but on the other hand, it gave students and teachers from the various international partners the possibility to meet in person and discuss with each other. The international collaboration was promoted by various means, such as group activities where students coming from multiple countries were asked to practice on simulation tools. To evaluate the factors the RQ implies, we administered, measured, and analyzed a survey that assessed both quantitative and qualitative aspects, in the spirit of the mixed method research approach (Creswell and Plano-Clark, 2017). This survey was organized into several questionnaires and evaluation forms; in this study, we mainly considered an initial questionnaire, administered at the beginning of the LTTA, and a final one administered at its end. For the quantitative facets, we made use of a 5-level Likert scale, where 1 stands for the lowest score and 5 for the highest one. We analyzed the numerical data using both descriptive and inferential methods; specifically, where it was possible to examine the differences between the two questionnaires' responses, we conducted the paired Wilcoxon signed-rank test intending to detect whether these increments or decrements were statistically significant or not (null hypothesis: they were not; alternative hypothesis: they were), nonetheless providing proper observations also in the case they were deemed not to be significant. In the qualitative responses, we looked for opinions and impressions about students' experience, both under the usefulness and the preference points of view, by also asking them how they thought such an initiative could be further boosted. We collected the data from the students participating in the event and responding to the two questionnaires of our interest: we had 52 responses to the initial one, and 38 to the final one, with all the students answering the second questionnaire except one having previously responded to the first. They were categorized by (institutional) country as follows; note that also a French student was involved, despite France not officially being part of the project:

Table 1. Distribution of the respondent students by country

| Country | Initial questionnaire | Final questionnaire | Both questionnaires |
|---|---|---|---|
| Hungary | 9 | 6 | 6 |
| Italy | 15 | 15 | 15 |
| Romania | 9 | 7 | 7 |
| Slovakia | 8 | 6 | 5 |
| Israel | 10 | 3 | 3 |
| France | 1 | 1 | 1 |

The majority of students were military (i.e., cadets), while five of them were civilians. The latter ones were studying in a program devoted to security and defence matters, giving them premises comparable to those the former ones had. The distribution of students between genders presented a moderate prevalence of males over females, as in the setting of the project is still expected, although the quota of women involved in these topics is increasing throughout time. The United Kingdom contributed to the LTTA with two instructors, but without sending either cadets or civilians as students.

## 5. RESULTS AND DISCUSSION

Let us start with some aspects that students were asked to rate both before and after the LTTA. Namely, before the LTTA we asked them to what extent they thought these aspects would be important for the success of the learning activities, while after the event we asked them to what extent they had paid attention to the aspects during the LTTA. The first aspect concerned *interacting with others regularly, especially with members of other cultures*; Table 2 shows how ratings changed from the first moment to the second one, by considering the 37 students answering both the initial and the final questionnaire:

Table 2. Rating of regular interaction with others, especially with members of other cultures

| Interaction | Before | After | Difference |
|---|---|---|---|
| Not at all (1) | 0 | 0 | 0 |
| Very little (2) | 0 | 0 | 0 |
| Somewhat (3) | 6 | 6 | 0 |
| Generally well (4) | 15 | 9 | -6 |
| To a great extent (5) | 16 | 22 | +6 |

We see that the scores, already quite high before the LTTA, on average further increased after the event: out of their total number, 9 increased, 3 decreased, and 25 remained the same. These steps up are reflected also in the mean score, increasing from 4.27 (standard deviation: 0.73) to 4.43 (standard deviation: 0.77). By pairing the data, we obtained a mean difference of 0.16 (standard deviation: 0.55), resulting in moderate statistical significance: the paired Wilcoxon signed-rank test, approximated normally, gave a z-score of 1.53, which yielded a p-value of 0.063. This means that the LTTA allowed students to confirm the positive impression they had regarding the importance of intercultural interaction, with several of them even having a better perception once the event had ended. The second aspect regarded *expressing interest and respect for the cultures different from yours, including their history, customs, and beliefs*; Table 3 unveils the relative results:

Table 3. Rating of expressing interest and respect for different cultures

| Expression | Before | After | Difference |
|---|---|---|---|
| Not at all (1) | 0 | 0 | 0 |
| Very little (2) | 1 | 0 | -1 |
| Somewhat (3) | 2 | 4 | +2 |
| Generally well (4) | 19 | 11 | -8 |
| To a great extent (5) | 15 | 22 | +7 |

The situation is comparable to what emerged regarding interaction: the scores were generally high beforehand, but the LTTA confirmed students' feelings and fueled in some of them even greater confidence. Indeed, the average score increased from 4.30 (standard deviation: 0.70) to 4.49 (standard deviation: 0.69), with 12 increasing scores and 6 decreasing, while 19 remained unchanged. The mean of the paired differences is 0.19 (standard deviation: 0.84), and the statistical significance is again moderate, with the Wilcoxon signed-rank test giving a z-score of 1.24, which corresponds to a p-value of 0.107. Therefore, it can be said that the event permitted students to confirm and improve also how they perceived the importance of expressing interest and respect for other cultures, thus fostering a comfortable environment also under a softer point of view. The third aspect considered *spending time observing, reading about, and studying the other cultures*, as Table 4 displays.

Table 4. Rating of spending time in observations, readings, and studies about other cultures

| Time spent | Before | After | Difference |
|---|---|---|---|
| Not at all (1) | 0 | 0 | 0 |
| Very little (2) | 0 | 5 | +5 |
| Somewhat (3) | 8 | 6 | -2 |
| Generally well (4) | 21 | 11 | -10 |
| To a great extent (5) | 8 | 15 | +7 |

Here, the situation is more balanced, if compared with the two previous aspects: while the scores were already high before the event also for this aspect, they did not increment in general after the LTTA, with their average exhibiting on the contrary a slight decrease from 4.00 (standard deviation: 0.67) to 3.97 (standard deviation: 1.07). At the same time, the number of increasing scores is a little higher than those decreasing, 13 versus 10 with 14 remaining unchanged. The paired data further confirm this uncertain trend: the mean of the differences is -0.03, and their standard deviation is 1.28, highlighting marked diversities in how students paid attention to this aspect with respect to their opinion about its importance *a priori*. Consequently, the variation is not statistically significant, but we can infer that the event confirmed a positive impression about time spent immersing oneself in different cultures, albeit with some degree of diversity. The fourth aspect contemplated *understanding ambiguous situations and making sense of new experiences*, and it is represented in Table 5.

Table 5. Rating of understanding ambiguous situations and making sense of new experiences

| Understanding | Before | After | Difference |
|---|---|---|---|
| Not at all (1) | 0 | 0 | 0 |
| Very little (2) | 0 | 0 | 0 |
| Somewhat (3) | 2 | 5 | +3 |
| Generally well (4) | 24 | 14 | -10 |
| To a great extent (5) | 11 | 18 | +7 |

These scores were high at the beginning too, and they became even higher after the event: their variation is however lower than interaction and expression of interest and respect. Indeed, their mean increased from 4.24 (standard deviation: 0.55) to 4.35 (standard deviation: 0.72), with the 14 increasing scores equalizing those staying the same and outnumbering the 9 decreasing ones. Pairing the data results in a mean difference of 0.11 (standard deviation: 0.84), which does not yield statistical significance of increment (Wilcoxon signed-rank test z-score: 0.70, p-value: 0.242). Nonetheless, the trend is slightly more definite than time spent immersing oneself in different cultures, with the possibility of the LTTA giving beneficial effects that would be detected in a larger setting; not to mention how, by starting from quite high scores, also having a confirm transporting aprioristic importance to actual attention is a positive outcome. The fifth and final aspect examined was *translating personal thoughts into other languages*, as shown in Table 6.

Table 6. Rating of translating personal thoughts into other languages

| Translation | Before | After | Difference |
|---|---|---|---|
| Not at all (1) | 0 | 1 | +1 |
| Very little (2) | 0 | 1 | +1 |
| Somewhat (3) | 2 | 4 | +2 |
| Generally well (4) | 23 | 10 | -13 |
| To a great extent (5) | 12 | 21 | +9 |

These ratings are generally analogous to those emerging in association with understanding: they were high from the beginning, and they tendentially further increased but with a low overall variation, also if compared with the first two aspects we took into account. Namely, their average increased from 4.27 (standard deviation: 0.56) to 4.32 (standard deviation: 0.97), by having 13 increasing and 8 decreasing scores (16 remained unchanged). The paired differences have a mean of 0.05 and a standard deviation of 1.03, which results in the increment being not statistically significant (Wilcoxon signed-rank test z-score: 0.40, p-value: 0.345), but in any case present in a context characterized by already strong scores. These insights extend to the final aspect the prospect of detecting benefits from the event, or similar ones, by involving a broader audience.

Table 7. Aspects versus differences before/after

| Aspects | -3 | -2 | -1 | 0 | +1 | +2 | +3 |
|---|---|---|---|---|---|---|---|
| **Interaction** | 0 | 0 | 3 | 25 | 9 | 0 | 0 |
| **Expression** | 0 | 1 | 5 | 19 | 10 | 2 | 0 |
| **Time spent** | 3 | 1 | 6 | 14 | 10 | 3 | 0 |
| **Understanding** | 0 | 1 | 8 | 14 | 14 | 0 | 0 |
| **Translation** | 1 | 2 | 5 | 16 | 12 | 1 | 0 |

Table 7 summarizes how scores increased or decreased by comparing them before and after the LTTA, for each aspect. While a quota of respondents between 40% and 50% depending on the aspect - except for "interaction" having the highest proportion of around 2/3 - did not change their rating, it can be seen that the number of students increasing the score is always higher than those decreasing. The fact hindering statistical significance for some of the aspects lies in a few answerers sharply diminishing the rating, by two or even three points. Indeed, if we consider sign tests, which do not take into account the differences' magnitude but only their either positive or negative nature, we obtain lower p-values for the last three aspects, although still not significant. This means that a small group of students failed to perform the transposition from the theoretical importance of those aspects to the implementation in their practical behavior. They are contrasted by a considerably larger number of peers that on the contrary made capital out of this opportunity, but their presence played a role in the outcomes of these analyses. Unfortunately, we were not able to detect the reasons why they did not manage to confirm their intentions with their deeds, since in the open-ended questions they did not write about these aspects.

We will consider now some qualitative features. After the LTTA, we asked students how they took advantage of the international interaction and collaboration with peer students and teachers from different countries and institutions. All the students answering the final questionnaire gave a response to this question; among the most relevant ones here are some worth mentioning:

- "Personally, I endeavored to maximize the opportunity to speak and collaborate with colleagues from other countries. Firstly, in the phase of mutual acquaintance, I aimed to interact as much as possible by asking questions about their culture and, most importantly, about their previous training experiences, so as to compare them with my own. Furthermore, during group activities, I sought to collaborate as much as possible with the objective of enhancing my ability to work in international teams."

- "From my point of view, it was a great experience to be a part of an international group. First of all, it allowed me to get comfortable with a different language. Then, I was able to compare the approaches of students and teachers from other countries in discussions, solving problems and tasks, with my approach. Finally, yet importantly, I managed to improve well the knowledge about the event topics."
- "I have benefited from different perspectives, cultural exchanges, and collaborative projects. Sharing ideas with colleagues and teachers around the world can improve understanding and promote a global perspective."

From these responses, we see how initially is important to organize a moment of interaction that can allow for comparisons between students coming from different backgrounds, both under the cultural and formative points of view. Then, after having secured some common understanding, a collaboration more centered on specific topics can be set up, thus enhancing the hard skills related to those themes as well as the soft skills resulting from collaborating internationally.

In a subsequent question, we asked students what they liked the most about the LTTA experience. Again, we had responses from the whole group of students, the most prominent ones being:

- "Firstly, I appreciated the fundamental theme addressed in this course, as I consider it to be a very relevant and useful topic for future applications. Moreover, I highly valued the high level of integration among the participants, which is beneficial for developing critical thinking and collaborative skills."
- "Getting to know different cultural thinking: as a student coming from the field of psychology, it was interesting to get to know different mindsets, and I believe that this understanding of how other people live and think can help us greatly in uniting our countries to peace by understanding and empathizing with them better. Indeed, this sharing of knowledge disrupts this bubble each of our countries lives in, and I can personally say that I emphasize and understand more cultures right now, with that helping me understand their thinking and choices."
- "Surely, I will carry with me several notions learned about the main topic. In my opinion, the most important aspect of this kind of event is the contact with personnel from foreign countries, to foster friendships and knowledge and to enhance intercultural connections. I admit that the two dinners organized were also greatly appreciated and very useful for breaking the ice and starting to relate with colleagues."

This intertwining between cultural exchange and shared learning had a positive effect in terms of both fulfilling the event's purposes and enhancing participants' appreciation. This provides another softer perspective regarding respect and understanding through observation, which is genuinely transversal.

Finally, the students were asked for possible suggestions organizers and staff may benefit from, also in relation to follow-up events. Some interesting answers were:

- "I would like to suggest organizing the course to include some of the topics covered in the MOOC modules in the face-to-face lessons as well, in order to enhance the level and quality of learning. Additionally, I would propose adding to the final exercise the opportunity to undertake roles with greater responsibility related to the military sphere, possibly introducing malfunctions and unforeseen events to test contingency plans."
- "Maybe lessons could treat some themes like current case studies in relation to the hybrid warfare. For example, many online presentations to listen to before the course could have been presented during the course."
- "About the MOOC, I would like to suggest the possibility of downloading the material provided by every module. The video lessons should be available with subtitles because many of us do not know the English language as a mother tongue does."

These responses show the opportunity to tighten the bond between the online and brick-and-mortar parts even further as required by the blended paradigm we followed, since some students thought that this link could be further reinforced. At the same time, a need emerged for having electronic resources available also outside the course: generally speaking, this could be a good practice, but only provided that no copyright issues arise, thus suggesting the use of Open Educational Resources where applicable.

We can relate these results with the theoretical framework and the research question by observing that our findings confirm the importance of social interactions among international students, taking place in a proper learning environment where digital tools play a role, also from the methodological point of view by considering novel forms of teaching and learning such as the blended one. A way to teach and learn that allows students to collaborate internationally and understand each other better from a cultural point of view is to include moments when peers can discuss, after they have started to study individually (possibly remotely and with the facilitation of Open Educational Resources), but preferably before discussing in person topics belonging to their specialized field; this is favored by designing courses with the features of the case. In this regard, digitally driven education has a positive impact since it helps to constitute a common base for the learners, by granting access to diverse perspectives and encouraging them to share insights from their backgrounds. It should be noted that most studies on this purview essentially result in qualitative outcomes, giving little or no space to quantitative indicators and scores. In our case, we were able to provide also statistically significant grounds in support of our proposed action, numerically corroborating the evidence the open responses gave us.

## 6. CONCLUSION

This research allowed us to explore how teaching and learning initiatives including e-learning and digital learning can benefit international collaboration and cultural understanding, thus generally answering the research question. We found out how a Learning, Teaching, and Training Activity, being held in presence as part of a project characterized by a significant online and especially blended approach was able to increment or confirm students' ratings of several aspects related to different cultures: regular interactions with their members, interest and respect for them, time worth spending for their observation and study. Moreover, the attention paid to the understanding of ambiguous situations as part of their new experiences as well as the use of a foreign language for their reflections were deemed as important factors worthy of observation. Students perceived the usefulness of constructively comparing backgrounds, which pave the way for a proficient collaboration in the specialized fields where they share domain expertise. This collaboration turned out to be a formative and appreciable practice. They furthermore recognized the importance of course design and made some suggestions; all the stakeholders would benefit from acknowledging them, since there is always room for improvement. As future work, it would be possible to expand the simulations, by structuring some exercises at the decision-making level: this would allow affording a more synthetic response based on cultural experiences and preferences, by also reflecting how such activity integration can help better cross-fertilization of policy response at a more strategic level and where national interest is more ingrained. Furthermore, the research may be expanded by considering other projects related to international strategic partnerships belonging to the Erasmus+ Key Action 2, such as DIGICODE (Marchisio et al., 2023; Marchisio Conte et al., 2024) and INVITE.

## ACKNOWLEDGEMENTS

The research activity described in this paper has been carried out within the Erasmus+ Key Action 2 Strategic Partnership N. 2021-1-HU01-KA220-HED-000032179 "Interdisciplinary Education and Training on Hybrid Warfare – HYBRID", and within the GNSAGA group of INDAM – "Istituto Nazionale di Alta Matematica Francesco Severi".

## REFERENCES

Bruhn, E., (2017). Towards a framework for virtual internationalization. *International Journal of e-Learning and Distance Education*, vol. 32, no. 1.

Caniglia, G. et al., (2018). The glocal curriculum: A model for transnational collaboration in higher education for sustainable development. *Journal of Cleaner Production*, vol. 171, pp. 368-376.

Creswell, J.W. and Plano-Clark, V.L., (2017). *Designing and Conducting Mixed Methods Research, 3rd edition*. Sage Publications, Thousand Oaks, USA.

de Wit, H., (2015). Globalisation and internationalisation of higher education. *International Journal of Educational Technology in Higher Education*, vol. 8, pp. 241-248.

Galluzzi, F. et al., (2021). Mathematics in higher education: a transition from blended to online learning in pandemic times. *Proceedings of IEEE 45th COMPSAC Conference*, pp. 84-92.

Hrnčiar, M., (2017). Keystones of Irregular Warfare. *International Conference Knowledge-Based Organization*, vol. 23, no. 1, pp. 150-154.

Huang, X. and Lajoie, S.P., (2023). Social emotional interaction in collaborative learning: Why it matters and how can we measure it? *Social Sciences & Humanities Open*, vol. 7, no. 1, paper 100447.

Lacaste, A.V. et al., (2022). Blended and collaborative learning: Case of a multicultural graduate classroom in Taiwan. *PLoS One*, vol. 17, no. 4, paper 0267692.

Marchisio, M. et al., (2020). Teaching Mathematics in Scientific Bachelor Degrees Using a Blended Approach. *Proceedings of IEEE 44th COMPSAC Conference*, pp. 190-195.

Marchisio, M. and Spinello, E., (2021). Internationalization for enhancing the European Security and Defence Higher Education. *15th International Conference on e-Learning (EL2021) – Held at the 15th Multi-Conference on Computer Science and Information Systems (MCCSIS2021)*, pp. 99-106.

Marchisio, M. et al., (2022). Teaching Mathematics to Non-Mathematics Majors through Problem Solving and New Technologies. *Education Sciences*, vol. 12, no. 1, paper 34.

Marchisio, M. et al., (2023). EU security and defence students' perception and use of digital competencies in higher education. *17th International Conference on e-Learning and Digital Learning (ELDL2023) – Held at the 17th Multi-Conference on Computer Science and Information Systems (MCCSIS2023)*, pp. 35-42.

Marchisio Conte, M. et al., (2024). Development of digital competencies in education through staff training events and international schools. *Proceedings of IEEE 48th COMPSAC Conference*, accepted.

Mihalova, G., (2006). E-learning as internationalization strategy in higher education. *Baltic Journal of Management*, vol. 1, no. 3, pp. 270-284.

Nascimbeni, F. et al., (2021). Patterns for higher education international cooperation fostered by Open Educational Resources. *Innovations in Education and Teaching International*, vol. 58, no. 3, pp. 361-371.

Ratiu, A. and Tudorache, P., (2021). Comprehensive integration of power instruments and civil-military capabilities for countering hybrid threats. *Proceedings of the 17th International Scientific Conference Strategic Changes in Security and International Relations (Strategies XXI)*, vol. 17, no. 1, pp. 104-110.

Sam, D.L. and Berry, J.W., (2010). Acculturation: When Individuals and Groups of Different Cultural Backgrounds Meet. *Perspectives on Psychological Science*, vol. 5, no. 4, pp. 472-481.

Sleeman, J. et al., (2020). Social media, learning and connections for international students: The disconnect between what students use and the tools learning management systems offer. *Australasian Journal of Educational Technology*, vol. 36, no. 4, pp. 44-56.

Szabó, A. et al., (2020). Social contact configurations of international students at school and outside of school: Implications for acculturation orientations and psychological adjustment. *International Journal of Intercultural Relations*, vol. 77, pp. 69-82.

Tudorache, P. et al., (2023). An innovative conceptual model for education and training on hybrid warfare. *Management & Marketing. Challenges for the Knowledge Society*, vol. 18, no. 3, pp. 234-250.

Vromans, P. et al., (2023). Intercultural learning in the classroom: Facilitators and challenges of the learning process. *International Journal of Intercultural Relations*, vol. 97, paper 101907.

Wiley, D., (2016). *An Open Educational Reader.* EdTech Books.

# A STUDY ON THE PROMOTION OF E-LEARNING TO THE PROFESSIONAL DEVELOPMENT OF RURAL TEACHERS IN GUANGXI, CHINA

Chenfei Xiang, Jiachao Wei, Yuchen Gou, Shan Liu and Liqing Huang
*School of Education science, Nanning Normal University*
*Nanning, Guangxi, 530299, China*

**ABSTRACT**

The rapid development of information technology has made E-Learning a new path of professional development for rural teachers. Through a survey of rural teachers in Guangxi, China, based on the E-Learning System' Theoretical Framework proposed by Manuela Aparicio et al, the study found that the teachers' needs, the social interaction situation of technology, and the Services in E-Learning three dimensions are positively related to teachers' professional development situation, with the social interaction of technology having the greatest impact on teachers' professional development. At the same time, E-Learning is also positively related to the three dimensions of teachers' professional development: professional knowledge, professional competence, and professionalism, with the strongest effect on the promotion of rural teachers' professional knowledge. Based on the above findings, the study makes recommendations to stimulate teachers' learning needs, enhance social interactivity, and increase service support in order to promote rural teachers' professional development.

**KEYWORDS**

Teachers' Professional Development, E- Learning, Rural Teachers, China

## 1. INTRODUCTION

The revitalization of rural education is currently the focus of world attention because it is related to the sustainable development and social progress in rural areas in China. However, Guangxi is a developing region in China. In the past, due to the inconvenient transportation caused by its geographical environment, its economic and commercial activities lagged behind those of coastal cities in China. In the context of China having solved absolute poverty, promoting regional development through education is a highly effective strategy. Nevertheless, the quality of education in Guangxi is comparatively low in China, let alone in some rural areas. Teacher quality is paramount to the advancement of education. Therefore, it is of immediate necessity to enhance the professional quality of rural teachers in Guangxi. Meanwhile, the implementation of policies to promote education digitalization and the construction of high-quality rural teacher teams has ushered in new development opportunities for rural education. In August 2020, the Ministry of Education issued the Opinions on Strengthening the Construction of Rural Teacher Teams in the New Era (MOEPRC,2020), proposing that we should cultivate high-quality and qualified teachers in rural areas in the new era. In 2017, the State Council in China issued a series of documents such as New Generation Artificial Intelligence Development Plan. Additionally, the report of the 20th CPC National Congress has also proposed to promote education digitalization and build a lifelong learning society. Therefore, In this context, the Teachers' Professional Development in rural areas in China is expected to achieve a breakthrough through E-Learning.

Gordon Kirk, a British educator, divides Teachers' Professional Development (TPD) into three aspects: professionalism, knowledge base and professional competence (Gordon Kirk, 1988). Besides, many Chinese scholars also point out that teacher's professional literacy has three dimensions: professional knowledge, professional competence, and professional concepts and ethics (Xudong Zhu, 2014). Among them, teacher's knowledge base is the specialized knowledge that teachers need to have to engage in education and teaching (Shulman, L.S.1987; Ma Yunpeng. et al, 2014), including four aspects of pedagogy, curriculum, subjects and teaching. Teacher's professional competence includes five factors, which are teacher's basic skills, teaching technique, nurturing, self-development and teaching innovation (Weiping Hu, Yipping Zhang, 2018). Teacher's professionalism is the driving force for Teachers' Professional Development, as well as the key element that distinguishes the educational sentiment of "educators" and "pedagogues". This definition is also used in this study. Considering that the geographical environment and learning atmosphere are relatively isolated in rural areas, E-Learning can break down the barriers to the professional development of teachers in rural areas and promote them to grow rapidly. E-Learning, which refers to the learning and teaching activities through the Internet or other digital content, makes full use of the learning environment provided by modern information technology with a new communication mechanism and rich resources to realize a new kind of learning (USDE, 2001). Manuela Aparicio et al. proposed E-Learning System's Theoretical Framework in 2016, which established People, Technologies, and Services in a three-dimensional system (Aparicio M, et al, 2016). This theory gives us a more comprehensive perspective to understand E-Learning.

Therefore, this study first tries to present the current situation of teachers in the E-Learning environment based on the theoretical framework of E-Learning System Theoretical Framework, to analyze the relationship between E-Learning and TPD of teachers in rural areas, and to propose a path for E-Learning to promote the professional development of rural teachers, so as to help them to use E-Learning more effectively.

# 2. METHODOLOGY

## 2.1 Participants

Due to the small size of rural schools in Guangxi, China, the number of teachers in these schools is usually limited, and the homogeneity within teacher groups is not high. Therefore, this study employed a cluster sampling method to ensure the validity of the sample while conserving manpower. A total of 12 rural schools in Guangxi were selected for this study, including 7 primary schools and 5 middle schools. Questionnaires were distributed to 583 rural teachers across these 12 schools, and after excluding invalid responses, a total of 546 effective questionnaires were collected. The response rate to the questionnaires was 93.65%. Meanwhile, cluster sampling was applied to select teachers in rural areas in Guangxi Province as the subjects. Among the valid subjects, 184 of them are male teachers, accounting for 33.70%, while 262 of them are female teachers, accounting for 47.99%. In the distribution of teachers' professional titles, there were 18 teachers without titles, 169 teachers of second-grade titles, 324 teachers of first-grade titles, and 35 teachers of senior titles, accounting for 3.30%, 30.95%, 6.41%, and 59.34% respectively.

## 2.2 Research Method

The E-Learning survey was conducted by a self-designed questionnaire, which is based on An E-Learning Theoretical Framework proposed by Manuela Aparicio et al. 2016. This theoretical framework is based on three principal dimensions: Users, Technology, and Services related to E-Learning (Weiping Hu, Yipping Zhang, 2018). Among them, Users refer to the stakeholders in E-Learning, including consumers, suppliers, education ministry, etc. However, in this study, only the user teachers will be investigated since the research object is teachers in rural areas. Technology refers to E-Learning content (including learning documents, digital audio and video, etc) and communication (including discussion area, forum, chat, social network, etc.). Services refer to the activities that learners will be through, such as some pedagogical models and instructional strategies during E-Learning process. Under this framework, each section includes corresponding questions with a total of 10. The Cronbach α and KMO for this questionnaire were 0.809 and 0.751 respectively.

The Teachers' Professional Development Questionnaire (TPDQ) includes three dimensions based on existing theories: Teacher's Knowledge Base, Professional Competence, and Professionalism(Xudong Zhu, 2014). Teacher's Knowledge Base includes not only subject content knowledge but also how to teach, nurture and serve, while Professional Competence includes both academic competence and educational professional competence. Additionally, Professionalism includes a teacher's professional awareness, thinking and general mental state. The questionnaire is with a total of 20 questions. The Cronbach's α of this part is 0.901 and the KMO is 0.820, which indicates that the overall questionnaire had good reliability and validity.

## 2.3 Data Analysis

This study used SPSS 17.0 for data analysis, starting with descriptive statistics for overall data and then analyzing the relationship between Teachers' Professional Development and E-Learning situation, which contains three dimensions of Users, Technology, and Services, by Pearson's correlation analysis.

# 3. RESULTS

## 3.1 The Descriptive Statistics on TPD of Teachers in Rural Areas in Guangxi

From the results of the study, the total score of TPD in rural areas in Guangxi was average (M=3.782). Meanwhile, during the same period, a survey was conducted among special-grade teachers in Jiangxi Province, China—the highest rank in Chinese teacher titles. The teachers achieved an overall average score of 4.5 after percentage points were converted, which was significantly higher than the score of rural teachers in Guangxi. There was not much difference in the performance of Knowledge Base(M=3.733), Professional competence (M=3.776) and Professionalism (M=3.815). Among overall statistics, Professionalism scored higher, followed by Professional competence, while the lower score was for Knowledge Base as shown in Table 1.

Table 1. Descriptive Statistics on E-Learning and the TPD in Rural Areas in Guangxi

| | Total Score on E-Learning | Total Score on TPD | Knowledge Base | Professional Competence | Professionalism |
|---|---|---|---|---|---|
| M | 3.751 | 3.782 | 3.733 | 3.765 | 3.815 |
| SD | 0.694 | 0.598 | 0.661 | 0.622 | 0.588 |

Notes:M:mean;SD:standard deviation;

The TPD did not have significant difference in terms of gender ($t=-1.833$, $p>0.1$), while there was a significant difference in terms of title ($F=11.890$, $p<0.001$). The professional title of Chinese teachers refers to the academic and professional ranking within the education system of China. Typically, primary and secondary school teachers are categorized as either senior teachers, first-grade teachers, or second-grade teachers, based on criteria such as their ideological and moral character, level of knowledge, professional abilities, and teaching performance. Novice teachers who have not yet met the criteria for formal evaluation are often referred to as 'unevaluated teachers.' The salary and other benefits that teachers receive can vary based on their professional title and years of experience. Among the statistics, senior teachers obtained the highest scores in terms of Professionalism (M=4.210), followed by second-grade teachers (M=4.070), untitled teachers (M=3.767), and last first-grade teachers (M=3.617). The result showed that the TPD of second-grade teachers is better than first-grade and the the first-grade teachers were the lowest, which is inconsistent with the common situation that teachers with higher titles will develop better. In further analysis, we found that first-grade teachers (M=4.069) did have higher scores compared to second-grade teachers (M=3.741) and untitled teachers (M=3.591) on Knowledge Base, while the Professionalism scores were slightly lower for first-grade teachers (M=3.657) compared to second-grade teachers (M=4.085) and untitled teachers (M=3.825). The reason why this result was presented in the study is that compared with teachers who have just joined the profession, it is believed that first-grade teachers have possibly entered a period of burnout in their career, and there is a bottleneck in the development of Professionalism. To some extent, a Chinese study can support our conjecture. It found significant differences in job burnout among teachers with different professional titles. First-level teachers, in particular, exhibited the most severe cases of job burnout. This is likely because they have a strong desire to advance further in their careers but have not yet achieved their goals, which leads to increased emotional exhaustion.(WU Xinchun, et al, 2019)

In further research, we found that there were also some differences among the TPD in rural areas in Guangxi in terms of education levels ($F=3.978$, $p<0.05$). Specifically, the average score of teachers with a master's degree (M=4.123) was higher than that of teachers with a bachelor's degree (M=3.847), teachers graduating from a college (M=3.669), and teachers graduating from secondary school degree or lower (M=3.445). This indicates that as the level of education increases, the level of the TPD increases accordingly.

However, in terms of Professionalism, teachers with a master's degree (M=3.789) were not different from teachers with a bachelor's degree (M=3.841) or even slightly lower than teachers with a bachelor's degree. This may be due to the fact that teachers with master's degrees do not show a clear advantage in professionalism because they have more professional expectations, but the actual teaching in rural ares is somewhat different from what they imagined.

In addition, we analyzed the relationship between the TPD and age in rural areas in Guangxi. The results showed that there was a significant difference in the age variable in TPD (F=3.216, p<0.05). Specifically, young teachers under the age of 40 (M=3.958) showed higher score in TPD t, while those between the ages of 40 and 50 (M=3.679) and those over the age of 50 (M=3.442) were relatively lower. This may be for the reason that young teachers generally have higher education and professional knowledge background, but are relatively easy to face burnout and weak in professionalism.

In conclusion, the overall TPD of teachers in rural areas in Guangxi is average. There is no significant difference in gender but there are some differences in terms of title, education level and age.

## 3.2 E-Learning Situation is Positive Related to TPD of Teachers in Rural Areas in Guangxi

As shown in Table 2, we conducted a correlation analysis between E-Learning and the three dimensions it contains (User, Technology, and Services), as well as the TPD with its three dimensions (Knowledge Base, Professional Competence, and Professionalism).

Table 2. Correlation Analysis between E-Learning and the TPD in rural areas in Guangxi

| | 1 | 2 | 3 | 4 | 5 | 6 | 7 |
|---|---|---|---|---|---|---|---|
| 1.Services | | | | | | | |
| 2.User | .770*** | | | | | | |
| 3.Technology | .753*** | .908*** | | | | | |
| 4.Knowledge Base | .790*** | .897*** | .909*** | | | | |
| 5.Professional-Competence | .775*** | .851*** | .852*** | .904*** | | | |
| 6.Professionalism | .714*** | .813*** | .847*** | .880*** | .877*** | | |
| 7.E-Learning | .844*** | .975*** | .965*** | .929*** | .881*** | .851*** | |
| 8.TPD | .792*** | .882*** | .893*** | .956*** | .964*** | .955*** | .914*** |

Notes : **p<0.01, ***p<0.001

The result showed that, firstly, the TPD of teachers in rural areas in Guangxi was positively correlated with their E-Learning situation (r=0.914, p<0.001), which indicated that the better E-Learning situation is, the higher TPD scores teachers will get. We believe that the higher professional development levels observed among teachers with robust professional spirit can be attributed, in part, to their greater propensity to proactively enhance their professional competence through E-Learning platforms. This indicates that such teachers are not only committed to their current practice but also open to embracing new technologies and methodologies, which are integral to the evolving educational landscape

Secondly, the three dimensions of E-Learning (User, Technology, and Services) are all positively related to TPD. Among them, communication within Technology had the highest correlation with TPD (r=0.893, p<0.001). This suggests a robust correlation between the communicative and collaborative efforts of educators within E-Learning environments and their ongoing professional growth.

Thirdly, E-Learning was positively correlated with each of the three dimensions of TPD (Knowledge Base, Professional Competence, and Professionalism), with the highest correlation coefficient between E-Learning and Knowledge Base (r=0.929, p<0.001). We hypothesize that E-Learning could significantly facilitate the acquisition of professional knowledge among educators, thereby enhancing their professional development and the quality of their educational practices.

The fourth,tthe correlation between the service situation of mobile learning and the professional knowledge of teachers is a 9-component dimension: “the Services-Knowledge Base,” “the Services-Professional Competence,” “the Services-Professionalism,” “the User-Knowledge Base,” “the User-Professional Competence,” “the User-Professionalism,” “the Technology-Knowledge Base,” “the

Technology-Professional Competence," and "the Technology-Professionalism." The coefficient was the highest in the "the Technology-Knowledge Base" correlation analysis ($r=0.909$, $p<0.001$). This finding suggests a strong correlation between the services that teachers themselves receive while engaging in mobile learning and the professional knowledge they possess.It may be because the diverse teaching methods and guidance strategies encountered during e-learning can influence the professional knowledge acquisition of teachers through m-learning. Additionally, it may also reflect that teachers are able to access content closely related to their areas of expertise more directly and efficiently when utilizing mobile learning tools. This strong correlation may further fuel teachers' enthusiasm and participation in mobile learning, creating a virtuous circle: high-quality services enhance teachers' learning experience, which, in turn, promotes their professional knowledge growth, thereby enabling them to make full use of mobile learning tools to garner increasingly valuable information.

## 4. DISCUSSION

Our study revealed that the overall TPD in rural areas of Guangxi is average. Among the evaluated aspects, Teacher's Knowledge Base is the weakest, followed by Professional Competence and Professionalism. Additionally, it was observed that some teachers who are experiencing burnout, as well as those with higher levels of education, are lacking in professionalism.

Additionally, the analysis above demonstrates a strong correlation between E-Learning and TPD. In particular, communication skills within Technology contexts show the highest correlation with TPD. Furthermore, teachers' professional knowledge exhibits the highest correlation coefficient with E-Learning performance. When examining the three-dimensional components, it becomes evident that teachers' professional knowledge is most closely associated with the facilitation of E-Learning services.

To achieve this, we call for the government to integrate mobile learning for rural teachers into the broader framework of educational informatization and to offer the necessary policy backing and financial assistance. Encouraging teachers to embrace E-Learning hinges on fostering their intrinsic motivation to utilize it, enhancing their capacity to amalgamate network communication technology into their teaching practices, establishing effective E-Learning collaborative groups, and refining the system for evaluating teachers' E-Learning outcomes (Yang Qiguang, Xu Tingting, 2021). Drawing on the insights provided by these findings, we can advance the E-Learning capabilities of rural teachers and foster their professional growth by focusing on three interconnected dimensions: increasing their engagement in E-Learning communities, addressing their specific E-Learning requirements, and expanding the range of E-Learning services available to them.

### 4.1 Improving the Interactivity of E-Learning: Facilitating the Expansion of Teacher's Knowledge Base

In the framework of E-Learning, Communication in Technology is a kind of social interaction, which is mainly reflected in online interactive communication. The findings show that focusing on the social interactivity of E-Learning is an important way to enhance teachers' professional knowledge. Therefore, the expansion of teacher's knowledge base can be promoted by enhancing social interactivity.

First, a sharing platform for teachers' E-Learning can be constructed. On this platform, teachers can share teaching resources, exchange teaching experience and insights, learn from each other and make progress together. At the same time, the digital sharing platform can be utilized to discuss cutting-edge knowledge and broaden teacher's knowledge levels. Second, we can implement an online collaborative teaching mode. Through online collaborative teaching, teachers can observe each other's classes and add rich interactive links, such as online discussions and group activities. The joint discussion of teaching methods enables teachers to gain more professional teaching knowledge.

## 4.2 Stimulating Teachers' E-Learning Needs: Promoting Teacher's Professionalism

Teachers, as the main body of E-Learning, first need to improve their E-Learning awareness. It has been argued that teachers' subjective norms, behavioral attitudes, perceived usefulness, perceptual behavioral control, self-efficacy, superior influence, and convenience in rural primary and secondary schools have a significant positive effect on E-Learning willingness(Zhang Jia,2022; Yang Kuizhi,2012). By stimulating their digital learning needs, they can realize their greater potential in education and teaching.

First, in terms of teacher training, increase the investment in digital education and provide teachers with rich online courses and field training, so that they can quickly master digital teaching methods in a short period of time. Second, in the process of education and teaching, teachers are encouraged to make full use of information technology, such as Cloud, big data and artificial intelligence, for innovative teaching. The application of these technologies can enrich the teaching methods and improve the teaching effect, and at the same time, it can also make teachers feel the charm of digital education and further stimulate their enthusiasm for learning. Lastly, the education sectors concerned should establish a sound incentive mechanism to reward teachers who have made outstanding achievements in education digitalization. This will help raise the status of teachers and make them attach more importance to digital education, thus promoting education and teaching reform.

## 4.3 Enhancing Services for E-Learning: Promoting Teacher's Professional Competence

To strengthen the strategic service of E-Learning is to expand the specific guidance in teachers' E-Learning. For example, we can focus on contextualization when conducting the teachers' online training courses so that we can activate the learning process of teachers' professional competence.

First, we should focus on personalized teaching. Since the teaching needs and abilities of each teacher are different, personalized factors should be fully considered when designing online training courses. Through the assessment of teachers' competence, teachers are provided with teaching resources and support that meet their needs. It also enables them to quickly find a learning style that suits them in the E-Learning environment. Second, we should innovate public E-Learning resources. Making full use of QQ, WeChat and other large social platforms can optimize the content of E-Learning resources for teachers in rural elementary schools. E-Learning provides learning convenience for learners with its unique characteristics of learning fragmentation. Therefore, fragmented reading content can be provided for teachers in terms of equipment support so that they can use trivial time to learn complete knowledge blocks. Third, we should build a diversified E-Learning model. In addition to optimizing personal E-Learning devices and innovating public E-Learning resources, we should also explore more E-Learning modes adapted to the needs of teachers. For example, online and offline training can be carried out to allow teachers to utilize E-Learning tools in actual teaching scenarios and improve their teaching techniques. At the same time, by organizing E-Learning seminars, salons and other activities, teachers are encouraged to share their E-Learning insights and practical experiences to enhance communication and cooperation among themselves.

Additionally, several limitations of the study should be acknowledged. First, due to limited funds and time, cluster sampling method was employed in the selection of samples, restricting the study to the Guangxi region of China. Consequently, the generalizability of the findings is constrained. Second, the study identified a strong correlation between e-Learning and TPD from a macro perspective. However, whether a causal relationship exists between them, and if so, the specific mediators through which they are influenced, has not been thoroughly investigated. Future research is anticipated to delve deeper into the underlying mechanisms based on this foundation.

## ACKNOWLEDGEMENT

The authors are grateful for the support by the 2019 Humanities and Social Sciences Research Planning Fund Project of the Ministry of Education of China: "Research on the Path of Improving the Subject Core Literacy Competence of Secondary School Teachers in Townships in Southwest Ethnic Areas and the Construction of Their Support Platforms(19YJA880064).Thanks to the Ministry of Education of China (MOE) for funding and support for this study; and thanks to all members of this research group for your hard work to complete this study.


## REFERENCES


Aparicio, M., et al, (2016), An E-Learning theoretical framework. *Educational Technology & Society*, 19 (1),292–307.

China's Ministry of Education and other six departments. (2020, Aug). *Opinions on Strengthening the Construction of Rural Teaching Staff in the New Era* . http://www.moe.gov.cn/srcsite/A10/s3735/202009/t20200903_484941.html

Hu, W., Zhang, Y. (2018). The theory and practice of teachers' professional capacity development. *Journal of Shaanxi Normal University (Philosophy and Social Science Edition)* (02), 139-145.

Jia, Z. (2022). A study on rural primary and secondary school teachers' willingness to mobile learning based on the theory of planned behavior. *Chinese Journal of ICT in Education,* 28(01),46-55.

Kirk, G. (1988). *Teacher Education and Professional Development.* Scottish Academic Press, Glasgow.

Kuizhi, Y. (2012). An investigation of learners' motivation in modern distance education. *Continuing Education Research,*(04):68-70.

Qiguang, Y., Xu, T., (2021). Mobile Learning for Teachers' Professional Development: Practice, Challenges and Responses. *Journal of Capital Normal University (Social Science Edition)*,(03),171-180.

Shulman, L. S. (1987), Knowledge and Teaching: Foundations of the New Reform. *Harvard Educational Review,*(1), 1-22.

US Department of Education, Office of Educational Technology. (2001, Dec). *E-Learning: putting a world class education at the finger tips of all children. National Educational Technology Development Plan*. www. Ceoforum.org.

Wu, X., et al, (2019). Overall characteristics and differential performance of teacher burnout in primary and secondary schools in China. Journal of South China Normal University (Social Science Edition),(01):37-42+189-190.

Yunpeng, M., et al, (2014). Teacher professional knowledge development and teacher education reform. *Journal of Teacher Education,*(1),31-45.

Zhu, X.D. (2014). On the construction of theoretical model of Teachers' Professional Development. *Educational Research* (06), 81-90.

# INTEGRATION OF ARTIFICIAL INTELLIGENCE IN TEACHER TRAINING: TRAINING TO INNOVATE IN THE DIGITAL AGE

Bruno F. Gonçalves
*Research Center in Basic Education, Polytechnic Institute of Bragança, Campus de Santa Apolónia, 5300-253 Bragança, Portugal*

**ABSTRACT**

Artificial intelligence (AI) is already part of people's daily lives in Portugal, since it is present in practically every economic sector of society, making it difficult for citizens not to use this type of emerging technology in all kinds of contexts. The education sector is precisely one example of a context where AI is used and should be taught in a pedagogical way, helping to clarify its proper use for both students and teachers. In this sense, it is essential that AI is integrated into teacher training so that these professionals are prepared to use this type of technology, integrating it into the teaching-learning process. The research is carried out through a systematic literature review, carried out between 2020 and 2024, which seeks to identify the benefits of integrating AI in teacher training, thus contributing to the discussion and reflection on the subject. The results show that, although there is a need for more literature in the area, there is recognition of the importance of providing teachers with digital skills in the area of AI through continuous training so that they can be in line with the new educational reality.

**KEYWORDS**

Artificial Intelligence, Teachers, Digital Technologies, Teaching-Learning Process, Teacher Training

## 1. INTRODUCTION

Although society is beginning to come into contact with AI at different levels, whether in its daily personal and family life or at a professional and academic level, much is said about its challenges and the contributions it brings to organizations and to life. of citizens. It is such an emerging topic, so appropriate to reality and so interesting that it is currently being talked about everywhere. It is talked about in the media (television programs, news, newspapers, magazines and radio); It is also talked about in the educational, academic and scientific communities (in meetings between educational agents, conferences and congresses and other scientific events, but also in debates and in the literature found in scientific repositories and bibliometric databases); There is also talk of the national and international political spectrum (in public forums, in the regulations created, reports, recommendations and in the legislation that is being produced by the various European and Portuguese bodies).

But what is AI after all? A set of sciences, theories and techniques whose purpose is to reproduce by a machine the cognitive abilities of a human being. Current developments aim, for instance, to be able to entrust a machine with complex tasks previously delegated to a human (Europa, 2021). Also UNICEF, in the same year as the Council of Europe, presents a definition of AI that seems interesting to us: Machine-based systems that can, given a set of human-defined objectives, make predictions, recommendations or decisions that influence real or virtual environments. AI systems interact with us and act on our environment, either directly or indirectly. Often, they appear to operate autonomously, and can adapt their behaviour by learning about the context (UNICEF, 2021). When we talk about AI systems, we are referring to software on computers or machines programmed to perform tasks that normally require human intelligence, for example learning or reasoning. Through the use of data, certain AI systems can be “trained” to make predictions, formulate recommendations or make decisions, sometimes without any human involvement (Comissão Europeia, 2022).

Considering these definitions of AI presented by international reference organizations, the question immediately arises: If AI can do all this, simplify life for citizens, contribute to the speed of processes in organizations and faster production, then no should the education sector look at AI as an opportunity rather than a problem? Our answer is very clear, Yes! Yes, because the use of AI systems can potentially improve teaching, learning and assessment, provide better learning outcomes and help schools run more efficiently (Comissão Europeia, 2022). The cited document (Comissão Europeia, 2022) also states that AI systems can be used in different ways to support teaching or facilitate learning, namely: (i) Teaching students — using AI to teach students (student-oriented); (ii) Student support — using AI to support student learning (student-oriented); (iii) Supporting teachers — using AI to support the teacher (teacher-oriented); (iv) Systems support — use AI to support system-level (system-oriented) diagnosis or planning. This means that the contribution that AI can make to education (or rather, it already makes) is truly important, especially when we are increasingly immersed in online environments, where digital technologies play an absolutely vital role in achieving goals. teaching-learning processes.

Noting the importance of AI in education highlighted in the literature (Baltar & Baltar, 2023; Beijing, 2019; Cardoso et al., 2023; da Indústria, 2018; Hsu & Ching, 2023; Limo et al., 2023; Tavares et al., 2020), it becomes evident that all educational actors must be prepared for the new era of the digital society in which we find ourselves (Web 4.0).

This preparation seems to be able to take place in the context of teacher training, through training actions aimed at the integration, use and application of AI in the teaching-learning process. Teacher training in AI is vital since if students have contact with AI, parents have contact with AI and society has contact with AI, it seems logical and absolutely indispensable that those who teach must also have skills in this type of technologies. This will lead to the development of new digital skills to be considered in the context of the European Digital Competence Framework for Educators (C & Y, 2017; Redecker, 2017), which provides a general framework of reference to support the development of educator-specific digital skills in Europe (Comissão Europeia, 2022). Some of these AI skills are related to: (i) The use of digital technologies for communication, collaboration and professional development; (ii) Acquisition, creation and sharing of digital resources; (iii) Managing and orchestrating the use of digital technologies in teaching and learning; (iv) Use digital technologies and strategies to improve assessment; (v) Use digital technologies to improve inclusion, personalization and active participation of students; (vi) Empower students to creatively and responsibly use digital technologies for information, communication, content creation, well-being and problem solving (Comissão Europeia, 2022). Each of these skills is very important for teachers to be able to reach students through the integration of AI in their classes, which is why training aimed at understanding and using this type of technology is important.

With the development of this research, we intend, through a systematic review of the literature, to identify the benefits of integrating AI in teacher training, thus contributing to the discussion and reflection on the subject.

The article has four sections, the first being the introduction where we try to provide a brief contextualization. Subsequently, we present the main methodological options adopted for the development of this research. In the subsequent sections, we present and discuss the results obtained and present the main conclusions obtained with the development of this investigative work.

## 2. METHODOLOGY

The research is carried out through a systematic literature review which seeks to identify the benefits of integrating AI in teacher training, thus contributing to the discussion and reflection on the subject.

For this, the systematic literature review is adopted as the investigative methodology to support the study. The systematic review of the literature will be carried out with the support of a pre-defined set of criteria that will be essential for the serration of information on the theme addressed, namely: (i) Time interval: 2020-2024; (ii) Keywords (PT): A IA na formação de professores; contributos da IA para a formação de professores; benefits of AI in teacher training; Keywords (EN): AI in teacher training; contributions of AI to teacher training; (iii) Documents: reference articles and doctoral theses; (iv) Search language: English and Portuguese; (v) Bibliometric databases: B-On, Scopus and Web of Science.

Based on the criteria previously established, the general framework of the documents found is presented:

Table 1. Systematic literature review - Articles found

| Id | Year | Title | Authors |
|---|---|---|---|
| 1 | 2020 | The role of artificial intelligence in teacher professional development. | Al-Zyoud, H. M. M. (Al-Zyoud, 2020) |
| 2 | 2021 | The responses to artificial intelligence in teacher integrated science learning training program | Gunawan, K. D. H., Kaniawati, I., & Setiawan, W. (Gunawan et al., 2021) |
| 3 | 2022 | Artificial intelligence and learning analytics in teacher education: A systematic review | Salas-Pilco, S. Z., Xiao, K., & Hu, X. (Salas-Pilco et al., 2022) |
| 4 | 2023 | Formação de professores para o uso de tecnologia: a inteligência artificial (IA) e os novos desafios da educação | Duque, R. de C. S., Monteiro, R. R., de Oliveira Filho, F. L. C., Loureiro, V. J. S., do Nascimento, I. J. B. M. F., Placido, R. L., da Silva , C. J., da Silva, J. M., Saraiva, M. do S. G., & Silva, A. M. de B. (Duque, Monteiro, et al., 2023) |
| 5 | 2023 | Formação de professores e a Inteligência Artificial: desafios e perspectivas | Duque, R. de C. S., Turra, M., dos Santos, A. A., Soares, L. G., Pascon, D. M., Bernardina, L. D., Peres, H. H. C., Barros, M. W. B., do Nascimento, I. J. B. M. F., Gomes, D. J. R. de A., Simões, G. S., & de Oliveira, E. A. R. (Duque, Turra, et al., 2023) |
| 6 | 2023 | Desafios e Oportunidades da Inteligência Artificial na Educação e na Formação | Santos, A. (Duque, Turra, et al., 2023) |
| 7 | 2023 | Redefining Teacher Training: The Promise of AI-Supported Teaching Practices | Kusmawan, U. (Kusmawan, 2023) |
| 8 | 2023 | Integrating AI tools in teacher professional learning: a conceptual model and illustrative case | Tammets, K., & Ley, T. (Tammets & Ley, 2023) |
| 9 | 2023 | Investigating Effective Ways to Use Artificial Intelligence in Teacher Education | Tunjera, N., & Chigona, A. (Tunjera & Chigona, 2023) |
| 10 | 2023 | A study on the need for AI integration in teacher education | Mahendran Maniam, D. K. (Mahendran Maniam, 2023) |
| 11 | 2023 | Introduction of Artificial Intelligence Literacy and Data Literacy in Computer Science Teacher Education | Olari, V., Zoppke, T., Reger, M., Samoilova, E., Kandlhofer, M., Dagiene, V., ... & Lucke, U. (Olari et al., 2023) |
| 12 | 2023 | The Role of Artificial Intelligence (AI) in Teacher Education: Opportunities & Challenges | Afiya Jamal (Jamal, 2023) |
| 13 | 2023 | Artificial Intelligence and the Future of Teaching and Learning | Miguel A. Cardona, Roberto J. Rodríguez, Kristina Ishmael (Cardona et al., 2023) |
| 14 | 2023 | The Opportunities and Challenges of Artificial Intelligence (AI) Application in Education: Literature Review | Wang Ruizhe, Jenny Chen Jie, Albattat Ahmad (Wang Ruizhe, Jenny Chen Jie, 2023) |
| 15 | 2023 | Conhecimento Poderoso e Inteligência Artificial (IA): Aliando Didaticamente Tecnologias para Educabilidades | Oliveira, E. A. D. S. (Oliveira, 2023) |
| 16 | 2023 | Formação docente e Inteligência Artificial | Aroma Bandeira, Jéfte Amorim, Pedro J. Florencio (Amorim et al., 2023) |
| 17 | 2024 | Use of Artificial Intelligence in Teacher Training | Wei Wu, Gulnara Burdina, Alena Gura (Wu et al., 2023) |

Nine documents were excluded because they were outside the lines of the research and, in this sense, seventeen documents were considered for the purposes of this research.

A content analysis was carried out as part of the systematic literature review, involving the careful selection of a set of relevant articles through relevant databases and specific inclusion criteria. A data extraction process was carried out, identifying variables of interest and developing a protocol for the consistent extraction of information. The results were interpreted in light of the review objectives and discussed regarding implications for the study area.

## 3. DISCUSSION OF RESULTS

The research was carried out through a systematic literature review which sought to identify all the articles that discuss the benefits of integrating AI in teacher training, thus contributing to the discussion and reflection on the subject.

Amorim et al. (2023) identifies a set of benefits of integrating AI into teacher training, namely: Possibility of more effective planning, implementation and evaluation; Development of new critical, philosophical and ethical skills and competencies for teachers; Reformulation of degree curriculum and continuing teacher training; Definition of new ethical and organizational frameworks; Adoption of transversal, intersectoral and multisectoral practices for the planning and governance of AI in education; Reformulation of assessment models and development of teaching-learning systems; Inclusion of media education in teacher training and school curriculum; Automation of management and planning processes for teaching work (Amorim et al., 2023).This perspective brings us to the importance of looking at AI as a way of reformulating teacher training, looking to the future.

Personalization of Learning, expansion of capabilities and personalized dynamic experience are also other benefits of AI. Thus, in personalizing learning, AI can be directed by educators to personalize learning, offering an adaptive approach that meets the individual needs of students. In expanding capabilities it appears that collaboration between humans and AI does not replace the role of the educator, but expands their capabilities, allowing them to focus on more complex and meaningful aspects of education. A dynamic and personalized experience can be realized with an effective collaboration between educators and AI systems that enhances the educational process, transforming it into a dynamic and personalized experience (Duque, Monteiro, et al., 2023). This demonstrates that AI cannot operate alone in the educational process, but rather in conjunction with educational agents.

Other authors (Duque, Turra, et al., 2023) point out that AI allows skills development, particularly with regard to the ability to analyze educational data, mastery of technological tools and ability to effectively integrate AI into pedagogical practice. They also point to the personalization of teaching since AI can be used as a support tool to improve the quality of teaching and student learning. They also highlight opportunities for innovation, as AI offers perspectives and opportunities in teacher training, such as personalizing teaching, analyzing educational data and adopting innovative practices (Duque, Turra, et al., 2023). In this sequence, Oliveira (2023) also tells us that AI enables the personalization of learning, helping teachers to personalize teaching according to the individual needs of each student, offering adapted resources and activities. It also enables data analysis, as it can assist teachers in analyzing educational data, providing valuable insights into student performance and guiding pedagogical interventions. They also highlight task automation, as AI can automate administrative and repetitive tasks, allowing teachers to spend more time planning lessons and supporting individual students (Oliveira, 2023).

AI also allows pedagogical and technological support, facilitating the work of teachers and students with intelligent tutors, identifying trends and educational strategies. It also promotes innovative learning environments, such as gamification, video-assisted learning and immersive training environments, improving the educational experience for both teachers and students (Santos, 2023). In this sense, AI is expected to contribute to improving the educational experience of teachers and students, promoting more individualized and fluid learning.

AI can be used to integrate technology more systematically into teacher education programs. It can also help in the creation of personalized education systems, adapted to the individual needs of students. AI can be used to assist in assessing student performance and providing personalized feedback (Cardona et al., 2023). This is fundamental for increasingly inclusive and close teaching. AI can still be used to facilitate collaboration between teachers, enabling the sharing of educational resources, effective practices and teaching experiences and to promote AI literacy among teachers, enabling them to understand and effectively utilize emerging technologies in their educational practices (Cardona et al., 2023).

AI can analyze large volumes of educational data to identify patterns, trends and areas for improvement in student performance, assisting teachers in making evidence-based decisions. Can provide more efficient and accurate assessment tools, allowing teachers to assess student progress more objectively and identify areas needing intervention. AI-based tutoring systems can offer additional support to students outside of class hours by providing personalized explanations, feedback and guidance. AI can help automate routine administrative tasks, freeing up time for teachers to focus on more meaningful activities such as lesson planning and developing innovative teaching strategies (Oliveira, 2023). The automation of these tasks is essential so that bureaucracy in teaching is reduced, processes are faster and employees can carry out other more complex tasks.

In addition to task automation, AI in teacher training helps identify the specific needs of each teacher, enabling personalized training programs; Enables the simulation of real-world teaching scenarios, assisting teachers in classroom management practices and effective instruction; Promotes teacher effectiveness and satisfaction, in addition to increasing the retention of these professionals; Facilitates data-driven decision-making, drives efficiency and fosters innovation in education; Prepares educators for an AI-driven world, empowering them to create engaging, personalized learning experiences for students (Mahendran Maniam, 2023). AI also allows you to visualize behaviors and interactions of pre-service teachers, predict dropouts and identify risk groups and support self-study of teachers in training. It also allows detecting discourse characteristics in textual data, evaluating the qualities of oral presentations through automatic scoring and assessing teaching competence through intelligent assessment systems (Salas-Pilco et al., 2022).

AI can help teachers perceive and understand what is happening in the classroom, assisting in decision-making and the development of professional competence (Tammets & Ley, 2023). AI can provide insights, integrate cues and knowledge-based reasoning into the official teacher qualification standard, promoting professional development. This type of technology can help teachers adapt teaching to student needs, allowing real-time adjustments according to student demands. AI can facilitate the integration of technological tools into teaching practice, promoting effective learning and aligning practices with competency frameworks (Tammets & Ley, 2023). In this regard, AI can also facilitate overcoming problems faced by teachers. Provides personalized pedagogical suggestions based on teachers' existing knowledge. Improves the effectiveness of lesson planning and the development of learning activities and provides real-time feedback to improve performance and teaching effectiveness; Enables the creation of more interactive and engaging learning environments and contributes to innovation and continuous improvement of teaching and learning methods (Gunawan et al., 2021). AI can play a crucial role in improving the quality of teacher education by providing access to high-quality educational resources and personalized learning materials. AI can provide feedback on areas in need of improvement, and provide opportunities for ongoing professional development (Jamal, 2023).

AI allows teachers to actively participate in a dynamic learning ecosystem, receiving feedback to constantly improve and democratizes access to teacher training, allowing even teachers in remote areas to access high-quality training modules; Granular and Real-Time. AI provides detailed information into aspects such as tone modulation, pace of content delivery, and student engagement levels, allowing teachers to improve their practices based on data-driven recommendations. Avatar technologies allow teachers to practice their teaching skills in simulated scenarios, receiving instant feedback and refining their practices based on AI-driven recommendations (Kusmawan, 2023). This is very interesting for teachers' commitment to their training and development as education professionals.

AI is also being used to improve teaching quality by helping teachers develop more effective and personalized teaching methods and enables the creation of individualized teaching plans for students, helping teachers meet the specific needs of each student. AI helps stay teachers up to date with changes in the school environment and perform their duties more effectively (Wang Ruizhe, Jenny Chen Jie, 2023).

In addition to these, Al-Zyoud (2020) highlight others such as: the development of AI-based educational software to raise teacher qualifications; the establishment of AI-based training trajectories for all education professionals; the provision of accurate databases in all areas of education, including human resources; Integration of databases into a global network to help advance teachers' careers (Al-Zyoud, 2020).

Integrating AI into teacher training can improve equal access to quality education in marginalized communities. Personalization of educational content through AI can make learning more relevant and meaningful for students, meeting their specific learning needs. Investing in teacher training programs that develop technological and pedagogical skills is essential to address the skills gap and maximize the benefits of AI in education (Tunjera & Chigona, 2023). Also the collaboration between educators, educational technology experts, and other stakeholders is critical to ensuring the effective and equitable integration of AI into education, focusing on issues such as data privacy, equitable access, and fairness (Tunjera & Chigona, 2023).

Finally, Wu et al. (2023) highlight another set of AI benefits: Improving students' knowledge through training based on AI Effective combination of technology with traditional learning approaches; Potential to improve educational quality and improve teacher training; Opportunity to refine educational processes, reducing time and resource costs associated with the preparation of pedagogical personnel; Contribution to improving educational quality and developing highly qualified educators (Wu et al., 2023).

The integration of AI into teacher training represents a significant advance in contemporary education, providing a series of benefits that go beyond the simple modernization of teaching methods. These results highlight the importance of incorporating AI into the educational process to promote a more personalized and effective approach to learning, adapting to students' individual needs.

Another relevant aspect is the preparation of educators to face an increasingly technological educational scenario. AI in teacher education not only empowers teachers to use technological tools and resources effectively, but also develops their critical, philosophical and ethical skills necessary to deal with the ethical and social challenges associated with the use of technology in education.

The results seem to highlight the crucial importance of innovating in education, adopting advanced practices and technologies, such as AI, to continually improve the teaching-learning process, prepare teachers for the challenges of the future and promote quality and more accessible education for all.

## 4. CONCLUSIONS

The results show that, although there is a need for more literature in the area, there is recognition of the importance of providing teachers with digital skills in the area of AI through continuous training so that they can be in line with the new educational reality.

The systematic review of the literature on the integration of AI in teacher training highlights a series of significant benefits that this integration can provide. From personalizing learning to automating administrative tasks, to developing critical and ethical skills, AI offers a vast field of opportunities to improve educational practice and, consequently, improve the quality of education.

Integrating AI into teacher education not only enables the creation of more adaptive and personalized educational experiences for students, but also empowers educators to become more effective in their practice. Collaboration between humans and AI systems expands teachers' capabilities, allowing them to focus on more complex and meaningful aspects of education. Furthermore, AI analysis of educational data provides valuable insights that can inform evidence-based pedagogical decisions and continually improve the teaching and learning process.

By investing in integrating AI into teacher training, educational institutions can prepare educators for an AI-powered world, empowering them to create engaging, personalized learning experiences for students. Furthermore, by addressing issues such as equity in access and use of technology, collaboration between all stakeholders can ensure that the benefits of AI are maximized and distributed fairly across all educational contexts.

Therefore, it is essential to recognize the strategic importance of integrating AI into teacher training as an effective means of improving the quality of education, empowering educators and preparing students to face the challenges of an increasingly technological and interconnected world. This integration not only benefits individual teachers and students, but also contributes to the development of more robust and effective education systems on a global scale.

Finally, it should be noted that AI is an opportunity for teachers to modernize their educational practices, taking the opportunity to update and refresh their digital skills. If teachers seize the opportunity to understand the importance of AI in making education more inclusive, then they will certainly be at the helm of innovation and will make an indispensable contribution to achieving a teaching-learning process that is appropriate to reality.

Challenges are part of change and are fundamental to transforming education. We must take advantage of them to change the current teaching context!

## ACKNOWLEDGEMENT

This work has been supported by FCT – Fundação para a Ciência e Tecnologia within the Research Center in Basic Education with reference UIDB/05777/2020 (https://doi.org/10.54499/UIDB/05777/2020).

## REFERENCES

Al-Zyoud, H. M. M. (2020). The role of artificial intelligence in teacher professional development. Universal Journal of Educational Research, 8(11B), 6263–6272.

Amorim, A., Bandeira, J., & Florencio, P. J. (2023). Formação docente e Inteligência Artificial.

Baltar, R., & Baltar, C. S. (2023). Professores serão substituídos pela inteligência artificial? Authorea Preprints.

Beijing, C. De. (2019). Consenso de Beijinng sobre a inteligência artificial e a educação. https://en.unesco.org/themes/ict-education

C, R., & Y, P. (2017). European Framework for the Digital Competence of Educators: DigCompEdu (Issue KJ-NA-28775-EN-C (print),KJ-NA-28775-EN-N (online)). Publications Office of the European Union. https://doi.org/10.2760/178382 (print),10.2760/159770 (online)

Cardona, M. A., Rodríguez, R. J., & Ishmael, K. (2023). Artificial Intelligence and the Future of Teaching and Learning: Insights and Recommendations.

Cardoso, F., Silva, N. Da, Bragion, R., Andrioli, M. G., & Chaves, P. (2023). Uso da inteligência Artificial na Educação: uma revisão exploratória e bibliográfica.

Comissão Europeia. (2022). Orientações éticas para educadores sobre a utilização de inteligência artificial (IA) e de dados no ensino e na aprendizagem. Luxemburgo: Serviço das Publicações da União Europeia, 2022.

da Indústria, C. N. (2018). TENDÊNCIAS EM INTELIGÊNCIA ARTIFICIAL NA EDUCAÇÃO NO PERÍODO DE 2017 A 2030.

Duque, R. de C. S., Monteiro, R. R., de Oliveira Filho, F. L. C., Loureiro, V. J. S., Placido, R. L., da Silva, C. J., da Silva, J. M., Saraiva, M. do S. G., & de Brito Silva, A. M. (2023). Formação de professores para o uso de tecnologia: A Inteligência Artificial (IA) e os novos desafios da educação. Caderno Pedagógico, 20(2), 838–852.

Duque, R. de C. S., Turra, M., dos Santos, A. A., Soares, L. G., Pascon, D. M., Dalla Bernardina, L., Peres, H. H. C., Barros, M. W. B., de Alencar Gomes, D. J. R., & Simões, G. S. (2023). Formação de professores e a Inteligência Artificial: desafios e perspectivas. CONTRIBUCIONES A LAS CIENCIAS SOCIALES, 16(7), 6864–6878.

Europa, C. da. (2021). Artificial Intelligence. Conselho Da Europa.

Gunawan, K. D. H., Kaniawati, I., & Setiawan, W. (2021). The responses to artificial intelligence in teacher integrated science learning training program. Journal of Physics: Conference Series, 2098(1), 12034.

Hsu, Y.-C., & Ching, Y.-H. (2023). Generative {Artificial} {Intelligence} in {Education}, {Part} {One}: the {Dynamic} {Frontier}. TechTrends. https://doi.org/10.1007/s11528-023-00863-9

Jamal, A. (2023). The Role Of Artificial Intelligence (AI) In Teacher Education: Opportunities & Challenges.

Kusmawan, U. (2023). Redefining Teacher Training: The Promise of AI-Supported Teaching Practices. J Adv Educ Philos, 7(9), 332–335.

Limo, F. A. F., Tiza, D. R. H., Roque, M. M., Herrera, E. E., Murillo, J. P. M., Huallpa, J. J., Flores, V. A. A., Castillo, A. G. R., Peña, P. F. P., Carranza, C. P. M., & Gonzáles, J. L. A. (2023). Personalized tutoring: {ChatGPT} as a virtual tutor for personalized learning experiences. Przestrzen Spoleczna, 23(1), 292–312. https://www.scopus.com/inward/record.uri?eid=2-s2.0-85162122474&partnerID=40&md5=1ec60b0b6e01c95f00a1ccedc6dc9873

Mahendran Maniam, D. K. (2023). A STUDY ON THE NEED FOR AI INTEGRATION IN TEACHER EDUCATION. Journal of Research Administration, 5(2), 10332–10342.

Olari, V., Zoppke, T., Reger, M., Samoilova, E., Kandlhofer, M., Dagiene, V., Romeike, R., Lieckfeld, A. S., & Lucke, U. (2023). Introduction of Artificial Intelligence Literacy and Data Literacy in Computer Science Teacher Education. Proceedings of the 23rd Koli Calling International Conference on Computing Education Research, 1–2.

Oliveira, E. A. dos S. (2023). Conhecimento Poderoso e Inteligência Artificial (IA): Aliando Didaticamente Tecnologias para Educabilidades. Sisyphus: Journal of Education, 11(3), 31–45.

Redecker, C. (2017). European framework for the digital competence of educators: DigCompEdu. Joint Research Centre (Seville site).

Salas-Pilco, S. Z., Xiao, K., & Hu, X. (2022). Artificial intelligence and learning analytics in teacher education: A systematic review. Education Sciences, 12(8), 569.

Santos, A. (2023). Desafios e Oportunidades da Inteligência Artificial na Educação e na Formação. RE@ D-Revista de Educação a Distância e Elearning, 6(2), e202311–e202311.

Tammets, K., & Ley, T. (2023). Integrating AI tools in teacher professional learning: a conceptual model and illustrative case. Frontiers in Artificial Intelligence, 6.

Tavares, L. A., Meira, M. C., & do Amaral, S. F. (2020). Inteligência Artificial na Educação: Survey. Brazilian Journal of Development, 6(7), 48699–48714. https://doi.org/10.34117/bjdv6n7-496

Tunjera, N., & Chigona, A. (2023). Investigating Effective Ways to Use Artificial Intelligence in Teacher Education. European Conference on E-Learning, 22(1), 331–340.

UNICEF. (2021). Policy guidance on AI for children. UNICEF. https://www.unicef.org/globalinsight/reports/policy-guidance-ai-children

Wang Ruizhe, Jenny Chen Jie, A. A. (2023). The Opportunities and Challenges of Artificial Intelligence (AI) Application in Education: Literature Review.

Wu, W., Burdina, G., & Gura, A. (2023). Use of Artificial Intelligence in Teacher Training. International Journal of Web-Based Learning and Teaching Technologies (IJWLTT), 18(1), 1–15.

# JUDOMATHS: AN INTELLIGENT SEMANTIC SYSTEM TO SUPPORT SKILLS-BASED LEARNING OF FRENCH THIRD-CYCLE MATHEMATICS

Clément Aubeuf
*ELLIADD Laboratory*
*University of France-Comté, France*

**ABSTRACT**

This article presents JudoMaths, a skills-based learning support system for mathematics exploiting semantic and artificial intelligence technologies with the aim of personalizing the learner's educational journey and helping the teacher in various recurring tasks. The system uses a semantic reasoner as well as ontologies and knowledge bases created for guiding the learner in the acquisition of mathematical skills from the Eduscol framework for the French third cycle. In addition to responding to a need expressed by teachers, the system and its technologies aim to improve student success and monitoring while relieving the teacher of certain automatable tasks. The semantic aspect paves the way for a formal description of mathematical skills and improves the interpretation and interoperability of the knowledge used by the system. However, certain limits are discernible, both ethical, technological and behavioral.

**KEYWORDS**

Artificial Intelligence, Education, Skill-Based Learning, Semantic Technologies

## 1. INTRODUCTION

For several years, artificial intelligence (AI) has increasingly penetrated our practices (Mitchell, 2021). Initially limited in terms of power and reflection due to its symbolic approach, the increase in the volumes of data produced and the progress made in the field have allowed the emergence of connectivist networks, making AI more and more efficient. to the point of creating a disruption in our social practices (Dessaire & al., 2023). Popularized during the Dartmouth conference in 1956, AI then infiltrated many fields including education through intelligent tutoring systems (ITS), the main goals of them are to adapt the educational resources offered to the learner and generate feedback (Rebah, 2023). This technology has changed teaching practices, but has also shown that human intervention in learning cannot be replaced, leading to a review of the knowledge-teacher-learner relationship by adding the notion of mediation (Seldon & al., 2018). At the end of November 2022, the new chatGPT tool shook the global media sphere by offering a multitude of services usually associated with human skills (Leiter & al., 2023). This technology quickly became as fascinating as it was frightening because it allows an uninitiated individual to produce seemingly correct knowledge on a subject even though he or she has no prior knowledge. This tool, which is only a reflection of the evolution of AI technologies, leads to a change in the way our digital practices, including education, are constructed today (Romero & al., 2023). This is, among others, the case of the Khan Academy with Khanmigo, which chose to develop and use an intelligent tutor based on a GPT-4 model and adapted for education (Khan, 2023). Rather than fighting against a technology that is in full democratization, it is therefore important to learn to use and identify the limits of these tools in order to maximize their potential in education, without forgetting that they can lead to a reduction of the quality of learning in the event of poor understanding and/or use (Ijaz & al., 2017). It is with this in mind, in addition to the observed drop in the level of mathematics in France (Mons & al., 2020) and following discussions with several primary school teachers, that JudoMaths, an intelligent skill-based tutoring system was designed, exploiting semantic and AI technologies in order to evolve practices concerning the learning of algebra in primary school. The interest in using such a system is twofold: on the one hand, it facilitates the representation of structured and standardized concepts, principles and relationships in order to improve the computer processing and sharing of knowledge (Munir et al., 2018). The reasoner then allows the personalization of learning through knowledge inference while authorizing subsequent developments on the ontologies used. On the other hand, an adaptive computer system takes advantage of the fun and attractive aspect of digital technology for young students, greatly improving attention capture (Pllana, 2020) and maintaining engagement (Diwan & al., 2023). The system also offers additional educational tools to the teacher such as the visualization of learner data, the automatic correction of certain exercises or the aggregation of relevant educational content when creating course material.

The stated objectives are therefore:

to provide personalized support to students by adapting exercises and concepts worked on in relation to their levels and progression;

to improve the capture and maintenance of attention for better absorption and retention of information;

to help teachers choose their teaching approaches through data visualization;

to automate some time-consuming tasks (e.g. corrections) of little educational interest.

This project is a continuation of the research carried out within the laboratory, the objective of which is to create a more personalized, more efficient and more inclusive learning environment thanks to the combined power of digital technology, AI and semantics. We can cite the creation of an Active and Semantic Learning System (Szilagyi, 2014), the uses of machine learning for education (Bouchereau & Roxin, 2022) or the AI-VT project (Carvalho & al., 2023).

## 2. METHODS

ResearchGate, ScienceDirect, arXiv, IEEExplore and Google were searched using the keywords "artificial intelligence", "education", "intelligent tutoring system", "chatGPT" and "semantic technologies". This research includes questions on learning and these cognitive mechanisms, learning by skills, e-learning, AI and its techniques, semantic technologies and standardized object descriptions, the functioning of intelligent tutors in education, the technical and ethical limits of these tools. Other topics were also covered such as the use of Natural Language Processing, recommendation systems, computational thinking, logic programming, theories of brain processes of information processing or even reflections and debates on evolution of technology, education and society. For all searches, the keywords were searched in the titles and abstracts of the articles present in the above-mentioned databases. The research protocol was restricted to articles meeting various inclusion criteria: the article must be published in a peer-reviewed journal between January 2010 and December 2021, and must report at least one analytical or empirical study. Dissertations and secondary data analyzes were excluded. Study observations must be based on average K-12 students, and must include a control group in order to be able to make a comparison (e.g. teacher-led class, human tutoring). The effectiveness of the system must be measured using standardized tests, and the study must provide the information necessary to take effect size into account. Finally, similar studies were discriminated and an examination of the references used in the selected articles was made, but no other publications were added during the process. Judomath was developed incrementally and is currently being tested for one term in a class including students from the French third cycle of education. It will be used for learning several mathematical concepts (e.g. multiplication rules, operations, fractions) in addition to the lessons given by the teacher. Tests and questionnaires will be carried out before, during and after use of the system to test its effectiveness, efficiency and its impact on user affect. The evolution of students' knowledge will be compared to a control class not using IT tools, but must be modulated by external factors (e.g. mood, average age, social level and background) that can impact the results without having a link directly with the quality of the tool proposed. The results will be analytical and empirical in nature and must take into account the different statistical phenomena linked to the size of the sample observed.

## 3. SYSTEM OVERVIEW

All of the expert knowledge of the system as well as the learner modeling is represented using ontologies, which make it possible to describe in a formalized and structured manner the knowledge of a set called a discourse domain. They are described using OWL (Web Ontology Language), a knowledge representation language based on RDF which allows the formal description of ontologies through its vocabulary (e.g. classes, instances, attributes) and its reasoning capacity logic through the use of semantic reasoners. Classes represent the entities or categories of elements constituting the domain, and are linked together through hierarchical (e.g. a subset) or non-hierarchical (e.g. a prerequisite) relationships describing how these concepts are linked together. Instances are specific elements of a concept, and represent the concrete data of the system. Classes and instances have attributes, which are properties specific to the object to which they are attached in order to list its specific characteristics, while rules describe the relationships between concepts, instances and attributes through logical statements in order to allow the inference of new knowledge and verification of the overall consistency of the ontology.

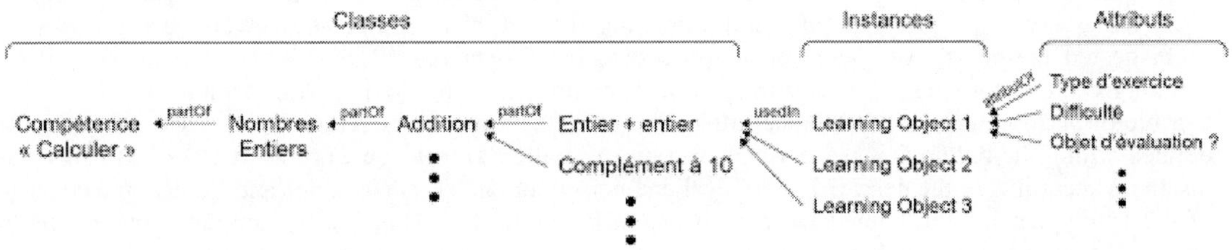


Figure 1. Representation of the ontology aiming the mathematical skill "Calculate"

All of these resources are identifiable using a URI (Uniform Resource Identifier), which is a unique name for locating a semantic resource across different systems or domains. Ontologies operate within a knowledge base, which can be divided into two parts: one representing the terminological aspect (TBox) describing classes and relationships structuring the knowledge domain, and another representing the assertional aspect (ABox) containing specific assertions and facts defined by the TBox.

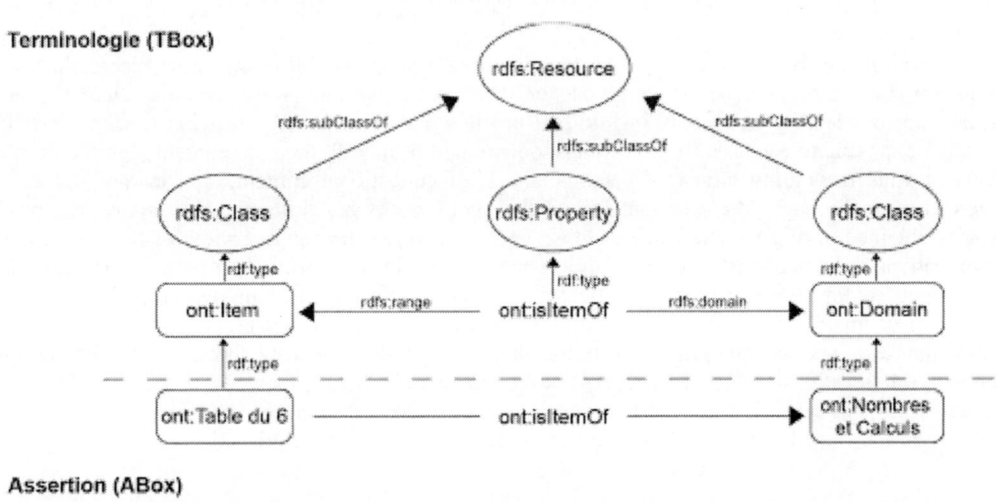


Figure 2. RDFS schema for table of 6

A knowledge base is an interoperable and structured repository of information on which it is possible to carry out logical reasoning, de facto allowing the adaptation of the learning system to the needs of the learner. Resources are described in RDF (Resource Description Framework), a standardized semantic model for describing resources using subject-predicate-object triples in order to explain the relationships between the different entities. The subject represents a resource (e.g. a concept, an object) and the predicate symbolizes the relationship or a property between the subject and the object, which can be another resource or a literal value (e.g. a number, a text). RDF is independent of applications or programming languages, which facilitates interoperability, and can be serialized in different formats (e.g. XML/RDF, Turtle, N-Triples). RDF data is used using SPARQL (SPARQL Protocol and RDF Query Language), a query language developed by the W3C for querying and manipulating structured semantic data. It supports several types of queries (e.g. select, construct, ask, describe) which can be made more complex using operators and allows direct update or delete operations on RDF data. It also allows query federation to integrate data from different sources, and has also a standardized protocol for web operations.

The system was developed using Apache Jena, a framework specifically designed to interface seamlessly with the previously mentioned technologies and featuring an extensive library of tools for robust development. Apache Jena enables the creation and manipulation of RDF models and incorporates a SPARQL query engine for querying and extracting data efficiently. This framework also integrates two sophisticated reasoners, one dedicated to processing RDF schemas (RDFS) and the other specialized in handling OWL ontologies. The semantic reasoner, or inference engine, is a vital component of the system, capable of performing advanced logical inferences on semantic data. Leveraging the rules and relationships defined within the ontologies, the reasoner can make implicit knowledge explicit. It classifies concepts and instances according to the defined hierarchical and non-hierarchical relationships and ensures the consistency of the ontologies and the knowledge base by identifying and resolving logical contradictions or modeling errors. This enhances the system's reliability and accuracy in handling complex semantic queries.

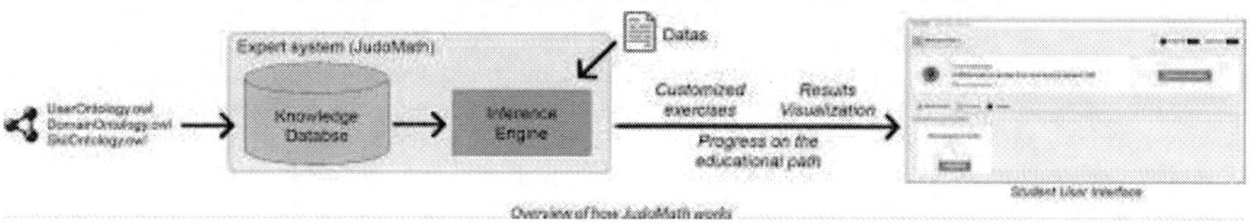


Figure 3. Overview of how JudoMaths works

Concerning the user interface, the system is structured in two distinct parts, accessible via a secure connection interface: one part is reserved for the teacher and the other for the student. The teacher, responsible for creating and modifying student profiles, has an interface allowing him to view data either globally or specific to each profile. It also has tools to add or modify the data contained in the ontologies and exercises, thus facilitating increased personalization of educational content. The learner, for his part, has access to a profile page where he can follow the evolution of his skills and personalize his profile using rewards obtained during his exercise sessions. This page also offers an overview of the progress made and the objectives to be achieved, thus reinforcing engagement in the learning process. In addition, the student has two options for his educational activities: the first is designed for training phases on a concept chosen by the student or guided by the system according to his needs, which allows precise adaptation to the learner's gaps or interests. The second option aims to test their level of skills on a specific concept (for example, whole numbers, fractions) or their general progress in the mathematical skills expected at the end of the third cycle. All exercises, whether training or testing, are procedurally generated based on defined educational parameters, such as difficulty, level of advancement and necessary prerequisites.

# 4. DISCUSSION

The benefits of using such a system are multiple and significant. First of all, the system effectively responds to a very specific problem formulated by the teachers involved in the project: reducing the workload associated with repetitive and unenriching tasks, such as the generation of exercises adapted to the learner's level. and their correction. Monitoring the evolution of skills acquired by students is greatly improved thanks to the continuous tracking of their progress, making it possible to highlight the difficulties encountered or to visualize in detail the learner's journey through the different educational modules. This continuous monitoring makes it possible to quickly identify warning signs of potential school dropouts, thus helping to reduce the digital divide often associated with the use of virtual tutoring systems (Pitchforth, 2021). The skills-centered approach forges a robust link between the theoretical notions studied and their practical application, at the same time promoting the development of transversal skills essential to academic and professional success, while stimulating the engagement and motivation of learners in their own knowledge acquisition process (Jääskä & al., 2022). In addition, the semantic aspect of the description of knowledge allows a precise and detailed representation of the expert knowledge to be assimilated, thanks to the creation of semantic relationships which clarify their meaning and their context. Presented in this form, this structure reflects the way in which the expert organizes and structures his knowledge, thus facilitating navigation and human assimilation of complex knowledge. The ontologies used are not only extensible, making it possible to adapt the system to changes in knowledge, but they also contribute to improving the interoperability of data between different computer systems, thus facilitating the dynamic generation of exercises adapted to the specific needs of learners.

However, some limitations to the use of this system are discernible. This tool must be used in addition to courses traditionally taught by the teacher, which requires significant time to adapt to the tool and adjustments in the organization of educational units to integrate this tool with existing educational resources. The adoption of a decision-making IT system can also raise challenges in terms of understanding decision-making processes, where the choices made by artificial intelligence can appear opaque, which can affect trust in a system whose mechanisms remain obscure for users (Vilani, 2018). The JudoMaths system partially overcomes this problem by operating according to clearly defined rules, where each decision is traceable and explicit, however the future introduction of machine learning techniques is envisaged and could potentially reintroduce this "black box" problem (Wiener, 1950). In-depth ethical questioning about the management of personal data is also crucial, given that the system, in addition to collecting information about the learning journey, aims to carry out detailed and individual modeling of learners. This data, although crucial for the effectiveness of the system, is particularly sensitive and must therefore be handled with the greatest caution, thus guaranteeing its protection and confidentiality.

Future research in our intelligent tutoring systems should focus on several key areas. First, enhancing the transparency of AI decision-making processes is essential to improve user trust and interaction, emphasizing the development of interpretable AI models. Additionally, there is a need for comprehensive studies to evaluate the long-term impacts of these systems on diverse learning styles and outcomes, possibly through adaptive learning technologies. Another critical research area is the ethical management of sensitive data, particularly concerning data privacy and security in educational settings involving minors. Finally, interdisciplinary collaborations between educators, technologists, and cognitive scientists could further optimize AI tools to better align with educational goals and facilitate their integration into classroom settings.

# 5. CONCLUSION

With recent advancements in the field of artificial intelligence (AI), it has become crucial for students and educators alike to learn how to harness this technology effectively. This proficiency will maximize AI's positive impact within educational spheres. The emergence of technologies like ChatGPT underscores the urgency of this need, illustrating that educational methodologies must evolve. Modern education should adopt approaches that integrate AI technologies with innovative mediation methods. These strategies are essential for ensuring the appropriate and thoughtful use of these new tools. An exemplary initiative in this realm is JudoMaths, an intelligent tutoring system tailored to provide personalized academic support to students. This system not only enhances the capture and maintenance of student attention but also automates numerous time-consuming tasks that hold little educational value. By relieving educators from these mundane duties, JudoMaths allows them to focus more on delivering high-quality, engaging instruction. Moreover, this system is a part of a larger, cutting-edge wave of research. This research is dedicated to creating highly personalized and efficient learning environments by synergizing artificial intelligence with semantic technologies. The goal is to develop systems that understand and adapt to individual learning preferences and needs, thereby optimizing the educational experience. This approach promises to revolutionize how educational content is delivered and assimilated, making learning more accessible, engaging, and effective for everyone involved.

While intelligent tutoring systems like JudoMaths offer significant advancements in the realm of education, it is clear that they will never fully replace the crucial role of a human teacher. The unique aspects of human intelligence, such as empathy, the ability to adapt to unforeseen circumstances, and efficient learning from nuanced data, are not easily replicated by machines. Human evolution has finely tuned the ways in which we transmit information, including through facial analysis, the assessment of attention levels, and the consideration of emotions and other external factors. Preserving this human element is essential, as it is one of the greatest strengths of our species—the profound ability to teach and learn. JudoMaths is designed not just as a tool but as a partner in education. It individualizes the learning journey and integrates seamlessly with the teacher's vision and decisions, thereby enhancing the educational process. This system exemplifies how personalization and automation can be used to support—rather than replace—the teacher, who is fundamental to effective learning. By incorporating these aspects, JudoMaths not only aims to enrich the educational experience but also to underscore and reinforce the central role of the teacher as a guide and mediator. This approach maximizes the benefits of AI technologies in education, ensuring they serve as valuable aids that enhance, rather than diminish, the irreplaceable human touch in teaching.

# REFERENCES


Bouchereau, A. & Roxin, I. (2022). Usages de l'apprentissage artificiel pour l'éducation : enjeux et dispositif de régulation. *Communication, Technologies et Développement*, 12. ⟨10.4000/ctd.7955⟩. ⟨hal-03942420⟩

Carvalho, F., Henriet, J., Greffier, F., Betbeder, M-L. & Di Pardo Leon-Henri, D. (2023). Deep learning for the detection of acquired and non-acquired skills in students' algorithmic assessments. *Journal of Education and e-Learning Research*, Vol. 10, No. 2, 111-118. ISSN: 2410-9991. DOI: 10.20448/jeelr.v10i2.4449

Desaire, H., Chua. E., Isom, M., Jarosova, R. & Hua, D. (2023). *ChatGPT or with over 99% academic scientistaccuracy using off? Distinguishing authorship the shelf machine learning tools*. arXiv: 2303.16352.

Diwan, C., Srinivasa, S., Suri, G., Agarwal, S. & Ram, P. (2023). AI-based learning content generation and learning pathway augmentation to increase learner engagement. *Computers and Education: Artificial Intelligence*, 4, pp. 100-110. ISSN : 2666-920X. DOI : https://doi.org/10.1016/j.caeai.2022.100110.

Ijaz, K., Bogdanovych, A. & Trescak, T. (2017). Virtual worlds vs books and videos in history education. *Interactive Learning Environments*, vol. 25, no. 7, pp. 904–929.

Jääskä, E., Lehtinen, J., Kujala, J. & Kauppila, O. (2022). Game-based learning and students' motivation in project management education. *Project Leadership and Society*, vol. 3, 100055. ISSN: 2666-7215. https://doi.org/10.1016/j.plas.2022.100055.

Khan, S. (2023). *Harnessing GPT-4 so that all students benefit. A nonprofit approach for equal access. Khan Academy. URL : https://blog.khanacademy.org/harnessing-ai-so-that-all-students-benefit-a-nonprofit-approach-for-*equal-access/

Leiter, C., Zhang, R., Chen, Y., Belouadi, J., Larionov, D., Fresen, V. & Eger, S. (2023). *ChatGPT: A Meta-Analysis after 2.5 Months*. arXiv: 2302.13795v1. February 20, 2023.

Mitchell, M. (2021). *Intelligence artificielle : Triomphes et Déceptions*. Dunod Editions. ISBN: 978-2-10-081372-8

Mons, N., Tricot, A., Chesné, J-F. & Botton, H. (2020). Numérique et apprentissages scolaires : dossier de synthèse. [Rapport de recherche] Centre national d'étude des systèmes scolaires (Cnesco); Conservatoire national des arts et métiers (Cnam). ⟨hal-03234543⟩

Munir, K. & Sheraz Anjum, M. (2018). The use of ontologies for effective knowledge modelling and information retrieval. *Applied Computing and Informatics*, Volume 14, Issue 2, pp. 116-126. ISSN: 2210-8327

Pitchforth, J. (2021). *Et si l'Intelligence Artificielle était source d'égalité dans l'enseignement supérieur ?* https://www.actuia.com/contribution/jamiepitchforth/et-si-lintelligence-artificielle-etait-source-degalite-dans-lenseignement-superieur/

Pllana, D. (2020). Keeping Students' Attention Active. *Education, Society and Human Studies*, 1, pp. 171. 10.22158/eshs.v1n2p171.

Rebah, H., Barthes, D. & Carnus, M.-F. (2023). Personal learning environment: instrument system for learning beyond the boundaries of the university. *Learning Environments Research*. 25. 10.1007/s10984-023-09457-x.

Romero, M., Heiser, L. & Lepage, A. (2023). *Enseigner et apprendre à l'ère de l'IA*. GTnum LINE #Scol_IA, Canop{é}. Livre blanc.

Seldon, A. & Abidoye, O. (2018). *The Fourth Education Revolution*, pp. 1–14, University of Buckingham Press, London, UK.

Szilagyi, I. (2014). *Technologies sémantiques pour un système actif d'apprentissage* [Thèse de doctorat, Université Franche Comté]. URL : https://www.theses.fr/2014BESA1008

Villani, C. (2018). Donner un sens à l'intelligence artificielle. *AI for Humanity Rapport*.

Wiener, N. (1950). *The human use of human beings: cybernetics and society*. Houghton Mifflin.

# GATEWAY TO SUCCESS: IMPROVING ACADEMIC ACHIEVEMENT THROUGH ONLINE MODULES

Linda van Ryneveld
*Comprehensive Online Education Services*
*University of Pretoria*
*South Africa*

**ABSTRACT**

This paper explores the efficacy of gateway modules in higher education, particularly focusing on their role in providing students with a second chance to excel academically. Gateway modules offer students who failed a pre-requisite module the opportunity to revisit and improve their understanding of foundational concepts, thereby enhancing their overall academic performance and ensuring that students can complete their degrees in minimum time. What sets these gateway modules apart is their provision during the extended university holidays, allowing students to engage in fully online modes of learning. By embracing this flexible approach, students can retake the module and demonstrate their proficiency from any location, facilitating a seamless continuation of their academic journey. Notably, successful completion of the module during the holiday period enables students to progress immediately to subsequent modules that require the failed module as a prerequisite. This alternative pathway not only mitigates the need to repeat an entire semester but also empowers students to stay on track towards timely degree completion.

**KEYWORDS**

Success Rates, Failure Rates, Fully Online Learning, Prerequisites, Student Success, And Throughput Rates

## 1. INTRODUCTION

In recent years, the challenge of high failure rates in foundational courses has prompted educational institutions to explore innovative teaching methodologies (Simpson, 2018). Traditional face-to-face and blended learning models, together with a myriad of contextual factors, often fail to address the diverse needs of students, leading to disengagement and academic underperformance (Edwards et al., 2020). A higher education contact institution in South Africa has responded to this challenge by implementing fully online Gateway modules, which are designed to provide students who have struggled with prerequisite modules, offered in-person and on campus, a second chance to succeed. This initiative aims to enhance student engagement and academic outcomes through a more flexible and immersive online learning environment.

The introduction of Gateway modules represents a strategic shift in educational delivery, targeting foundational in-person modules that traditionally exhibit high failure rates. By transitioning these modules to an online format, the initiative seeks to enhance student engagement and success by leveraging the benefits of digital learning environments. This approach is particularly pertinent given the increasing demand for flexible educational models that accommodate varying learning styles and schedules. The timeliness of the online offering further aims to increase students' ability to complete their studies in minimum time.

The primary problem addressed by this research is therefore the persistent high failure rates in pre-requisite modules, which impede student progress and strain academic systems. High failure rates not only impede student progress but also congest academic systems with timetable clashes and administrative burdens due to repeat enrollments. The research aspires to address this issue by introducing fully online, intensive four-week modules known as Gateway modules, which provide a highly structured, self-directed learning opportunity for students to deeply engage with course content and achieve mastery at a foundational level. The research question that guides this study is therefore focused on how fully online Gateway modules impact student performance and satisfaction compared to traditional learning models.

The main aim of this paper is to evaluate the efficacy of fully online Gateway modules in improving student performance and satisfaction. By analysing data from the initial implementations of these modules at the higher education institution where the study was undertaken, the paper seeks to determine whether the transition to online learning can effectively mitigate the challenges associated with high failure rates in critical pre-requisite modules.

This paper provides an analysis of empirical data from the Gateway modules implemented in 2023 and 2024, insights from a literature review on online learning effectiveness, and a discussion on the broader implications for educational practice. The paper presents findings on module pass rates, student retention, and qualitative feedback, comparing these outcomes with those of traditional learning models. Additionally, the paper will outline strategic recommendations for future educational practices, with a focus on sustaining and enhancing the effectiveness of online learning environments.

## 2. LITERATURE REVIEW

The landscape of higher education is punctuated with discussions about student success and failure rates, particularly within foundational courses that serve as prerequisites for advanced learning. Success rates in these courses are critical as they often dictate the trajectory of a student's academic journey, with failure rates presenting a significant barrier to progression, or throughput rates, within degree programs (Kreisberg & Hsin, 2021).

Fully online learning has emerged as a response to the challenges in higher education, offering a more flexible and accessible alternative to traditional in-person instruction. Literature suggests that online environments can cater to diverse learning styles and potentially reduce failure rates among students who might struggle with the rigid schedules of traditional classrooms (Palloff & Pratt, 2013). This is particularly relevant for modules that are identified as prerequisites, where the mastery of content is crucial for advancement in academic programs (Shaw et al., 2016).

A body of research has begun to examine the efficacy of fully online courses in improving success rates. Studies indicate that online learning can improve throughput rates by offering remedial opportunities that are not bound by geographic and time constraints, thus providing a strategic advantage for students at risk of academic delay (Mathew et al., 2021). However, the effectiveness of such interventions varies based on the time of delivery, course design, the extent of interactivity, and the support structures in place.

Despite its advantages, online learning also presents unique challenges. The retention and success rates in fully online programs have been a concern, with some studies suggesting that the absence of a physical learning community may impact student engagement negatively (Li et al., 2023). Conversely, recent developments in digital pedagogy have begun to address these issues by incorporating elements such as adaptive learning technologies and community-building tools, which are shown to foster engagement and potentially improve success rates (Sato et al., 2023).

Regarding prerequisites, fully online learning offers an avenue for students to revisit foundational concepts and solidify their understanding, which is fundamental for their academic success. Research indicates that well-structured online prerequisite modules can enhance knowledge retention and prepare students better for subsequent coursework (Trip et al., 2024).

The relationship between fully online learning and student throughput rates is a complex one. While online learning offers flexibility and accessibility, it requires a high level of self-regulation and motivation from students. Studies have reported that students who engage with online content regularly and participate actively in online discussions are more likely to succeed (Cavinato et al., 2021). This suggests that student engagement is a pivotal factor in the success rates of online learning and, consequently, affects throughput rates in higher education programs (Qureshi et al., 2023).

In the South African Higher Education context, winter and summer schools typically refer to short, intensive academic sessions of a week or two, that are offered during university vacation periods, namely winter and summer breaks. These sessions were originally designed to provide students with opportunities to catch up on or retake courses they have failed or missed during the regular academic year, to improve grades, or to accelerate their studies by earning extra credits. This format was intended to help students stay on track for graduation by making up for lost time or enhancing their understanding of challenging subjects (Mccombs et al., 2011). However, numerous criticisms can be raised against summer and winter schools, including intensity and pace, the quality of the instruction, access and equity, and their long-term effectiveness in terms of mastery of the learning material (Katzir et al., 2013).

Building upon the foundation laid by the literature, which underscores the transformative potential of fully online learning for enhancing student success and managing failure rates, the Gateway module initiative, as an alternative to the traditional summer and winter schools, emerges as a practical embodiment of these theoretical frameworks. This initiative was specifically designed to address the low pass rates in foundational prerequisite courses, acknowledging that such barriers have historically delayed academic progress by at least a year. By enabling students to retake the full curricula of essential prerequisite modules in a fully online format during university breaks, the Gateway modules aim to streamline the path to academic advancement.

The following section delves into the inaugural iteration of the fully online Gateway modules, evaluating their structure, content delivery, and impact on the academic trajectory of students. It examines how this strategic online intervention aligns with the empirical findings reviewed, seeking to not only mitigate the time lost in traditional settings but also to capitalise on the strengths of digital learning to foster a deeper and more accessible educational experience.

# 3. GATEWAY MODULES

In January 2023, the first fully online Gateway module was offered over four weeks in the South African summer recess period. The first-year, 10-credit Marketing module was normally offered as a semester module that is taken over 14 weeks in a contact mode, on campus. The module is a prerequisite for all the Marketing modules in the second year of study. The module was identified as problematic as the module pass rate was low resulting in large numbers of first-year students not being able to continue with their full load of second-year modules the following year.

During the second semester of 2022, the academic department developed a highly structured fully online version of the module in collaboration with a learning design specialist. The module covered the entire curriculum in full and included a variety of learning materials and learning activities, including annotated PowerPoints, interactive videos, quizzes and discussion forums. Special attention was given to the implementation of the principles of best practices for online learning. In this way, it was ensured that students had a positive online learning experience that required them to extensively engage with the curriculum in a highly structured, semi-synchronous manner.

Multiple continuous assessment opportunities were included, for students to complete in a self-directed manner so that they could learn from the detailed feedback provided. The aim was to actively engage students with the learning materials and to keep them on track throughout the four weeks. The continuous assessment opportunities, some of which were assessed by the computer and others by the lecturer, contributed to the semester mark that constituted 70% of the final mark. A final in-person exam was written on campus and counted 30% towards the final mark obtained for the module. The Gateway module was repeated for a second time in January 2024, and was taken by students who failed the in-person module during the last semester of 2023.

The objective of implementing the fully online Gateway module was to allow students to retake the failed module during their university break. This strategic approach ensured that upon their return to campus for the subsequent academic year, they were equipped to proceed directly with their second-year modules for which the Gateway course served as a prerequisite. Consequently, this circumvented the delay often experienced with the traditional academic model, where failing a module could prolong a student's educational timeline by at least a year, as they were previously required to pass the failed module in a subsequent semester before advancing in their program.

The Gateway modules, delivered entirely online, provided the flexibility for students to engage with their coursework from any location during the holiday period, eliminating the need for them to return to campus or arrange for transportation and special accommodation before the academic year commenced. Likewise, lecturers could administer these modules from their holiday destinations, allowing for a seamless teaching and learning experience irrespective of their physical return to campus.

# 4. METHODOLOGY

The research methodology used in this study is a case study approach, specifically a single-case design, focusing on the implementation and evaluation of a specific Gateway module at a traditional contact university in South Africa. This methodology allowed for an in-depth examination of the intervention, its context, and its outcomes (Yin, 2018). The study employed a mixed-methods design, combining quantitative data from student performance and engagement metrics with qualitative data from student surveys, interviews, and observations (Cresswell, 1999; Chandra et al., 2011). The data collection methods used in this study included document analysis, student performance data, student surveys, interviews, and observations. Data was analysed employing descriptive statistics, a thematic analysis and content analysis. In reporting the data participants were given a number and a unique code, for example, P33 indicated Participant 33. The research approach described here enabled a comprehensive understanding of the Gateway module's effectiveness, student experiences, and potential areas for improvement.

# 5. RESULTS AND FINDINGS

The Gateway module initiative has shown promising results in addressing academic challenges and enhancing student success rates. The fully online, intensive four-week modules provided a structured, self-directed learning opportunity that allowed students to collaborate with one another, deeply engage with module content, and achieve mastery at a foundational level. The educational framework and methodology of these modules blended synchronous and asynchronous interactions, maximising accessibility, and flexibility for students. Comparatively, traditional remedial approaches such as winter or summer schools, as defined in the South African context, often do not offer the breadth or depth needed to ensure mastery of foundational knowledge due to time constraints and the condensed nature of these modules.

The Gateway module has demonstrated improved pass rates compared to traditional summer school formats (See Table 1). In 2023, the Gateway module had a pass rate of 73.8%, and in 2024, the pass rate was slightly lower at 70.5%, but still significantly higher than the summer school's 63.2% pass rate in 2021.

Table 1. Number of students who attended the online Gateway offering

| *Variables* | *Total enrolled* | *Dropped (after module start)* | *Failed* | *Passed* | *Module Pass rate* |
|---|---|---|---|---|---|
| Gateway 2022 | **42** | 1 | 10 | 31 | **73.80%** |
| Gateway 2023 | **44** | 0 | 13 | 31 | **70.50%** |

From the students' comments on whether they would recommend this course to other students, valuable insights came to light. One student said:

*Yes, I really would, because this months' worth of work made me understand everything that I didn't understand in the whole of the semester last year* [P18].

Another mentioned:

*I would recommend this course to other students because it is very engaging. The real-life examples given by the lecturer helped me to better understand the content and it also helped me to engage more with the work. The way the course was presented also encouraged me to work in a timely manner because of the deadlines. Overall, the module was very engaging and I understood more of the work better as I was able to read the content by myself and also because the lecturer provided narrated videos to better explain the content* [P30].

It seems as if the students valued the fact that they could learn at their own pace [P33] and that a lot of assistance was available [P20].

*I learnt a lot more in this course than I did in class. There was a lot of help available to us to help us understand the course [P19].*

The biggest challenges experienced were related to time constraints and access to the internet. One student mentioned:

*I was studying in an environment that did not have electricity and a stable connection. This became problematic as the essay type of questions had limited time allocated to it [P18].*

Another stated:

*I was slow at typing, which affected my marks on essay questions* [P11].

Students commented on the way that the study material was presented. They valued the principles followed in the structuring of the module content. One student said that

*videos are of great help when trying to understand concepts* [P24],

while another mentioned that

*I have learned that I need to break the content down bit by bit to grasp more and understand more* [P18].

Time management was mentioned as one skill that they have learned [P10],[P7]. The conscious efforts during the design of the module to link the theory with real-world scenarios paid off. One student stated:

*I learned how to apply myself to real situations in the world and generate business ideas in my head* [P3].

One student offered valuable comments as a summary of their feedback:

1. *This online module is way better than the F2F classes reason being that we are given a lot of practice questions to better understand the content being taught.*

*2. The due dates put on the small activities push me to continuously do the work and submit it right on time, doing the work makes me understand the content and practice* [P33].

The facilitator of the Gateway module also commented that the structured approach followed in the Gateway module made it much easier to facilitate the module. They indicated that students could manage the workload seeing that it was presented over four weeks and students didn't have competing modules to pay attention to as well. They did indicate that the department would investigate the possibility of an online exam as it would be more convenient for the facilitator and the students.

The key findings in the study include an improvement in the module pass rate, an increase in the level of student satisfaction, effective skills development, increased academic performance and positive student testimonials and feedback.

The Gateway module showed improved module pass rates compared to traditional summer school formats. Students expressed high levels of satisfaction with the module's format and delivery, rating various aspects such as facilitation, content presentation, and feedback above 4 out of 5, indicating a positive reception to the highly structured online learning environment. Despite slightly lower ratings in skill development areas in 2024 compared to 2023, students valued the development of critical thinking, reasoning, and problem-solving skills through the module. There was a noticeable improvement in the average final marks of students attending the Gateway module, indicating a positive impact on student learning outcomes. Students highlighted the benefits of the Gateway module's approach, such as the ability to learn at their own pace from the comfort of their own homes, and the practical application of concepts made possible by the careful attention that was paid to the design of the online module. The continuous assessment component plays a crucial role in keeping students engaged and providing them with timely feedback to improve their learning outcomes. However, challenges such as time constraints and access to stable internet connections were noted as affecting the learning experience.

These findings suggest that the Gateway module effectively addressed academic challenges, enhanced student success rates, and provided a valuable alternative to traditional educational models.

## 6. CONCLUSION

The study on the Gateway module delved into the innovative approach introduced to combat high failure rates in challenging courses. By implementing fully online, intensive four-week modules, the University aimed to address the limitations of traditional educational models that struggle to accommodate diverse learning paces and styles, often resulting in high failure rates and academic congestion. The Gateway modules were designed to provide a highly structured, self-directed learning experience, allowing students to deeply engage with course content and achieve mastery at a foundational level.

This educational initiative aligns with the broader trend towards flexible, digital learning environments that cater to diverse student needs and promote academic success. The methodology of the Gateway modules combined synchronous and asynchronous interactions to maximise accessibility and flexibility, enabling students to engage with the material at a deeper cognitive level outside traditional classroom settings. This pedagogical shift towards hybrid learning environments has been associated with improved student outcomes.

In comparison to traditional remedial approaches like winter or summer schools, the Gateway modules offered a more comprehensive learning experience over a condensed four-week period. This approach aligns with competency-based education principles, emphasising mastery of course material over seat time. Technological integration played a crucial role in the success of the Gateway modules, with advanced educational technologies like the Learning Management System (LMS) supporting various forms of content delivery, including video lectures and interactive learning activities.

Financial and operational considerations are essential for the sustainability of the Gateway modules, requiring initial investment in module development and ongoing costs for facilitation and assessment. The financial model associated with Gateway modules aims to balance upfront design and development costs with long-term benefits in terms of improved student retention and success. The study's findings revealed improvements in pass rates, student engagement, satisfaction, skill development, and academic performance, indicating the effectiveness of the Gateway module initiative in addressing academic challenges and enhancing student success rates.

While the Gateway module initiative shows promise, challenges such as ensuring equitable access to technology and managing program scalability remain. Future research should focus on evaluating the effectiveness of Gateway modules in improving student outcomes and expanding this model to other challenging modules across different disciplines.

In conclusion, the Gateway module proposal represents a strategic response to high failure rates in higher education, leveraging digital technologies and pedagogical innovations to transform the educational landscape and make learning more accessible, flexible, and aligned with students' individual needs and circumstances. Further empirical studies are necessary to validate the efficacy of this model and optimise its integration into the broader educational system.

## ACKNOWLEDGEMENT

We would like to express our sincere gratitude to all the stakeholders involved in the design, development, and facilitation of the Marketing Gateway module. Special thanks also go to the student administration team for their indispensable support throughout the administrative processes. Their collective efforts have been crucial to the success of this initiative.

## REFERENCES

Cavinato, A.G., Hunter, R.A., Ott, L.S. and Robinson, J.K., (2021). Promoting student interaction, engagement, and success in an online environment. *Analytical and Bioanalytical Chemistry*, 413, pp. 1513–1520.

Chandra, S., Theng, L. G., & Lwin, M. O. (2011). Learning management systems: A review of the literature. *International Journal of Instructional Media*, 38(1), 5-21.

Creswell, J.W., (1999). Mixed-method research: Introduction and application. In *Handbook of educational policy* (pp. 455-472). Academic press.

Edwards, R. L., et al., (2020). Exploring student engagement factors in a blended undergraduate course. Available online: https://doi.org/10.5206/cjsotl-rcacea.2020.3.8293.

Katzir, T., Goldberg, A., Aryeh, T.J., Donnelley, K., & Wolf, M., (2013). Intensity vs. Duration: Comparing the Effects of a Fluency-Based Reading Intervention Program, in After-School vs. Summer School Settings. *Journal of education and training studies, 1*, 61-73.

Kreisberg, A.N. and Hsin, A., (2021). The higher educational trajectories of undocumented youth in New York City. *Journal of Ethnic and Migration Studies*, 47(17), pp.3822-3845.

Li, L., Zhang, R. and Piper, A.M., (2023). Predictors of student engagement and perceived learning in emergency online education amidst COVID-19: A community of inquiry perspective. *Computers in Human Behavior Reports*, 12, p.100326.

Mathew, V., Abduroof, A.I. and Gopu, J., (2021). Digital Transformation of Higher Education: Opportunities and Constraints for Teaching, Learning and Research. In *Transforming higher education through digitalization* (pp. 145-171). CRC Press.

Mccombs, J. S., Augustine, C. H., Schwartz, H. L., Bodilly, S. J., Mcinnis, B., Lichter, D. S., & Cross, A. B., (2011). Effectiveness of Summer Learning Programs. In *Making Summer Count: How Summer Programs Can Boost Children's Learning* (pp. 27–36).

Palloff, R.M. and Pratt, K., (2013). *Lessons from the virtual classroom: The realities of online teaching.* John Wiley & Sons.

Qureshi, M.A., Khaskheli, A., Qureshi, J.A., Raza, S.A. and Yousufi, S.Q., (2023). Factors affecting students' learning performance through collaborative learning and engagement. *Interactive Learning Environments,* 31(4), pp.2371-2391.

Sato, S.N., Condes Moreno, E., Rubio-Zarapuz, A., Dalamitros, A.A., Yañez-Sepulveda, R., Tornero-Aguilera, J.F. and Clemente-Suárez, V.J., (2023). Navigating the New Normal: Adapting Online and Distance Learning in the Post-Pandemic Era. *Education Sciences*, 14(1), p.19.

Shaw, M., Burrus, S. and Ferguson, K., (2016). Factors that influence student attrition in online courses. *Online Journal of Distance Learning Administration*, 19(3), pp.211-231.

Simpson, O., (2013). Student retention in distance education: are we failing our students? *Open learning: The Journal of Open, Distance and e-learning*, 28(2), pp.105-119.

Tripp, B., Cozzens, S., Hrycyk, C., Tanner, K.D. and Schinske, J.N., (2024). Content Coverage as a Persistent Exclusionary Practice: Investigating Perspectives of Health Professionals on the Influence of Undergraduate Coursework. *CBE—Life Sciences Education*, 23(1), p.ar5.

Yin, R. K., (2018). *Case study research and applications: Design and methods.* Sage publications.

# PARTICIPATORY PROCESSES IN THE CONTEXT OF THE DEVELOPMENT AND IMPLEMENTATION OF (DIGITAL) LEARNING OFFERINGS: A CASE STUDY ON THE SUPPORT OF INTERNSHIP PHASES IN BUSINESS MANAGEMENT STUDIES

Yvonne Bönninger, Anne Florence Merkle, Elke Mählitz-Galler, Gunnar Voß,
Beate von Velsen-Zerweck and Michael A. Herzog
*Magdeburg-Stendal University of Applied Sciences*
*Osterburger Str. 25, 39576 Stendal, Germany*

**ABSTRACT**

This paper guides academic developers on the participatory development of digital teaching and learning resources within internship processes. It highlights opportunities, challenges, experiences, and success factors. The study examines the development of digital teaching and learning resources on the Practice Innovation Platform, implemented at the Department of Economics, allowing intensive support of students' practical experiences. The Internship Experience Report (IER) outlines various development process phases as an example. The participatory development involved all stakeholders (academic developers, teachers, students, administration, and platform developers) in an agile setting. The Design-Based Research approach (Fishman et al. 2013; Russell et al. 2013) provides a collaborative and iterative framework. This mixed-methods case study illustrates how qualitative and quantitative methods can improve teaching and learning resources for the target audience. Three guided interviews with the internship supervisor, qualitative workshop formats, quantitative questionnaires with students, and additional stakeholder perspectives in four different qualitative settings give insight into the process. Intensive collaboration with internship supervisors and a high level of interaction with all stakeholders is crucial for successfully developing and accepting these learning offerings. Trust-building measures and personal dialog is fundamental here.

**KEYWORDS**

Participation, Teaching and Learning Innovation, Internship, Digital Platform, Academic Development, Business Management Studies

## 1. INTRODUCTION

Since April 2023, the Department of Economics has been testing a digital platform to support the internship phases of students in the undergraduate program in sustainable business administration (cf. Voß et al. 2024). This Practice Innovation Platform (PIP) serves as an interface for networking between companies and the university. It provides targeted teaching and learning offerings for the preparation, during the internship, and the reflection (3rd - 5th semester). The development of the platform and the teaching-learning formats was interdisciplinary and participatory. All relevant internal and external stakeholders were involved from the beginning. Existing literature about participatory (university) development processes usually focuses on individual status groups such as students (Denker et al. 2021; Rink 2020; Lamprecht 2020) or lecturers (Voogt et al. 2014). This paper highlights all relevant stakeholders - teachers, students, administration, companies, and platform developers - and the interdependencies in the joint development process.

This article aims to provide academic developers with guidance for such processes in developing digital teaching and learning offerings for internship phases. Chapter 2 outlines the research questions and the methodology used in this case study. Chapter 3 presents the theoretical background to participation and the collaborative design of teaching and learning materials. An organizational development model is used to visualize the collaboration of academic developers with the accompanying research and the relevant stakeholders. Chapter 4 illustrates the development phases and shows the agile procedure within the framework of the DBR approach, using the Internship Experience Report - a reflection task for students. Chapter 5 discusses the empirical results that lead to our recommendations for the participatory design of internship teaching and learning offerings. The study concludes by summarizing key findings and reflecting on limitations and our experiences.

## 2. PROBLEM STATEMENT AND METHODOLOGY

Teaching in higher education cannot be improved by digitalization alone. Instead, it relies on the involvement of all stakeholders and design principles to enhance educational practices and foster a shift in the learning culture (cf. Getto et al. 2018, p.18). The cooperative relationships between university stakeholders are fundamental here and strengthened through equal joint work and discussions (cf. Merkt 2014, p.33). Through joint development, stakeholders establish a relationship with the developed product and foster acceptance. Students are not seen as consumers of a product designed by teachers or academic developers but are involved in the interactive design process aiming for co-creation (Kasnakoğlu and Mercan 2022). Throughout the entire development process of the IER, academic developers acted as a link between all status groups and closely accompanied the process in an agile approach.

A problem analysis revealed that the students needed support throughout the internship process (Voss et al. 2024, Mählitz-Galler et al. 2019), especially during reflection. So far, an initial reflection of the internship by students has been carried out through a reflection module in the 5th semester and writing an Internship Report (IR). The report includes around 20 pages and is unavailable digitally to other students. Younger semesters rate it as too long, difficult to access, and uninteresting. The newly developed IER using the PIP responds to student criticism and complements the traditional IR.

Table 1. Overview of the surveys and exchange formats

| Stakeholder | Date (DD/MM/YY) | Method | Setting | Tools | n |
|---|---|---|---|---|---|
| M.A.-Students (Risk Management) | 05/06/22 | Workshop/ Group discussion | Course | Miroboard, posters | 11 |
| Administration (Office for Regional Contacts/ International Office) | 02/03/23 | Group discussion | Virtual meeting | Miroboard, Zoom | 2 |
| Internship supervisor | 29/03/23 | Guided interview | Virtual meeting | Audio recording Zoom, Miroboard | 1 |
| Internship supervisor | 20/04/23 | Concept meeting/ Interview | Virtual meeting | Audio recording Zoom, Miroboard | 1 |
| BA-Students Sustainable Business Administration, Course: Preparation for internship | 19/07/23 -10/08/23 | Quantitative survey | Online-Survey | Soscisurvey | 7 |
| BA-Students Sustainable Business Administration, Course: Reflection on internship | 05/12/23 | Qualitative survey: experience reports, group discussion | Virtual meeting | Miroboard, Zoom | 9 |
| Internship supervisor | 07/02/24 | Guided interview | Virtual meeting | Audio recording Zoom, Miroboard | 1 |
| Platform developer | 03/01/24 -09/02/24 | Discussion and planning | E-Mail | E-Groupware | 1 |
| Platform users: Department of Economics | 24/01/24 -31/01/24 | Quantitative survey | Online-Survey | Soscisurvey | 21 |

The following research questions were considered in the development process:

1. How can all stakeholders be actively involved in developing (digital) teaching-learning offerings for the internship process?
2. What opportunities and challenges arise in participatory collaboration?
3. Which recommendations for action (does and don'ts) arise from an academic developer perspective for such organizational development processes?

The DBR approach serves as a research and development framework. The cyclical process of four phases (problem analysis, solution development, evaluation, and reflection) enables continual review and improvement of the learning materials (DBRC 2003; Reeves 2006; Reinmann 2005). Table 1 provides an overview of the qualitative and quantitative methods used (mixed methods).

The initial needs were surveyed by students and professors in qualitative settings. Chapter 4 presents original quotes from the qualitative survey, illustrating the development process of the IER. The evaluation focused on testing the "internal structures of an intervention; how it is used in practice" regarding alpha and beta testing based on software engineering (McKenney and Reeves 2014, p. 143). A quantitative survey (n=21) was conducted regarding the use of the learning offerings by students of the Sustainable Business Administration course from all semesters. It was analyzed using descriptive statistical methods such as frequency analyses. Additionally, a group discussion was conducted with students and lecturers of the 5th semester to assess the work with the developed IER. The results of the group discussion were paraphrased and clustered thematically. The evaluation will help to improve the didactic design of teaching and learning materials in the next iteration of the re-design phase.

# 3. PARTICIPATION AS A DRIVING FORCE FOR ORGANIZATIONAL DEVELOPMENT

## 3.1 Academic Developers and Organizational Development

The continuous development of teaching should align with the latest concepts of high-quality university teaching and study conditions that support academic success. This remains the daily responsibility of teachers, researchers, and professionals in the third space at German universities who are involved in improving teaching and learning (cf. Mayrberger 2020, p. 322). This objective is pursued by the BMBF-funded project "H2D2 – competent didactic and digital teaching and learning", which aims to "test, evaluate and further develop innovative (digital) ideas for the design of teaching and learning" (cf. H2.de). Five different university-wide laboratories were established to improve teaching and learning quality comprehensively. This approach addresses the issue that innovation often occurs on a project basis without structural anchoring in the university (cf. ibid.). The Practice Innovation Lab (PIL) deals with the internship phases of students, initially in the Department of Economics, and is accompanied and researched by the Evaluation Lab. The synergy of these two labs initiates organizational development processes based on the social field needs. "[…] organizational development should also help to change an organization from within, for example, when employees report dissatisfaction or processes need to be redesigned" (Kauffeld 2019, p.75). Organizational development processes can be considered at different levels (see Merkt 2014). Individuals are located at the micro level, while the "structure and culture of an institution" is located at the meso level (cf. ibid. p.35). Here, the establishment of a "professional community of teachers [...] is crucial for the sustainable anchoring of a supportive teaching-learning culture" (cf. ibid. p.40). In the present case, the professional community of educators is understood broadly. This includes the role as teachers and is also part of the university's organizational units (departmental council, senate, university management). Additionally, there are administration members, such as the Office for Regional Contacts, the International Office, and IT and media technology staff. The macro level is embodied by stakeholders outside the university, such as the companies and the platform developer.

### 3.2 The Participatory Process

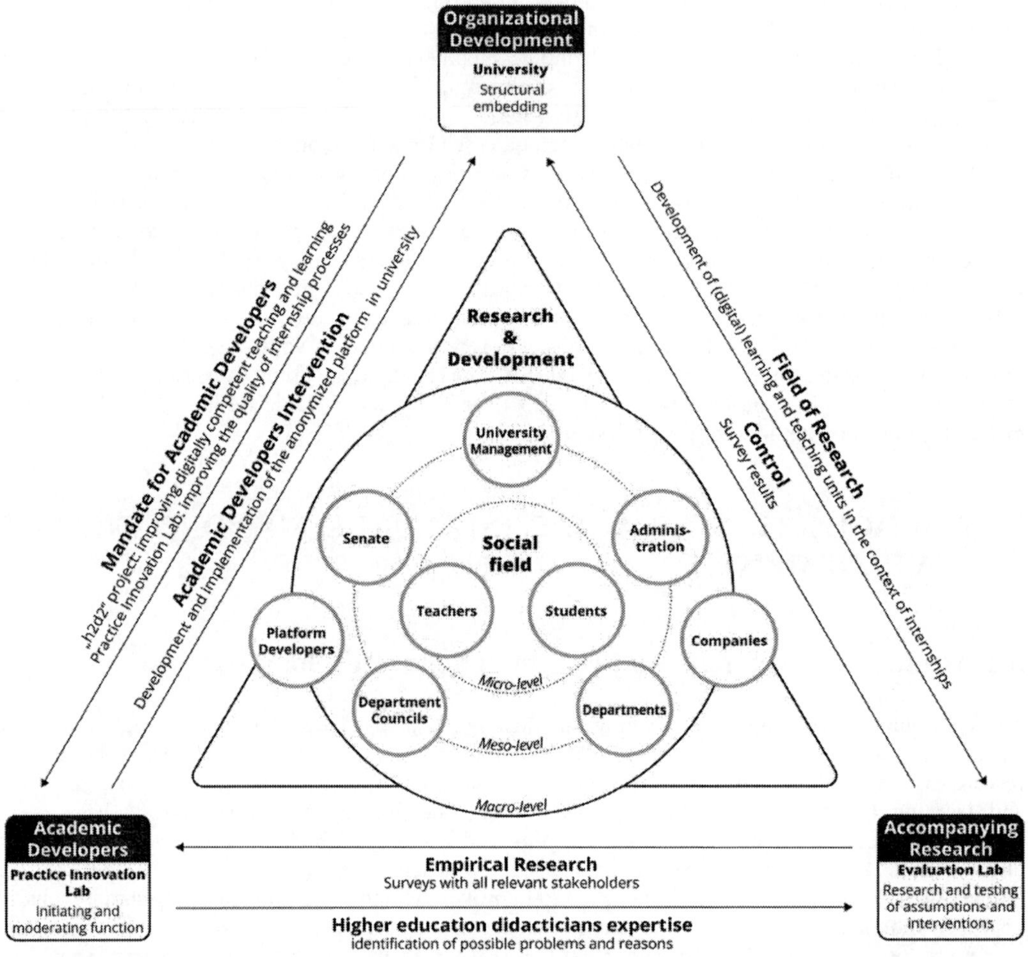


Figure 1. Diagram of the process flow (based on Merkt 2014, p.31)

A model for university development, organizational development, and accompanying research was developed at Magdeburg-Stendal University of Applied Sciences. "The integrative model aims to relate the measures and processes in the three fields of action mentioned above to each other through systematic feedback [...] so that they can be aligned with the core objective of academic development [..]to support students' learning processes and generate synergy effects in their impact" (cf. Merkt 2014, p.31). This model was expanded for this article, including the micro, macro, and meso levels to describe the participatory working methods and processes throughout the project (see Figure 1).

The accompanying research verifies the assumptions of academic development, and the results serve as a basis for controlling organizational development (cf. Merkt 2014, p. 32). The structural, university-wide anchoring of the PIP is a long-term goal pursued jointly by the PIL and E-Value Lab in a bottom-up process. The identified problem areas were derived from the social field and were not specified by the university management. The academic development and accompanying research fields of action interlock by providing each other with impulses for further development and jointly initiating processes to improve the quality of teaching and learning. The development processes involve all stakeholders in the social field at the micro, meso, and macro level. The working method is outlined below based on the development of the IER.

# 4. PARTICIPATORY DEVELOPMENT OF TEACHING-LEARNING MATERIALS - A MODEL PROCESS USING THE EXAMPLE OF THE IER

The following table shows the individual development phases, including a rough estimation of working hours for the team, which consists of four people involved to varying degrees. It is important to mention that the initial needs assessments (step 1) and the evaluation (step 5) concerned the whole development of the PIP with its teaching and learning offerings, not only the IER.

## 4.1 First Iteration

Table 2. Step 1 – Needs assessments

| Stakeholder | ~ 100 hours (academic developers) |
|---|---|
| Students | - "Digital experience reports clearly arranged on one page"<br>- "More experience reports from other students"<br>- "Evaluation of internships" |
| Teachers and Internship Supervisor | - "Maybe we [...] make a difference between experience report and internship report. [...] How was it in general? How did I prepare myself? What are the visa requirements? [...] And the other is really the work assignment, the internship report. So maybe that would also get around the fact that you're not allowed to make these internship reports public [...] if you have to hand in an experience report [...] as well - in addition." (Interview 1).<br>- "The report is such a compulsory task. Actually, the interesting thing is these questions around it [...] You don't get that from the report."<br>- "So far, [...] if you want to know something about England or France, you go to [...] and have all the reports picked out and then you read through them. And I think that's kind of super old-school." |
| Administration | - "Responsibilities must be clarified."<br>- "Specifications and contents of an IER" |

Table 3. Step 2 – Development of the concept

| Stakeholder | ~ 32 hours (academic developers) |
|---|---|
| Academic Developers | - Digital internship experience report on PIP<br>- Exchange of experiences across semesters; direct networking and communication between readers and authors<br>- Multimedia forms of presentation (text, image, and video files) -> promotion of digital skills<br>- Assessment options<br>- Individual visibility settings<br>- Filter option by location, semester, subject group, and keywords<br>- Documentation of competence growth (see Stratmann et al. 2019) |

Table 4. Step 3 – Technical implementation

| Stakeholder | ~ 22 hours (academic developers) + 8 hours (platform developer) |
|---|---|
| Platform Developer | - Development of a new project type for the IER<br>- New filters in the project archive -> search for IER by location, assessments and skills acquired.<br>- Filter by project type, subject group, semester, and keywords already available |

| | |
|---|---|
| Academic Developers | - Draft sample IER with key questions on work culture, environment and reflection, sample images for a "photo gallery on the working atmosphere" at the internship location -> Key questions: "Do you have any pictures of your journey to work? Of your workplace? Where did you spend your breaks?" and for the "Life photo gallery" with pictures of life in the city and favorite places<br>- Description of the steps for creating your own project |

Table 5. Step 4 – Implementation of the prototype in the course: Reflection on the internship phase

| Stakeholder | ~ 2 hours (academic developers) |
|---|---|
| Teachers and Internship Supervisor | - Implementation of the prototype IER in the course: Reflection on the internship phase |

Table 6. Step 5 – Evaluation

| Stakeholder | ~ 68 hours (academic developers) |
|---|---|
| Students | Quantitative survey:<br>- 80% of respondents from Sustainable Business Administration degree program<br>- Survey of satisfaction among PIP users (n=14) five-point scale: Five people = very satisfied, six people = satisfied, and three people = partly satisfied<br>- Information on satisfaction with uploading the report (n=9) five-point scale: Three people = very satisfied, three people = satisfied, and three people = neutral<br><br>Evaluation of IER:<br>- "Without the questions, I wouldn't have known what to put in."<br>- "It takes away some of the excitement when you read the good experiences of others."<br>- "You would have a better overview of what kind of internships you can do because sometimes you can't really assess the possibilities yet." |
| Teachers and Internship Supervisor | - On the topic of key questions: "We have to define together what is important for reflection and attractive for others." |

## 4.2 Second Iteration

Table 7. Step 6 – Further development

| Stakeholder | ~ 16 hours (academic developers) |
|---|---|
| Academic Developers | - Draft of a IER template with the tested key questions and sample images |

Table 8. Step 7 – Technical Implementation

| Stakeholder | ~ 4 hours (platform developer) |
|---|---|
| Platform Developer | - Implementation of a template (sections with headings, guiding questions for editing, sample images, and explanations for editing the template) |

Step 8 - Implementation of the further developed offer in the course:
In the coming winter semester 2024/25, students will create their IER as part of the course using the template.

## 5. DISCUSSION

A development process for digital teaching and learning offerings in the context of internship phases, which actively involves all university stakeholders, provides many opportunities to improve the quality of teaching and learning. Problem areas can be identified directly in the social field, and interventions can be initiated. Pooling the different perspectives of the various stakeholders prevents "blind spots" in problem analysis and introduces new points of view. When they work together, the need for change becomes visible to all stakeholders, and transformation processes can be approached from different starting positions.

However, participatory collaboration presents several challenges. Adequate resources (see Chapter 4) and tactful communication are crucial. Bringing together stakeholders from different cultures and logic of action requires understanding, empathy, and motivation on the part of academic developers. Concerns and fears of change must be taken seriously, and commitments must be adhered to create trust. The implementation of the developed offerings remains voluntary by the teachers.

At the beginning of the project, we hypothesized that the teaching and learning offerings on the PIP would be utilized more due to the stakeholders' intrinsic motivation and that the added value would be recognized more rapidly. It turned out that neither the platform nor the digital teaching-learning offerings were self-selling, especially to the students. Therefore, it is crucial to implement the developed materials in courses and to motivate and involve all stakeholders continuously. The participatory process took much longer and required more resources than we initially assumed (see Chapter 4). DBR (McKenney and Reeves 2014) is well-suited for continuous testing and further development of interventions due to its agile approach with multiple iterations. Unfinished and prototype-like interventions can be revised and improved repeatedly. The IER was first implemented as a sample project with critical questions, which had to be recreated from scratch by students before a template was integrated into the PIP. The evaluation (step 5) showed that these questions successfully achieved their goal by stimulating reflective thinking among the students. However, the initial creation of a project without a template led to difficulties. This impacted the evaluation, requiring the analysis to differentiate between content-related and technical aspects.

A change in learning culture can only work if the interventions are supported by all those involved. In this specific case, numerous negotiation processes took place between the platform developer and the academic developers regarding the platform's further development.

It is important to note that the evaluation's validity is limited by the low level of participation, which is probably caused by students' voluntary attendance outside of lectures. Further surveys and research are necessary to validate or refute the results of the participatory development process of a teaching-learning offer in other subject cultures.

## 6. CONCLUSION

Successful (organizational) development processes thrive on continuous exchange at an appropriate level of intensity. Academic developers have an initiating and moderating function in this process and must neither overburden nor underburden the stakeholders. The iterative approach of the DBR enables the continuous review of innovations. It allows a multifocal view of the process and target group-oriented development: frequent feedback and constant information flow support stakeholders' interest and active participation. The collaborative approach can potentially drive long-term structural changes but requires significant resources. The following recommendations for action should serve as a guide based on our experiences and reflections.

## 7. RECOMMENDATIONS FOR ACTION

1. Identify key persons, involve all status groups in the development processes from the outset, and take problems/concerns/fears seriously.
2. Academic developers have a supporting role: consider the needs and ideas of the teaching staff.
3. Participatory development processes are resource- and time-intensive. Keep the burden on stakeholders to a minimum.

4. Digital innovations need a technical introduction by academic developers. As multipliers, teaching staff initially need technical support to feel confident and can pass on knowledge to students.
5. Internal and external university networks create awareness and trust for the project and the project staff. Use university-wide events (Science Day, university forum, regional advisory board, etc.).
6. Use demonstrations (e.g., prototypes) of interventions in presentations (e.g., in the faculty council).
7. Carry out evaluations in the course. Mails have a low response rate.

Including external stakeholders (companies) in the PIP is relevant for networking, supporting the transfer of theory and practice for students, and promoting early insights into professional fields. The PIL will increasingly involve companies in the next project phase, using their perspective to develop further offerings.

## REFERENCES


Denker, F. et al., (2021). Partizipative Hochschulentwicklung für den digitalen Wandel – Leitlinien des studierendenzentrierten Change Management. *Digitalisierung in Studium und Lehre gemeinsam gestalten*. Springer VS, Wiesbaden, pp. 233–247. DOI:10.1007/978-3-658-32849-8_14.

Fishman, B. J. et al., (2013). Design-Based Implementation Research: an emerging model for transforming the relationship of research and practice. *In Teachers College Record*, vol. 115, no. 14, pp. 136–156. DOI:10.1177/016146811311501415.

Getto, B. et al., (2018). *Digitalisierung und Hochschulentwicklung. Proceedings zur 26. Tagung der Gesellschaft für Medien in der Wissenschaft E.V.* Waxmann, Münster, New York. DOI:10.25656/01:16860.

Kasnakoğlu, B. T. and Mercan, H., (2022). Co-creating positive outcomes in higher education: are students ready for co-creation?. *Journal of Marketing for Higher Education*, Vol. 32, no. 1, pp. 73–88. DOI:10.1080/08841241.2020.1825031.

Kauffeld, S. et al., (2019). Organisationsentwicklung. *Arbeits-, Organisations- und Personalpsychologie für Bachelor*. Springer, Berlin, Heidelberg, pp. 73–104. DOI:10.1007/978-3-662-56013-6_4.

Lamprecht, M., (2020). Lehre und Lernen mitbestimmen. *Zukunft Lernwelt Hochschule*. De Gruyter, Berlin, Boston, pp. 96–102. DOI:10.1515/9783110653663-010.

Mählitz-Galler, E. et al., (2019). Qualitätssicherung des Auslandspraktikums im grundständigen Studiengang Betriebswirtschaftslehre (Poster). *HRK-Nexus Tagung*. Ulm. URL: https://www.hrk-nexus.de/fileadmin/redaktion/hrk-nexus/07-Downloads/07-01-Tagungen/07-01-82_Praktika_2019/HS_Magdeburg_Stendal.pdf.

Mayrberger, K., (2020). Agilität als Motor für Transformationsprozesse in der Lehrentwicklung - Digitalisierung von Lehren und Lernen partizipativ gestalten, erproben und verankern. *Vom E-Learning zur Digitalisierung. Mythen, Realitäten, Perspektiven*. Waxmann, Münster, New York, pp. 320–337. DOI:10.25656/01:21739.

McKenney, S. and Reeves, T.C., (2014). Methods of evaluation and reflection in design research. *In Zeitschrift Für Berufs- Und Wirtschaftspädagogik*, vol. 27, pp. 141–153. URL: https://research.ou.nl/en/publications/methods-of-evaluation-and-reflection-in-design-research.

Merkt, M., (2014). Hochschuldidaktik, Organisationsentwicklung und Begleitforschung an der Hochschule Magdeburg-Stendal – ein integrativer Ansatz. *Hochschuldidaktische Weiterbildung an Fachhochschulen. Lernweltforschung*. Springer VS, Wiesbaden, pp. 27–48. DOI: 10.1007/978-3-658-01497-1_3.

Reeves, T.C., (2006). Design research from a technology perspective. *Educational Design Research*. Routledge, London, pp. 64–78. DOI: 10.4324/9780203088364-13.

Reinmann, G., (2005). Innovation ohne Forschung? Ein Plädoyer für den Design-Based Research-Ansatz in der Lehr-Lernforschung. *In Unterrichtswissenschaft*, vol. 33, no. 1, pp. 52–69. DOI: 10.25656/01:5787.

Rink, S., (2020). Organisation von Hochschule mitgestalten: Studierendenpartizipation als zentrale Perspektive. *Zukunft Lernwelt Hochschule*. Berlin, Boston, De Gruyter, pp. 57–64. DOI: 10.1515/9783110653663-006.

Russell, J. et al., (2013). Theories and Research Methodologies for Design-Based Implementation Research: Examples from Four Cases. *In Teachers College Record*, vol. 115, no. 14, pp. 157–191. DOI:10.1177/016146811311501401.

Stratmann, J. et al., (2009). Lernerfolg und Kompetenz bewerten. Didaktische Potenziale von Portfolios in Lehr-/Lernkontext. *In MedienPädagogik*, vol. 18, pp. 1–19. DOI:10.21240/mpaed/18/2009.12.18.x.

The Design-Based Research Collective, (2003). Design-Based Research: an Emerging paradigm for educational inquiry. *In Educational Researcher*, vol. 32, no. 1, pp. 5–8. DOI:10.3102/0013189x032001005.

Voogt, J. et al., (2015). Collaborative design as a form of professional development. *In Instructional Science*, vol. 43, pp. 259–282. DOI:10.1007/s11251-014-9340-7.

Voß, G. et al., in press 2024. Wissenstransfer und Zusammenarbeit: Die Praxis-Innovation-Plattform als digitale Brücke zwischen Hochschule und Praxis. *In Die Hochschullehre*, Jahrgangssammlung 2024.

# PREDICTING MOBILE LEARNING APPS' USER RATING BASED ON THEIR USABILITY EVALUATION BY GENERATIVE ARTIFICIAL INTELLIGENCE (AI)

Victor K. Y. Chan
*Macao Polytechnic University*
*Rua de Luís Gonzaga Gomes, Macao*

**ABSTRACT**

This article explores the prediction of mobile learning apps' user ratings from various online app stores by means of the apps' usability evaluation by generative artificial intelligence (AI). The generative AI robot employed in the study was Meta Llama, which was requested to award rating scores to the eight major usability dimensions, namely, (1) content/course quality, (2) pedagogical design, (3) learner support, (4) technology infrastructure, (5) social interaction, (6) learner engagement, (7) instructor support, and (8) cost-effectiveness of 17 currently popular mobile learning apps. The apps' user ratings were basically the average of those "star" ratings from the two leading online apps stores, viz., Apple App Store and Google Play. Multiple regression of such a user rating (as the dependent variable) on the Llama rating scores for the eight dimensions or their subset (as the independent variables) was performed. It was found that multiple regression of such a user rating on the Llama rating scores for the five usability dimensions (3) learner support, (4) technology infrastructure, (5) social interaction, (7) instructor support, and (8) cost-effectiveness constituted the final prediction model ($R^2 = .596$, the $F$-test's $F$ statistic = 2.947 with its $p$-value = $0.069 < 0.1$) with the respective regression coefficients for the above five Llama rating scores being -2.150, .355, 2.578, -2.502, and .223 and the corresponding $t$-tests' $p$-values being $.008 < 0.05$, .729, $.043 < 0.05$, .238, and .669.

**KEYWORDS**

Mobile Learning Apps, M-Learning Apps, User Rating, Usability, Generative Artificial Intelligence, AI

# 1. INTRODUCTION

With the rapid advancement of mobile technologies, mobile learning (m-learning) has emerged as a transformative approach to education and training. M-learning leverages the portability, accessibility, and interactivity of mobile devices to provide learners with flexible and personalized learning experiences. The widespread adoption of m-learning has been accompanied by a plethora of m-learning apps, each catering to diverse learning needs and preferences. As the name implies, m-learning apps are mobile forms of learning apps, which are in general apps capable of assisting education (Singer, 2021; Wells, 2023). Typical examples include Khan Academy, Adobe Learning Manager, etc.

## 1.1 Acceptance of and User Satisfaction with M-Learning Apps

The success and effectiveness of m-learning apps heavily rely on user satisfaction. User satisfaction encompasses various aspects, including the app's functionality, user interface (UI), content quality, overall user experience, to name just a few. Several studies have investigated acceptance of and user satisfaction with m-learning apps and the related features and factors underlying such acceptance and user satisfaction, providing valuable insights into users' perceptions and preferences.

For instance, a study by Sophonhiranrak (2021) focuses on and reports a systematic review on analyzing the features of m-learning as well as the barriers and influencing factors of using mobile devices in learning. It pays particular attention to clarifying the factors, techniques, and strategies that enhance learners' experiences in using mobile devices. The results of the review indicated that mobile devices can be used as learning tools for tasks such as submitting homework, reflecting on immediate learning experiences, and sharing ideas. The review suggests that instructors consider three main components in m-learning: learners' and instructors' readiness, learning management, and supporting systems.

Another research by Al-Emran (2019) has objectives that are threefold. First, it analyzes students' perceptions of integrating Knowledge Management (KM) factors in m-learning systems through a preliminary study. Second, it develops a new model by extending the Technology Acceptance Model (TAM) with the KM factors as external variables. Third, it validates the proposed model through the development of m-learning application and the use of statistical analyses methods. Data were collected through a questionnaire survey from 735 information technology undergraduate students in two different universities in two different countries, namely Universiti Malaysia Pahang (UMP) in Malaysia and Al Buraimi University College (BUC) in Oman. The results suggest that knowledge acquisition, application, and protection have a positive effect on perceived ease of use and perceived usefulness of m-learning systems in both samples. Nonetheless, knowledge sharing was found to be partially supported in both samples. Furthermore, perceived usefulness and perceived ease of use were found to be significant determinants of the behavioral intention to use m-learning systems. The overall results contributed to the then existing literature by validating and extending the TAM with the KM factors in two different contexts (i.e., UMP and BUC) and provided various implications to the theory, research, and practice in respect of acceptance of and satisfaction with m-learning.

## 1.2 User Ratings of M-Learning Apps on Online App Stores

User ratings (i.e., "star" ratings) on online app stores, such as the Apple App Store and Google Play, serve as a valuable source of information about users' perception and acceptance of and satisfaction with m-learning apps. These ratings provide insights into the overall user experience, including the app's functionality, reliability, and educational value.

A study by Singh and Suri (2022) analyzed the reviews and "star" ratings of m-learning apps. Word frequency analysis, sentiment analysis, and content analysis were performed on 2,000 reviews of four highly rated, downloaded, and reviewed m-learning apps on Google Play. The results revealed that the most frequently mentioned topics included mobile applications, courses, and teaching. The majority of the reviews had a positive sentiment. Trust, anticipation, and joy were found to be the prominent emotions. It was uncovered that seven identified sub-themes (technical quality, customer support quality, content quality, teaching quality, usefulness, compatibility, and learner influencing others), learners' sentiment, and length of review influenced "star" ratings of apps. The "star" ratings and emotions of joy, trust, disgust, and sadness were moderately associated. The association was strong between the emotions trust and joy and between sadness and fear. Besides, there was a significant difference between "star" ratings based on the apps' sector (public or private) and based on the apps' education level (school or higher).

## 1.3 Evaluation of M-Learning Apps' Usability by Generative AI

Generative AI robots, also known as generative models or generative adversarial networks (GANs), capitalize on a type of artificial intelligence that can contrive new and original content such as images, music, or text. These robots are trained on large datasets and use complex algorithms to generate content that is similar to the training dataset but also unique in its own right. Famed examples of generative AI robots include OpenAI's ChatGPT (Zhang et al, 2023), Claude+ (de Souza et al., 2023), Google PaLM, (Anil et al., 2023), and Meta Llama (Oxford Analytica, 2023), which are capable of generating human-like text with a high degree of coherence and creativity. (Wang et al., 2017)

For the evaluation of m-learning apps, generative AI robots can be programmed or otherwise instructed to analyze the apps' such perspectives or dimensions as (1) content/course quality, (2) pedagogical design, (3) learner support, (4) technology infrastructure, (5) social interaction, (6) learner engagement, (7) instructor support, and (8) cost-effectiveness, among many others. Thereby, a robot can generate an objective and standardized rating score for each dimension of each app, serving the purpose of the app's evaluation.

In recent years, researches have studied the application of generative AI robots to education. One seminal one (Eysenbach, 2023) is an interview with ChatGPT, which provides a snapshot of the current capabilities of ChatGPT. It also illustrates the vast potential for medical education, research, and practice but also hints at current problems and limitations. In the interview, ChatGPT generated some ideas on how to use chatbots in medical education. It also illustrated its capabilities to generate a virtual patient simulation and quizzes for medical students, critiqued a simulated doctor-patient communication, attempted to summarize a research article (which turned out to be fabricated), commented on methods to detect machine-generated text to ensure academic integrity, generated a curriculum for health professionals to learn about AI, and helped to draft a call for papers for a new theme issue to be launched in JMIR Medical Education on ChatGPT. The interview also highlights the importance of proper "prompting." Although the chatbot did make occasional mistakes, it admitted these when challenged. The well-known disturbing tendency of large language models to hallucinate became evident when ChatGPT fabricated references. The interview provides a glimpse into the capabilities and limitations of ChatGPT and the future of AI-supported medical education.

## 1.4 Relationship between M-Learning Apps' User Rating and their Usability Evaluation by Generative AI

In the context of m-learning apps, both user ratings and usability evaluation by generative AI target at quantifying users' perception and acceptance of and satisfaction with particular apps. Therefore, it is both logical and intuitive to expect consistency between them. If they really concur appreciably, generative AI will be able to supplant user ratings to an extent, obviating large-scale surveys or the deployment of specialized information systems to collect user ratings from real human- users.

Regrettably, the author is not aware of any existing literature specifically verifying such consistency between user ratings of m-learning apps and the corresponding usability evaluation by generative AI. This is exactly the gap that this article is to fill. In particular, on the one hand, via the two leading online app stores Apple App Store and Google Play, this article is to survey the "star" ratings (as the user ratings) of a considerable number of popular m-learning apps on the market. On the other hand, it is to utilize one of the most quintessential generative AI robots, Meta Llama, to evaluate the various usability perspectives or dimensions of these apps by means of analyzing global users' textual comments on the apps' dimensions as appear on the web. The ultimate purpose is to determine whether the user ratings are consistent with the evaluations by the robot.

# 2. METHODOLOGY

## 2.1 Data and Materials

For user ratings, the present study drew on the two leading online app stores, viz., the Apple App Store and Google Play. The user rating of any particular m-learning app was derived from averaging the "star" rating of the app as displayed on each of these two online app stores. In case of such a "star" rating being absent from either of the two online app stores, the user rating equated to the "star" rating from the other online app store. In case of such "star" ratings being absent from both the two online app stores, the corresponding app was excluded from any further analysis.

For the usability evaluation, the generative AI robot Meta Llama as accessed through the AI portal poe.com was adopted. Eight major dimensions were utilized to evaluate any m-learning app's usability (Albelbisi, 2020; Hew and Cheung, 2014; Khalil and Ebner, 2014; Kizilcec et al., 2013; Liyanagunawardena, et al., 2013). They were (1) content/course quality, (2) pedagogical design, (3) learner support, (4) technology infrastructure, (5) social interaction, (6) learner engagement, (7) instructor support, and (8) cost-effectiveness, which were rated by Meta Llama. (1) Content/course quality measures the overall quality and relevance of the course content, including the course design, instructional strategies, and assessment methods. (2) Pedagogical design refers to the design of the courses, including the teaching methods, assessment strategies, and learning outcomes. (3) Learner support includes the support provided to learners throughout the courses on the m-learning app. It is essential to ensure that learners have access to adequate support, including technical support and academic support. (4) Technology infrastructure measures the technological capabilities of the m-learning app, including its ability to deliver course content, interact with learners, and manage learner data as well as how user-friendly, accessible, and reliable it is. Technology infrastructure may be assessed by metrics like an m-learning app's uptime, speed of content delivery, compatibility with different devices, and security measures. (5) Social interaction refers to the opportunities for learners to interact with each other and with the instructors. It is essential to ensure that learners have opportunities to collaborate, discuss, and share ideas with each other. (6) Learner engagement measures the level of interaction between learners and the course content. It can be appraised by such metrics as the average time spent on the course content, the number of interactions with the course materials, the number of forum posts and comments by learners, and the average completion rate of the m-learning app's courses. (7) Instructor support measures the level of support provided to learners by the instructors. Instructor support is important because it fosters a sense of community and increases learner engagement. It can be gauged by metrics such as the response time to learners' queries, the quality of responses to learners' queries, the availability of instructors during course hours, and the frequency of instructor-led sessions. (8) Cost-effectiveness measures the cost of delivering the course content and the benefits derived from it. Cost-effectiveness is important because it determines the viability of an m-learning app as a mode of delivering education. It can be translated into metrics, namely, the cost per learner, the return on investment of the m-learning app, the cost savings compared to traditional modes of education delivery, and the revenue generated by the m-learning app. (Chan, 2023)

Despite such formal "definitions" of these eight dimensions above, how Llama defined and interpreted these eight dimensions, how much Llama's own definitions and interpretations differed from these formal "definitions," and how the models microscopically rated these eight dimensions for each app were all beyond the scope and control of this study. As a matter of fact, such "hows' have always been inside the black-box of AI and thus in the domain of AI's "explanability," which has always been a yet-to-accomplish "Holy Grail" in AI research (Adadi and Berrada, 2018). As long as a particular model produced its ratings for these eight dimensions of the apps in question, this study just took these ratings as what they were for the purpose of prediction of the apps' user ratings macroscopically regardless of these microscopic "hows."

The data collection took place in December 2023 and started off with a search for some popular m-learning apps by the Google search engine using the keywords "mobile learning apps." Eighteen apps were outputted as follows:

"Khan Academy, Quizlet, Photomath, EdApp, Kahoot!, iSpring Learn, Adobe Learning Manager, Connecteam-All-In-One, Dayforce, 360Learning, TalentCards, Adobe Connect, Moodle, Socrative Student, TalentLMS, LearnUpon, Trainual, BrainPOP Jr,"

which were referred to by the search engine as "Applications / M-Learning: From sources across the web." In essence, this search "empowered" the Google search engine to shortlist the currently popular m-learning apps.

Subsequently, the generative AI robot Llama was instructed to award rating scores to the eight usability dimensions of the above 18 mobile learning apps. Specifically, the instruction was to ask for a rating score on a scale of 1 to 10 (1 being the worst and 10 the best) for each of the 18 apps in each of the eight dimensions (1) content/course quality, (2) pedagogical design, (3) learner support, (4) technology infrastructure, (5) social interaction, (6) learner engagement, (7) instructor support, and (8) cost-effectiveness. The instruction purposefully specified that the scores be derived from global users' textual comments on these eight dimensions of the apps as appear all around the web. In order words, the robot was directed to derive its scores from global users' textual comments appearing all around the web instead of simply quoting any similar scores already posted somewhere on the web or elsewhere. Also, the 18 apps were explicitly enumerated in the instruction. Llama was employed in this study just experimentally. No superiority or inferiority of its was thereby endorsed.

The robot responded with the rating scores for all the eight dimensions of all the 18 apps mentioned. It is worth noting that these rating scores were real-number values as expressly given by the Llama, and thus the eight underlying usability dimensions were observed variables as opposed latent/unobserved variables (or constructs). Also, given BrianPOP Jr. focusing on the children's market (BrianPOP Jr., 2023) versus the general education market of the remaining 17 apps, the former was eliminated from further analysis.

## 2.2 Analysis

Multiple regression of the user rating (as the dependent variable) on the Llama rating scores for all the eight usability dimensions (as eight independent variables) was performed over all the 17 m-learning apps. Having said that, the tolerance level of any particular Llama rating score (i.e., any particular independent variable) had to be greater than 0.0001 for the Llama rating score to be included in the final multiple regression model. A tolerance level less than 0.0001 would signify rather serious multicollinearity between the corresponding independent variable and the remaining independent variables. For the final multiple regression model, the coefficient of determination $R^2$ was examined to estimate the proportion of the user rating's variation being explained by the Llama rating scores' (or their corresponding usability dimensions') variations. Likewise, the $F$-test outcome was noted to verify whether it was worthwhile to include the Llama rating scores (or their corresponding usability dimensions) in the final model for predicting the user rating. The regression coefficients corresponding to the Llama rating scores (or their corresponding usability dimensions) were recorded. For the $t$-test to test whether each regression coefficient differed from zero, the outcome was scrutinized to validate whether each Llama rating score (or its corresponding usability dimension) impinged on the user rating with statistical significance.

To avert multicollinearity, individual simple regression of the user rating (as the dependent variable) on the Llama rating score (as the only independent variable) for each of the usability dimensions in the above final multiple regression model followed. Each such simple regression model's $R^2$, $F$-test, and $t$-test for the regression coefficient therein indicated whether the corresponding Llama rating score substantially predicted the user rating.

To further discern the effects of the Llama rating scores on the user rating, hierarchical regression of the user rating (as the dependent variable) on the Llama rating scores (as the independent variables) for the usability dimensions in the above final multiple regression model that turned out to have statistically significant correlation coefficients. These Llama rating scores for the usability dimensions entered the hierarchical regression in increasing order of their $p$-values of the $t$-tests on the corresponding regression coefficients in the above final multiple regression.

## 3. RESULTS

The final multiple regression model was found to comprise the Llama rating scores for five usability dimensions, namely, (3) learner support, (4) technology infrastructure, (5) social interaction, (7) instructor support, and (8) cost-effectiveness. The model statistics included: the $R^2 = .596$ and the $F$-test's $F$ statistic = 2.947 with its $p$-value = $0.069 < 0.1$. This implied that 59.6% of the user rating's variation was explained by the variation of the five Llama rating scores for these five usability dimensions in the final model and that at the 10% significance level, the final model was superior to an alternative one with no Llama rating score for any usability dimension at all. In terms of predictability and explainability, the final model turned out to be moderately acceptable. Table 1 depicts the regression coefficients with their 95% confidence intervals (CIs), alongside their $t$ statistics with their $p$-values and degrees of freedom of the corresponding $t$-tests, in the final model.

Table 1. The regression coefficients with their 95% confidence intervals (CIs), alongside the $t$ statistics with their $p$-values and degrees of freedom of the corresponding $t$-tests, in the final multiple regression model

| Dimension Underlying the Llama Rating Score (as an Independent Variable) | Regression Coefficient / 95% CI | $t$ ($p$-value) / Degrees of Freedom |
|---|---|---|
| (3) Learner support | -2.150 / [-3.604, -.695] | -3.293 (.008**) / 11 |
| (4) Technology infrastructure | .355 / [-1.868, 2.578] | .356 (.729) / 11 |
| (5) Social interaction | 2.578 / [.093, 5.064] | 2.311 (.043*) / 11 |
| (7) Instructor support | -2.502 / [-6.944, 1.941] | -1.255 (.238) / 11 |
| (8) Cost-effectiveness | .223 / [-.905, 1.352] | .441 (.669) / 11 |
| * $p < 0.05$ ** $p < 0.01$ | | |

Despite five Llama rating scores for five usability dimensions being in the final multiple regression model, only the regression coefficients of the two Llama rating scores (or the corresponding two usability dimensions) (3) learner support and (5) social interaction were revealed to impinge on the user rating negatively and positively at the 1% ($p < 0.01$) and 5% ($p < 0.05$) significance levels respectively, upon taking into account multicollinearity among the five Llama rating scores.

When it comes to each individual simple regression model of the user rating on the Llama rating score for each of the five dimensions in the final multiple regression model, Table 2 delineates the $R^2$, the $F$ statistic with its $p$-value, and the $t$ statistic with its $p$-value and degrees of freedom of the $t$-test on the corresponding regression coefficient.

Table 2. The $R^2$, the $F$ statistic with its $p$-value, and the $t$ statistic with its $p$-value and degrees of freedom of the $t$-test on the regression coefficient of each individual simple regression model

| Simple regression of the Llama user rating on | $R^2$ | $F$ ($p$-value) | $t$ ($p$-value) / Degrees of Freedom |
|---|---|---|---|
| (3) Learner support | .101 | 1.576 (.230) | -1.256 (.230) / 15 |
| (4) Technology infrastructure | .007 | .101 (.755) | .318 (.755) / 15 |
| (5) Social interaction | .049 | .727 (.408) | .852 (.408) / 15 |
| (7) Instructor support | .023 | .329 (.575) | .574 (.575) / 15 |
| (8) Cost-effectiveness | .017 | .239 (.633) | -.489 (.633) / 15 |

None of these individual simple regression models ended up with an appreciable $R^2$ or an $F$ statistic with statistical significance. Nor did any of these models' regression coefficients give rise to a $t$ statistic with statistical significance. In other words, the variation of none of these models' Llama rating scores (or their corresponding usability dimensions) per se individually offered sufficient predictability and explainability of the user rating's variation. Similarly, none of these five Llama rating scores (or their corresponding usability dimensions) alone individually bore adequately on the user rating.

From Table 1, in the final multiple regression model, there were two usability dimensions, namely, (3) Learner support and (5) Social interaction ending up with significant $p$-values of the $t$-tests on the regression coefficients corresponding to these two dimensions' Llama rating scores, and the $p$-value corresponding to (3) Learner support was less than that corresponding to (5) Social interaction. Therefore, hierarchical regression of the user rating (as the dependent variable) on the Llama rating scores (as the independent variables) for these two dimensions were performed. These Llama rating scores for these two dimensions entered the hierarchical regression in the above increasing order of their $p$-values of the $t$-tests on the corresponding regression coefficients in the final multiple regression model. The hierarchical regression's final model statistics included: $R^2 = .363$ and the $F$-test's $F$ statistic $= 3.711$ with its $p$-value $= 0.053 < 0.1$. This implied that 36.3% of the user rating's variation was explained by the variation of these two Llama rating scores for these two dimensions and that at the 10% significance level, this hierarchical regression's final model was superior to an alternative one with no Llama rating score for any usability dimension at all. Table 3 delineates the hierarchical regression's intermediate results whilst Table 4 illustrates the regression coefficients with their 95% confidence intervals (CIs), alongside their $t$ statistics with their $p$-values and degrees of freedom of the corresponding $t$-tests, in this hierarchical regression's final model. The $p$-value of the $F$ statistic for the change in $R^2$ regarding (3) learner support's entry into the regression model was not statistically significant whereas that regarding (5) social interaction's entry into the model was significant at the 5% significance level ($p < 0.05$). Nevertheless, the regression coefficients of both these two Llama rating scores (or the corresponding two usability dimensions) (3) learner support and (5) social interaction were confirmed to bear on the user rating negatively and positively at the 5% ($p < 0.05$) significance level.

Table 3. The hierarchical regression's intermediate results: $R^2$, change in $R^2$ ($\Delta R^2$), the $F$ statistic for $\Delta R^2$, its $p$-value, and its degrees of freedom

| Dimension Underlying the Llama User Rating Entered | $R^2$ | $\Delta R^2$ | $F$ for $\Delta R^2$ ($p$-value) / (Degrees of Freedom) |
|---|---|---|---|
| (3) Learner support | .101 | .101 | 1.576 (.230) / (1, 14) |
| (5) Social interaction | .363 | .262 | 5.355 (.038*) / (1, 13) |
| * $p < 0.05$ | | | |

Table 4. The regression coefficients with their 95% confidence intervals (CIs), alongside the $t$ statistics with their $p$-values and degrees of freedom of the corresponding $t$-tests, in the hierarchical regression's final model

| Dimension Underlying the Llama User Rating Entered | Regression Coefficient / 95% CI | $t$ ($p$-value) / Degrees of Freedom |
|---|---|---|
| (3) Learner support | -1.325 / [-2.454, -.195] | -2.533 (.025*) / 14 |
| (5) Social interaction | .665 / [.044, 1.286] | 2.314 (.038*) / 14 |
| * $p < 0.05$ | | |

# 4. CONCLUSION

The present article attempts to establish a model to predict m-learning apps' user rating based on their usability evaluation by the generative AI robot Meta Llama. In the final multiple regression model consisting of the Llama rating scores for five usability dimensions (3) learner support, (4) technology infrastructure, (5) social interaction, (7) instructor support, and (8) cost-effectiveness, 59.6% of the user rating's variation is explainable by the Llama rating scores for these five usability dimensions, and at the 10% significance level, the model defeats an alternative one with no Llama rating score for any usability dimension at all. Such a model's predictability and explainability are moderately acceptable. Nevertheless, only the regression coefficients of the two Llama rating scores (or the corresponding two usability dimensions) (3) learner support and (5) social interaction impact the user rating with statistical significance. Furthermore, for each individual simple regression model of the user rating on the Llama rating score for each of the five usability dimensions, the variation of none of these models' Llama rating scores (or their corresponding usability dimensions) per se individually provides substantial predictability and explainability of the user rating's variation. When it comes the hierarchical regression of the user rating on the Llama rating scores for the usability dimensions in the above final multiple regression model that turned out to have statistically significant correlation coefficients (i.e., (3) learner support and (5) social interaction), the *p*-value of the $F$ statistic for the change in $R^2$ regarding (3) learner support's entry into the regression model is not statistically significant whereas that regarding (5) social interaction's entry into the model is significant at the 5% significance level ($p < 0.05$). Even so, in the hierarchical regression's final model, 36.3% of the user rating's variation is explainable by the Llama rating scores for the two usability dimensions (3) learner support and (5) social interaction.

Theoretically, such confounding findings may be due to multicollinearity among the Llama rating scores for the five usability dimensions. Whist the final multiple regression model predicts the user rating fairly well, such multicollinearity may have channeled the effects of the five Llama rating scores (or the corresponding five usability dimensions) on the user rating through only the two Llama rating scores (or the corresponding two usability dimensions) for (3) learner support and (5) social interaction, which thus appear to be the only two Llama rating scores (or the only corresponding two usability dimensions) affecting the user rating. Also, the variation of none of the five Llama rating scores (or the corresponding five usability dimensions) per se individually predicts and explains the user rating's variation well because the user rating hinges on all the five Llama rating scores, though interrelated through multicollinearity, instead of any one of them individually. In addition, there may exist interdependence between the usability dimensions (3) learner support and (5) social interaction.

This study itself is not without its critics. First, rating scores accorded by any generative AI robots based on users' textual comments on m-learning apps are susceptible to imperfection due to the variety of users' textual comments taken into consideration by the robots, the subjectivity, bias, and variation unavoidably in the users' textual comments, etc. (Chan, 2023). Second, the usability of m-learning apps was evaluated only by the generative AI robot Meta Llama, which might not be representative of the profuse robots available globally. Therefore, the findings may not be generalizable to all generative AI robots in general. Third, this robot Llama, like any of its "peers," was primarily trained on datasets up to some years back, so even the rating scores from Llama in December 2023 might lag behind the then latest m-learning apps, their versions and the then corresponding latest user ratings from the online app stores. In fact, there might be some time discrepancy between the then most current user ratings collected from the online app stores and such Llama rating scores. Therefore, the prediction model of the user rating might be plagued by some inherent prediction error in that obsolete Llama rating scores were used to predict the then most current user ratings. Therefore, it is to the benefit of the research circle to further extend this study by including more generative AI robots, in particular, those having been trained on the most current datasets in order to even out the adverse effects of the first two points above and to head-on redress the third.

In summary, whereas the interplay between the user rating and the Llama rating scores was rather intricate and beleaguered by multicollinearity, the above multiple regression model at least succeeds in predicting the user rating (and thus user satisfaction) of an m-learning app by utilizing the Llama rating scores for the five usability dimensions (3) learner support, (4) technology infrastructure, (5) social interaction, (7) instructor support, and (8) cost-effectiveness of the app. Generative AI robots are thus considered promising as a nascent method to insightfully interpret global users' textual comments at scale, to rate the omnibus usability dimensions of each m-learning app using such comments, and to ultimately predict real human-user ratings of the app.

## REFERENCES


Adadi, A. and Berrada, M., (2018). Peeking Inside the Black-Box: A Survey on Explainable Artificial Intelligence (XAI). *In IEEE Access 6*, 52138-52160. https://doi.org/10.1109/ACCESS.2018.2870052

Albelbisi, N. A., (2020). Development and Validation of the MOOC Success Scale (MOOC-SS). *In Education and Information Technologies*, Vol. 25, No. 5, pp. 4535-4555. https://doi.org/10.1007/s10639-020-10186-4

Al-Emran, M. N. H., (2019). *Extending the Technology Acceptance Model with Knowledge Management Factors to Examine the Acceptance of Mobile Learning*. PhD Thesis, Faculty of Computer Systems and Software Engineering, Universiti Malaysia Pahang, Pahang, Malaysia. https://core.ac.uk/download/pdf/333902136.pdf

Anil, R. et al., (2023). *PaLM 2 Technical Report*. https://doi.org/10.48550/arXiv.2305.10403

BrianPOP Jr., (2023). Homepage, https://educators.brainpop.com/contact-us. Last accessed December 14, 2023.

Chan, V. K. Y., (2023). Evaluating Popular MOOC Platforms by Generative Artificial Intelligence (AI) Robots: How Consistent are the Robots? *Proceedings of the 20th International Conference on Cognition and Exploratory Learning in Digital Age (CELDA 2023)*. Madeira Island, Portugal, pp. 329-336.

de Souza, B. C. et al., (2023). Are the New AIs Smart Enough to Steal your Job? IQ Scores for ChatGPT, Microsoft Bing, Google Bard and Quora Poe. https://papers.ssrn.com/sol3/papers.cfm?abstract_id=4412505

Eysenbach, G., (2023). The Role of ChatGPT, Generative Language Models, and Artificial Intelligence in Medical Education: A Conversation with ChatGPT and a Call for Papers. *In JMIR Medical Education*, Vol. 9. https://doi.org/10.2196/46885

Hew, K. F. and Cheung, W. S., (2014). Students' and Instructors' Use of Massive Open Online Courses (MOOCs): Motivations and Challenges. *In Educational Research Review*, Vol. 12, pp. 45-58. https://doi.org/10.1016/j.edurev.2014.05.001

Khalil, H. and Ebner, M., (2014): MOOCs Completion Rates and Possible Methods to Improve Retention - A Literature Review. *Proceedings of EdMedia 2014--World Conference on Educational Media and Technology*. Tampere, Finland, pp. 1305-1313. https://www.learntechlib.org/primary/p/147656/

Kizilcec, R. F. et al., (2013): Deconstructing Disengagement: Analyzing Learner Subpopulations in Massive Open Online Courses. *Proceedings of the Third ACM International Conference on Learning Analytics and Knowledge*. pp. 170-179. https://doi.org/10.1145/2460296.2460330

Liyanagunawardena, T. R. et al., (2013). MOOCs: A Systematic Study of the Published Literature 2008-2012. *In The International Review of Research in Open and Distributed Learning*, Vol. 14, No. 3, pp. 202-227. https://doi.org/10.19173/irrodl.v14i3.1455

Oxford Analytica, (2023). Meta LLaMa Leak Raises Risk of AI-Linked Harms. *In Expert briefings*. Emerald, Leeds, England, UK. https://doi.org/10.1108/OXAN-ES276597

Singer, N., (2021). Learning Apps Have Boomed in the Pandemic. Now Comes the Real Test. *In The New York Times*, March 17, 2021. https://www.nytimes.com/2021/03/17/technology/learning-apps-students.html

Singh, Y. and Suri, P. K., (2022). An Empirical Analysis of Mobile Learning App Usage Experience. *In Technology in Society*, Vol. 68, 101929. https://doi.org/10.1016/j.techsoc.2022.101929

Sophonhiranrak, S., (2021). Features, Barriers, and Influencing Factors of Mobile Learning in Higher Education: A Systematic Review. *In Heliyon*, Vol. 7, No. 4. https://doi.org/10.1016/j.heliyon.2021.e06696

Wang, K. et al., (2017). Generative Adversarial Networks: Introduction and Outlook. *In IEEE/CAA Journal of Automatica Sinica*, Vol. 4, No. 4, pp. 588-598. https://doi.org/10.1109/JAS.2017.7510583

Wells, R., (2023). 10 Mobile Learning Apps, Sites and Tools for Busy Professionals. *In Forbes*, November 8, 2023. https://www.forbes.com/sites/rachelwells/2023/11/08/10-mobile-learning-apps-sites-and-tools-for-busy-professionals/?sh=117f68c35941

Zhang, C. et al, (2023). A Complete Survey on Generative AI (AIGC): Is ChatGPT from GPT-4 to GPT-5 All You Need? https://arxiv.org/abs/2303.11717

# EVALUATION OF THE IMPACT OF ACTIVE LEARNING USING AUGMENTED REALITY ON CADETS' KNOWLEDGE

Jadranko Tuta[1] and Ljerka Luic[2]
[1]*University of Zagreb, Croatian Military Academy "Dr. Franjo Tudman"*
*Ilica 256b, 10000 Zagreb, Croatia*
[2]*University North*
*Trg dr. Zarka Dolinara 1, Koprivnica, Croatia*

**ABSTRACT**

The work is based on the conducted scientific research with the aim of evaluating active learning using augmented reality on the knowledge of cadets. The application of augmented reality in higher education teaching highlights the increased effectiveness of teaching, the development of digital literacy and digital competences that have a great impact on learning and teaching in university education, which is the subject of this paper, which considers information practices and digital perspectives of cadet education. The goal of the research is focused on the identification of information objects Located, Games, Research of the information package Active learning, and on the basis of them, the existence, relationship and intensity of connections of the created information constructs Perception, Knowledge were examined and determined by assessing the impact of active learning using augmented reality. In the implementation of cadet training, a problem related to the application of active learning in classes was observed. The research was conducted using the field method and quantitative analysis with the following methods: experiment, survey questionnaire and knowledge test with cadets of undergraduate and graduate university military studies in Military Engineering and Military Leadership and Management. The material on which the mentioned methods were carried out consists of cadets, students of the first, second, third, fourth and fifth years of study, which, according to the current availability, included N = 270 of them. In the experimental part of the research, data were collected using augmented reality technology through the HaubicAR application, quantitative analysis with methods: an experiment, a Google Forms questionnaire tool and a knowledge test, which are integrated in the HaubicAR application, and descriptive statistical analysis. The results obtained in this research support the claim that it is necessary to propose and create a model for teaching cadets using augmented reality technology and to develop a model for the serial development of an application in order for cadets to learn by applying new technologies to be used in learning through practice on virtual models, game-based learning and research-based learning in military studies. In light of this, as a result of the research, an active learning model was developed using AR technology, which is operationally applied in the implementation of cadet education at the University of Defense and Security "Dr. Franjo Tudman" in Croatia.

**KEYWORDS**

Active Learning, Augmented Reality, Cadet, Evaluation, Knowledge, Model

## 1. INTRODUCTION

The information model forms the basis of modeling in information and communication sciences. It explains information flows and shows the structure of information constructs. In the information model, information and data are abstracted as information objects that are classified according to their functions. Furthermore, the information model explains the structure and types of information packages through the logical model of information. They later appear in the functional model, where they represent the content that is transferred and, if necessary, transformed, as part of the communication between its functional entities (Stančić, 2005). One of the XR technologies (Extended Reality) that displays information in 3D space is Augmented Reality (AR). Augmented reality (hereafter AR) uses computer-generated sensory signals in the real world that allow the user to simultaneously experience digitized elements of the virtual and real world (Memarsadeghi and Varshney, 2020). AR technology is recognized as one of the most important new technologies for the development of active learning in higher education (Saltan and Arslan, 2017). The fact that some universities use augmented reality and virtual reality (VR) in teaching is a motivation for their more frequent use in innovative ways that expand educational dimensions (Swensen, 2016). Furthermore, AR and VR technologies provide the possibility of overlaying images, text, audio, video and audio-video on existing AR and VR structures, on space or images in real time and in communication with students (Greenwood and O'Neil, 2016). The use of AR and VR applications in teaching in higher education is compatible with certain pedagogical approaches that fit into the framework of active learning: constructivist learning, situated learning, game-based learning, research-based learning (Pantelić and Vukovac, 2017; Castellanos and Pérez, 2017). Compatibility with constructivist learning is reflected in the student's interaction with the learning environment and in building connections with the student's prior knowledge. In accordance with the approach of situated learning, AR has a special importance because it enables, in real time and environment, the display of virtual objects that students would otherwise not be able to encounter and that would be difficult to visualize. In a game-based learning approach, students can role-play an AR game that prepares them for everyday life, while in inquiry-based learning, students explore and experiment with virtual models. Furthermore, by applying augmented reality, it is possible to achieve a learning experience outside the classroom, placing it in the context of the connection between reality and the learning situation in which students participate, whereby any physical space can become a stage for active learning (Bower et al., 2014; Amaguana et al., 2018). Authors Akçayir and Akçayir emphasize the potential of using augmented reality in teaching. Among other things, they cite the positive effects of the use of AR, which are reflected in a better understanding of the teaching content and point out that the main topics are learned through the effective design of active learning (Akçayir and Akçayir, 2017).

The results of the research conducted at the Croatian Military Academy "Dr. Franjo Tudman" in cooperation with the Faculty of Electrical Engineering and Computing of the University of Zagreb show that XR technologies have the potential to improve the education process in Croatia as well. The insights reached by the researchers confirmed the hypothesis that XR technologies greatly contribute to active learning because they arouse curiosity, encourage active participation and increase motivation to learn (Tuta et al, 2019). The authors He, Han and colleagues (He et al, 2017) estimate that it is possible to improve education by applying AR, especially in supporting familiarization with equipment such as dealing with emergency situations during which students are unable to attend practical classes. A recent example of current research on the application of AR in higher education points to the importance of teaching effectiveness and the ability of students to develop digital literacy (Eshet-Alkalai, 2012) and digital competencies that have a great impact on learning and teaching in university education (Saez-Lopez et al. , 2020). For several years now, augmented reality has been recognized as one of the technologies with the greatest impact on university education. This is especially important with immersive AR and VR media because they depend on parallel effects in the real world. Most of these effects are triggered by the student's actions as an AR user, especially in situations where during teaching the student needs to look in a certain direction or use controls to manipulate a virtual object (Richards, 2017). The application of augmented reality in the higher education system has a significant impact on modern education, not only on the acquisition of knowledge and skills by students, but also on the transfer of knowledge and skills by teachers to students (Karlsson, 2015). Furthermore, from the review of previous research, it is possible to conclude that learning in an AR environment coincides with the concepts of educational theory, one of which apostrophizes learning as a result of the connection between a stimulus and a response to that stimulus. An AR-based learning platform provides students with a choice of modeling tools and scenarios, designed to be easy to use (Lu and Liu, 2015; Martin-Gonzalez et al, 2016). In recent years, learning by interacting with virtual objects integrated in augmented reality has become a modern way of digital learning. The author Radu also points out in his research that these practices promote higher performance in learning, are related to their creative, motivational and motor potential and the power of an immersive sense of experience (Radu, 2014). Several studies (Man and Madhusudan, 2023; Tekedere and Göker, 2016; Cózar-Gutiérrez and Sáez-López, 2016; Cabero and Barroso, 2016), whose goal is to analyze the use of augmented reality in an educational context, highlight positive examples of the application of AR in teaching, where the motivation to learn is held to be the most prominent factor. In their work, Cabero and Barroso analyze published studies on the implications of using AR technologies and conclude that AR facilitates the understanding of complex phenomena and concepts; promotes contextualization and enrichment of information; enables individualization of practical teaching and adaptation to different forms of digital intelligence; provides students with the ability to interact through AR, manipulate real objects; favors ubiquitous and contextualized learning by turning any physical space into a stimulating academic environment; facilitates the development of constructivist teaching and/or learning methodologies; promotes the development of graphic skills through the perception of spatial content and 3D objects; favors learning through practice (experiential learning) and increases motivation and improves learning success. Newer research suggests that scientific studies and reports that suggest that XR technology trends should be introduced into teaching as soon as possible, with the aim of changing teaching practice, should be accepted in university settings (O'Flaherty and Phillips, 2015; Whitton and Langan, 2018; Ding, 2019).; Riman, et al, 2020). In addition to the above, modern research introduces new technologies into the university education of students and teachers, who must adapt to the challenges and demands of the modern digital society, taking into account the perception and emerging trends that today's students, tomorrow's professionals will encounter in their immediate work environment (Abid et al., 2022). The mentioned research partly answers the problem related to the application of active learning in teaching students. However, they do not provide a solution on how to systematically validate the learning outcomes achieved by applying AR in teaching, nor do they accept the information package of active learning and its information objects, the conceptualization of which was considered in this research, the goal of which is to assess the impact of active learning on learning outcomes.

## 2. BACKGROUND OF THE RESEARCH

The modern world is experiencing changes in various parts of life, and the last decade of the twentieth century and the beginning of the twenty-first century saw tremendous progress, especially in the field of new technologies. The field of education stands out as the most significant area affected by new information and communication technologies, which resulted in the development of different types of active learning; constructivist learning, situated learning, game-based learning, research-based learning. Letina (2016: 8) states in his work "Active forms of learning imply a student who is not only a passive observer and receiver of information, but an engaged participant in the teaching process who actively contributes to the overall development of his being".

Active learning is a method that enables the improvement of education through the application of AR, especially in supporting familiarization with equipment, such as dealing with emergency situations during which students cannot attend practical classes. Enthusiasm for the application of active learning has developed in parallel with the introduction of new technologies. The method of active learning has undergone significant development with the development of new technologies.

Due to the rapid growth in today's world, thanks to scientific and technological progress, it is essential to develop the educational process parallel to the innovations of new technologies and different types of active learning. Consequently, many university institutions around the world have implemented active learning using new technologies, which has contributed to the modern way of digital learning. Therefore, these practices promote higher performance in learning and are related to their creative, motivational and motoric potential (Radu, 2014). The application of available new technologies and accessible digital content successfully ensured the possibility of high-quality education for all members of society. The rapid development of new technology has created innovative teaching experiences through the impact of active learning using new technologies on student knowledge. Furthermore, it provides students with the ability to interact through AR, manipulate real objects; it favors ubiquitous and contextualized learning by turning any physical space into a stimulating academic environment with the aim of changing teaching practice (Miralles et al, 2019).

Therefore, it was important to determine what correlations exist between independent variables (gender, field of study and year of study) and dependent variables (immersive experience, knowledge of XR technologies, perception of the use of augmented reality in teaching, prior knowledge of AR applications and experience in using the *HaubicAR* application). Furthermore, the aim of the paper is to create an information model for assessing the impact of active learning using augmented reality and to determine the best predictors for the further development of technological possibilities of future application solutions based on new technologies using augmented reality.

### 2.1 A Research Problem

Immersive technologies are rapidly changing the field of education. Among them, augmented reality (AR) is promising as a resource, especially for education in the field of Science, Technology, Engineering, Arts and Mathematics (STEAM). However, few teachers directly apply this new medium in the classroom and thus, only a select few students benefit from AR-enriched offerings (Jesionkowska et al, 2020). Curricula are overloaded and faculties generally lack development resources, so there is no room for experimentation. Active learning using new technologies, including (XR), virtual reality (VR) and augmented reality (AR), is increasingly used in most parts of the world as a viable alternative to conventional education. It is believed that the potential of information and communication technology, and especially active learning using new technologies, should greatly improve the educational process and contribute to the development of active learning. New technologies in higher education bring increased and positive effectiveness to teaching and learning by providing students and teachers with flexibility, accessibility, more opportunities for participation and collaboration, and better outcomes (U. S. Department of Education, 2017). Furthermore, digital literacy and digital competence of students are being developed, which have a huge impact on learning and teaching in university education. Any change in teaching and learning strategies is always evaluated according to its impact on academic performance. From the literature review, the authors noted that there is limited research on the effects of active learning using augmented reality on cadet knowledge. Most research on active learning has focused on issues, challenges, attitudes and perceptions. The mentioned research partially answers the problem related to the application of active learning in teaching students (Letina, 2016), but they do not provide a solution on how to systematically validate the learning outcomes achieved by applying AR in teaching, nor do they accept the information package of active learning and its information objects, whose conceptualization will be considered in the conducted research, the goal of which is to assess the impact of active learning on learning outcomes. The rationale for this research arose from this limited research on evaluating the impact of active learning using augmented reality (AR) on cadet knowledge. The potential for improving the education process will serve as an incentive for this research to fill an existing important research gap in the literature.

## 2.2 Research Hypotheses

By evaluating the impact of three different forms of active learning (situated learning, learning through play and exploratory learning) using augmented reality AR on the knowledge of cadets, the perception and level of knowledge of cadets was investigated. The aim of the study is to test the hypotheses of students' perceptions towards the application of AR in classes and the form of AR active learning that provides a higher level of knowledge. On the basis of the set research goal, the established factors of the information objects Located, Games, Research of the information package Active learning, the intensity of the connections between the created information constructs Perception and Knowledge was investigated, by assessing the impact of active learning using augmented reality, and the following hypotheses were tested:

H1: The perception of cadets towards the application of augmented reality, AR in teaching is related to their immersive experience and knowledge of AR technology.

H2: Cadets who were taught through all forms of AR active learning (Situated, Games, Research) achieve a statistically significantly higher level of knowledge compared to cadets of the other two groups.

# 3. METHODOLOGY

The design of the experiment represents the basic procedures aimed at conducting the research. It also provides useful, precise information that will answer research questions or hypotheses. The research was conducted using an experimental method with a quantitative approach that was descriptive and predictive in nature. In order to achieve this, Spearman's correlation coefficients were used for data analysis, which were used to examine the connection between encountering the term VR and the term AR in previous education with the cadets' perception of the application of augmented reality technology in classes and the connection between the frequency of using virtual reality technology in everyday life.

## 3.1 Design

As Woodwell (2013) explains, descriptive research offers answers to reasonable questions to make a decision about current events. In this capacity, descriptive research often involves the use of statistics to delineate the nature and form of data and the distribution of measured data. Descriptive statistics (such as field of study, year of study, age group, gender, use of technology (VR) and (AR), mean scores, mean scores, and Spearman's correlation) were used in this research. Creswell (2022) explained that the process of analyzing sample-based data can be facilitated by descriptive statistics by drawing inferences from a given population. In this research, descriptive statistics were used for hypothesis testing. Maxwell (2012) described the research method as an approach that provides a context within which appropriate techniques and methods can be adopted and developed to achieve the overall purpose of the research. The examination of the cadets' perception of the application of AR technology in teaching was assessed with a Likert scale, and the principal component analysis (PCA) method was used for the factor structure of the construct of the cadets' perception of the application of AR technology in teaching. Verification of the suitability of the correlation matrix for factorization was calculated by Kaiser-Meyer-Olkin (KMO) and Bartlett's test indicating suitability for factorization. For both criteria, the determination of the number of factors was calculated by Kaiser-Gutman. Kruskal-Wallis H-test and Post hoc analysis (Dunn-Bonferroni) were used to test the statistical significance of differences in the time spent solving the knowledge pretest and knowledge subtest between cadets from three groups. Quantitative data were analyzed using SPSS statistical software, and descriptive statistics were also presented. Complete statistical processing was performed in the IBM SPSS 25 program.

## 3.2 Participants

The experiment included 270 respondents, cadets attending the Croatian Military Academy "Dr. Franjo Tudman" which were divided into three groups, where each group had 90 respondents. The experiment took place in a classroom in groups of ten subjects. Each group was assigned different forms of active learning through a sample auditory exercise on tablets in the HaubicAR application on the same assigned topic "Technical and operational characteristics of the D-30 122 mm Howitzer", common to all three groups of subjects.

## 3.3 Subjects' Tasks for all Three Groups

Before the start of the experiment, each group of subjects was presented with the purpose and goal of conducting the auditory exercise. The respondents participated in the experiment, which was divided into three groups; Group 1 – *Situated learning*, Group 2 – *Situated learning* + *Games* and Group 3 – *Situated learning* + *Games* + *Research*. Group 1 - *Situated learning* was assigned the task of studying digital material on a given topic in PDF format. Through situated learning, respondents are enabled to learn an embedded topic in augmented reality through the activity of educational experiences in a real environment. The subjects had a certain amount of time to study the digital material in order to take the knowledge test after the experiment. Group 2 – *Situated learning* + *Games* had a quiz game. The respondents used immersive learning on a given topic through a quiz game where authentic sources and contextually relevant information were incorporated. The subjects had a certain amount of time to play the quiz in order to take the knowledge test after the experiment. Group 3 – *Situated learning* + *Games* + *Research learning* had a D-30 122mm Howitzer in the form of a digital object in augmented reality where they explore and experiment with it. The respondents used immersive learning through visualization that is contextual to the given research topic. The respondents had a certain amount of time for research in order to take the knowledge test after the experiment.

## 3.4 Data Collection

The data was collected through the *HaubicAR* application, in which the Google Forms tool is integrated, with which all data were collected: data collected by the method of a survey questionnaire, by the method of a pretest of knowledge acquired before exposure to AR active learning, by the method of an experiment and by the method of a subtest of knowledge acquired after active participation in a certain form of AR active learning. Data collection in the form of surveys, pretest, experiment and subtest of knowledge was carried out through a sample auditory exercise that lasted up to 60 minutes. Data collection was carried out with cadets in the first year of study in the compulsory course "Military Topography", in the second year of study in the compulsory course "Tactics", in the third year of study in the compulsory course "Organization and Technology of Military Techniques", in the fourth year of study in the compulsory course "Infantry Tactics" and fifth year of study from the compulsory course "Fundamentals of the use of tactical units and services".

## 3.5 Data Preparation and Processing

Data processing for the purposes of this work included checking the questionnaire in Google Forms, editing the data and coding within the *HaubicAR* application, which was done before starting the experiment. In this research, data was collected and analyzed using a quantitative method. After preparing the data from the *HaubicAR* application, the obtained data were analyzed using the statistical package IBM SPSS 25. Based on the highlighted important points, the observed data were categorized and discussed in an appropriate manner. The authors observed the causes arising from the data, and the findings were interpreted by analyzing the results collected using the following methods: an experiment, a survey questionnaire and a knowledge test. The Cronbach Alpha coefficient was applied to the research to determine the validity and reliability of the instruments used. After the end of the experiment, all collected data were accessed for processing on the given link. By clicking on the link, the data is downloaded from the format file (.csv), i.e. in a table-structured format. Given that there was a possibility that the respondent accessed an activity more than once, for example, he started and closed the application several times, those activities of the respondent were considered an error in the examination within the format file (.csv) and were corrected before the actual preparation of the data for processing. When the application was free of errors related to the subject or errors of a technical nature, the data were ready for further processing. In order to prepare the data for processing, the respondents were divided into groups (Groups 1, 2, 3). Each group included 90 respondents, which is a total of 270 respondents. Furthermore, sorting according to groups enables high-quality data processing of respondents through activity format files (.csv): ID of respondent, Activity, Start of activity, End of activity, Duration of activity. Furthermore, editing of the data table (transposition of data, definition of variables, creation of an SPSS database with data), descriptive statistics for the variables that measured the use and knowledge of VR/AR on the total sample, scoring of tasks on the knowledge pretest and knowledge subtest, implementation of appropriate analysis that will be tested (+ prerequisites for analyses, tabular presentation of results, interpretation of performed analyses): the effect of the form of AR active learning on the knowledge of cadets, the difference between groups solving the pretest, the difference between groups solving the subtest, factor analysis and reliability testing for the perception scale (six particles), examining the relationship between perception and knowledge. After processing the data, the results from the SPSS database were delivered in tabular form and an interpretation was written in the MS Word program.

# 4. RESULTS AND DISCUSSION

According to Pallant (2016), Spearman's correlation, post hoc tests and two-factor analysis of variance (group x measurement) are used to break down multiple comparisons with Bonferroni correction indicating the mean score of subjects. Furthermore, the Kruskal-Wallis H test and Post hoc analysis (Dunn – Bonferroni) is used for the time spent solving the knowledge test when there are three different groups of subjects or conditions, to suggest whether there is a statistically significant difference in their mean scores. Below is an overview of the results that were reached based on the conducted research, which was set up on an experimental model.

## 4.1 Cadets' Perception of the Application of AR in Teaching

To examine the cadets' perception of the application of augmented reality technology in teaching, a scale of six items was used, i.e. statements on the basis of which respondents assessed their agreement/disagreement with the statements offered, which were formulated using a five-point Likert scale (1 - I completely disagree to 5 – I completely agree). In order to examine the factor structure of the construct of the cadets' perception of the application of augmented reality technology in teaching, the method of principal components (Principal Component Analysis) was carried out, which included these six particles listed in order of factor size and communality: A6 The application of augmented reality technology in teaching would contribute to the quality cadet learning outcomes; A2 The application of augmented reality technology would improve the implementation of military professional practice; A7 Application of augmented reality technology in cadet training would increase interest in military studies; A4 The use of augmented reality technology in teaching would improve the digital skills of cadets; A3 Frequent use of augmented reality technology in classes could endanger the health of cadets; A5 Using augmented reality technology in class will not improve my digital skills significantly. In order to check the suitability of the correlation matrix for factorization, the Kaiser–Meyer–Olkin (KMO) value of 0.795 and Bartlett's test ($\chi2= 362.128$, df = 15, $p < 0.001$) were calculated. Given that KMO values greater than 0.06 are acceptable and that a significant Bartlett's test indicates suitability for factorization, these prerequisites are met. Based on both criteria for determining the number of factors to be retained (Kaiser-Guttman criterion - retention of factors whose eigenvalue is greater than 1 and Cattell's scree-test - graphic representation with eigenvalues on the ordinate and the number of factors on the abscissa, based on which the number of factors to be retained is determined by the inflection point) one factor is extracted. Table 1 lists the results of the performed analysis: factor saturations, eigenvalue communalities and percentage of explained variance. The extracted factor explained 45.26% of the variance. Two particles (A3 – Frequent use of augmented reality technology in teaching could endanger the health of cadets and A5 – Application of augmented reality technology in teaching will not significantly improve my digital skills) have lower communalities and are negatively saturated (refer to a negative opinion about the application of augmented reality technology reality in teaching).

Table 1. Results of the factor analysis of the cadets' perception of the use of AR in teaching

| | Faktor<br>1 | Communality |
|---|---|---|
| A6 Application of augmented reality technology in teaching would contribute to the quality of cadet learning outcomes. | .82 | .67 |
| A2 The application of augmented reality technology would improve the implementation of military professional practice. | .77 | .59 |
| A7 Application of augmented reality technology in cadet training would increase interest in military studies. | .71 | .50 |
| A4 The use of augmented reality technology in teaching would improve the digital skills of cadets. | .69 | .48 |
| A3 Frequent use of augmented reality technology in classes could endanger the health of cadets. | -.49 | .24 |
| A5 Applying augmented reality technology in class will not significantly improve my digital skills. | -.49 | .24 |
| Eigenvalue | 2.72 | |
| % explained variances | 45.26 | |

Furthermore, table 2 lists the descriptive statistics for the perception scale: number of subjects – N, minimum score – Min, maximum score – Max, arithmetic mean – M, standard deviation – SD, symmetry index – Sk and distribution flattening index – Ku. The average score on the scale obtained for the total sample ($M = 4.25$, $SD = 0.51$) indicates a relatively positive perception of cadets about the application of augmented reality technology in teaching. The variable is slightly negatively asymmetric ($Sk = – 0.50$), which indicates a greater representation of higher scores on the scale.

Table 2. Descriptive statistics for the scale of cadets' perception of the application of AR technology in teaching

| N | Min | Max | M | SD | Sk | Ku |
|---|---|---|---|---|---|---|
| 268* | 2.67 | 5.00 | 4.25 | 0.51 | – 0.50 | – 0.31 |

*two extreme results were excluded

In order to examine the difference in the average score on the scale of perception of the application of augmented reality technology in teaching with regard to the use of AR (Do you use AR?: Yes, No), a t-test for independent samples was conducted, and the results are shown in Table 3 (number of respondents – N, arithmetic mean – M, standard deviation – SD, t-ratio, degrees of freedom – df, level of significance – p, Cohen's d as a measure of effect size). A statistically significant difference was obtained ($t = 2.19$, $df = 266$, $p < 0.05$). Cadets who use AR have a significantly more positive perception of the application of augmented reality technology in classes ($M = 4.46$, $SD = 0.47$) compared to cadets who do not use AR ($M = 4.23$, $SD = 0.51$).

Table 3. Result on the scale of perception about the application of AR technology in teaching - t-test results

| Do you use AR? | N | M | SD | t-test | | | Cohenov d |
|---|---|---|---|---|---|---|---|
| | | | | t | Df | p | |
| No | 243 | 4.23 | 0.51 | 2.19 | 266 | 0.03 | 0.45 |
| Yes | 25 | 4.46 | 0.47 | | | | |

## 4.2 Integrative Results by Conducting a Knowledge Test

The knowledge of the cadets was tested with a knowledge test consisting of 11 questions. The result on the knowledge test was obtained by adding up the achieved points, whereby a correct answer to an individual question carried one point, and an incorrect answer zero points. Therefore, the theoretical range of points is from a minimum of zero to a maximum of eleven points. Table 4 shows the descriptive statistics for the results of the knowledge test in the initial measurement and in the final measurement.

Table 4. Descriptive statistics for the points achieved on the knowledge test in two measurements

| | N | Min | Max | M | SD | Sk | Ku |
|---|---|---|---|---|---|---|---|
| Points in initial measurement | 270 | 0.00 | 11.00 | 5.37 | 1.63 | – 0.01 | – 0.05 |
| Points in the final measurement | 270 | 4.00 | 11.00 | 8.90 | 1.77 | – 0.59 | – 0.49 |

In the total sample of cadets (N = 270), the average score (M) on the knowledge test in the initial measurement was 5.37 points (SD = 1.63), and in the final measurement it was 8.90 points (SD = 1.77). From the results (95% confidence intervals) shown in Table 5, it is evident that there were no differences in the average results achieved by cadets from different groups on the initial knowledge test. However, the results achieved by cadets from different groups on the final knowledge test differ significantly. At the same time, cadets from the *Situated learning* group achieved the lowest number of points on the final knowledge test (M = 7.72, SE = 0.15, 95 % CI [7.42, 8.02]).

Table 5. Result on the knowledge test in the initial and final measurement for three groups of cadets

| Group | Measurement | M | SE | 95 % IP |
|---|---|---|---|---|
| *Situated learning* | Initially | 5.54 | 0.17 | [5.20, 5.88] |
| | Final | 7.72 | 0.15 | [7.42, 8.02] |
| *Situated learning + Games* | Initially | 5.12 | 0.17 | [4.78, 5.46] |
| | Final | 8.72 | 0.15 | [8.42, 9.02] |
| *Situated learning + Games + Research* | Initially | 5.43 | 0.17 | [5.10, 5.77] |
| | Final | 10.27 | 0.15 | [9.97, 10.56] |

According to the average number of points, cadets from the *Situated learning + Games* group follow (M = 8.72, SE = 0.15, 95 % IP [8.42, 9.02]). The best result, i.e. the highest average points, on the final knowledge test was achieved by cadets from the *Situated learning + Games + Research.* group (M = 10.27, SE = 0.15, 95 % IP [9.97, 10.56]). Also, it is evident from the results that each group of cadets achieved a significantly higher average number of points on the final knowledge test compared to the average number of points achieved on the initial knowledge test.

## 4.3 Discussion

### 4.3.1 Elaboration of hypothesis H1

In order to prove the hypothesis H1, the connection between encountering the term AR in previous education and the cadets' perception of the application of augmented reality technology in classes with regard to the use of AR was examined. Analysis of the results shows that cadets who use AR have a significantly more positive perception of the application of augmented reality technology in classes (M = 4.46). Furthermore, they have a strong relationship with perception compared to cadets who do not use AR (M = 4.23), therefore they have a weak relationship with perception about the application of AR technology in teaching. The statistical significance of the differences in the average score on the perception scale between cadets from three groups/conditions (*Situated learning*, *Situated learning* + *Games* and *Situated learning* + *Games* + *Research*) was tested and a one-way analysis of variance was performed. The result of the analysis of variance is not statistically significant ($F2, 265 = 0.61$, $p > 0.05$, $\eta p2 = 0.01$), that is, it indicates that cadets from different groups do not differ significantly in their perception of the application of augmented reality technology in classes. Cadets from Group 2 *Situated learning* + Games (M = 4.29) have a strong connection in their perception of the application of technology. Cadets from Group 3 *Situated learning* + *Games* + *Research* (M = 4.24) have a weaker relationship, while cadets from Group 1 *Situated learning* (M = 4.21) have a weak relationship.

This satisfies and confirms the statement from hypothesis H1: The perception of cadets towards the application of augmented reality, AR in classes is related to their immersive experience and knowledge of AR technology.

### 4.3.2 Elaboration of hypothesis H2

In order to prove the hypothesis H1, the difference in the results of the cadets on the knowledge test was examined in two measurements (pretest and subtest) with regard to three forms of AR active learning (three groups/conditions: *Situated learning* , *Situated learning* + Games, *Situated learning* + *Games* + *Research*). On the pretest of knowledge from all three forms of AR active learning, the cadets achieved a significantly lower average number of points in the pretest (M = 5.37) than in the subtest (M = 8.90). It can be seen that there were no differences in the average results achieved by cadets from different groups on the pre-test of knowledge. The cadets of Group 2 *Situated learning* + Games on average achieved the lowest number of points on the knowledge pretest (M = 5.12), followed by the average number of points followed by the cadets of Group 3 *Situated learning* + *Games* + *Research* (M = 5.43) and the best result, i.e. the highest average points on the knowledge pretest were achieved by cadets from Group 1 *Situated learning* (M = 5.54). However, the results achieved by cadets from different groups on the basis of knowledge differ significantly in the results achieved. Therefore, the cadets from Group 1 *Situated learning* on average achieved the lowest number of points on the final knowledge test (M = 7.72). According to the average number of points, the cadets from Group 2 *Situated learning* + Games (M = 8.72) follow. The best result, i.e. the highest average points, on the final knowledge test was achieved by cadets from Group 3 *Situated learning* + *Games* + *Research* (M = 10.27).

This satisfies and confirms the statement from hypothesis H2: Cadets who were taught through all forms of AR active learning (*Situated learning, Games, Research*) achieve a statistically significantly higher level of knowledge compared to cadets of the other two groups.

## 5. CONCLUSION

The purpose of the work was to evaluate the impact of active learning using augmented reality on the knowledge of cadets using the *HaubicAR* application. Furthermore, by conducting original (representative) research, it was necessary to determine the level of learning outcomes of cadets in augmented reality, the frequency of dealing with new technologies virtual reality (VR) and augmented reality (AR), previous qualifications, active learning in new technologies, the influence of different forms of learning on knowledge, cadets' perception towards application (AR) in class, time spent per day in augmented reality for active learning and its impact on their academic success. Furthermore, the goal of the research was achieved since the following information objects were identified: *Situated learning*, *Games*, *Research* of the information package

Active learning and based on them, the existence, relationship and intensity of connections of the created information constructs *Perception* and *Knowledge* were examined and determined by assessing the impact of active learning using augmented reality. In addition to the above, the subject research determined the predictors of academic success of cadets through active learning using augmented reality. The applied goal of the research focused on implementing the application of augmented reality in cadet training was achieved through the implementation of an experiment, the results of which indicate the possibility of implementing the subject form of active learning at any related higher education institution. By evaluating the results obtained in the experimental part, using the information construct Research of the information package Active learning, the most significant determinant that determines the success of the outcome of the information constructs Perception and Knowledge using augmented reality is presented.

## ACKNOWLEDGEMENT

Authors express their gratitude to University North in supporting the publication of this article through scientific research project "d-Learning: Digital Competences for Innovation Culture" funds.

## REFERENCES

Abid, H. et al. (2022). Understanding the role of digital technologies in education: A review, *Sustainable Operations and Computers,* Volume 3, pp. 275-285.

Akçayır, M.; Akçayır, G. (2017). Advantages and challenges associated with augmented reality for education: A systematic review of the literature. *Educational Research Review*. Volume 20, pp. 1–11.

Amaguana, F. et al. (2018). Simulation System Based on Augmented Reality for Optimization of Training Tactics on Military Operations. *5th International Conference on Augmented Reality, Virtual Reality, and Computer Graphics*, Otranto, Italy, pp. 394-403.

Bower, M. et al. (2014). Augmented Reality in education – cases, places and potentials. *Educational Media International*, Vol. 51, No. 1, pp. 1–15.

Cabero, J.; Barroso, J. (2016). The educational possibilities of Augmented Reality. *New Approaches Educational Research*. Vol. 5. No. 1. pp. 44-50.

Cózar-Gutiérrez, R.; Sáez-López, J.M. (2016). Game-based learning and gamification in initial teacher training in the social sciences: An experiment with MinecraftEdu. *International Journal of Educational Technology in Higher Education*, Vol. 13. No. 1., pp. 2-11.

Creswell, J. W. (2022). Research design: Qualitative, quantitative, and mixed methods approaches. *Journal of Electronic Resources in Medical Libraries*, Volume 19, Issue1-2, pp. 54–55.

Ding, L. (2019). Applying gamifications to asynchronous online discussions: A mixed methods study. *Computers in Human Behavior*, 91, pp. 1–11.

Eshet-Alkalai, Y. (2012). Thinking in the Digital Era: A Revised Model for Digital Literacy, *Issues in Informing Science and Information Technology*, Volume 9, pp. 267-276.

Greenwood, T.A., P. O'Neil, M. (2016). Harnessing the Potential of Augmented and Virtual Reality for Military Education, *Workshop Proceedings of the 12th International Conference on Intelligent Environments (IE), Intelligent Environments*, Volume: 21, pp. 249-254.

He, J., et al. (2017). The research and application of the augmented reality technology, *Proceedings of 2017 IEEE 2nd Information Technology, Networking, Electronic and Automation Control Conference (ITNEC)*, Chengdu, China pp. 496-501.

Jesionkowska, J., Wild, F., Deval, Y. (2020). Active Learning Augmented Reality for STEAM Education—A Case Study. 10(8), 198. *Journals Education Sciences,* Volume 10 Issue 8, Pp. 2-15.

Karlsson, M. (2015). *Challenges of designing Augmented Reality for Military*, Institutionen för informatik, Publisher, UMEA Universitet, Umea, Sweden.

Letina, A. (2016). Strategije aktivnog učenja u nastavi prirode i društva, *Školski vjesnik : časopis za pedagogijsku teoriju i praksu*, Vol. 65 No.1, pp. 1–26.

Lu, S. J., & Liu, Y. C.. (2015). Integrating augmented reality technology to enhance children's learning in marine education. *Environmental Education Research*, 21(4), pp. 525-541.

Mani, D. R., Madhusudan J. V. (2023). Augmented Reality as a Performance Enhancement Technology in Primary Education: A Systematic Review. *International Journal of Advanced Virtual Reality*, Vol. 1 No. 1. Page: 20-25.

Martin-Gonzalez, A. (2016). Usability evaluation of an augmented reality system for teaching Euclidean vectors. *Innovations in Education and Teaching International*, Volume: 53 Issue: 6 , pp. 627-636.

Maxwell, J. A. (2012). *Qualitative research design: An interactive approach*. Publisher, Sage publications, Inc. 2455 Thousand oaks, California, USA.

McCrindle, M. (2014). *The ABC of XYZ: Understanding the Global Generations*. Published by McCrindle Research Pty Ltd, published by UNSW Press, Lexington Drive Bella Vista, Australia.

Memarsadeghi, N., Varshney, A. (2020). Virtual and Augmented Reality Applications in Science and Engineering. *IEEE Computer Society Computing in Science & Engineering*, vol. 22, pp. 4-6.

O'Flaherty, J.; Phillips, C. (2015). The use of flipped classrooms in higher education: A scoping review. *Internet Higher Educ.*, 25, pp. 85–95.

Pantelić, A., Plantak, Vukovac, D. (2017). The Development of Educational Augmented Reality Application: A Practical Approach, *ICERI 2017 Conference*, Seville, Spain, pp. 8745-8752.

Radu, I. (2014). Augmented reality in education: A meta-review and cross-media analysis. *Springer, Pers. Ubiquitous Comput.* 18, pp.1533–1543.

Richards, J., et al. (2017). *Virtual, Augmented, And Mixed Realities In Education*, Springer Nature Singapore, Gateway East, Singapore.

Riman, N., et al. (2020). *Emerging Extended Reality Technologies For Industry 4.0,* edition first published 2020 by John Wiley & Sons, Inc., 111 River Street, Hoboken, NJ 07030, USA.

Saez-Lopez, JM., et al. (2020). Augmented Reality in Higher Education: An Evaluation Program in Initial Teacher Training, *MDPI Journals Education Sciences* Volume 10 Issue 2 Volume: 10 Issue: 2, pp. 1-12.

Saltan, F., Arslan, O. (2017). The Use of Augmented Reality in Formal Education: A Scoping Review, *Eurasia Journal Of Mathematics Sciene And Technology Education*, Volume: 13, Special Issue, pp. 503-520. https://doi.org/10.12973/eurasia.2017.00628a.

Stančić, H. (2005). Teorijski model postojanog očuvanja autentičnosti elektroničkih informacijskih objekata, *Doktorska disertacija*, Sveučilište u Zagrebu , Filozofski fakultet, pp. 25-38.

Swensen, H. (2016). Potential of Augmented Reality in sciences education a literature review, *Conference: 9th International Conference of Education, Research and Innovation (ICERI)*: Seville, Spain, pp. 2540-2547.

Tekedere, H.; Göker, H. (2016). Examining the effectiveness of augmented reality applications in education: A meta-analysis. *International Journal of Environmental & Science Education*, Vol. 11, No. 16, pp. 9469–9481.

Tuta, J., et al. (2019). A conceptual model of augmented virtual and reality in cadet training, *European Conference on Electrical Engineering and Computer Science, (EECS)*, Athens, Greece, pp. 128 – 133.

Whitton, N.; Langan, M. (2018). Fun and games in higher education: An analysis of UK student perspectives. Teach. Higher Educ., 23, pp, 1000–1013.

Woodwell, D. (2013). *Research Foundations: How Do We Know what We Know?* ISBN-13:9781483306742. SAGE Publications. Thousand oaks, California. USA.

# GREEN IT ADOPTION BY SOFTWARE DEVELOPERS: KEY FACTORS FROM THE THEORY OF PLANNED BEHAVIOR PERSPECTIVE

Yassine Talas and Hajer Rabii
*Amaris Research Unit, France*

**ABSTRACT**

Green Information Technology (Green IT) could play a major role in reducing the environmental impact of IT industry. At the software level, developers are at the forefront of Green IT while researchers have worked to theorize their integration of sustainable IT practices. However, these efforts do not suggest how to make adoption mechanisms operational. Hence, this exploratory study relies upon the Theory of Planned Behavior to investigate based on three components (attitude, subjective norms, perceived behavioral control) how to enhance, in a more practical manner, the adoption of Green IT by software developers. In this regard, we conducted a qualitative research approach by means of interviews to understand how Green IT is perceived by software developers and understand the main factors that would accelerate a practice change. Our findings suggest that the following 7 key factors are needed to ensure Green IT adoption: awareness raising, team-level action, company-supported initiative, referent role, coding best practices, key performance indicators and trainings. Guidance on how to implement these key factors are provided based on interviews' insights. These findings provide valuable and actionable information to help decision-makers plan and deploy Green IT efficiently, and to enlighten software developers that are newcomers to Green IT.

**KEYWORDS**

Green Information Technology, Sustainable Information Technology, Software Developers, Theory of Planned Behavior

## 1. INTRODUCTION

Green IT refers to the process of developing, operating and disposing of IT in a manner that minimizes harm to the environment (Dalvi-Esfahani et al., 2020; Elliot, 2007). This process involves a set of principles, methodologies and tools at the software and the hardware levels. Both academia and practice provide clear evidence that software Green IT can contribute to more energy efficiency and thus to less carbon emissions (Capra et al., 2012; Katal et al., 2023; Kravets & Egunov, 2022; Ournani et al., 2021; Verdecchia et al., 2021). Approaches in software Green IT include virtualization, cache management, coding and algorithmic efficiency, data compression techniques, etc. Since the early 2010's, multiple studies called for investigating factors that could impact Green IT adoption in companies (Dedrick J., 2010; Melville, 2010). In 2011, Jenkin et al. pointed out that companies were still in infancy stages of awareness and adoption of Green IT, mainly due to four gaps: knowledge, practice, opportunity, and knowing-doing gaps. Unfortunately, this observation was corroborated by industry 7 years later, according to the "We Green IT" study (Trudu et al., 2018). In this paper, we make two assumptions that may explain these gaps from a research standpoint. The first one is the lack of understanding in Green IT adoption mechanisms at the individual level. Indeed, early studies in the field highlighted the importance to tackle the issue of Green IT adoption at the "employee" level (Jenkin et al., 2011), while only few studies explored it (Asadi et al., 2017; Yoon, 2018). The second is related to the scarcity of detailed insight and context-specific studies. Ribeiro et al. (2021) pointed out the need for a more complex and multi-scale approach to uncover Green IT adoption mechanisms. Tanveer (2021) and others tackled this issue in a rather generic and aggregated manner. They did neither consider the specificities related to the position of IT professionals who apply Green IT in their work line, nor the difference in Green IT practices, for instance between software and hardware related practices.

Our exploratory work tries to fill these gaps while addressing one specific context at the individual level: Green IT adoption by software developers. Since developers are mainly concerned with software issues, the focus of our empirical investigation is solely placed on software-related Green IT practices (i.e. software Green IT) during development activities. Our goal is to understand the key factors that could lead to the adoption of Green IT by software developers. This question is explored based on the Theory of Planned Behavior, which was already used to predict behavior in relation to Green aspects (Maichum et al., 2016).

## 1.1 Literature Review

### 1.1.1 Introduction to the Theory of Planned Behavior

The Theory of Planned Behavior (TPB) is one of the most popular social-psychological models for the prediction of behavior (Ajzen, 1991). The TPB helps clarifying an individual's decision-making process (Han, 2015) and is grounded in the idea that the intention to engage in a behavior is a key predictor of whether the behavior will occur. In turn, the intention to use the behavior is dependent on three components: (i) attitude towards the behavior, (ii) subjective norm to perform behavior and (iii) perceived behavioral control over behavior. First, attitude towards behavior refers to the beliefs about the positive or negative consequences an individual might experience when they perform behavior (Passos et al., 2013). Second, subjective norms encompass the perceived opinions of persons or groups whose beliefs may be important for the individual (Mathieson, 1991). In other words, subjective norms refer to a person's perception of social pressure coming from external environment to perform or not a given behavior (Fishbein & Ajzen, 2010). Third, perceived behavioral control refers to the individual's perceptions of the presence or absence of requisite means and tools necessary to perform the behavior, that reflects the ease or difficulty of performing the behavior (Ajzen, 1991; Mathieson, 1991; Orbell et al., 1997). Mishra et al. (2014) explored Green IT adoption factors among IT professionals using the theory of reasoned action, a psychological theory explaining behavior based on individual intention, similar to TPB. Du Buisson & Naidoo (2014) found a correlation between perceived behavioral control and intention to practice green computing for IT professionals. Ojo & Fauzi (2020) conducted a survey on IT professionals and found out that belief and attitudes towards Green IT contribute to their engagement in basic Green IT practices. Dalvi-Esfahani et al. (2020) showed that Green IT attitude, perceived behavioral control and personal norms positively impact intention to practice Green IT. Finally, the findings from Yoon, 2018 imply that environmental beliefs, personal norms (i.e. the feeling of individual moral obligation), descriptive norms (i.e. subjective perceptions about what most people do) and perceived usefulness play an important role in the use of Green IT.

### 1.1.2 Conceptual Model

There is a lack in understanding how IT professionals and decision-makers can take advantage of the TPB and its components to enhance Green IT adoption in software developers' day-to-day work. In this regard, our study aims to uncover key factors associated to TPB that should be leveraged to promote intention to apply and adopt Green IT. Figure 1 presents the conceptual model adopted in our research paper. The dotted sections in the figure represent the gaps in the literature that we aim to address.

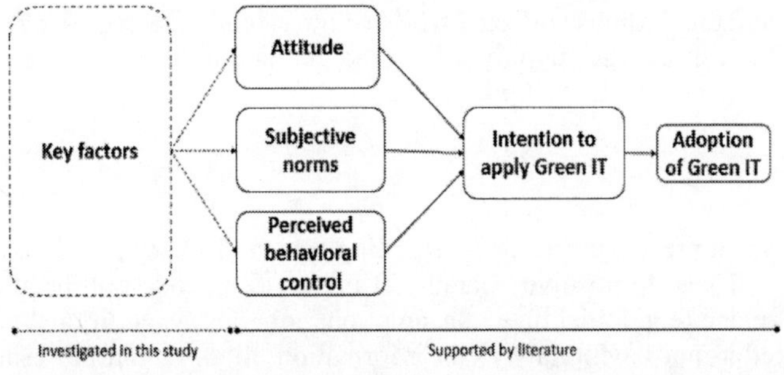


Figure 1. Conceptual model of the TPB and the adoption of Green IT

## 2. METHODOLOGY

Based on the initial conceptual model presented in Figure 1, we started by collecting empirical data to be coded and analyzed. More than 15 IT professionals (software developers and IT project managers) contributed thoroughly to the iterated construct of an interview guideline, comprised of several questions. The questions targeted all software developers, without requiring specific knowledge or application of Green IT. They aimed at collecting the different ideas, opportunities, and risks the interviewees foresee if they were to apply Green IT within their work line. The questions also explored their expected needs to ensure Green IT adoption, in line with key constructs from the TPB. For example, the following questions were asked: "For you, what is the ideal working environment to apply Green IT", "What is your motivations to engage in Green IT initiatives?", "What extra knowledge do you require to apply Green IT?", etc. Two pilot interviews with software developers were conducted for pretesting and adjustments.

All of our interviewees are employees of a French leading IT consulting company. They were assigned to different client projects covering various sectors such as energy, telecommunication and public transport. This diversity leads to more credibility and increased potential of transferability. Interviewees were selected randomly, based on their profile that shows "software developer". Sample size (N=20) was determined based on the recommended range from 20 to 30 interviewees that was suggested by Creswell & Poth (2018). The interviewed developers were mainly experienced, as 70% of them have more than 5 years' experience. This fact helped us collect detailed insight as the interviews were based on all their past experience. The interviewees had different backgrounds and expertise, essentially front-end, back-end and full stack. During the interviews, participants were asked to answer questions based on their experiences in the projects they were involved in. With the consent of interviewees, all interviews have been taped and transcribed.

We conducted 20 separate semi-structured interviews of software developers between October 2022 and January 2023. The duration of each interview was about 30 to 45 min. Resources were sent beforehand to the interviewees in order to introduce our research, to present Green IT and its practices, and also to remind of the link between software development and environmental impact. These resources consisted of a prepared slideshow and additional reading material. Especially that, upon initial contact, none of the interviewees reported to have ever purposely practiced Green IT.

Following interview transcriptions, data coding was conducted anonymously to link the transcribed textual data with TPB components and the identified Green IT adoption key factor. The coding followed a two-cycle approach, as suggested by Saldaña (2013). In the first cycle, a structural coding method was selected as it is suitable for exploratory investigations using semi-structured interviews. This coding method allows us to label data with investigated questions. Then, for the second cycle coding, axial coding was conducted to regroup initial codes into key factors and associate them to the theory of planned behavior. Dovetailapp, a dedicated research analysis tool, was used to organize and support data coding. An Excel gathering and organizing all collected data was also used. To enhance reliability, two software developers followed a double coding approach to ensure proper and common understanding of the assigned codes (Miles et al., 2014).

## 3. FINDINGS

In this section, we present the findings collected from the interviews. We organize them according to the three components of the TPB and provide support with some quotations from the interviewees. A more detailed selection of quotations is available in the Appendix 1.

### 3.1 Attitude

All of the interviewees had preconceived ideas regarding Green IT. They either thought that applying Green IT in their work line will lead to positive operational impact (e.g. code optimization, productivity, product quality) or to a negative one (e.g. initial time consumption, complexity, additional project constraints, stress). The interviewees regretted not having any exact information about Green IT, especially its impact from a developer's perspective. As one of the interviewees said to explain this mixed situation: "*The missing word today is awareness raising*".

Indeed, if the awareness of software developers is not properly raised regarding different elements depicting Green IT, the risk is that their attitude would be established on biased information or common beliefs. All interviews shed light on the high importance of awareness raising. Interviewees specified how they expect this awareness raising to be conducted. They highlighted the importance of continuous communication, instead of an isolated approach.

According to interviewees, awareness raising needs to cover different aspects such as "*numbers*", "*stakes*", "*importance*", "*operational impacts*" and "*productivity and performance*" at multiple levels such as "*environment*", "*company*", "*project*" and "*developer*". For the latter, one interviewee stated that "*it is really important to show that even developers can play an important role, a role that not everyone of us is aware of*".

Based on these insights, we make the following proposition:

**Proposition 1:** Continuous awareness raising, covering different aspects, and at multiple levels, is a key factor of attitude, and is important to achieve higher adoption of Green IT by developers.

## 3.2 Subjective Norms (SN)

First, at the project level, the interviewees unanimously agreed that if they were part of a team whose members already practice Green IT, they would naturally feel "*obligated*", "*encouraged*", "*reassured*" or "*motivated*" to adopt the same practices.

Second, at the company level, feedback from the interviewees indicated that it is not only about Green IT, but about the prerequisites the company can implement, the resources it needs to afford (support, trainings…), the understanding of the challenges it can add for developers (time constraints, necessary training…) and the coherence between its practices and strategy on one hand, and the Green IT initiative on the other hand.

According to the developers, the company needs to set the example, by globally adopting and promoting a sustainable culture and engaging in environment friendly initiatives such as sorting waste. This can contribute to the important feeling of sharing a common vision, which the developers expect. Another interesting requirement that was brought up by many developers, is the necessity to define a role of Green IT referent at the project level.

Based on the previous insights, we make the following propositions:

**Proposition 2:** Making software Green IT a team-level action is a key factor of SN, that is important to achieve higher adoption of Green IT by developers.

**Proposition 3:** Making software Green IT part of a global company-supported initiative and effort is a key factor of SN, that is important to achieve higher adoption of Green IT by developers.

**Proposition 4:** Creating a software Green IT referent role at the project level is a key factor of SN, that is important to achieve higher adoption of Green IT by developers.

## 3.3 Perceived Behavioral Control (PBC)

For this TPB component, three main points stood out repeatedly during the interviews: coding best practices, key performance indicators (KPIs) and training.

### 3.3.1 Coding Best Practices

First, interviewees agreed that coding best practices should be made easy to comprehend and to implement. According to some interviewees, the best practices should be implemented in a way that they are smoothly and gradually integrated in the regular pathway and processes of the developer.

This approach can help developers "*know what is working for them and what is not*". In other words, it would help them assess progressively the application of the best practices. Plus, some interviewees highlighted that these best practices should be "*grouped*", "*sorted from the most to the least important*" for Green IT (i.e. prioritized), and presented with enough detail and precision. Many interviewees stressed their expectation to have access to dynamic best practices. In other words, Green IT best practices are meant to evolve, and thus should be regularly updated. Some of these interviewees think it is important for the developers to be included in this updating process.

### 3.3.2 Key Performance Indicators (KPIs)

The following statement sums up the current feeling of developers, as KPIs are one of the major challenges of Green IT: "*I know how much carbon emissions I cause when I take my car to work, but I don't know how much carbon emission I am responsible for when I code*". Indeed, all interviewees stressed the importance to have live information on "*exactly what is the impact of each best practice*" and "*understand the real impact on the environment*" through Green IT KPIs. As expressed by another interviewee, "*it is a major source of motivation*". Such KPIs are important in favor of Green IT adoption, because for instance: "*Three months after Green IT application, developers need to know concretely what has changed, otherwise, for them, it will be just a marketing maneuver*". KPIs are also important tools to help developers compare development techniques and choices based on which offers the best Green IT performance.

### 3.3.3 Training

Interviewees agreed that training should be "*regular*" and "*continuous*". All interviewees suggest that a key success factor is to get "*practical*" rather than theoretical trainings. A related view shared by the interviewees, is that the format of Green IT trainings should be different than mainstream trainings. Many suggested that the "*gamification*" of trainings could be beneficial.

One important aspect that was brought up by several interviewees is the necessity to adapt and personalize trainings and best practices depending on the developers' profiles and the development tools and technologies. Based on the above insights, we make the following propositions:

**Proposition 5:** Coding best practices, integrated in the work path, grouped and sorted, dynamic, and progressively implemented, are key factors of PBC, and are important to achieve higher adoption of Green IT by developers.

**Proposition 6:** Availability of live Green IT KPIs at the global and at each coding best practice level, is a key factor of PBC, and is important to achieve higher adoption of Green IT by developers.

**Proposition 7**: Personalized, practical, gamified and continuous trainings are key factors of PBC, and are important to achieve higher adoption of Green IT by developers.

Figure 2 sums up our findings.

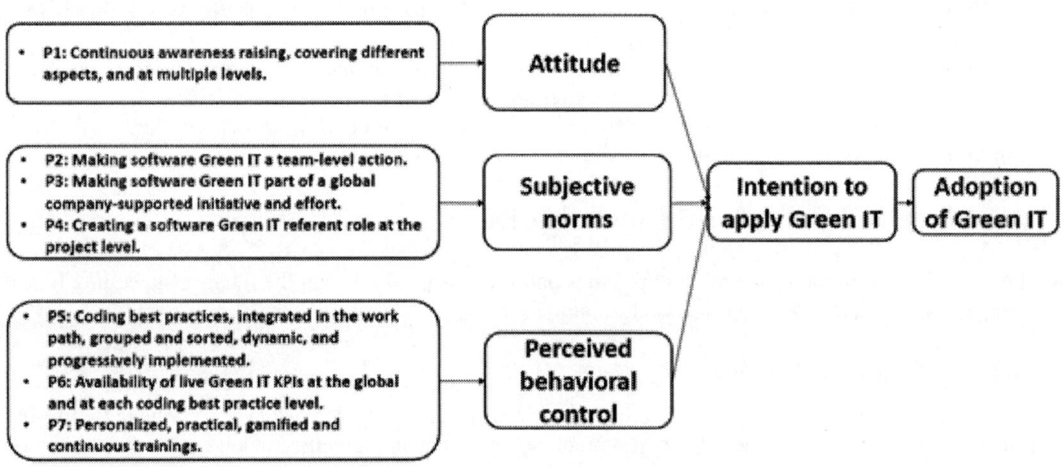


Figure 2. Propositions: TPB and Green IT

# 4. CONCLUSION

Our findings are essentially coherent with previous research conducted in other contexts. We relied on a qualitative empirical approach, which is different than quantitative existing studies that are related to this work. Our findings extend previous research by providing some guidance on how to make the identified key factors operational and somehow ready for use. For instance, in Proposition 1, not only we highlight that awareness raising is important for Attitude, but we also specify how it should be implemented (namely: continuous, covering different aspects and at multiple levels). Our attitude findings regarding the important role of awareness raising (proposition 1) are in accordance with existing research. Puvaneswaran & Alsdorf (2021) attempted to understand Green IT capability maturity. They explicitly suggest that "sensitizing employees" might be as effective as investments in Green IT. Our subjective norms findings (propositions 2, 3 and 4) are in line with previous research that highlighted the importance for Green IT adoption of governance (Patón-Romero et al., 2021), organizational support (Mouakket & Aboelmaged, 2022; Ojo & Fauzi, 2020), resource commitment (Lei & Ngai, 2013; Mouakket & Aboelmaged, 2022), greening of organizational culture (Deng & Ji, 2015; Molla et al., 2014; Schmermbeck, 2019), championing of the adoption process via designated referents (Hernandez, 2018) and alignment of green IT initiative with company goals (Deng & Ji, 2015; Puvaneswaran & Alsdorf, 2021). Past studies suggest a lack of technical skills and knowledge as barriers to Green IT adoption (Dalvi-Esfahani & Nilashi, 2022). We assume that our propositions on perceived behavioral control (5, 6, and 7) can equip software developers with the required skills and knowledge to ease the perceived technological complexity of Green IT (Dangelico, 2017). Our findings on Green IT's key performance indicators (Proposition 6) align with Patón-Romero et al. (2021), who identified the absence of specific metrics to evaluate Green IT practices' performance as a key improvement opportunity in two case studies. Finally, Shahzad et al. (2022) suggested that some facilitating conditions positively affect green behavioral intention. Their definition of these conditions include training on Green IT, which is coherent with proposition 7 of our findings. Similarly, the findings from Sayeed & Onetti (2018), indicate that Green IT adoption requires changes in human resources capabilities by performing trainings.

This research tried to explore mechanisms of Green IT adoption by software developers. Seven propositions were formulated as answers to our research question. Our findings indicate that some key factors might be leveraged along with their specific criteria in order to enhance Green IT adoption. These factors can be linked to the TPB. Also, our research confirms the suitability of the TPB to understand Green IT adoption. Furthermore, this study suggests that it might be necessary to adopt empirical human-centered approaches (e.g. interviews, action research) to better investigate complex mechanisms of Green IT adoption and to capture detailed insight that surveys cannot. From the developers' perspective, our findings can guide and help those discovering Green IT to better understand what can facilitate adoption. Our propositions could be viewed as a checklist or as a starting point for self-thinking on their own needs to better engage in Green IT practices. From a managerial standpoint, our study encourages IT managers, chief information officers and other decision-makers to acknowledge not only technical considerations (e.g. tools), but also non-technical aspects (e.g. organizational support) in order to enhance Green IT adoption by software developers. In this sense, this research provides these decision-makers with determined key factors to put in place or address when planning Green IT practices. Moreover, this study offers some guidance on how to implement these key factors. This guidance is based on the needs expressed by software developers themselves, and therefore is important to consider. Furthermore, our study calls decision-makers in companies (e.g. chief information officers) for a multiscale initiative (company, project, team, individual) to disseminate Green IT practices. For such an initiative to succeed, we assume that it might be important to consider the needs and knowledge of all relevant stakeholders.

Although our study provides meaningful implications for researchers and practitioners, it comes with some limitations. As we position this research as exploratory, we cannot conclude that saturation of data has been totally obtained and our findings cannot be generalized. Our study was not able to uncover other possible key factors that might have been hindered by our sample size. As such, this study did not explore contextual characteristics that could have mediated or changed the effects of our propositions (e.g. company size, project type, team size…). Another limitation of our research is that the understanding of Green IT by interviewees was not evaluated to ensure a common baseline for analysis although this bias was mitigated by the common resources shared with interviewees. Also, our research did not investigate the possible interactions between the components of the TPB (i.e. Attitude, subjective norms, perceived behavioral control), as other research suggested. Finally, all interviewees have never practiced Green IT purposely, even though they applied some of the common best practices. Therefore, collected data was based on their perception, experiences, and expected needs, rather than on current practices.

Future studies are required to explore a more comprehensive model of key factors, while prioritizing the latter to facilitate implementation. To expand and enrich this exploratory study, we plan to conduct in-depth case studies to challenge our initial findings and possibly uncover other key factors. The resulting model will be then tested in real company settings, by applying the different propositions within a pilot project.

## REFERENCES


Ajzen, I. (1991). The theory of planned behavior. *Organizational Behavior and Human Decision Processes*, *50*(2), 179–211. https://doi.org/10.1016/0749-5978(91)90020-T

Asadi, S., Hussin, A. R. C., & Dahlan, H. M. (2017). Organizational research in the field of Green IT: A systematic literature review from 2007 to 2016. *Telematics and Informatics*, *34*(7), 1191–1249. https://doi.org/10.1016/j.tele.2017.05.009

Capra, E., Francalanci, C., & Slaughter, S. A. (2012). Is software "green"? Application development environments and energy efficiency in open source applications. *Information and Software Technology*, *54*(1), 60–71. https://doi.org/https://doi.org/10.1016/j.infsof.2011.07.005

Creswell, J. W., & Poth, C. N. (2018). *Qualitative Inquiry and Research Design: Choosing Among Five Approaches* (4th ed.). Sage Publications.

Dalvi-Esfahani, M., & Nilashi, M. (2022). *Analyzing the Barriers Affecting the Adoption of Green Computing: A Hybrid Method Approach*. https://doi.org/10.21203/rs.3.rs-2348009/v1

Dalvi-Esfahani, M., Alaedini, Z., Nilashi, M., Samad, S., Asadi, S., & Mohammadi, M. (2020). Students' green information technology behavior: Beliefs and personality traits. *Journal of Cleaner Production*, *257*. https://doi.org/10.1016/j.jclepro.2020.120406

Dangelico, R. M. (2017). What Drives Green Product Development and How do Different Antecedents Affect Market Performance? A Survey of Italian Companies with Eco-Labels. *Business Strategy and the Environment*, *26*(8), 1144–1161. https://doi.org/https://doi.org/10.1002/bse.1975

Dedrick J. (2010). Green IS: Concepts and issues for information systems research. *Communications of the Association for Information Systems*, *27*(1), 11. https://doi.org/10.17705/1CAIS.02711

Deng, Q., & Ji, S. (2015). Organizational green IT adoption: Concept and evidence. *Sustainability (Switzerland)*, *7*(12), 16737–16755. https://doi.org/10.3390/su71215843

Du Buisson, W., & Naidoo, R. (2014). Exploring factors influencing it workers' green computing intention at a South African firm. *ACM International Conference Proceeding Series*, *28-September-2014*, 148–156. https://doi.org/10.1145/2664591.2664609

Elliot, S. (2007). Environmentally Sustainable ICT: A Critical Topic for IS Research? *PACIS 2007 Proceedings*, 114.

Fishbein, M., & Ajzen, I. (2010). *Predicting and changing behavior: A reasoned action approach* (1st ed.). Psychology Press.

Han, H. (2015). Travelers' pro-environmental behavior in a green lodging context: Converging value-belief-norm theory and the theory of planned behavior. *Tourism Management*, *47*, 164–177. https://doi.org/10.1016/j.tourman.2014.09.014

Hernandez, A. A. (2018). Exploring the factors to green IT adoption of SMEs in the Philippines. *Journal of Cases on Information Technology*, *20*(2), 49–66. https://doi.org/10.4018/JCIT.2018040104

Jenkin, T. A., McShane, L., & Webster, J. (2011). Green information technologies and systems: Employees' perceptions of organizational practices. *Business and Society*, *50*(2), 266–314. https://doi.org/10.1177/0007650311398640

Katal, A., Dahiya, S., & Choudhury, T. (2023). Energy efficiency in cloud computing data centers: a survey on software technologies. *Cluster Computing*, *26*(3), 1845–1875. https://doi.org/10.1007/s10586-022-03713-0

Kravets, A. G., & Egunov, V. (2022). The Software Cache Optimization-Based Method for Decreasing Energy Consumption of Computational Clusters. *Energies*, *15*(20), 7509. https://doi.org/10.3390/en15207509

Lei, C. F., & Ngai, E. W. T. (2013). Green IT Adoption: An Academic Review of Literature. *PACIS 2013 Proceedings*, 95.

Maichum, K., Parichatnon, S., & Peng, K. C. (2016). Application of the extended theory of planned behavior model to investigate purchase intention of green products among Thai consumers. *Sustainability (Switzerland)*, *8*(10). https://doi.org/10.3390/su8101077

Mathieson, K. (1991). Predicting user intentions: Comparing the technology acceptance model with the theory of planned behavior. *Information Systems Research*, *2*(3), 173–191. https://doi.org/10.1287/isre.2.3.173

Melville, N. P. (2010). Information systems innovation for environmental sustainability. *MIS Quarterly*, *34*(1), 1–21. https://doi.org/10.2307/20721412

Miles, M. B., Huberman, M. A., & Saldana, J. (2014). *Qualitative Data Analysis: A methods sourcebook* (3rd ed.). Sage Publications.

Mishra, D., Akman, I., & Mishra, A. (2014). Theory of Reasoned Action application for Green Information Technology acceptance. *Computers in Human Behavior*, *36*, 29–40. https://doi.org/10.1016/j.chb.2014.03.030

Molla, A., Abareshi, A., & Cooper, V. (2014). Green IT beliefs and pro-environmental IT practices among IT professionals. *Information Technology and People*, *27*(2), 129–154. https://doi.org/10.1108/ITP-10-2012-0109

Mouakket, S., & Aboelmaged, M. (2022). Drivers and outcomes of green information technology adoption in service organizations: an evidence from emerging economy context. *Journal of Science and Technology Policy Management*, *13*(4), 898–924. https://doi.org/10.1108/JSTPM-09-2020-0137

Ojo, A. O., & Fauzi, M. A. (2020). Environmental awareness and leadership commitment as determinants of IT professionals engagement in Green IT practices for environmental performance. *Sustainable Production and Consumption*, *24*, 298–307. https://doi.org/10.1016/j.spc.2020.07.017

Orbell, S., Hodgkins, S., & Sheeran, P. (1997). Implementation Intentions and the Theory of Planned Behavior. *Personality and Social Psychology Bulletin*, *23*(9), 945–954. https://doi.org/10.1177/0146167297239004

Ournani, Z., Belgaid, M. C., Rouvoy, R., Rust, P., & Penhoat, J. (2021). Evaluating the Impact of Java Virtual Machines on Energy Consumption. *Proceedings of the 15th ACM / IEEE International Symposium on Empirical Software Engineering and Measurement (ESEM)*, 1–11. https://doi.org/10.1145/3475716.3475774

Passos, C., Cruzes, D. S., Hayne, A., & Mendonca, M. (2013). Recommendations to the adoption of new software practices: A case study of team intention and behavior in three software companies. *International Symposium on Empirical Software Engineering and Measurement*, 313–322. https://doi.org/10.1109/ESEM.2013.31

Patón-Romero, J. D., Baldassarre, M. T., Rodríguez, M., Runeson, P., Höst, M., & Piattini, M. (2021). Governance and Management of Green IT: A Multi-Case Study. *Information and Software Technology*, *129*. https://doi.org/10.1016/j.infsof.2020.106414

Puvaneswaran, S., & Alsdorf, H. (2021). Why People and Processes Enable Sustainability and Not the Latest Technology: On Achieving Green IT/IS Capability Maturity. *PACIS 2021 Proceedings*, 163.

Ribeiro, M. P. L., Tommasetti, R., Gomes, M. Z., Castro, A., & Ismail, A. (2021). Adoption phases of Green Information Technology in enhanced sustainability: A bibliometric study. *Cleaner Engineering and Technology*, *3*. https://doi.org/10.1016/j.clet.2021.100095

Saldaña, J. (2013). *The coding manual for qualitative researchers* (2nd ed.). Sage Publications.

Sayeed, L., & Onetti, A. (2018). The relationship between organizational resources and Green IT/S adoption: A RBV approach. *Journal of International Technology and Information Management*, *27*(2), 43–72. https://doi.org/10.58729/1941-6679.1331

Schmermbeck, H. (2019). On Making a Difference: Towards an Integrative Framework for Green IT and Green IS Adoption. *Proceedings of the 52nd Hawaii International Conference on System Sciences*, 2045–2054.

Shahzad, M., Qu, Y., Rehman, S. U., & Zafar, A. U. (2022). Adoption of green innovation technology to accelerate sustainable development among manufacturing industry. *Journal of Innovation and Knowledge*, *7*(4). https://doi.org/10.1016/j.jik.2022.100231

Tanveer, B. (2021). Sustainable software engineering-have we neglected the software engineer's perspective? *Proceedings - 2021 36th IEEE/ACM International Conference on Automated Software Engineering Workshops, ASEW 2021*. https://doi.org/10.1109/ASEW52652.2021.00059

Trudu, O., Pontal, A., & Bordage, F. (2018). *Etude We GREEN IT: Quelle démarche Green IT pour les grandes entreprises françaises ?*

Verdecchia, R., Lago, P., Ebert, C., & De Vries, C. (2021). Green IT and Green Software. *IEEE Software*, *38*(6), 7–15. https://doi.org/10.1109/MS.2021.3102254

Yoon, C. (2018). Extending the TAM for Green IT: A normative perspective. *Computers in Human Behavior*, *83*, 129–139. https://doi.org/10.1016/j.chb.2018.01.032

# APPENDIX

Appendix 1. Selection of quotations from interviews

| TBP | Quotations |
|---|---|
| A | *"the missing word today is awareness raising".* |
| | *"If companies make all the necessary Green IT tools available to developers, and if these developers are not initially aware enough about Green IT, will they really use those tools? It's an important thing to consider".* |
| | *"If I am not informed enough, I will be 'scared' (reluctant) to engage in Green IT".* |
| | *"Green IT awareness should be even part of important project meetings. Awareness raising should be made and repeated before, at the start, the middle and the end of projects. To make it successful, it's not an email sent one time that will help…".* |
| | *"the more you say it, the more people will understand it… It has to be continuous, throughout the whole career path in one company. It's only this way that this issue can move forward".* |
| | *"it is really important to show that even developers can play an important role, a role that not everyone of us is aware of"* |
| SN | *"If it was a team race, and I was the only one running, while my colleagues and my company are not… It wouldn't work".* |
| | *"Green IT is all about community… If it is a teamwork, it can be done more easily, and I would feel motivated every day to do it".* |
| | *"If there is a team of 6 developers using Green IT practices, then joined by me as a 7th developer… I would definitely try to use the same practices. And it would all be beneficial".* |
| | *"A necessary condition for Green IT success, is when the whole company is putting effort to make it work".* |
| | *"it should be a collective action"* |
| | *"It's not only about raising awareness… What tools and resources is the company willing to make available? Is the company ready to change its strategy and processes?"* |
| | Developers *"…would go to this referent for any questions about Green IT principles and good practices". This referent would have "advanced Green IT knowledge", would "help developers on a daily basis" and "validate their Green IT choices".* |
| PBC | *"The company should introduce best practices in my everyday life at work, without me feeling that these practices are cumbersome"* |
| | *"It would be more comfortable for me to receive goals and best practices to achieve and apply progressively"* |
| | *"It is important to know what requires lots of coding skills, and what requires less, and to have an indication of what is feasible for me"* |
| | *"It is important for me to know exactly what do they mean by optimizing the requests"* |
| | *Best practices need to "make the link between Green IT and their tasks"* |
| | *"We need, as developers, to contribute to the evolution of the best practices, it is rather important"* |
| | *"I know how much carbon emissions I cause when I take my car to work, but I don't know how much carbon emission I am responsible for when I code".* |
| | *"understand the real impact on the environment"* |
| | *"It is very important to monitor Green IT, through quantifiable data on what is working, and what is not, and indications on where are the problems* |
| | About Green IT KPIs: *"it is a major source of motivation"* |
| | *"Three months after Green IT application, developers need to know concretely what has changed, otherwise, for them, it will be just a marketing maneuver".* |
| | About trainings: *"Between an if, foreach or a while loop, which is the best for Green IT ?"* |
| | *"When such a change is introduced, we can easily change our habit first, but then go back to the old habits after a while".* |
| | *"I need to know practically what are the coding best practices? how can I use them? when to use them?"* |
| | *"IT is extremely vague, generic best practices are good to take, but it is very important to receive specific training and best practices".* |
| | About trainings: *"Green IT for mobile developers is different than Green IT for backend developers"* |

# STUDYING THE ENERGY CONSUMPTION OF INSTAGRAM FUNCTIONALITIES ON PC AND TABLET

Elisa Jiménez, Carlos Pulido, Coral Calero, Mª Ángeles Moraga, Félix García and Alberto Gordillo
*Institute of Technology and Information Systems, University of Castilla-La Mancha.*
*Camino de Moledores, Ciudad Real, 13005, Spain*

**ABSTRACT**

(1) Background: The internet's widespread use has made social media platforms (SMPs) an integral part of everyday life, transforming global communication and information sharing. Instagram, one of the leading SMPs, has 500 million daily users and offers various functionalities, such as stories and posts. However, the environmental impact of using these platforms remains largely unexplored.
(2) Methods: Using the FEETINGS framework, this paper investigates the energy consumption of Instagram by conducting measurement studies of various functionalities on both tablets and PCs.
(3) Results: The study found significant differences in energy consumption across different Instagram features. Sharing videos in stories and reels was identified as one of the most energy-intensive activities, and device choice was found to influence overall energy consumption, with PCs consuming more energy than tablets.
(4) Conclusions: The study highlights the importance of optimising Instagram use for greater energy efficiency. Prioritising energy-efficient features and considering device choice can mitigate environmental impact and promote sustainable digital practices.

**KEYWORDS**

Social Media, Instagram, Energy Consumption, Software Sustainability

## 1. INTRODUCTION

The internet has become an integral part of our lives, providing us with endless possibilities for entertainment and information. This is evidenced by the 66.2% of the world's population accessing it by 2024 (Kemp, 2022, p. 202)

The internet has transformed the way we carry out many everyday activities, including the way we interact with others.

Social networking sites (SNS) are digital communication platforms that connect users worldwide via the internet. The first social network, SixDegrees (BBC News World, 2019), was created in 1997 and allowed users to locate other members of their network and create lists of friends, based on the theory of 6 degrees of separation (Kirvan, 2022). Since then, social networking applications have become increasingly popular, among people of different ages. Depending on their profile, age, and interests, users choose one social network over another. Modern social networking sites have added new features, such as real-time chatting and the ability to share content, and have even diversified to include specific activities, such as job searching or partner seeking.

For this reason, social media is one of the most widespread online activities worldwide. According to (Dixon, 2023), currently over 5.17 billion people use them, and this number is expected to increase to 5.85 billion by 2027. However, it is unclear whether the development and subsequent use of these platforms have an impact on the environment (BEREC, 2022). Therefore, it is crucial for users to be aware of the impact of their activities. This awareness is also aligned with the United Nations Sustainable Development Goals (SDGs) (United Nations, 2015), particularly SDG 13, which focuses on taking urgent action to combat climate change and its impacts.

To assess the energy consumption associated with social media use and aid users in optimizing their usage, this study analyses various features of Instagram (Instagram, n.d.) from their energy consumption point of view. Instagram was launched in 2010 as a photo-sharing platform, but over time, the social network has evolved and now offers additional features such as stories, short videos, and messaging with other users. As a result, it has become one of the most popular social networks among young people worldwide, with 500 million daily users (Khoros, 2022).The findings obtained in the presented study will be used to provide recommendations for the use of this social network.

The research questions to be answered are:

**RQ1:** Which Instagram features result in higher consumption?

**RQ2:** Does the inclusion of multiple parameters in the same functionality lead to an increase in consumption?

**RQ3:** Is the consumption for performing the same functionality equivalent on both PC and tablets?

To address the research questions, two studies were conducted, one on tablet and the other on PC, measuring the different combinations available on Instagram for uploading and viewing content. As not all features are available on PCs, we have defined different actions for each device.

To explain the results of our study, this document is structured as follows: Section 2 describes the methodology used to perform the measurements, Section 3 presents and analyses the results obtained from both studies, and finally, Section 4 presents the conclusions.

## 2. MATERIALS AND METHODS

To measure software energy consumption, the GSMP (Green Software Measurement Process) (Mancebo, Calero, & García, 2021) is used, which is a specific process for measuring the energy consumption of software. This process is part of the FEETINGS framework (Framework for Energy Efficiency Testing to Improve Environmental Goals of the Software) (Mancebo, Calero, Garcia, et al., 2021), which aims to accurately capture, analyze, and interpret software energy consumption data.. FEETINGS consists of two other components: A conceptual component (GSMO), which contains the terminology related to the measurement of software energy consumption, and a technological component, which includes EET (Energy Efficiency tester) (Mancebo et al., 2018), a hardware device for measuring the software energy consumption; and ELLIOT (Gordillo & Mancebo, 2022), a tool to analyse the huge amount of data captured by EET. In this study, we have chosen to use FEETINGS, which collects actual energy consumption data rather than estimates as other tools do (e.g. RAPL (Khan et al., 2018)), which ensures more accurate and reliable results.

GSMP (Mancebo, Calero, & García, 2021), the process used, consists of seven phases (Figure 1). Initially, the requirements and software system to be evaluated are defined. The next two phases focus on configuring and preparing the measuring environment. The fourth phase is the measurement of energy consumption. Finally, the data obtained is analysed and reported in the last phases.

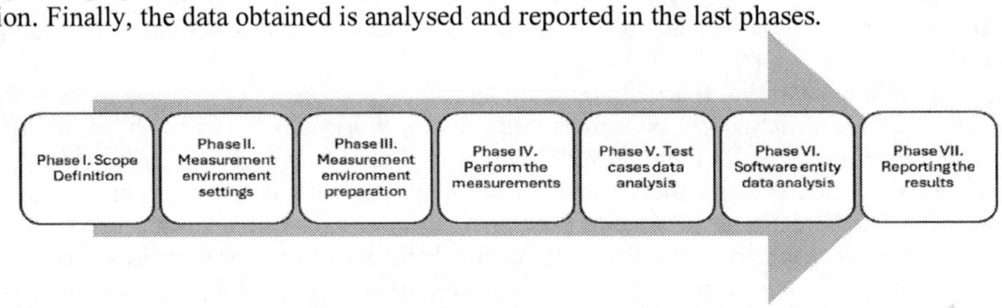


Figure 1. GSMP phases for evaluating the energy efficiency of a software

The scope of the present study (*Phase I*) is to analyse how the use of Instagram features affects consumption, including sharing one's own content and viewing external content. We have considered the following Instagram features to measure the consumption of one's own content:

- Stories are audiovisual contents that include photos, videos, and images, and are available to the public for only 24 hours.
- Posts consist of sharing photos or videos without a predetermined duration.

In both cases, a publication can be uploaded using a combination of different additional elements available on the social network, such as music or effects.

And for the external content viewing, we have considered the following three features:

- Stories: This section of Instagram allows us to see the content posted by our contacts for a period of 24 hours.
- Reels: Reels are vertical full-screen videos with a maximum duration of 90 seconds. Randomly selected videos provided by the application are available in the Reels section.
- Content visualisation: Involves viewing different types of content.

According to the considered Instagram features, the actions to be analysed for the tablet and PC studies are shown in Table 1 and Table 2 respectively. It is worth emphasizing that for Action13 to Action15, measurements will be made by randomly viewing content for a period of 120 seconds. It is important to note that not all the selected Instagram features will be available on the PC, as Instagram is a social network designed for use on tablets. Therefore, only the actions shown in Table 2 will be analysed in the PC study.

Table 1. Actions of the study in Tablet

| Software Entity | TestCase | Description |
|---|---|---|
| Stories | Action1 | Storie with a photo. |
| | Action2 | Storie with a photo and text. |
| | Action3 | Storie with a photo and music. |
| | Action4 | Storie using a combination of photo, music, and text. |
| | Action5 | Storie with a photo that has a static effect and background music. |
| | Action6 | Storie with a photo that has a dynamic effect and background music. |
| | Action7 | 15-second video storie. |
| Posts | Action8 | Post a photo. |
| | Action9 | Post a photo and comment. |
| | Action10 | Post a photo with effect and comment. |
| | Action11 | Post a 15-second video. |
| | Action12 | Post a 15-second reel. |
| Content visualisation | Action13 | Visualising stories. |
| | Action14 | Visualising reels. |
| | Action15 | Visualising general (standard use). |

Table 2. Actions of the study in PC

| Software Entity | TestCase | Description |
|---|---|---|
| Posts | Action16 | Post a photo. |
| | Action17 | Post a photo and comment. |
| | Action18 | Post a 15-second reel. |

During *Phase II*, we chosen EET on its version for PC and for Tablet, as measuring instruments. The specifications of the two Devices Under Test (DUTs) used (Tablet and PC) are detailed in Table 3.

Table 3. DUT specifications for Tablet and PC

| DUT specifications for **Tablet** | |
|---|---|
| **Model** | Lenovo TB-X306F |
| **Screen** | HD IPS (1280 x 800) de 25,65 cm (10,1") y 400 nits |
| **Processor** | MediaTek P22T (4 2,3 GHz + 4 1,8 GHz) |
| **RAM** | 2 GB LPDDR4x |
| **O.S.** | Android 10 |

| DUT specifications for **PC** | |
|---|---|
| **Monitor** | Philips 170s6fs LCD |
| **Motherboard** | ASUS Prime B460-Plus |
| **Processor** | Intel i7 10700 2900MHz |
| **RAM** | 2 modules of 16GB Kingston Hiperx Fury DDR4 |
| **Graphics card** | Sapphire ATI Radeon X1950 GT, 256mb RAM DDR3 |
| **Hard disk** | Western Digital Blue 500GB SSD |
| **Power supply** | 360 PS5805 – 580W |
| **O.S.** | Ubuntu 20.4 LTS |

While the tablet EET captures the power consumption of the entire tablet, the PC EET can measure the power consumption of various parts of the PC. However, in this study we decided to analyse just the power measurement of the entire PC, to align another aspect with the tablet study and be able to make the comparison between the two. In the *Phase III* we decided to run each test case 10 times to assure the accuracy of the measurements. Considering that both EETs can capture 100 samples per second, 10 measurements are a good sample size to mitigate the impact of outliers. Once the measurements have been taken (*Phase IV*), it is possible to analyse the data obtained using ELLIOT (Gordillo & Mancebo, 2022), which ensures greater reliability of the analysis obtained. The data analysed by ELLIOT can then be interpreted to answer our research questions (*Phases V and VI*). It should be noted that the consumption data analysed are those obtained after subtracting the baseline, i.e. the consumption of the operating system and hardware devices in the background.

# 3. RESULTS

This section presents and analyses the results obtained during the measurement of energy consumption in the two studies (Tablet and PC). The energy consumption data obtained from the measurements can be found in https://github.com/EliJR/InstagramEnergyConsumption.

To answer the research questions, the following comparisons will be presented: Actions on Stories, Actions on Posts, Actions on Content Visualisation, Actions on Stories vs. Actions on Posts and Actions on Tablet vs. Actions on PC (see Table 1 and Table 2).

## 3.1 Tablet

Firstly, the study was carried out on a tablet to find out which features lead to higher energy consumption. Table 4 shows the time and consumption results obtained after performing the measurements. These values, provided by ELLIOT after removing the baseline (idle DUT consumption), represent the average of all the time data captured by EET during the execution of the corresponding feature. The energy consumption has been calculated by using the time and the data power consumption average, also provided by ELLIOT.

Table 4. Tablet actions time and energy consumption

| | Time (s) | Energy consumption (J) |
|---|---|---|
| Action1 | 18.062 | 27.403 |
| Action2 | 29.972 | 28.126 |
| Action3 | 65.845 | 96.050 |
| Action4 | 74.474 | 128.928 |
| Action5 | 80.288 | 139.176 |
| Action6 | 137,549 | 218,789 |
| Action7 | 45.000 | 82.647 |
| Action8 | 17.770 | 25.952 |

| | Time (s) | Energy consumption (J) |
|---|---|---|
| Action9 | 29.619 | 39.907 |
| Action10 | 32.011 | 44.058 |
| Action11 | 54.590 | 103.604 |
| Action12 | 100.117 | 207.732 |
| Action13 | 120.000 | 193.849 |
| Action14 | 120.000 | 198.405 |
| Action15 | 120.000 | 213.588 |

### 3.1.1 Stories

This section aims to compare the difference in consumption when posting an Instagram storie with different content. The results show that Action1 (post a storie with a photo), obtained the best result. Table 5 shows the percentage of increase in consumption and the time required to add more parameters when posting stories (Action1 vs. Action2-Action6). In most cases, there is a significant difference in the amount of energy consumed, which increases by almost 700% compared to Action1. Similarly, in terms of time, the percentage increase compared to Action1 is also very high, reaching a maximum of approximately 662% of increase. Figure 2 shows the relationship between the number of parameters added and the corresponding increase in time and energy consumption required for publishing a photo.

Table 5. Time and energy consumption comparison: Action1 vs. Action2-Action6

| | Action1 time with respect Action n (%) | Action1 energy consumption with respect Action n (%) |
|---|---|---|
| Action2 | 65,943 | 2,641 |
| Action3 | 264,562 | 250,512 |
| Action4 | 312,337 | 370,492 |
| Action5 | 344,526 | 407,890 |
| Action6 | 661,561 | 698,417 |

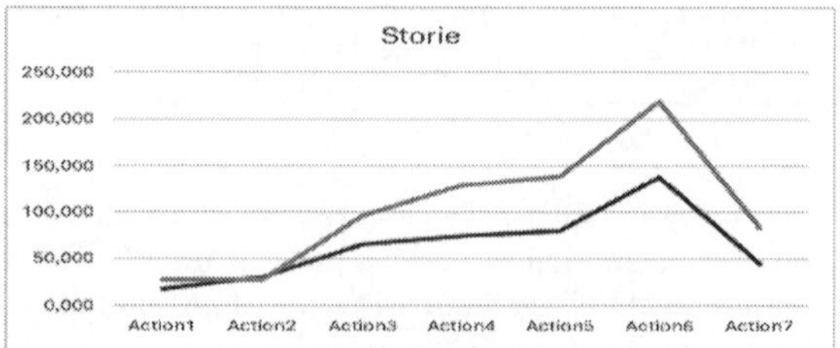


Figure 2. Evolution of time and consumption across stories actions

On the other hand, publishing a 15-second video storie (Action7) results in an increase in both time and consumption compared to a storie with just a photo and text (Action2). However, it involves less time and consumption compared to other cases (Action3-Action6). Table 6 displays the percentages of increase and savings of Action7 in comparison to the other Actions.

Table 6. Time and energy consumption comparison: Action7 vs. Action1-Action6

| | **Action7 time with respect Action n (%)** | **Action7 energy consumption with respect Action n (%)** |
|---|---|---|
| Action1 | 149.148 | 201.599 |
| Action2 | 50.141 | 193.839 |
| Action3 | -31.658 | -13.955 |
| Action4 | -39.577 | -35.897 |
| Action5 | -43.952 | -40.617 |
| Action6 | -67.285 | -62.225 |

Therefore, it can be concluded that:

- Adding more parameters to a storie increases both the time and consumption required.
- Even a short video (such as the 15-second example given) is more time-consuming than a photo.

### 3.1.2 Posts

As a continuation of the analysis, we will compare the difference in time and energy consumption between the content posts functionalities. As in the previous case, the best results in terms of time and energy consumption are obtained by the simplest post, which contains only a photo. Table 7 shows the percentage increase in these variables that Action9-Action12 entail compared to Action8. In this case, publishing a video or a reel takes more time and consumes more energy than publishing photos in any of their variants. The increase in consumption can be up to 700.46% in the most extreme case (Action12).

Figure 3 illustrates the increase in time and energy consumption for the different types of posts.

Table 7. Time and energy consumption comparison: Action 8 vs. Action9-Action12

| | **Action 8 Time with respect Action8 (%)** | **Action 8 Energy consumption with respect Action8 (%)** |
|---|---|---|
| Action9 | 66.682 | 53.774 |
| Action10 | 80.145 | 69.769 |
| Action11 | 207.205 | 299.220 |
| Action12 | 463.409 | 700.458 |

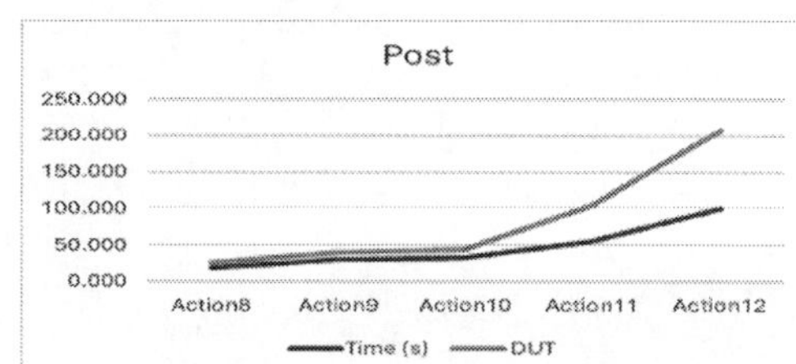


Figure 3. Evolution of time and consumption across post actions

Therefore, it can be concluded that:

- Including additional parameters in images increase their consumption.
- Videos in any of their variants consume more energy than images.

### 3.1.3 Visualisation

This section compares the energy consumption required to randomly view different content for 120 seconds. The action with the best consumption results was viewing only stories (Action13). Table 8 displays the percentage increase in consumption compared to Action13. Figure 4 illustrates the variations in energy consumption among the different display actions.

Table 8. Energy consumption comparison: Action13 vs. Action14-Action15

| | **Energy consumption with respect Action13 (%)** |
|---|---|
| Action14 | 2.350 |
| Action15 | 10.182 |

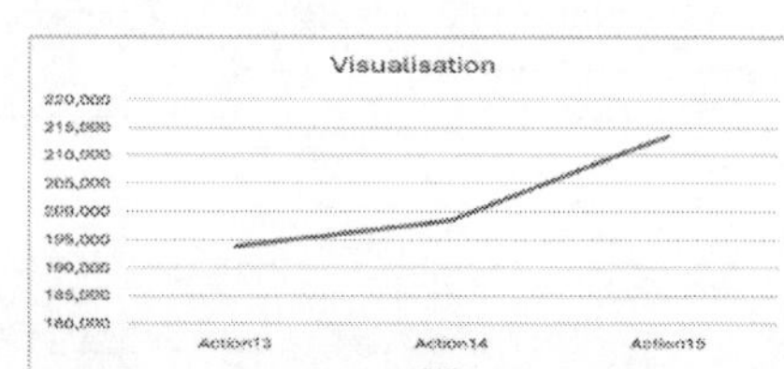


Figure 4. Evolution of consumption across visualization actions

It can be concluded that:

- Displaying specific content separately is better than displaying content alternately due to the considerable differences between some of the actions.

### 3.1.4 Post vs. Storie

To conclude the analysis of tablet functionalities, this section compares the time and consumption differences between sharing content as a story or a post. Due to varying available parameters, comparisons include: Action8 vs. Action1; Action9 vs. Action2; and Action11 vs. Action7. Table 9 displays the percentage difference when sharing the same content in post or storie format.

There is minimal difference in time and effort between sharing a photo as a story or a post, with posts requiring slightly less time. However, adding text significantly increases the posting time by almost 42%. Similarly, posting a video takes 25% more time. Conversely, there is only a small-time difference for sharing photos and photos with text, with a 1-2% decrease in time when posting. Videos, however, show a 21% increase in posting time compared to stories. Figure 5 illustrates the time and consumption differences for the three comparisons.

Table 9. Comparison rates between posts and stories

| | Time (s) | Energy consumption (%) |
|---|---|---|
| Action8 vs. Action1 | -1.615 | -5.296 |
| Action9 vs. Action2 | -1.177 | 41.884 |
| Action11 vs. Action7 | 21.311 | 25.358 |

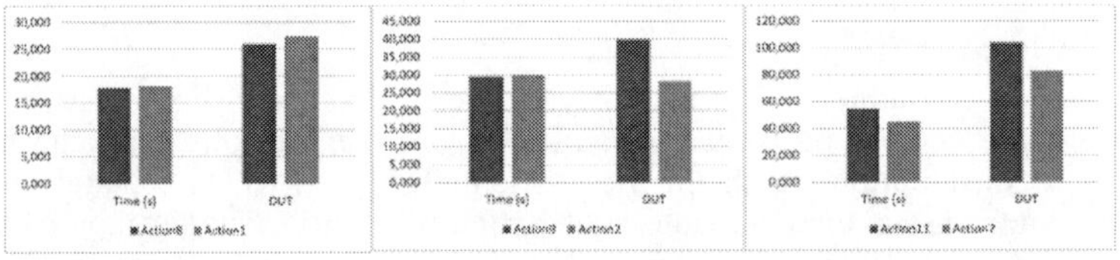

Figure 5. Difference in time and consumption between post and storie

In two out of the three actions, sharing content in post format resulted in higher consumption compared to storie format. As the remaining actions represent only a minimal percentage, it can be concluded that:

- Sharing content in the form of stories seems to be more effective than permanent posts.
- Sharing a photo without effects seems to be more efficient to do as a permanent post.

## 3.2 PC

As already mentioned, the functionalities of Instagram for PCs are very limited, as it is a social network designed to be used on tablets. Table 10 shows the results obtained after measuring the only three actions available on PCs (see Table 2).

Action16 achieved the best results among the actions by uploading a photo without any additional content. Table 11 presents the percentage increase in time and consumption for the remaining actions in comparison to Action16. The data presented in Figure 6 illustrates that adding text to a photo results in a 90% increase in consumption and a 63% increase in time.

Posting a video results in the highest consumption, with a 189% increase compared when posting a photo. Similarly, Action18 results in a 151% increase in time compared to posting a photo. Figure 6 shows the increase in time and consumption in the different PC actions.

Table 10. Time and energy consumption of PC actions

| | Time (s) | Energy consumption (J) |
|---|---|---|
| Action16 | 18.951 | 1937.193 |
| Action17 | 30.893 | 3681.768 |
| Action18 | 47.599 | 5608.286 |

Table 11. Time and energy consumption comparison: Percentage increase Action17-18 vs. Action16

| | Time with respect Action16 (%) | Energy consumption with respect Action16 (%) |
|---|---|---|
| Action17 | 63.013 | 90.057 |
| Action18 | 151.165 | 189.506 |

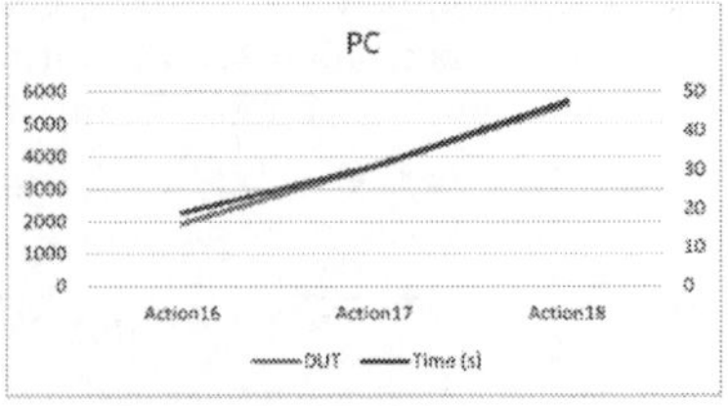


Figure 6. Evolution of time and consumption across PC actions

Based on the analysis of the obtained data, it can be concluded that:

- Publishing a photo on a PC, either alone or with text, seems to consume less energy and time compared to publishing a video.
- Publishing a photo without any additions requires the least amount of time and energy.

## 3.3 Tablet vs PC

Finally, to see whether the same functionality consumes more on PC or Tablet, this section compares the results of both. As the features available on PCs are less than those available on tablets, the following comparisons are made: Action16 vs. Action8; Action17 vs. Action9; Action18 vs. Action11.

Table 12 shows the percentage increase in PC functionality over tablet functionality. As shown, although the time differences are small, the percentage differences in power consumption are significant. This is because the power required by the components of a PC is much greater than that of a tablet.

Table 12. Comparison of increase and savings of PC compared to Tablet

| | **PC time with respect Tablet (%)** | **PC energy consumption with respect Tablet (%)** |
|---|---|---|
| Action16 vs. Action8 | 6.648 | 7364.625 |
| Action17 vs. Action9 | 4.301 | 9125.882 |
| Action18 vs. Action11 | -12.807 | 5313.176 |

Based on this data, it can be concluded that:

- Using Instagram on a tablet seems to be more energy efficient than on a PC.

## 4. CONCLUSIONS

In this paper we have presented two studies aimed to analyse the energy consumption of Instagram's features on Tablet and PC. We have compared the energy consumption required by several actions to be performed on Instagram both on Tablet and PC. The findings allow us to provide recommendations, (see Figure 7) for users to make more environmentally responsible use of the social network. Due to space limitations, threats to the validity can be consulted in the provided repository.

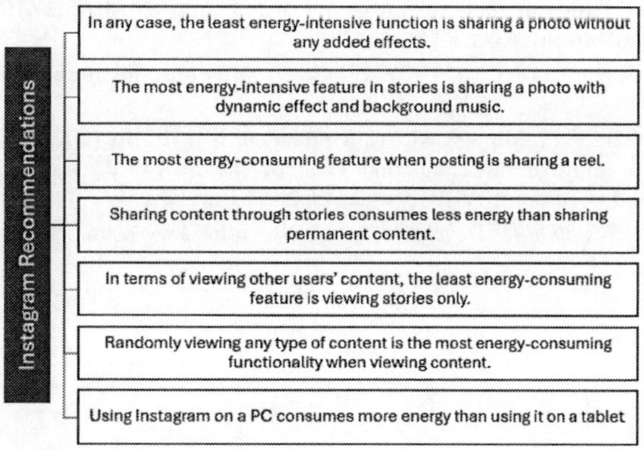


Figure 7. Instagram recommendations for users

## ACKNOWLEDGEMENT

This work was supported by projects: OASSIS (PID2021-122554OB-C31/ AEI/10.13039/ 501100011033/FEDER, UE); EMMA (Project SBPLY/21/180501/000115, funded by CECD (JCCM) and FEDER funds); SEEAT (PDC2022-133249-C31 funded by MCIN/AEI/ 10.13039/501100011033 and European Union NextGenerationEU/PRTR); PLAGEMIS (TED2021-129245B-C22 funded by MCIN/AEI/ 10.13039/501100011033 and European Union NextGenerationEU/PRTR). Financial support for the execution of applied research projects, within the framework of the UCLM Own Research Plan, co-financed at 85% by the European Regional Development Fund (FEDER) UNION (2022-GRIN-34110).


## REFERENCES


BBC News World. (2019, June 8). Six Degrees: What it was like and who created the internet's first social network, inspired by the theory of 'six degrees'. *BBC News*. https://www.bbc.com/mundo/noticias-48558989

BEREC. (2022). *BEREC Report on Sustainability Assessing BEREC's contribution to limiting the impact of the digital sector on the environment*. https://www.berec.europa.eu/en/document-categories/berec/reports/berec-report-on-sustainability-assessing-berecs-contribution-to-limiting-the-impact-of-the-digital-sector-on-the-environment

Dixon, S. (2023). *Number of social media users worldwide from 2017 to 2027*. https://www.statista.com/statistics/278414/number-of-worldwide-social-network-users/

Gordillo, A., & Mancebo, J. (2022). Elliot: Gestión Y Análisis De Datos De Consumo De Software. In *Calidad y sostenibilidad de sistemas de información en la práctica* (1st ed.).

Instagram. (n.d.). *Instagram*. Retrieved 8 March 2023, from https://about.instagram.com/es-la

Kemp, S. (2022, October 20). Digital 2024: October global statshot report. *Datareportal*. https://datareportal.com/reports/digital-2024-global-overview-report

Khan, K., Hirki, M., Niemi, T., Nurminen, J., & Ou, Z. (2018). RAPL in Action: Experiences in Using RAPL for Power Measurements. *ACM Transactions on Modeling and Performance Evaluation of Computing Systems (TOMPECS)*, *3*. https://doi.org/10.1145/3177754

Khoros. (2022). The 2024 Social Media Demographics Guide. *Khoros, LLC*. https://khoros.com/resources/social-media-demographics-guide

Kirvan, P. (2022, July). Six degrees of separation. *WhatIs.Com*. https://www.techtarget.com/whatis/definition/six-degrees-of-separation

Mancebo, J., Arriaga, H. O., García, F., Moraga, M. Á., García-Rodríguez de Guzmán, I., & Calero, C. (2018). EET: A Device to Support the Measurement of Software Consumption. *2018 IEEE/ACM 6th International Workshop on Green And Sustainable Software (GREENS)*, 16–22.

Mancebo, J., Calero, C., & García, F. (2021). GSMP: Green Software Measurement Process. In *Software Sustainability* (pp. 43–67). Springer.

Mancebo, J., Calero, C., Garcia, F., Moraga, M. A., & Guzman, I. G.-R. de. (2021). FEETINGS: Framework for Energy Efficiency Testing to Improve Environmental Goal of the Software. *Sustainable Computing: Informatics and Systems*, *30*, 100558. https://doi.org/10.1016/j.suscom.2021.100558

United Nations. (2015). *Sustainable Development Goals*. https://www.un.org/sustainabledevelopment/development-agenda/

# ASSESSMENT OF HYGROTHERMAL PROPERTIES OF ADOBE BRICKS. RESEARCH EXPERIENCE AT KEENE STATE COLLEGE

Fernando del Ama Gonzalo, Paul Fowler and Zachary Punch
*Keene State College*
*229 Main St. 03431 Keene, NH, USA*

**ABSTRACT**

Throughout history, soil-based building materials have been a popular choice for construction. However, in recent years, a greater focus has been put on using eco-friendly materials for new construction, such as adobe and rammed earth, to promote sustainable architecture. Concrete and fired clay brick industries impact the ecological footprint due to their use of clinker and the firing process. Adobe is a promising alternative because it has the potential to significantly reduce the ecological footprint of conventional building materials. In addition, the use of fibers for adobe reinforcement is a critical factor in improving adobe performance. By using residues from the paper and pulp industry to replace traditional fibers, the ecological footprint can be further reduced. As a part of an undergraduate research project, this article intends to explain a methodology to assess the thermo-hydrometrical properties of adobe bricks by building prototypes on the Keene State College campus. The second goal is to measure comfort conditions by analyzing heat sources in a full-scale prototype. The probes and sensors will provide data on the prototype throughout the year to assess comfort, energy consumption, and indoor air quality.

**KEYWORDS**

Adobe Bricks, Hygrothermal Properties, Simulated and Measured Data, Undergraduate Research

## 1. INTRODUCTION

New energy efficiency standards and life-cycle analysis have different effects on building materials. Improving the energy efficiency of buildings and mitigating greenhouse gas emissions are new strategies that construction stakeholders must consider (Ciancio et al. 2020). As sustainable building materials and environmental concerns become prevalent, Adobe Blocks have emerged as a top choice for eco-friendly materials due to their ability to maintain indoor air quality, regulate indoor temperatures, and minimize environmental impact (Hao et al. 2020). While traditionally used in rural areas of developing countries, research has shown that these blocks have some limitations in terms of strength and durability (Imada et al. 2021). Developing Adobe blocks will help from many environmental perspectives, especially regarding global warming and climate change, and can also help achieve minimal energy and resource utilization (Brito et al. 2023). Some studies have highlighted the significance of analyzing the properties of adobe brick walls and testing for durability, life expectancy, and carbon footprint (Mellaikhafi et al. 2022). A control mixture will create bricks from 60% sand, 30% clay, and 10% other additives (Thi et al. 2021). These different materials will be appropriately hydrated to avoid the material sucking water from the brick mixture as the bricks dry. One of the project's goals is to set the optimum moisture in terms of proportion between sand, clay, and fibers by simultaneously analyzing the mixture's physical and hygrothermal properties. Different types and amounts of fibrous materials (straw, timber waste, and chipped wood from local vines) will be tested in varying percentages of fiber-to-brick volume.

In Northern European and American regions, roofs and walls must be able to amplify insulation and conserve energy, whereas in hot and arid climates, they must regulate heat transfer. While most studies have relied on one-dimensional heat transfer simulations, novel methodologies have emerged to enhance analysis and calculation. Nonetheless, further advancements in multi-dimensional modeling techniques are crucial to attain more accurate results that reflect the performance of diverse roofing technologies in varying weather conditions (Tootkaboni et al. 2021). A two-dimensional hygrothermal simulation has been devised to enhance the accuracy and realism of results (Costantine et al. 2020). When calculating hygrothermal models that affect building envelope performance and energy consumption, the three critical factors of water surface exposure, absorption, and intrusion must be considered (Gaur et al. 2019). Despite this, limited studies are available that provide accurate results in multi-dimensional hygrothermal models and evaluate the effects of moisture on adobe walls (Libralato et al. 2021). After reviewing various modeling options, WUFI was the most suitable software for the simulation due to the need for walls to be simulated together for accurate comparison (WUFI). This tool precisely calculates heat and moisture exchanges between the building envelope and the surroundings, utilizing a comprehensive precipitation database and calibrated moisture transport to measure vapor diffusion and liquid water transport (Frasca et al. 2018). In each time step, WUFI can accurately calculate the hygrothermal loads based on equations utilizing the Thomas algorithm. In addition to its primary functions, WUFI Pro is capable of indoor air modeling and can simulate space heating and cooling. The last stage of this research was to build a small-scale mockup to assess the accuracy of the WIFI model compared to the measured results in the prototype.

Undergraduate research is an essential activity of higher education. However, sometimes, it is a challenge to include those activities in a tight fifteen-week academic calendar. As sustainable building materials and environmental concerns become prevalent, undergraduate research can become a meaningful educational experience to engage students in this topic (Arikan et al., 2022). Shavers and Mitchell (2019) highlight the importance of undergraduate research in university education today and underscore the need for educators and students to develop essential skills such as information search, assimilation and retention, organization, analysis, and decision-making. Current trends in undergraduate research involve incorporating the construction of prototypes to compare simulated and measured data (Fowler et al. 2023). These practices require faculty mentors to guide students through reading relevant literature, providing valuable feedback, designing, and implementing independent work, and developing proficient communication skills (Beier et al. 2019). Additionally, undergraduate research provides an opportunity to disseminate practical strategies throughout academia (Engberg and Fox, 2011). This project continued the experience of instructors and students in the Keene State College Architecture program in implementing high-impact practices, such as hands-on and project-based activities and undergraduate research.

The main aim of this project was to assess the impact of moisture absorption on adobe walls constructed in extremely cold climates, such as the weather conditions found in New Hampshire, US. Another objective was to compare the accuracy of simulated data with the data obtained from sensors embedded in prototypes throughout the year. Additionally, the project sought to evaluate the ability of traditional adobe walls to offer comfort and energy efficiency. Lastly, the project aimed to demonstrate how research can be used as a teaching tool to enhance the learning experiences of undergraduate students.

## 2. DESCRIPTION OF THE EDUCATIONAL EXPERIENCE

This project delved into Adobe structures' hygrothermal properties and durability performance. The study emphasized the importance of further testing in indoor air quality criteria. We can significantly address environmental concerns and minimize energy and resource utilization by developing Adobe Blocks. The educational experience includes three phases: firstly, the manufacture of the adobe bricks; secondly, the simulation of the hygrothermal behavior of the material and comparison with a small mockup; and finally, the construction of a full-scale prototype where data on indoor thermal parameters and indoor air quality will be taken throughout a year.

### 2.1 Materials and Methods

To properly construct the digital model, it was crucial to identify the main hygro-thermal parameters of the construction material using software inputs as a starting point. Next, we focused on gathering technical details about the Adobe bricks, paying particular attention to density and moisture retention. For a traditional building, the next step would be to determine the number of occupants during different periods and the characteristics of the equipment used inside the building that could impact internal gains, such as lighting or electronic devices. However, for this case study, our main objective was to replicate the impact of a heat source on indoor air parameters, including air temperature and relative humidity. To achieve this, we collected comprehensive thermal data between April and May 2024 using indoor and outdoor sensors to ensure we captured all necessary data on the internal environmental conditions throughout the study. Once we had all the relevant information on the mockup, we needed to determine the most appropriate modeling strategy, the goals we wanted to achieve, and the level of precision required for the subsequent simulation. However, significant challenges appeared due to the strong fluctuations in outdoor temperature and relative humidity data throughout the study, which made it challenging to meet comfort requirements and analyze the building's energy consumption.

The adobe blocks were made by manually mixing materials in the correct proportions, with each batch of 4 bricks mixed individually to ensure consistency. The selected proportion was 60% sand, 30% clay and 10% of chipped wood from local vines. This process follows the same procedure used in real-world applications as documented in other studies (Charai et al. 2022). The mixture was then poured into wooden molds, and after production, the density of the blocks was measured at 1.45 g/cm$^3$. Figure 1 depicts the mixture being poured into the molds.

Figure 1. Fabrication of the adobe bricks

### 2.1.1 Simulation Tools

The simulation, conducted over a two-year period, utilized the WUFI Pro software developed by the Fraunhofer Institute for Building Physics (Antretter et al. 2011). This software enables the comprehensive evaluation of heat and moisture transfer within building envelopes, making it a widely accepted tool for assessing new materials and calculation methods.

Using WUFI, two scenarios were modeled for the mockup: one with a heat source inside and one without it. These simulations were conducted to determine the annual temperature and relative humidity performance of the prototype's internal and external faces and the water content within the adobe wall. The energy-saving potential of each scenario was estimated and compared to the baseline building design. The simulations were run in 1-hour increments for two years, allowing for an analysis of the impact of moisture on energy loads and the feasibility of designing a non-cooling building. The WUFI software allows users to test water surface exposure, absorption, and intrusion to replicate rainfall and moisture in wall and roof materials. Additionally, a precise localized weather model has been created to account for the influence of urban context on microclimate conditions.

### 2.1.2 Small-Scale Mockup

The first approach involves making a mock-up with adobe walls and a plywood roof with an electric heating source. In this research phase, the main concern is testing the functionality of the measurement devices. By prototyping the experience, students can compare simulated with measured data and understand what interactions may occur and what external data dependencies may disrupt the intended flow of the experience. This phase can also reveal potential problems with asset positioning, scale, and size. Figure 2 illustrates the size of the mockup and the position of the internal heat source and the probes that measured internal and external temperature, relative humidity, and the water content inside the adobe walls. To model the mockup accurately, it was necessary to start by identifying the construction materials and their key characteristics, using the building materials and drawings as a guide. The role of the heating system in the modeling process was crucial, as it directly affects the indoor air parameters. Therefore, the technical details of the heating system, including its operation and limitations, need to be collected in full. The goal of this case study was to replicate the effect of indoor air parameters, such as air temperature and humidity. To achieve this, an exhaustive thermal data collection took place between April and May of 2024, using both indoor and outdoor sensors to ensure comprehensive coverage of the environmental conditions throughout the study. With all the necessary information in hand, the next step was to determine the most appropriate modeling strategy, develop the simulation, establish goals, and define the required level of precision.

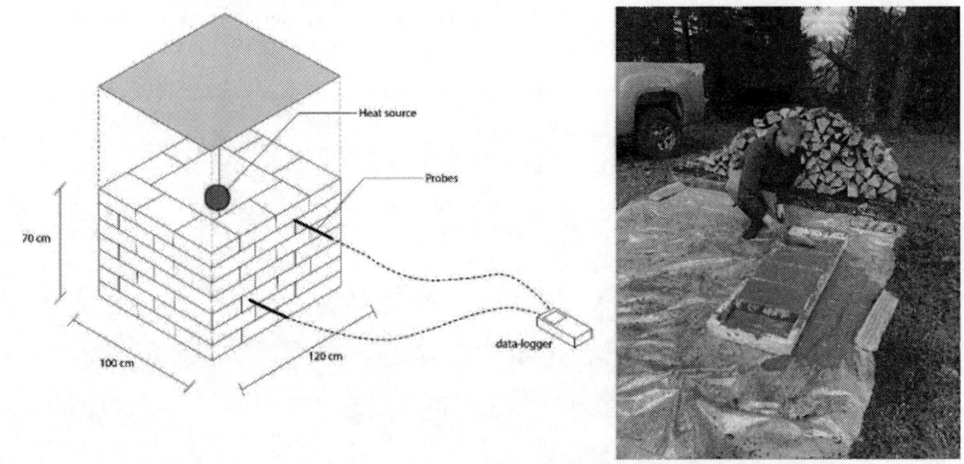


Figure 2. Description of the mockup with heat source and probe positioning

### 2.1.3 Large-Scale Prototype

The second stage will consist of making a full-scale prototype to test indoor and outdoor conditions. Figure 3 shows the dimensions of the prototype.

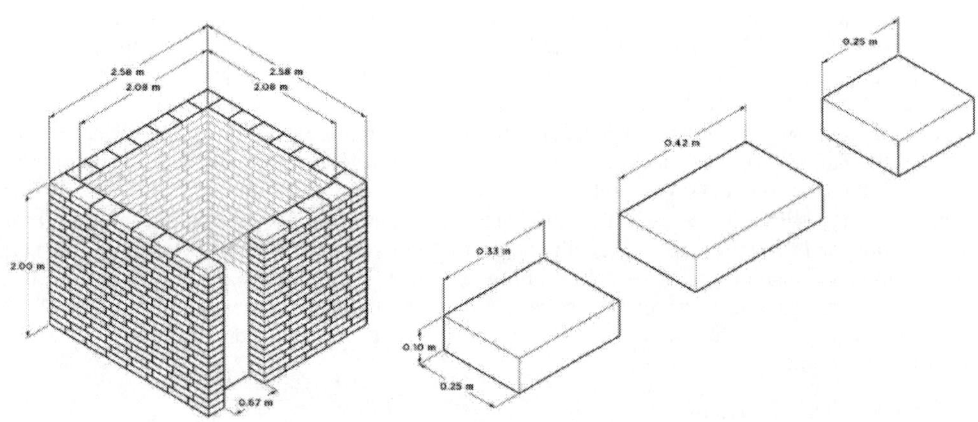


Figure 3. Description of the large-scale prototype

WUFI Pro plays a significant role in the simulation process, offering users the flexibility to select a gross, net, or mean internal volume for the thermal zone. This feature allows for the use of surface areas related to either side of a wall, enhancing the accuracy of the simulation. As noted in Section 2.2, the necessary experimental weather data for thermal dynamic simulation includes external air humidity and temperature, global direct solar radiation on a horizontal surface, diffuse solar radiation, and wind speed intensity and direction. To determine the simulation inputs, certain values such as outdoor temperature will need to be introduced. Finally, to ensure accuracy, verification will be conducted on indoor temperature and relative humidity.

### 2.1.4 Measurement Tools

The selected sensor was the General Tools & Instruments' MMD950 Pin and Deep Sensing Moisture meter. It is effective in detecting moisture just beneath the surface of specific materials. However, when dealing with hard materials such as adobe bricks, the pin measurements primarily indicate the surface moisture content. To measure the moisture inside the brick, we used a 5" Deep Wall Moisture Probe. This report aims to inform about the capabilities and limitations of this sensor and probes, and the method used for measuring moisture in hard materials.

## 2.2 Results

This section's objective was to establish a technique for aligning the data logger readings with the simulation outcomes. By incorporating the environmental data of the site and its vicinity into the model, the climatic aspects were accounted for. This included the measurements and technical features of the structure and the heat source within the prototype. To replicate the indoor and outdoor environment of the model, WUFI necessitated a climatic file. The specific file employed for the simulation included Keene Dillant-Hopkins Airport's pertinent details.

The performance of insulated walls and rooftops varies according to the specific season they are installed in. However, adobe bricks are significantly influenced by the prevailing climate conditions. The bricks' capacity to absorb moisture affects the insulation and indoor environment. Studies have explored the impact of surface moisture in measurements such as evaporation efficiency, ratio, and water vapor conductivity. Increasing the water content of adobe bricks can improve their thermal conductivity and specific heat. Table 1 shows the physical parameters of adobe bricks. Those parameters were measured and introduced in the simulation software as inputs.

Table 1. Parameters of adobe bricks

| Property | Unit | Value |
|---|---|---|
| Bulk density | $kg/m^3$ | 1514 |
| Porosity | $m^3/m^3$ | 0.42 |
| Initial water content | $kg/m^3$ | 4.3 |
| Water absorption coefficient | $kg(m^2 s^{1/2})$ | 0.0467 |

Figure 4 shows the evolution of temperature and relative humidity in both internal and external faces of the adobe brick. During the heating period, the outdoor mean air temperature showed a range of 0°C to 11°C, with the minimum dropping to -18°C. Notably, the warmest days were observed in April and October, reaching a maximum temperature of 23°C. The designed heat source ensured comfortable indoor conditions (between 18°–20°C) during operating hours from 8 a.m. to 5 p.m. In the cooling period of the year, July had the highest mean temperature of exterior air at 24°C, with a peak temperature of 33°C. The beginning days of May had the lowest temperature at 6°C. The indoor temperature of the wall during this period peaked at 26°C in July, while the mean temperature was still at 25°C without any cooling device.

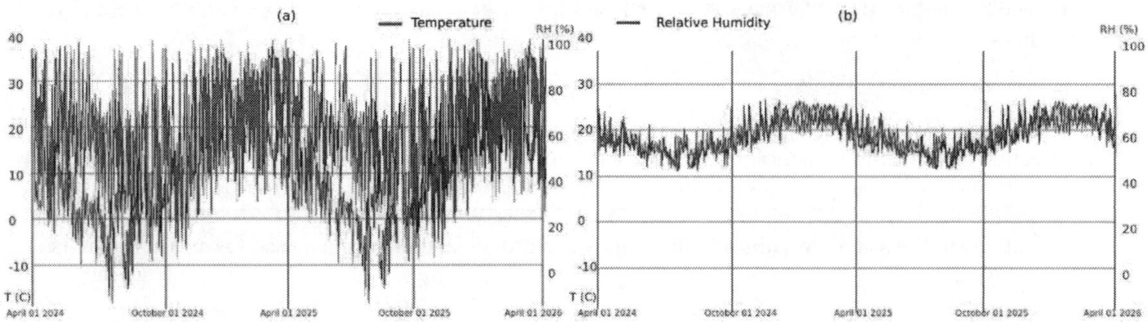

Figure 4. Temperature and relative humidity in the adobe wall; (a) external face, (b) internal face

A primary concern lies in the high porosity of the blocks, causing them to absorb a considerable amount of water and ultimately compromising their durability in areas with elevated humidity levels. As seen in Figure 5, the simulated water retention values over two years illustrate this issue. According to Table 1, the initial water content was measured at 4.3 kg/m3. Despite the material's high water retention abilities and minimal values of 3 kg/m3, the impact was notably more severe during winter and autumn, with a peak of 12 kg/m3 at the start of spring.

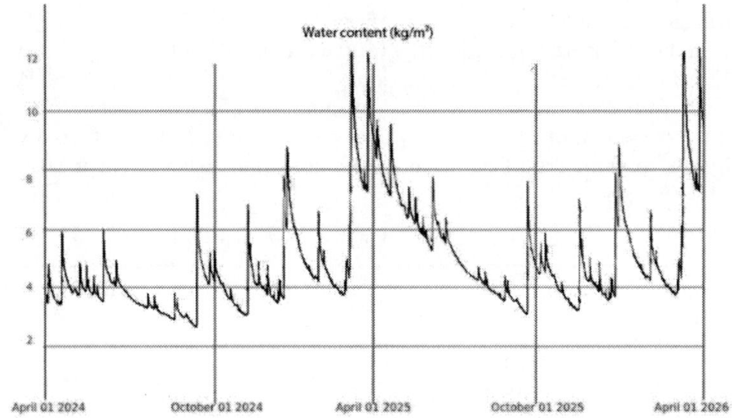

Figure 5. The water content of the simulated prototype over two years

According to the water absorption results, the adobe blocks used in this study, which were treated with sand and reinforced with fibers of local vines, exhibit remarkable stability in the presence of water. This outcome is highly promising and underscores the efficacy of the stabilized blocks in effectively absorbing moisture from the indoor surroundings.

The next step was to build the small-scale mockup and test its changing conditions throughout the year. Figure 6 illustrates the measured results between May 8 and May 29. Figure 6(a) shows the humidity as a percentage of weight, whereas Figure 6(b) shows the temperature and relative humidity evolution inside the mockup. The humidity measurements were carried out at three different points of the adobe bricks: at the exterior face (Hum OUT), at the interior face (Hum IN), and at the mid-depth of the brick (Hum MID). The moisture levels in the brick's exterior surface fluctuated notably due to the varying weather conditions in the chosen area. The water content remained consistent on the interior surface of the brick and followed a similar trend within the brick, although the water content within the brick was noticeably higher than on the interior surface.

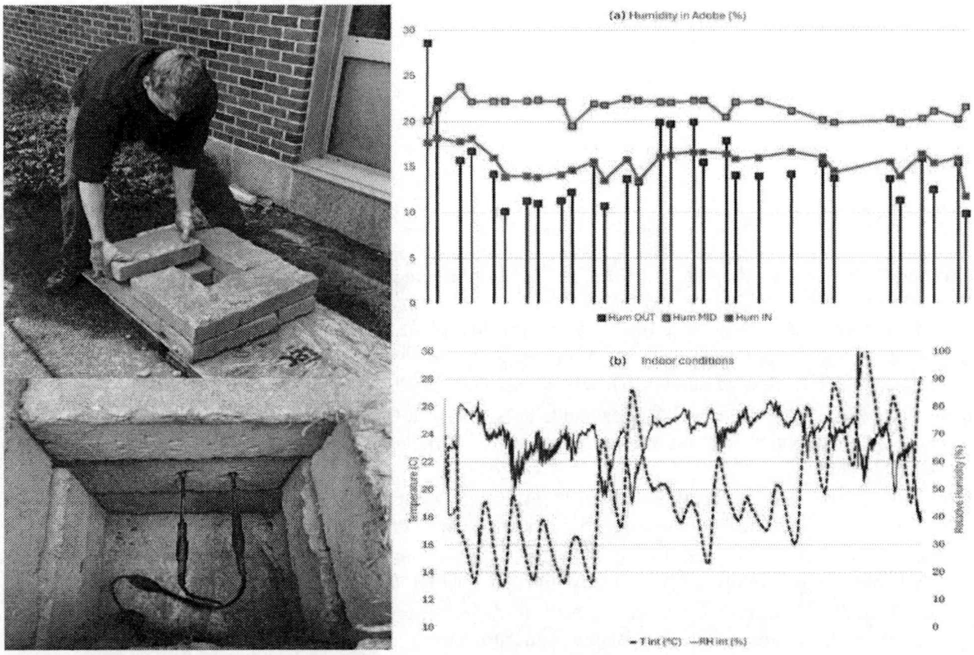


Figure 6. Description of the small-scale mockup: (a) water content of brick in percentage of weight; (b) temperature and relative humidity inside the mockup; (c) position of the deep probes

Undergraduate research activities have a lasting influence. It involves a systematic approach that encompasses documenting important information and analyzing and interpreting it according to the established protocols of specific academic and professional fields. Engaging in research is a pivotal way to enhance students' capacity for critical thinking and problem-solving, equipping them with the skills necessary to tackle real-world challenges.

# 3. CONCLUSION

This article showed the results of an educational experience at Keene State College in New Hampshire, US. Through project-based activities, students were introduced to a construction process that involved four key steps: assessment, planning, execution, and re-evaluation. While the first three steps could be effectively taught within the classroom, the fourth step was crucial in addressing unforeseen challenges and factors that arose during the process. These challenges were often related to new methods and materials that required innovative hands-on techniques to help navigate. Undergraduate research experiences allowed students to delve deeper into their research topic and gain valuable skills in conducting, managing, and presenting their work in non-traditional academic settings such as conferences and international journals.

The project started by making adobe bricks with local materials to be tested in prototypes located on the Keene State College campus. The results section showed a methodology to match simulated and measured hygrothermal parameters.

This research confirmed that, by developing a digital model in WUFI, it is possible to get simulated results for indoor temperature and relative humidity of a simple structure made of adobe walls. Several iterations were necessary to simulate, first, the thermal parameters of the envelope, second, the influence of air leakage, and finally, a heat source that mimicked the heat source and its impact on relative humidity. Comparing simulated and measured results sugges that using a thermal model for the building envelope could be a useful method to assess the hygrothermal behavior of the envelope and comfort inside buildings. The results stated in this report are preliminary outcomes of a more ambitious project that will include the construction of a full-scale prototype. This prototype will be the topic of future undergraduate research projects in the Architecture Keene State College program.

## ACKNOWLEDGEMENT

The authors wish to thank the University System of New Hampshire's Summer Undergraduate Research Fellowships (SURF) program for materials and student grants for this research.

## REFERENCES

Antretter, F.; Sauer, F.; Schöpfer, T.; Holm, A. (2011). Validation of a hygrothermal whole building simulation software. *In Proceedings of Building Simulation 2011: 12th Conference of International Building Performance Simulation Association*, Sydney, Australia, 14–16 November 2011; Volume 14, p. 16.

Arikan, S., Dochy, F., & Segers, M. (2022). Framing the Effects of High-Impact Practices from a High-Impact Learning Perspective. A Review of Studies. *Creative Education*, 13, 2994-3025. doi: 10.4236/ce.2022.139190.

Beier, M.E., Kim, M.H., Saterbak, A., Leautaud, V. Bishnoi, S., Gilberto J.M. (2019) The effect of authentic project-based learning on attitudes and career aspirations in STEM. *Journal of Research in Science Teaching*, 56(1), 3-23, 10.1002/tea.21465

Brito, M.R.; Marvila, M.T.; Linhares, J.A.T., Jr.; Azevedo, A.R.G.d. (2023). Evaluation of the Properties of Adobe Blocks with Clay and Manure. *Buildings*, 13, 657. https://doi.org/10.3390/buildings13030657

Ciancio, V.; Salata, F.; Falasca, S.; Curci, G.; Golasi, I.; de Wilde, P. (2020). Energy demands of buildings in the framework of climate change: An investigation across Europe. *Sustain. Cities Soc.* 60, 102213.

Charai, M.; Salhi, M.; Horma, O.; Mezrhab, A.; Karkri, M.; Amraqui, S. (2022). Thermal and Mechanical Characterization of Adobes Bio-Sourced with Pennisetum Setaceum Fibers and an Application for Modern Buildings. *Constr. Build. Mater.* 326, 126809.

Costantine, G.;Maalouf, C.;Moussa, T.; Kinab, E.; Polidori, G. (2020). Hygrothermal evaluation of hemp concrete at wall and room scales: Impact of hysteresis and temperature dependency of sorption curves. *J. Build. Phys.* 44, 183–224. doi:10.1177/1744259119896380.

Engberg, M. E., & Fox, K. (2011). Exploring the relationship between undergraduate service-learning experiences and global perspective-taking. *Journal of Student Affairs Research and Practice*, 48(1), 85–105.

Fowler P, Del Ama Gonzalo F, Newell S, Poolman J, Montero Burgos MJ and González Lezcano RA; (2023). Assessment of indoor air quality and comfort by comparing an energy simulation and actual data in Native American shelters. *Front. Built Environ.* 9:1202965. doi: 10.3389/fbuil.2023.1202965

Frasca, F.; Cornaro, C.; Siani, A. (2018). *Performance assessment of a heat and moisture dynamic simulation model in IDA ICE by the comparison with WUFI Plus*. IOP Conference Series: Materials Science and Engineering; IOP Publishing: Bristol, UK; Volume 364, p. 012024.

Gaur, A.; Lacasse, M.; Armstrong, M. (2019). Climate Data to Undertake Hygrothermal and Whole Building Simulations Under Projected Climate Change Influences for 11 Canadian Cities. *Data* 4, 72.

Hao, L.; Herrera, D.; Troi, A.; Petitta, M.; Matiu, M.; Pero, C.D. (2020). Assessing the impact of climate change on energy retrofit of alpine historic buildings: Consequences for the hygrothermal performance. *IOP Conf. Ser. Earth Environ. Sci.*, 410, 012050.

Imada, P.K.; de Almeida Paes, L.; Vitiello, P.R. (2021). Economic feasibility study of construction with raw earth block (adobe). *In Proceedings of the Google for Education, Rio de Janeiro, Brazil*; pp. 1–16.

Libralato, M.; De Angelis, A.; D'Agaro, P.; Cortella, G.; Saro, O. (2021). Multiyear Hygrothermal Performance Simulation of Historic Building Envelopes. In Proceedings SBE21 Sustainable Built Heritage, Bolzano-Bozen, Italy, 14–16 April.

Mellaikhafi, A.; Tilioua, A.; Benallel, A. (2022). Thermal Performance Assessment of a Wall Built with Earth-Based Adobes and Reinforced with Pinnate Leaves Fibers. Mater. Today Proc. 58, 1535–1540.

Shavers, C. L., & Mitchell, R. B. (2019). High-Impact Educational Experiences in Colleges of Business. *e-Journal of Business Education and Scholarship of Teaching*, 13, 1-12.

Thi, N.; Trang, M.; Anh, N.; Ho, D.; Babel, S. (2021). Reuse of Waste Sludge from Water Treatment Plants and Fly Ash for Manufacturing of Adobe Bricks. *Chemosphere*, 284, 131367.

Tootkaboni, M.P.; Ballarini, I.; Zinzi, M.; (2021). Corrado, V. A Comparative Analysis of Different Future Weather Data for Building Energy Performance Simulation. *Climate*, 9, 37

WUFI (en). Available online: https://wufi.de/en/ (accessed on 22 April 2024).

# RESEARCH ON AN ABSORPTION AIR-CONDITIONING SYSTEM MADE FROM SUSTAINABLE MATERIALS

Yuxian Zhou, Wenquan Wang and Xiuwei Li*
*School of Energy and Power Engineering, Nanjing University of Science and Technology, Nanjing 210094, China*

**ABSTRACT**

Tremendous $CO_2$ emission from the combustion of fossil fuels continues to plague human beings with climate change. The only way to protect the earth's environment is to drive all facilities with renewable energy. It is especially true for air-conditioning system, which is such an energy hog taking up over 30% of the energy supply. However, the current technology of renewable energy driven air-conditioning system is less than satisfactory at efficiency and cost. To improve, this paper presents a more efficient and cost-effective air-conditioning system that can be driven by different renewable energy. By coupling refrigeration cycle with a battery cycle, the new system realizes energy storage when producing cooling effect. In this way, it overcomes the unreliability of renewable energy. Cost has also been cut by fabricating the core components with the carbon materials derived from three biomass waste, corn cobs, corn stalks, and waste cigarette butts. Investigation has been made on the preparation and characterization of the derived materials. Analysis has been conducted on the performance of the air-conditioning system made from those materials. The new system could be a competitive potential air-conditioner in a coming sustainable low carbon future.

**KEYWORDS**

Air-Conditioning, Battery, Renewable Energy, Performance

## 1. INTRODUCTION

According to the International Energy Agency (IEA), more than 300 billion tons of $CO_2$ have been emitted as a result of burning fossil fuels in the past decade. These emissions are surpassing nature's carbon cycle and leading to significant environmental consequences, such as global warming and ocean acidification (Goeppert et al., 2012). Currently, approximately 30% to 40% of primary energy is consumed by air-conditioning systems in buildings (Wang et al., 2016; Granqvist et al., 2018; Gaiet al., 2017). Promoting air-conditioning systems powered by renewable energy sources could effectively mitigate the adverse impacts of fossil fuel consumption. One successful commercialized example is the absorption air-conditioning system, which can be directly driven by renewable energy. However, a drawback of the present absorption systems is their reliance solely on thermal power from renewable sources, sacrificing selectivity and flexibility. So far, there is no definitive solution for developing an air-conditioning system that is environmentally friendly, efficient, and compatible with various renewable energy sources. Another significant environmental threat arises from the exponential increase in global waste production over the past century (United Nations Framework Convention on Climate Change, 2008). The World Bank projects that current waste generation levels are expected to triple by 2100. Proper waste management practices are crucial for alleviating environmental pressures and ensuring sustainable development.

* Corresponding author: good3000best@163.com

This paper introduces a conceptual scheme for a novel system that utilizes everyday waste as its key component. The proposed system, named the air-conditioning battery system (ACBS), employs a new refrigeration cycle by integrating capacitive deionization (CDI) technology into its design. At the heart of this system lies its electrode, the properties of which determine the unit's capacity and significantly influence the performance of the ACBS. In this study, corn cobs, corn stalks, and discarded cigarette butts have been selected to produce micro/mesoporous carbon materials for electrode fabrication. Coefficient of Performance (COP) values have been analyzed for the ACBS employing various electrode materials.

## 2. BODY OF PAPER

### 2.1 Working Principle of CDI

Figure 1 shows the CDI units work in a battery way propelling the refrigeration cycle: the charging unit produces refrigerant and stores energy in the electrodes (capacitors), the discharging unit produces concentrated absorbent solution and releases energy. The charging unit stores energy either in the form of electric double layer (EDL) with non-faradic electrode or in the form of ions insertion with faradic electrode like that in Li ion battery (Sharma et al., 2008; Yu et al., 2013; Zhao et al., 2012; Cho et al., 2014; Li et al., 2015; Pernia et al., 2014; García-Quismondo et al., 2016; Gamaethiralalage et al., 2021; Hou et al., 2013; Zhao et al., 2018; Yang et al., 2021). The releasing energy from the discharging unit can be recovered to drive the charging unit through an external circuit. With the battery mode, the two CDI units alternatively stores energy by charging and drives the other units by discharging in the refrigeration cycle. Assume the recovery ratio of the discharging energy is 100%, the system only requires a start energy injection, then the energy would swing between the units and keep the cycle going continuously without any extra energy supply. With real recovery ratio below 100%, little extra energy is required to make up the energy loss during charging and discharging. The battery method undoubtedly upgrades the performance of the air-conditioning system. It also assists ACBS with inheriting the energy storage characteristic of the traditional absorption system by producing extra strong absorbent solution when energy is rich. The performance of an air-conditioning system is normally evaluated with the coefficient of performance (COP), which is the ratio of cooling capacity to energy input. Considering cost and sustainability, investigation has been made on ACBS with electrodes made from three daily life biomass waste materials (corn cobs, corn stalks, waste cigarette). The procedures of preparation are introduced below.

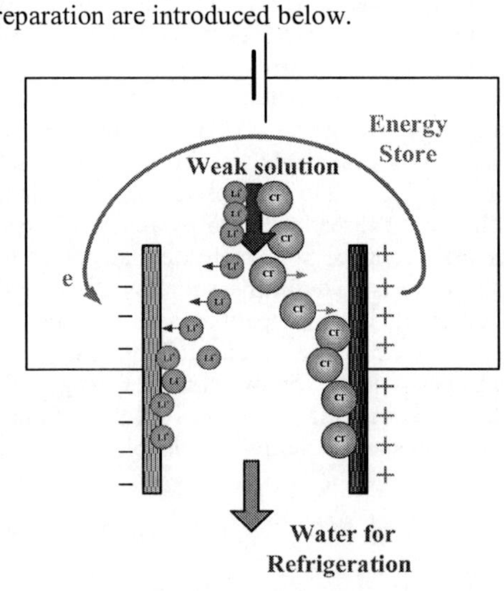


Figure 1. The battery mode of ACBS

## 2.2 Performance of Electrode

### 2.2.1 Preparation of Electrode Materials

The preparation method is basically same for three waste materials and the treatment of corn cobs is taken as an example. Firstly, the corn cobs were cleaned with deionized water and then dried at 80°C for 12h. The dried corn cobs were then crushed into powder with a crusher. Secondly, the corn cob powder was heated at 400°C for 2h, the heating rate was 5°Cmin$^{-1}$, and the whole process occurred under $N_2$ atmosphere. The carbonized corn cob powder was then washed with deionized water and dried at 80°C for 3h. Thirdly, the carbonized corn cob powder was mixed with $K_2CO_3$ in a mortar (the ratio of carbon to $K_2CO_3$ is 1:5). Subsequently, the mixture was heated at 700, 800 or 900 °C for 2h, the heating rate was 5°Cmin$^{-1}$, and the whole process occurred under $N_2$ atmosphere. Finally, the black mixture was washed with a 10% HCl solution to remove the inorganic impurities and then washed with deionized water until the pH value is neutral. The black mixture was then dried at 80°C for 3h. The obtained porous carbon materials derived from corn cobs were named CBC-X, where X is the activation temperature. The porous carbon derived from corn stalks and waste cigarette butts were named CS-X and WCB-X, respectively.

Figure 2. Schematic illustration of preparation process of CBC

### 2.2.2 Performance Test of ACBS

The performance of ACBS has been investigated using a conventionally designed CDI setup with CS-700 and WCB-800 as the electrode materials (Figure 3). The testing procedure is presented in the literature (Zhao et al., 2018; Yang et al., 2021). Carbon materials based symmetrical electrodes (28 cm$^2$ per electrode) were stacked across an insulation film. A nickel foam was used as the current collector. The electrodes were connected to the positive and negative terminal of an electrochemical workstation with connecting wires. The charging voltage of 0.8~1.2 V was applied to the CDI cell. A total feed volume of 100 mL of the absorbent solution (The absorbent solution is usually LiBr solution, LiCl solution and $CaCl_2$ solution) was pumped into the CDI cell with a peristaltic pump. The deionization result or ion adsorption capacity (*Ms*) is calculated with Equation (1):

$$M_s = \frac{(C_0 - C_t) * V}{m} \quad (1)$$

$C_0$ and $C_t$ (mg L$^{-1}$) represent the start and end concentrations, $V_l$ (L) is the solution volume, and $m$ (mg g$^{-1}$) is the weight of active material.

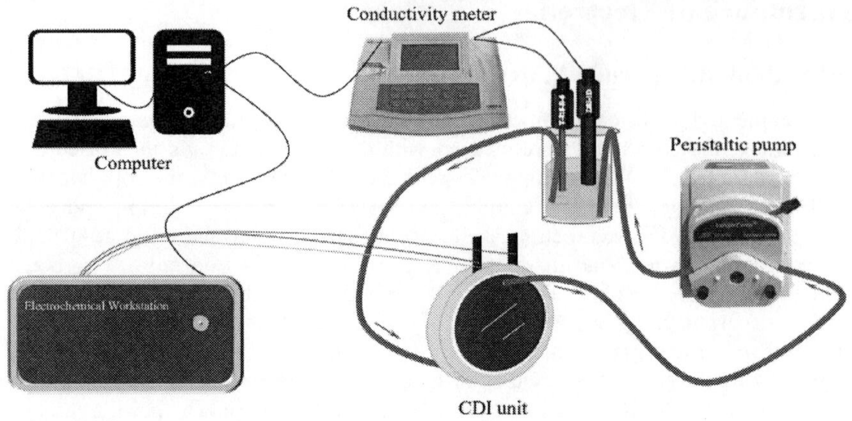


Figure 3. Schematic illustration of performance test

## 2.3 Results and Discussion

### 2.3.1 Morphology, Structural and Textural Characterizations of all Samples

The SEM images of CBC (Figure 4 (a-c)), CS (Figure 4 (d-f)), and WCB (Figure 4 (g-h)) demonstrates the porosity changes in the bio-char due to the high-temperature activation process, this is caused by the activation of $K_2CO_3$ at high-temperature. The dense pore structure provides ion storage and transport channels, which is conducive to ion transfer and adsorption. In addition, a large number of particles are attached to the surface of the pore structure, and the fiber network formed by such particles is beneficial to the accumulation and transport of ions. After high-temperature activation, the pore walls of CBC and WCB are thick, while the pore wall of CS is thin. This may be related to the structure of the raw materials. To further analyze the structure of the carbon materials, TEM has been used to characterize the samples. As shown in Figure 5, the carbon skeleton of all samples are preserved after high-temperature activation. The activation reaction is intense at high temperature. The larger the weight ratio of $K_2CO_3$ to bio-char, the higher the activation temperature, the intenser the reaction is. WCB-900 could not be prepared at 900℃ because the reaction was too violent in the activation process (Figure 4 (i)).

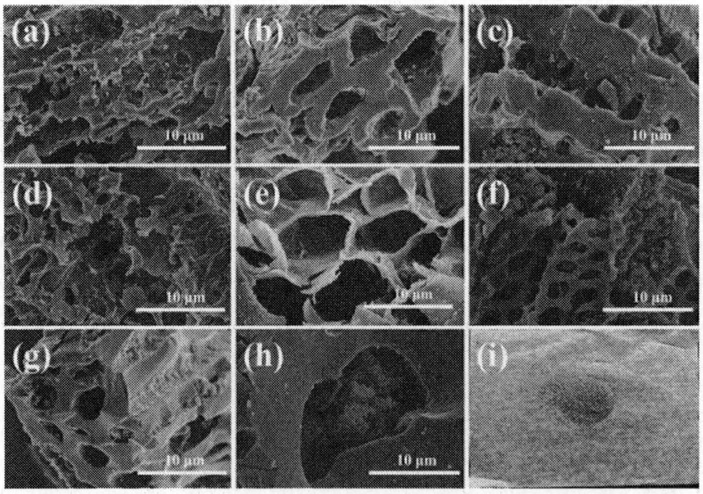


Figure 4. SEM images of bio-char: (a) CBC-700, (b) CBC-800, (c) CBC-900; (d) CS-700, (e) CS-800, (f) CS-900; (g) WCB-700, (h) WCB-800, (i) the appearance of WCB-900

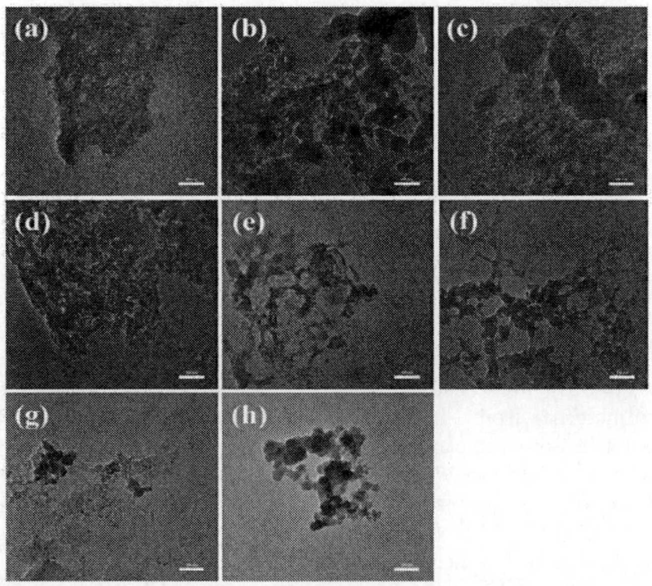


Figure 5. TEM images of bio-char: (a) CBC-700, (b) CBC-800, (c) CBC-900; (d) CS-700, (e) CS-800, (f) CS-900; (g) WCB-700, (h) WCB-800

### 2.3.2 Performance Analysis of ACBS

Experiments have been performed with the CS-700 electrode and the results are shown in Figure 6. When the concentration of LiBr solution increases from 0.1% to 55%, the ion adsorption capacity increases from 8.70 $mgg^{-1}$ to 89.50 $mgg^{-1}$ and the energy consumption increases from 28.8J to 278.4J. When the concentration of LiCl solution increases from 0.1 % to 40 %, the ion adsorption capacity increases from 5.60 $mgg^{-1}$ to 49.70 $mgg^{-1}$ and the energy consumption increases from 44.4J to 330J. When the concentration of $CaCl_2$ solution increases from 0.1% to 45%, the ion adsorption capacity increases from 6.70 $mgg^{-1}$ to 70.30 $mgg^{-1}$ and energy consumption increases from 32.4J to 301.2J. Compared with the CBC-800 electrode, the CS-700 electrode shows lower ion adsorption capacity because the CBC-800 has larger specific surface area, superior meso/microporous structure, as well as a right chemical composition. The performance of ACBS with the CS 700 electrode is shown in Figure 7: When the concentration of the LiBr solution ranges from 40% to 55%, the COP decreases from 2.80 to 1.63; When the concentration of the LiCl solution ranges from 30% to 40%, the COP decreases from 2.15 to 1.40; When the concentration of the $CaCl_2$ solution ranges from 30% to 45%, the COP decreases from 3.40 to 1.77.

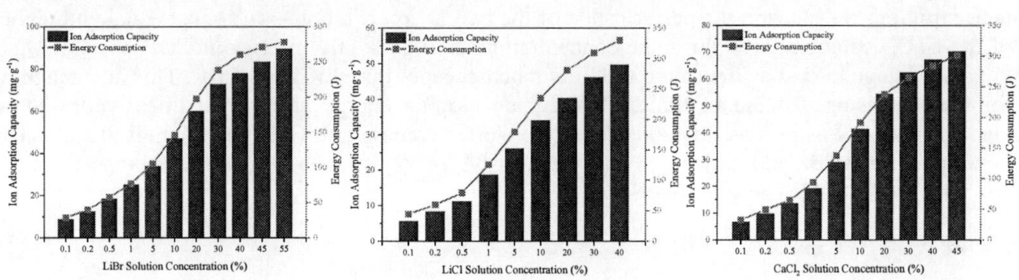


Figure 6. The ion adsorption capacity and energy consumption of CS-700 electrodes in different concentrations and solutions at 1.2 V

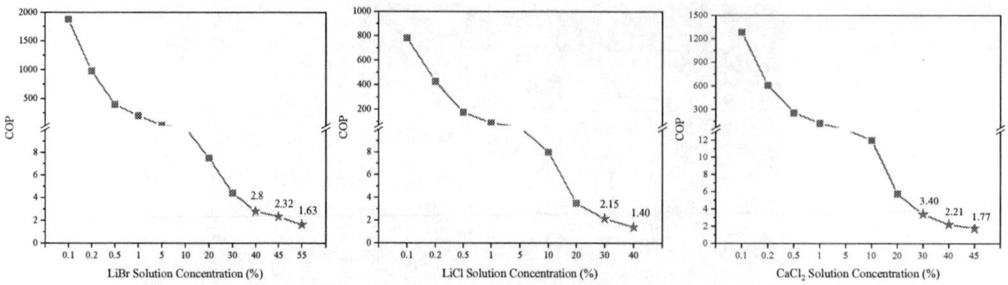


Figure 7. Actual COP of the cooling system with the CS-700 as the electrode materials

Figures 8 and 9 show the results with the WCB-800 electrode. When the concentration of LiBr solution increases from 0.1% to 55%, the ion adsorption capacity increases from 3.20 $mgg^{-1}$ to 62.30 $mgg^{-1}$ and the energy consumption increased from 15.6J to 252J. When the concentration of LiCl solution increases from 0.1% to 40%, the ion adsorption capacity increases from 2.30 $mgg^{-1}$ to 28.80 $mgg^{-1}$ and the energy consumption increases from 30.6J to 278.4J. When the concentration of $CaCl_2$ solution increases from 0.1% to 45%, the ion adsorption capacity increases from 2.80 $mgg^{-1}$ to 44.60 $mgg^{-1}$ and the energy consumption increases from 22.8J to 255.6J. When the concentration of the LiBr solution ranges from 40% to 55%, the COP decreases from 2.30 to 1.25. When the concentration of the LiCl solution ranges from 30% to 40%, the COP decreases from 1.48 to 1.0. When the concentration of the $CaCl_2$ solution ranges from 30 % to 45 %, the COP decreases from 2.45 to 1.33.

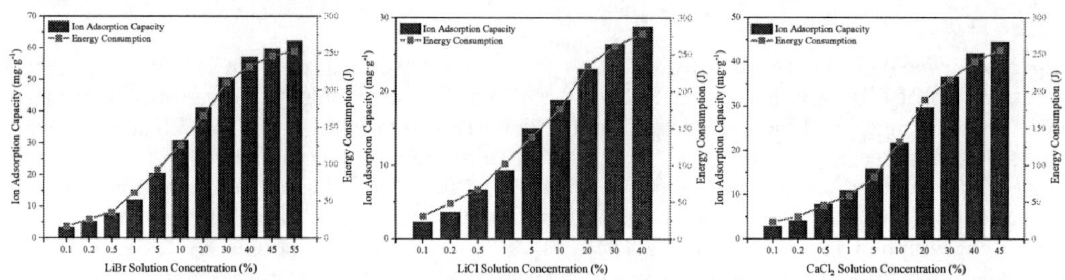


Figure 8. The ion adsorption capacity and energy consumption of WCB-800 electrodes in different concentrations and solutions at 1.2 V

With same electrode, COP varies with solution kind and concentration. For the two absorbent, the LiBr solution has better performance with higher COP under same mass concentrations, its COP value is almost twice that of the LiCl solution. But if the COP is evaluated with mole concentration instead of mass concentration, we can find the performance of the two Li ion solutions are almost equal and higher than that of the $CaCl_2$ solution given the same concentration. Being slightly inferior in COP, the $CaCl_2$ solution has huge advantage in cost as the solute $CaCl_2$ is much cheaper than LiCl and LiBr. This advantage will expand for the increasing demand of lithium salt for developing energy storage facilities while the resource of element Li is not as rich as Ca. Considering cost-effectiveness, the $CaCl_2$ solution as absorbent would be a better plan for ACBS and may provide an inexpensive energy storage measure for air-conditioning system in the coming renewable energy driving era.

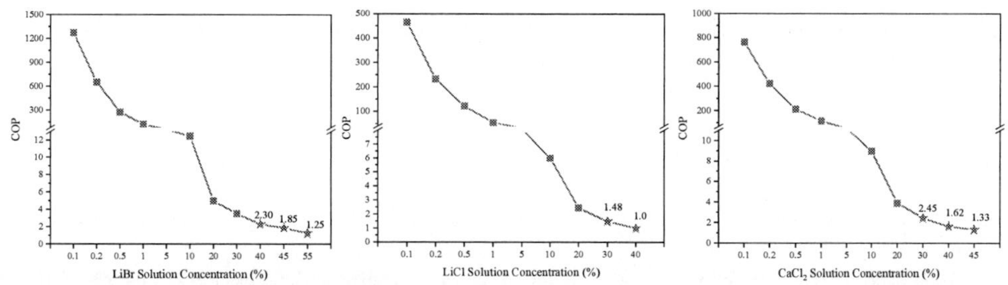


Figure 9. Actual COP of the cooling system when WCB-800 was used as the electrode materials

# 3. CONCLUSION

The air-conditioning battery system (ACBS) introduced in this paper represents a groundbreaking innovation that integrates the functions of a refrigerator and a battery. This system achieves a unique synchronization and fusion of refrigeration and battery cycles, incorporating two distinct energy storage mechanisms: one based on concentration differences in the working fluid and the other utilizing Electric Double Layer (EDL) or Faradic battery principle. These advancements have been proven to transform the ACBS into an environmentally friendly and highly efficient air-conditioning system, well-suited for the anticipated widespread adoption of renewable energy sources in the near future. The ACBS exhibits remarkable environmental benefits, offering a dual impact on environmental protection. The CDI electrode, a pivotal component of the ACBS derived from everyday waste materials, not only facilitates a significant reduction in $CO_2$ emissions through its integration with a renewable energy-powered air-conditioning system but also mitigates environmental pressures associated with waste disposal. Moreover, the utilization of these materials underscores the economic viability and sustainability of the ACBS, making it a compelling choice for green and resource-efficient building systems.

In the course of this study, micro/mesoporous electrode materials with a substantial specific surface area were successfully synthesized using corn cobs, corn stalks, and discarded cigarette butts to create highly efficient electrodes. It was observed that the optimal characteristics of the resulting material could be achieved through proper chemical activation at temperatures ranging from 700 to 800°C. The type and concentration of the absorbent solution also emerged as crucial factors influencing performance outcomes. When comparing solutions with the same mole concentration, it was found that LiBr and LiCl solutions exhibited nearly equivalent performance level, which is superior to those achieved with $CaCl_2$ solution. However, given its significantly better cost-effectiveness, $CaCl_2$ solution emerges as a favorable option and could potentially serve as an optimal absorbent for the ACBS.

# 4. SUMMARY

Great sustainability can be achieved in that only four to five kilograms of the wastes materials are required for making an ACBS air conditioner with two thousand watts cooling capacity. It not only solves the wastes disposal problems but also reduces the $CO_2$ emission. The products from the materials are green and cheap, saving cost of the air conditioning. In the research, the micro/mesoporous electrode materials with high specific surface area have been successfully prepared using corn cobs, corn stalks and waste cigarette butts for highly efficient electrodes. CBC-800 had the largest specific surface area (1201.90 $m^2$ $g^{-1}$) and electrochemical capacitance (141.4 F $g^{-1}$) in the derived materials. Also, the COP with CBC-800 reached the highest 12 in the experiments.

While the ACBS has demonstrated a COP comparable to that of conventional vapor compression air-conditioning systems, further research is needed to identify more sustainable materials and enhance the overall performance of the system. Conducting a thorough analysis of the refrigeration cycle and energy storage characteristics will be pivotal in optimizing the system. It is essential to subject the ACBS to testing under various operating conditions before considering its practical application. This iterative process of refinement and evaluation will be crucial in advancing the efficiency and sustainability of the ACBS for potential real-world implementation.

## REFERENCES


Cho, K. T., et al. (2014). Facile Synthesis of Highly Electrocapacitive Nitrogen-Doped Graphitic Porous Carbons. *J. Phys. Chem. C*, 118, 9357-9367.

G. Gai, G., et al. (2017). Electrochromic materials and devices for energy efficiency and human comfort in buildings: A critical review. *Adv. Sustainable Syst.*, 1, 1700074.

Gamaethiralalage, J. G., et al. (2021). Recent advances in ion selectivity with capacitive deionization. *Energy Environ. Sci.*, 14, 1095.

García-Quismondo, E., et al. (2016). New operational modes to increase energy efficiency in capacitive deionization systems. *Environ. Sci. Technol*, 50, 6053-6060.

Goeppert, A., et al. (2012). Air as the renewable carbon sources of the future: an overview of $CO_2$ capture from the atmosphere. *Energy Environ. Sci.*, 5, 7833-7853.

Granqvist, C. G., et al. (2018). Electrochromic materials and devices for energy efficiency and human comfort in buildings: A critical review. *Electrochim. Acta*, 259, 1170-1182.

Hou, C., et al. (2013). A comparative study of electrosorption selectivity of ions by activated carbon electrodes in capacitive deionization. *Desalination*, 314, 124-129.

Li, K. B., et al. (2015). Carbonaceous Electrode Materials for Supercapacitors. *Electrochim. Acta*, 174, 596-600.

Pernia, M., et al. (2014). New Control Strategy of an Up-Down Converter for Energy Recovery in a CDI Desalination System. *IEEE Transactions on Power Electronics*, 29, 3573-3581.

Sharma, R. K., et al. (2008). Growth and characterization of carbon-supported $MnO_2$ nanorods for supercapacitor electrode. *Physica B: Condensed Matter*, 403, 1763-1769.

United Nations Framework Convention on Climate Change. (2008). Kyoto Protocol Reference Manual On Accounting Of Emissions And Assigned Amount.

Wang, Y., et al. (2016). Switchable Materials for Smart Windows. *Annu. Rev. Chem. Biomol. Eng.*, 7, 283-304.

Yang, J. Q., et al. (2021). A direct energy reuse strategy for absorption air-conditioning system based on electrode regeneration method. *Renewable Energy*, 168, 353-364.

Yu, G., et al. (2013). Hybrid nanostructured materials for high-performance electrochemical capacitors. *Nano Energy*, 2, 213-234

Zhao, H., et al. (2018). Research on actual performance and energy recovery characteristic of capacitive deionization regeneration method for absorption air-conditioning system. *Energy Conversion and Management*, 171, 1549-1559.

Zhao, X., et al. (2012). Dual-heteroatom-modified ordered mesoporous carbon: Hydrothermal functionalization, structure, and its electrochemical performance. *J. Mater. Chem.*, 22, 4963-4969.

# EFFECTS OF LEARNING ENVIRONMENTS ON STUDENTS' SKILLS DEVELOPMENT: A COMPARATIVE VIEW OF EIGHT EUROPEAN COUNTRIES

Ana Tecilazić[1] and Biljana Birač[2]
[1]*Algebra University, Zagreb, Croatia*
[2]*University of Georgia, USA*

**ABSTRACT**

The aim of this study is to investigate the effects of learning environments during higher education on graduates' perceptions of their own field-specific skills and communication skills one year after graduation. Astin's Input-Environment-Outcome (I-E-O) model was used as a broad conceptual framework supplemented by social and learning theories. A logistic regression was applied on a sample of 11008 BA and MA graduates who graduated in the academic year 2016/2017 from higher education institutions in eight European countries (Austria, Croatia, Czech Republic, Germany, Greece, Lithuania, Malta, and Norway) that participated in the pilot survey EUROGRADUATE 2019. The results of the analysis show variation in relationships between students' inputs, environment, and outcomes across the eight countries. The most consistent finding across all countries relates to the teaching and learning modes, as both programmes with high extent of lecture-based teaching and problem-based learning have positive statistically significant relationships with field-specific and communication skills in nearly all the participating countries.

**KEYWORDS**

Graduate Tracking, Transversal Skills, Learning Environments, Problem-Based Learning, Field-Specific Skills, Communication Skills

## 1. INTRODUCTION

A stimulating learning environment is important for students to achieve successful academic results. Previous empirical studies support modern pedagogical theories and show that engaging learning practices and activating modes of teaching and learning such as problem-based learning, internships or study-related work, volunteering and studying or working abroad during studies are seen as effective ways of improving students' skills and competences during their studies (Carini, Kuh and Klein, 2006; Campbell and Norton, 2007; Hu and Wolniak, 2010; Biggs and Tang, 2011; Hoidn and Klemenčič, 2021; Kirby and Thomas, 2022; Børte, Nesje and Lillejord, 2023). One of the aims of higher education is to equip students with skills and knowledge that will help them to find meaningful employment after graduation, to develop personally and to be active citizens. To achieve these goals, graduates need to be equipped with both field-specific knowledge and skills and general (transversal) skills. This paper draws on the findings of the 2019 Eurograduate Pilot Survey, conducted in eight European countries, and explores the relationship between an engaging learning environment and graduates' self-perceived field-specific and communication skills, one year after graduation. While previous studies analysed the relationship between learning environments and self-perceived level of field-specific and communication skills on the whole Eurograduate sample (Meng *et al.*, 2020), this study examines these relationships for each country separately. This allows for a more nuanced view of various factors that impact positive self-perceived academic outcomes, so that the study contributes to a better understanding of the impact of learning environments on skills development.

## 2. CONCEPTUAL FRAMEWORK

To explore the relationship between learning environments and skills development Astin's (Magolda and Astin, 1993) Input-Environment-Outcome (I-E-O) model was used as a broad conceptual framework and complemented by social and learning theories. Astin's model describes how student inputs and the learning environment interact to produce outcomes over time. The inputs relate to the characteristics of students when they enter the institution and outputs are measured by the perceived level of graduates' competences one year after graduation.

The *inputs* used in this study are sex, age at graduation, and educational and social background of parents. Previous studies and results of the Eurograduate pilot analysis (Furnham, Hosoe and Tang, 2001; Meng *et al.*, 2020) have found that men tend to self-evaluate their intelligence and skills to a greater extent than females and that more mature graduates perceive their skills and knowledge at a higher level than younger graduates. Parents' education, and graduates' migrant background were included as inputs because they are indicative of graduates' cultural and social capital. Bourdieu's theory of social reproduction (Bourdieu, 1977) explains that students from families with higher educational levels often inherit cultural knowledge, language proficiency, and academic expectations, all of which may contribute to their academic success, and affect their skills development. Similarly, Boudon's wide rational choice theory (Boudon, 1974) acknowledges the importance of the social context and the influence of referent groups, such as peers and family, in shaping individuals' choices when taking their educational paths. These theories have been supported by studies that proved that social origin and family background affect educational outcomes (OECD, 2015; Salmi, 2020; Cahalan *et al.*, 2022). Migrant background could also affect one's social capital, which consists of social networks and relationships. Studies demonstrate that people with migrant background of low social origin have fewer chances to access higher education if they would not benefit from alternative access to higher education, such as taking into account students' prior learning and work experience (Griga and Hadjar, 2014; Giudici, Trappolini and Vicari, 2021; Guichard *et al.*, 2024). Graduates' cultural capital measured by the educational and migrant background of parents' affects academic outcomes. Moreover, the graduates' social background increases the chance to study abroad (Meng *et al.*, 2020) and improves the odds of benefitting from connections with teachers, peers, or community members, which can facilitate access to information, mentorship, and internship opportunities (Engle and Tinto, 2008; Jack, 2020). In consequence, better connections with teachers enhance motivation and academic achievements (Komarraju, Musulkin and Bhattacharya, 2010; Parnes *et al.*, 2020).

The *environment* encompasses various factors related to the institution and the experiences students may have during their studies. This study focuses on various educational modes and experiences, including experiential learning, while considering the type of higher education institution, field of study, and enrollment status of students. Experiential learning is an important approach to skills development, grounded in the belief that learning is an active, dynamic process. Experiential learning theory, proposed by Kolb (Kolb, 1984), emphasizes that learning occurs through interaction with the environment. As argued by constructivists, when individuals engage with real-world situations, they create knowledge by applying theoretical concepts to practical scenarios (Ahmad *et al.*, 2015; Tecilazić, 2022). Literature shows that different teaching and learning modes applied in study programmes affect skills development. There is a growing criticism towards traditional learning styles (Gerritsen, 1999; Hoidn and Klemenčič, 2021; Kirby and Thomas, 2022; Børte, Nesje and Lillejord, 2023), where students are treated as recipients of information provided by the teachers. On the other side, contemporary pedagogical approaches emphasise student-centred learning and teaching approaches that foster students' engagement in the creation of their knowledge (Damşa and Lange, 2019; Hoidn and Klemenčič, 2021; Graham and Longchamps, 2022; Kirby and Thomas, 2022). These approaches value and promote more activating modes of learning and teaching, such as problem-based learning (Boud and Feletti, 2013). However, the Eurograduate survey demonstrated that both are needed: activating learning methods (such as problem-based learning) where the student is expected to be an active constructor of their knowledge as well as more traditional lecture-based learning. According to the graduates responses, "lectures form a crucial part in higher education" and should not be replaced fully by an activating learning environment, but rather a good balance has to be found between the two approaches, project, and problem-based learning with lecture style (Meng *et al.*, 2020).

Researchers have explored the positive attributes of high-impact practices (AACU, 2007) such as study abroad, study-related work experience, and internships on perceived skills development. Study abroad has an impact on the development of intercultural communicative competence and on personal and educational growth (Lee and Song, 2019). Students with experience from study abroad demonstrate higher levels of language skills, intercultural competence, and disciplinary knowledge (Berg, Connor-Linton and Paige, 2009). Another study compared students who spent a study period abroad to those who stayed on campus as well as knowledge of study abroad students before departure and upon return and the results show greater knowledge of mobile students on five cognitive dimensions (Cisneros-Donahue *et al.*, 2012). Study-related work experience or internships may have positive effects on the graduates' academic and labour market outcomes (Andrews and Higson, 2008). These studies usually explore transversal (employability) skills developed through work-integrated learning (Freudenberg, Brimble and Cameron, 2011) and show that internships are important for finding a first employment (Tecilazić, 2021, 2023) but also for proactively navigating the world of work and self-managing long-term career development (Helyer and Lee, 2014; Hora, Chen and Her, 2020).

*Outcomes* describe the students' characteristics after exposure to the learning environment during the studies, and we explored which experiences in higher education are related to the development of field-specific and communication skills. Graduates are equipped to find employment if they possess specific skills required by employers and relevant academic knowledge. Field-specific competencies of graduates are the most direct outcome of study programmes. However, employers' surveys demonstrate that it is not sufficient to be knowledgeable in a specific discipline, but that there is an increasing demand for generic (transversal) competencies in the labour market (Tymon, 2013; AACU, 2015) and skills that increase graduates' employment prospects (McGunagle and Zizka, 2020). Feedback from employers indicate a need to include a focus on generic skills to complement disciplinary expertise (De La Harpe, Radloff and Wyber, 2000; AACU, 2015; McGunagle and Zizka, 2020).

The results of the Eurograduate survey have shown that several factors have a positive effect on perceived field-specific competence: field of study, age, sex, and type of higher education institution. Graduates in the fields of Education and Agriculture, more mature individuals, male graduates, and those who have completed their studies at universities show a higher level of perceived field-specific competence. In addition, the proportion of graduates reporting a high level of field-specific competence increases by almost 2% if the degree programme is more focused on problem-based learning and teaching, while internships and periods abroad increase field-specific competence, and voluntary work has a negative impact (Meng *et al.*, 2020). The findings of the Eurograduate study demonstrate that the factors that positively influence the self-assessment of communication skills tend to be non-study-related factors such as age, length of study, language of study, and social background. Mature graduates who have completed programmes of longer duration, who have studied in a language other than the language of the country, graduates with parents without an academic background and with a migration background have a higher level of perceived communication skills than others (Meng *et al.*, 2020). Building on the findings of the Eurograduate pilot survey, this paper looks at the factors determining developments of field specific and communication skills in eight European countries.

## 3. RESEARCH OBJECTIVE AND METHODS

Based on the theoretical perspectives of cultural and social capital, activating learning theories and building on previous empirical studies, this study aims to analyse how the learning environment of students affects the development of field-specific and communication skills. Specifically, this study is guided by the following research questions:

1. Do lecture-based and problem-based teaching and learning affect the development of graduates' self-perceived communication and field-specific skills in eight European countries?
2. Do high-impact practices such as study abroad, internships, and volunteering affect the development of graduates' self-perceived communication and field-specific skills in eight European countries?
3. What graduate or academic characteristics affect the development of graduates' self-perceived communication and field-specific skills in eight European countries?

While the Eurograduate analysis examined similar relationships on the whole sample, controlling for country, in this analysis we applied the statistical method of logistic regression to eight countries that participated in the pilot survey EUROGRADUATE 2019 separately. Given that the number of respondents to the survey varies greatly across countries, this approach should provide a more nuanced analysis of the higher education learning environments in each country.

## 3.1 Sample and Instrument

The data used for this paper were collected through the EUROGRADUATE – first pilot European graduate tracking survey conducted across eight European countries (Austria, Croatia, Czech Republic, Germany, Greece, Lithuania, Malta, Norway) and over the period October 2018 – February 2019 (Meng *et al.*, 2020). The data are presented in Table 1, and they include Bachelor-level and Master-level graduates from the academic year 2016/2017. The final dataset was weighted based on population data and contains responses from a total number of 11008 graduates as follows: 1120 (10,2%) from Austria, 4277 (38,9%) from Croatia, 1015 (9,2%) from the Czech Republic, 914 (8,3%) from Germany, 866 (7,9%) from Greece, 1150 (10,4%) from Lithuania, 506 (4,6%) from Malta and 1160 (10,5%) from Norway. The instrument used in the survey was a standardised online questionnaire, designed to measure different aspects of transition from higher education to labour market. Data were centrally collected and anonymised by the Deutsche Zentrum für Hochschul und Wissenschaftsforschung (German Centre for Higher Education Research and Science Studies) that made the data available in accordance with the GDPR regulations.

Table 1. Data Sample

| Country | Number | Percent |
|---|---|---|
| Austria | 1120 | 10,2 |
| Croatia | 4277 | 38,9 |
| Czech Republic | 1015 | 9,2 |
| Germany | 914 | 8,3 |
| Greece | 866 | 7,9 |
| Lithuania | 1150 | 10,4 |
| Malta | 506 | 4,6 |
| Norway | 1160 | 10,5 |
| **Total** | **11008** | **100,0** |

## 3.2 Variables

The two outcome variables measured in this analysis indicate the self-assessment of graduates' "field-specific skills" and "communication skills" originally measured with the 1-5 Likert scale, ranging from 1 ('not at all') to 5 ('to a very high extent') with respect to the estimated level of a particular skill at the time of the survey. These variables were recoded into dummy variables measuring the self-perception of having these skills developed one years after graduation or not. Therefore values "1", "2" and "3" were recoded into a new value "0" and the values "4" and "5" to form a new value "1". Based on the theoretical and conceptual framework presented and the results of the Eurograduate study, several independent variables were included in the analysis as factors that determine the self-perception of skills development. Two variables are chosen as indicative of the cultural capital of graduates: "educational background of parents" as a continuous variable with three values ("parents without higher education degrees", "one parent with a higher education degree" or "both parents with higher education degrees") and "migrant background of parents" as a dummy variable with two values: "1=yes" and "2=no"). Seven variables indicate the learning environments during the studies: teaching and learning modes prevalent in the degree programme and differentiated between "lecture-based programme" and "problem-based learning" both continuous variables with the 1-5 Likert scale ranging from 1 ('not at all') to 5 ('to a very high extent'); experience of study abroad measured as a dummy variable with two values: "1=yes" and "2=no"); activities of "volunteering in the higher education institution", "volunteering outside higher education institution", "internship or study-related work experience in the country" and "internship or study-related work experience abroad" all dummy variables with two values: "1=yes" and "2=no".

## 3.3 Analyses

In order to establish a relationship between different types of learning experiences and self-assessed set of skills (field-specific, communication skills), a logistic analysis was carried out, separately for each country and separately for each skill as an outcome variable. The analysis was controlled by several variables: "sex" (m/f), "age at graduation" (5 categories), "status of enrolment" (full-time/part-time), "type of institution" (university/non-university), "qualification level" (BA/MA) and 10 fields of study according to the broad ISCED classification: (1) "Education", (2) "Arts and Humanities", (3) "Social Sciences, Journalism and Information", (4) "Business, Administration and Law", (5) "Natural Sciences, Mathematics and Statistics" (6) "Information and Communication Technologies", (7) "Engineering, Manufacturing and Construction", (8) "Agriculture, Forestry, Fisheries and Veterinary", (9) "Health and Welfare", (10) "Services".

# 4. RESULTS AND DISCUSSION

The results of the analysis show variation in relationships between students' inputs, environment, and outcomes across the eight countries. The most consistent finding across all countries relates to the teaching and learning modes, as both programmes with high extent of lecture-based teaching and problem-based learning have positive statistically significant relationships with field-specific and communication skills in nearly all the participating countries. This finding is consistent with previous Eurograduate findings related to the importance of both traditional, lecture-based teaching methods and innovative, problem- and project-based learning modes (Meng *et al.*, 2020). While detailed results are presented in Tables 2 and 3, we highlight and discuss findings related to learning environments and high-impact practices.

## 4.1 Field-Specific Skills

The most consistent, statistically significant relationship found in all participating countries is between lecture-based programmes and field-specific skills. Graduates in all eight countries who perceived their study programmes to have a high extent of lecture-based teaching have greater odds of rating their field-specific skills high. A similar finding is present in the relationship between problem-based programmes and field-specific skills, with a positive statistically significant relationship found in all participating countries except for Malta and Norway. These findings highlight the importance of combining both traditional and innovative teaching and learning methods.

Internships in the country of study have a statistically significant positive relationship in reference to field-specific skills only on graduates in Croatia and Malta (*Exp(B)*= 1,175, *Exp(B)*= 2,190, $p<0{,}05$, respectively) while in Germany, that relationship is negative (*Exp(B)*= 0,580, $p<0{,}05$). These results differ from previous findings and indicate there might be qualitative differences in internship provisions.

Fields of study have statistically significant effects in Austria, the Czech Republic, Greece, and Croatia. Austrian and Croatian graduates in the fields of Social Sciences, Journalism, and Information have lower odds of reporting a high level of field-specific skills (*Exp(B)*=0,478, *Exp(B)*=0,574, $p<0{,}05$ respectively) than graduates in the field of Education. This also holds for Croatian graduates in the fields of Business, Administration, and Law (Exp(B)=0,537, p<0,05) and Engineering, Manufacturing, and Construction (*Exp(B)*=0,430, $p<0{,}05$) and the Czech graduates in the field of Services (*Exp(B)*=0,438, $p<0{,}05$). In Greece, graduates in the fields of Business, Administration, and Law have higher odds of reporting a higher level of field-specific skills compared to graduates in the field of Education (*Exp(B)*=2,484, $p<0{,}05$). Although previous Eurograduate survey analysis found that graduates in the Education field report higher levels of field-specific skills, this analysis shows that this is not the case in all participating countries.

Study abroad experience only makes an impact in Croatia as graduates who completed a study abroad have a higher likelihood to self-assess their own level of field-specific skills higher compared to graduates with no study experience abroad (*Exp(B)*=1,477, $p<0{,}05$).

## 4.2 Communication Skills

Environment variables appear to have a strong influence on the self-perceived development of communication skills. This is visible for teaching modes, fields of study, and experiences outside of study programmes (high-impact practices). Similar to findings related to field-specific skills, graduates who attended study programmes with a high extent of lecture-based teaching in all countries except for Malta have greater odds of rating their communication skills high, holding other variables constant. The same relationship also holds between problem-based programmes and communication skills, with a positive statistically significant relationship found in all participating countries except for Norway.

Fields of study can also affect the development of communication skills differently. Differences among the broad field of studies are particularly visible in Austria, Germany, and Croatia, where most graduates from other fields have lower odds of reporting high levels of communication skills compared to graduates in the field of Education, holding other variables constant. This also holds for graduates in the fields of Arts and Humanities (*Exp(B)*=0,390, *p*<0,05), and Engineering, Manufacturing and Construction (*Exp(B)*=0,321, *p*<0,05) in the Czech Republic, and for graduates in the fields of Business, Administration and Law (*Exp(B)*=0,130, *p*<0,05), and Services (*Exp(B)*=0,087, *p*<0,05 in Malta. Only in Greece do graduates in the fields of Arts and Humanities have greater odds of reporting higher communication skills compared to graduates in the field of Education (*Exp(B)*=2,654, *p*<0,05). Even though the finding is not consistent across all countries and fields, it appears that programmes in the field of Education provide more opportunities for the development of communication skills.

The analysis also shows that volunteering experiences, and internships are positively affecting the development of communication skills, with some exceptions. Graduates who volunteered in their higher education institution have a higher likelihood to self-assess their own level of communication skills higher compared to graduates who did not volunteer in Austria (*Exp(B)*=1,669, *p*<0,05), Germany (*Exp(B)*=1,938, *p*<0,05), and Greece (*Exp(B)*=2,086, *p*<0,05), keeping other variables constant. The same finding holds for volunteering experiences outside of higher education institutions in Austria (*Exp(B)*=1,416, *p*<0,05), Greece (*Exp(B)*=1,595, *p*<0,05), and Croatia (*Exp(B)*=1,274, *p*<0,05). Graduates who conducted internships in the country also have higher odds of rating their communication skills better than graduates who did not participate in internships in Austria (*Exp(B)*=1,441, *p*<0,05), the Czech Republic (*Exp(B)*=1,409, *p*<0,05), and Croatia (*Exp(B)*=1,246, *p*<0,05). This also holds for Czech graduates who went abroad to participate in an internship compared to graduates who did not (*Exp(B)*=1,963, *p*<0,05). Interestingly, the graduates with internships in the country from Greece have lower odds of rating their communication skills higher than graduates with internships (*Exp(B)*= 0,607, *p*<0,05). When it comes to study abroad, only the graduates with study abroad experience from the Czech Republic (Exp(B)=2,136, p<0,05) and Germany (Exp(B)=2,214, p<0,05) have a higher likelihood of self-assess their own level of communication skills higher compared to graduates with no study experience abroad. Differently, the graduates in Malta who have spent a study period abroad have a lower likelihood to self-assess their own level of communication skills higher than those who stayed on campus (*Exp(B)*=0,383, *p*<0,05).

## 5. CONCLUSION

This study aimed to examine which factors during study programs (lecture-based teaching, problem-based learning, participation in high-impact practices) affect the self-perceived development of field-specific and communication skills in eight European countries that participated in the Eurograduate pilot survey. Our findings indicate that both lecture-based teaching and problem-based learning affect skills development in nearly all examined countries. This finding is aligned with previous studies conducted on the overall Eurograduate sample and highlights the importance of combining both methods in study programmes. On the other hand, while participation in high-impact practices fosters the development of both field-specific and communication skills, this relationship was not found in most of the countries. The only high-impact practice that was positively associated with the development of both communication and field-specific skills in half of the participating countries was an in-state internship. This finding indicates there might be important qualitative differences related to volunteering and study abroad experiences. Our country-by-country analysis approach offers more nuanced insights that can be used by national higher education institutions to investigate further the reasons behind these findings.

The study's limitations arise from the inherent constraints of the Eurograduate survey's predefined variables. Firstly, graduates were asked to assess the application of specific learning and teaching modes in their study programmes which may overlook effects or nuanced interactions among different modes. Secondly, the evaluation of field-specific and communication skills relied solely on graduates' self-perception, which may introduce biases or inaccuracies. Consequently, the findings may be influenced by graduates' subjective interpretations of these concepts and their confidence levels, rather than providing a comprehensive understanding of their proficiency. In considering future research, employers' perspectives might be included in order to gain more holistic understanding of the graduates' skills and competencies.

Finally, it's important to clarify that our study does not aim to establish a causal model. Instead, our focus lies in identifying factors that may influence graduates' experiences and outcomes, laying the groundwork for further research of the complex dynamics between higher education and labour market. Regarding external validity, it is essential to acknowledge the contextual limitations inherent in the study's scope, particularly within the context of the eight participating countries. While the findings offer valuable insights into the experiences of graduates within these specific national contexts, the results should not be extrapolated beyond these boundaries.

Table 2. Field-specific skills

| | Austria | | Czech Rep. | | Germany | | Greece | | Croatia | | Lithuania | | Malta | | Norway | |
|---|---|---|---|---|---|---|---|---|---|---|---|---|---|---|---|---|
| | Exp(B) | S.E. | Exp(B) | S.E. | Exp(B) | S.E. | Exp(B) | S.E. | Exp(B) | S.E. | Exp(B) | S.E. | Exp(B) | S.E. | Exp(B) | S.E. |
| Sex (f) | **0,676** | 0,177 | 0,845 | 0,179 | **0,328** | 0,217 | 0,853 | 0,190 | **0,828** | 0,082 | 0,930 | 0,213 | 0,661 | 0,370 | 0,869 | 0,158 |
| Age at graduation | 1,134 | 0,098 | 1,042 | 0,111 | 0,891 | 0,143 | **1,210** | 0,085 | **1,161** | 0,048 | 1,167 | 0,132 | 1,102 | 0,222 | 0,933 | 0,076 |
| Status (part time) | - | - | **1,862** | 0,301 | 1,057 | 0,534 | - | - | **1,326** | 0,116 | 1,551 | 0,359 | 4,098 | 0,872 | 1,710 | 0,312 |
| Type of HE (non-uni) | 1,051 | 0,225 | 1,093 | 0,301 | 0,869 | 0,265 | 1,159 | 0,259 | 0,928 | 0,099 | 1,139 | 0,259 | 1,206 | 0,545 | 1,259 | 0,159 |
| Level of qualif. (MA)* | 0,898 | 0,187 | 1,252 | 0,170 | 1,467 | 0,230 | **1,735** | 0,245 | 1,160 | 0,081 | **1,711** | 0,248 | **0,314** | 0,587 | **1,882** | 0,174 |
| Arts and Humanities | 0,893 | 0,367 | 0,624 | 0,367 | - | - | 2,056 | 0,414 | 0,847 | 0,206 | 1,332 | 0,520 | 0,804 | 0,962 | 1,689 | 1,375 |
| Social Sci., Jour., Inform. | **0,478** | 0,307 | 0,706 | 0,346 | - | - | 1,205 | 0,399 | **0,574** | 0,212 | 1,225 | 0,496 | 0,600 | 0,955 | 1,182 | 1,382 |
| Business, Admin., La | 0,636 | 0,289 | 0,727 | 0,311 | - | - | **2,484** | 0,397 | **0,537** | 0,180 | 0,955 | 0,446 | 0,320 | 0,877 | 1,004 | 1,371 |
| Nat. Sci., Math., Stat. | 0,594 | 0,356 | 0,843 | 0,427 | - | - | 1,601 | 0,431 | 0,764 | 0,234 | 2,257 | 0,597 | - | - | 0,795 | 1,363 |
| ICT | 2,173 | 0,689 | 1,071 | 0,490 | - | - | 2,238 | 0,617 | 0,734 | 0,235 | 3,101 | 0,700 | 1,165 | 1,148 | 1,030 | 1,398 |
| Engineer., Manuf., Constr. | 0,656 | 0,319 | 0,809 | 0,342 | - | - | 2,107 | 0,405 | **0,430** | 0,191 | 1,550 | 0,466 | 0,847 | 1,071 | 1,034 | 1,396 |
| Agri., For., Fish., Vet. | 1,132 | 0,969 | 0,416 | 0,462 | - | - | 0,288 | 0,706 | 0,647 | 0,245 | 1,419 | 0,653 | - | - | 0,627 | 1,372 |
| Health and Welfare | 0,733 | 0,354 | 0,813 | 0,364 | - | - | 1,471 | 0,432 | 0,873 | 0,205 | 1,874 | 0,489 | 0,780 | 0,968 | 0,364 | 1,618 |
| Services | 0,421 | 0,620 | **0,438** | 0,403 | - | - | 1,367 | 0,561 | 0,705 | 0,221 | 0,879 | 0,814 | 0,173 | 1,199 | 1,390 | 1,368 |
| Educationa background | 0,895 | 0,110 | 0,892 | 0,105 | 0,951 | 0,119 | 0,843 | 0,108 | 1,090 | 0,052 | 0,835 | 0,120 | 0,920 | 0,269 | 1,022 | 0,092 |
| Migration Backgrounc | 1,038 | 0,232 | 1,088 | 0,277 | 1,726 | 0,370 | 1,175 | 0,443 | 1,004 | 0,134 | 0,477 | 0,646 | 1,494 | 0,827 | **0,514** | 0,261 |
| Lecture-based programme | **1,353** | 0,075 | **1,247** | 0,074 | **1,297** | 0,087 | **1,415** | 0,056 | **1,527** | 0,022 | **1,408** | 0,071 | **1,590** | 0,131 | **1,615** | 0,047 |
| Problem-based learning | **1,295** | 0,073 | **1,181** | 0,067 | **1,351** | 0,087 | **1,259** | 0,071 | **1,113** | 0,030 | **1,349** | 0,073 | 1,159 | 0,146 | 0,896 | 0,061 |
| Study Abroad | 0,933 | 0,212 | 1,429 | 0,244 | 1,050 | 0,244 | 1,404 | 0,301 | **1,477** | 0,159 | 0,956 | 0,284 | 0,915 | 0,536 | 0,875 | 0,190 |
| Volunteerir in HEI | 0,847 | 0,230 | 1,340 | 0,248 | 1,383 | 0,226 | 1,206 | 0,220 | **1,261** | 0,095 | 1,033 | 0,239 | 0,827 | 0,410 | 0,991 | 0,161 |
| Volunteerir outside HE | 1,169 | 0,176 | 0,989 | 0,187 | **0,659** | 0,193 | 1,318 | 0,202 | 0,952 | 0,095 | **0,512** | 0,223 | 0,541 | 0,386 | 1,048 | 0,158 |
| Internship i the country | 0,850 | 0,178 | 1,206 | 0,168 | **0,580** | 0,227 | 1,121 | 0,196 | **1,175** | 0,079 | 0,763 | 0,234 | **2,190** | 0,380 | 1,039 | 0,167 |
| Internship abroad | 0,705 | 0,221 | 0,920 | 0,260 | 0,788 | 0,252 | 1,433 | 0,322 | 1,187 | 0,177 | 0,987 | 0,270 | 1,457 | 0,826 | 0,750 | 0,269 |

$p<0,05$; Referent variable for the fields of study is “Education”.

*Except for Malta where referent variable is the short-cycle

Table 3. Communication skills

| | Austria | | Czech Rep. | | Germany | | Greece | | Croatia | | Lithuania | | Malta | | Norway | |
|---|---|---|---|---|---|---|---|---|---|---|---|---|---|---|---|---|
| | Exp(B | S.E. | Exp(B | S.E. | Exp(B | S.E. | Exp(B | S.E. | Exp(B | S.E. | Exp(B | S.E. | Exp(B | S.E. | Exp(B | S.E. |
| Sex (f) | 0,855 | 0,166 | **1,478** | 0,174 | 0,686 | 0,215 | 1,088 | 0,196 | 1,107 | 0,085 | 1,184 | 0,199 | 0,862 | 0,325 | 1,123 | 0,156 |
| Age at graduatic | 1,098 | 0,89 | 0,990 | 0,105 | 1,217 | 0,148 | 1,139 | 0,086 | **1,183** | 0,051 | 1,181 | 0,118 | **1,569** | 0,211 | 1,087 | 0,076 |
| Status (part-tim | - | - | 1,690 | 0,288 | 1,531 | 0,514 | - | - | **1,296** | 0,122 | 1,161 | 0,328 | 1,397 | 0,645 | 1,202 | 0,304 |
| Type of HEI (nor uni) | 1,322 | 0,206 | 1,362 | 0,323 | 1,437 | 0,248 | 1,355 | 0,262 | 0,977 | 0,104 | 1,418 | 0,246 | 0,544 | 0,445 | 1,038 | 0,158 |
| Level of qualif. (MA) | 1,390 | 0,174 | 1,180 | 0,166 | 0,839 | 0,219 | **2,176** | 0,253 | **1,236** | 0,085 | 1,458 | 0,230 | 0,542 | 0,454 | 1,264 | 0,169 |
| Arts and Humanit s | 0,604 | 0,370 | **0,390** | 0,365 | **0,080** | 0,725 | **2,654** | 0,447 | 0,824 | 0,221 | 0,637 | 0,498 | 0,194 | 1,006 | 1,718 | 1,277 |
| Social Sci., Jou Inform. | **0,492** | 0,332 | 0,760 | 0,359 | **0,133** | 0,769 | 0,930 | 0,412 | 0,650 | 0,229 | 0,992 | 0,477 | 0,172 | 1,016 | 0,744 | 1,282 |
| Business Admin., Law | **0,410** | 0,310 | 0,698 | 0,316 | **0,068** | 0,750 | 1,151 | 0,404 | **0,677** | 0,195 | 0,868 | 0,433 | **0,130** | 0,952 | 1,186 | 1,272 |
| Nat. Sci. Math., Stat. | **0,253** | 0,354 | 0,505 | 0,413 | **0,146** | 0,780 | 1,137 | 0,444 | **0,528** | 0,244 | 1,494 | 0,565 | 0,250 | 1,184 | 0,866 | 1,264 |
| ICT | **0,321** | 0,435 | 0,936 | 0,474 | **0,048** | 0,817 | 0,889 | 0,611 | **0,271** | 0,242 | 0,880 | 0,609 | 0,127 | 1,071 | 0,790 | 1,299 |
| Engineer Manuf., Constr. | **0,259** | 0,323 | **0,321** | 0,330 | **0,046** | 0,757 | 1,259 | 0,412 | **0,366** | 0,204 | 0,893 | 0,448 | 0,141 | 1,037 | 0,335 | 1,301 |
| Agri., Fo Fish., Ve | 0,262 | 0,763 | 0,510 | 0,466 | **0,048** | 0,995 | 0,358 | 0,652 | **0,533** | 0,257 | 0,837 | 0,622 | 0,345 | 1,852 | 0,878 | 1,273 |
| Health a Welfare | **0,349** | 0,367 | 0,603 | 0,366 | 0,261 | 0,803 | 1,333 | 0,447 | 0,871 | 0,222 | 0,960 | 0,468 | 0,224 | 1,006 | 6,279 | 2,091 |
| Services | 0,890 | 0,758 | 0,649 | 0,411 | - | - | 0,490 | 0,589 | **0,610** | 0,234 | 0,491 | 0,775 | **0,087** | 1,232 | 1,636 | 1,270 |
| Educatio al backgrou d | 0,921 | 0,105 | 1,050 | 0,104 | 1,061 | 0,117 | 1,018 | 0,111 | 1,020 | 0,054 | 0,853 | 0,115 | 1,619 | 0,248 | 1,162 | 0,091 |
| Migratio Backgrou d | 0,930 | 0,218 | 1,031 | 0,283 | 1,893 | 0,344 | 0,934 | 0,455 | 1,217 | 0,145 | 0,525 | 0,637 | **0,131** | 0,644 | 1,238 | 0,273 |
| Lecture-based programm e | **1,277** | 0,072 | **1,201** | 0,073 | **1,502** | 0,086 | **1,607** | 0,059 | **1,618** | 0,023 | **1,251** | 0,068 | 1,240 | 0,112 | **1,483** | 0,045 |
| Problem-based learning | **1,226** | 0,068 | **1,235** | 0,066 | **1,185** | 0,083 | **1,135** | 0,074 | **1,071** | 0,032 | **1,400** | 0,069 | **1,308** | 0,123 | 0,891 | 0,061 |
| Study Abroad | 1,108 | 0,217 | **2,136** | 0,264 | **2,214** | 0,256 | 1,848 | 0,321 | 1,163 | 0,162 | 1,239 | 0,279 | **0,383** | 0,458 | 1,075 | 0,187 |
| Voluntee ng in HE | **1,669** | 0,251 | 1,462 | 0,257 | **1,938** | 0,237 | **2,086** | 0,234 | 1,039 | 0,099 | 1,328 | 0,233 | 1,307 | 0,374 | 1,275 | 0,160 |
| Voluntee ng outsid HEI | **1,416** | 0,170 | 1,032 | 0,189 | 1,106 | 0,194 | **1,595** | 0,212 | **1,274** | 0,101 | 1,067 | 0,217 | 1,699 | 0,383 | 1,092 | 0,156 |
| Internshi in the country | **1,441** | 0,165 | **1,409** | 0,165 | 1,192 | 0,206 | **0,607** | 0,205 | **1,246** | 0,082 | 0,910 | 0,216 | 0,608 | 0,315 | 1,125 | 0,165 |
| Internshi abroad | 0,853 | 0,222 | **1,963** | 0,300 | 0,760 | 0,246 | 1,460 | 0,339 | 1,194 | 0,184 | 1,224 | 0,261 | 3,765 | 0,771 | 0,771 | 0,271 |

p<0,05; Referent variable for the fields of study is "Education".

*Except for Malta where referent variable is the short-cycle

## REFERENCES

AACU (2007) College learning for the new global century: a report from the National Leadership Council for Liberal Education & America's Promise. Washington, D.C.: Association of American Colleges and Universities.

AACU (2015) Falling short? College learning and career success. Association of American Colleges and Universities, Washington DC.

Ahmad, C.N.C. et al. (2015) 'Relationship between Constructivist Learning Environments and Educational Facility in Science Classrooms', The Proceedings of 6th World Conference on educational Sciences, 191, pp. 1952–1957. Available at: https://doi.org/10.1016/j.sbspro.2015.04.672.

Andrews, J. and Higson, H. (2008) 'Graduate Employability, "Soft Skills" Versus "Hard" Business Knowledge: A European Study', Higher Education in Europe, 33(4), pp. 411–422. Available at: https://doi.org/10.1080/03797720802522627.

Berg, M., Connor-Linton, J. and Paige, R. (2009) 'The Georgetown Consortium Project: Interventions for Student Learning Abroad', Frontiers: The Interdisciplinary Journal of Study Abroad, 18, pp. 1–75. Available at: https://doi.org/10.36366/frontiers.v18i1.251.

Biggs, J.B. and Tang, C.S. (2011) Teaching for quality learning at university: what the student does. 4th edition. Maidenhead, England New York, NY: McGraw-Hill, Society for Research into Higher Education & Open University Press (SRHE and Open University Press imprint).

Børte, K., Nesje, K. and Lillejord, S. (2023) 'Barriers to student active learning in higher education', Teaching in Higher Education, 28(3), pp. 597–615. Available at: https://doi.org/10.1080/13562517.2020.1839746.

Boud, D. and Feletti, G. (2013) The Challenge of Problem-based Learning. 0 edn. Routledge. Available at: https://doi.org/10.4324/9781315042039.

Boudon, R. (1974) Education, opportunity, and social inequality: changing prospects in Western society. New York: Wiley (Wiley series in urban research).

Bourdieu (1977) 'Cultural Reproduction and Social Reproduction', in J. Karabel & A. H. Halsey (Eds.), Power and Ideology in Education (pp. 487-511). New York: Oxford University Press.

Cahalan, M.W. et al. (2022) Indicators of Higher Education Equity in the United States: 2022 Historical Trend Report. Washington, DC: The Pell Institute for the Study of Opportunity in Higher Education, Council for Opportunity in Education (COE), and Alliance for Higher Education and Democracy of the University of Pennsylvania (PennAHEAD).

Campbell, A. and Norton, L. (eds) (2007) Learning, teaching and assessing in higher education: developing reflective practice. Exeter: Learning Matters (Teaching in higher education).

Carini, R., Kuh, G. and Klein, S. (2006) 'Student Engagement and Student Learning: Testing the Linkages*', Research in Higher Education, 47, pp. 1–32. Available at: https://doi.org/10.1007/s11162-005-8150-9.

Cisneros-Donahue, T. et al. (2012) 'Assessing the Academic Benefit of Study Abroad', Journal of Education and Learning, 1. Available at: https://doi.org/10.5539/jel.v1n2p169.

Damşa, C. and Lange, T. de (2019) 'Student-centred learning environments in higher education: From conceptualization to design', Uniped, 42(01), pp. 9–26. Available at: https://doi.org/10.18261/issn.1893-8981-2019-01-02.

De La Harpe, B., Radloff, A. and Wyber, J. (2000) 'Quality and Generic (Professional) Skills', Quality in Higher Education, 6(3), pp. 231–243. Available at: https://doi.org/10.1080/13538320020005972.

Engle, J. and Tinto, V. (2008) 'Moving Beyond Access: College Success for Low-Income, First-Generation Students', Pell Institute for the Study of Opportunity in Higher Education.

Freudenberg, B., Brimble, M. and Cameron, C. (2011) 'WIL and Generic Skill Development: The Development of Business Students' Generic Skills Through Work-Integrated Learning', Asia-Pacific Journal of Cooperative Education, 12, pp. 79–93.

Furnham, A., Hosoe, T. and Tang, T.L.-P. (2001) 'Male hubris and female humility? A crosscultural study of ratings of self, parental, and sibling multiple intelligence in America, Britain, and Japan', Intelligence, 30(1), pp. 101–115. Available at: https://doi.org/10.1016/S0160-2896(01)00080-0.

Gerritsen, R.J. (1999) 'Do Students Study More Thoroughly Within a Problem-Based Learning Course', in J. Hommes et al. (eds) Educational Innovation in Economics and Business IV: Learning in a Changing Environment. Dordrecht: Springer Netherlands, pp. 127–141. Available at: https://doi.org/10.1007/978-94-015-9181-2_8.

Giudici, C., Trappolini, E. and Vicari, D. (2021) 'The academic performance of students with a migrant background: evidence from a cohort enrolled at Sapienza University of Rome', Genus, 77(1), p. 38. Available at: https://doi.org/10.1186/s41118-021-00145-3.

Graham, C. and Longchamps, P. (2022) Transformative education: a showcase of sustainable and integrative active learning. London New York: Routledge, Taylor & Francis Group. Available at: https://doi.org/10.4324/9781003276203.

Griga, D. and Hadjar, A. (2014) 'Migrant background and higher education participation in Europe: The effect of the educational systems', European Sociological Review, 30(3), pp. 275–286.

Guichard, E. et al. (2024) 'When National Origins Equal Socio-economic Background: The Effect of the Ethno-class Parental Background on the Education of Children Coming of Age in Switzerland', Journal of International Migration and Integration [Preprint]. Available at: https://doi.org/10.1007/s12134-024-01129-w.

Helyer, R. and Lee, D. (2014) 'The Role of Work Experience in the Future Employability of Higher Education Graduates', Higher Education Quarterly, 68(3), pp. 348–372. Available at: https://doi.org/10.1111/hequ.12055.

Hoidn, S. and Klemenčič, M. (eds) (2021) The Routledge international handbook of student-centered learning and teaching in higher education. London New York: Routledge (Routledge international handbooks).

Hora, M., Chen, Z. and Her, P. (2020) 'Problematizing college internships: Exploring issues with access, program design and developmental outcomes', 21, pp. 235–252.

Hu, S. and Wolniak, G.C. (2010) 'Initial Evidence on the Influence of College Student Engagement on Early Career Earnings', Research in Higher Education, 51(8), pp. 750–766.

Jack, A.A. (2020) The privileged poor: how elite colleges are failing disadvantaged students. First Harvard University Press paperback edition. Cambridge, Massachusetts London, England: Harvard University Press.

Kirby, L.A.J. and Thomas, C.L. (2022) 'High-impact teaching practices foster a greater sense of belonging in the college classroom', Journal of Further and Higher Education, 46(3), pp. 368–381. Available at: https://doi.org/10.1080/0309877X.2021.1950659.

Kolb, D. (1984) Experiential Learning: Experience As The Source Of Learning And Development, Journal of Business Ethics.

Komarraju, M., Musulkin, S. and Bhattacharya, G. (2010) 'Role of Student–Faculty Interactions in Developing College Students' Academic Self-Concept, Motivation, and Achievement', Journal of College Student Development, 51, pp. 332–342. Available at: https://doi.org/10.1353/csd.0.0137.

Lee, J. and Song, J. (2019) 'Developing Intercultural Competence through Study Abroad, Telecollaboration, and On-campus Language Study', Language, Learning and Technology, 23, pp. 178–198.

Magolda, M. and Astin, A. (1993) 'What Matters in College: Four Critical Years Revisited', Educational Researcher, 22. Available at: https://doi.org/10.2307/1176821.

McGunagle, D. and Zizka, L. (2020) 'Employability skills for 21st-century STEM students: the employers' perspective', Higher Education, Skills and Work-Based Learning, 10(3), pp. 591–606. Available at: https://doi.org/10.1108/HESWBL-10-2019-0148.

Meng, C. et al. (2020) Eurograduate pilot survey: design and implementation of a pilot European graduate survey. LU: Publications Office. Available at: https://data.europa.eu/doi/10.2766/149071 (Accessed: 30 May 2021).

OECD (2015) 'How do differences in social and cultural background influence access to higher education and the completion of studies?', (35). Available at: https://doi.org/10.1787/5jrs703c47s1-en.

Parnes, M.F. et al. (2020) 'Academic Outcomes Among Diverse Community College Students: What Is the Role of Instructor Relationships?', Community College Review, 48(3), pp. 277–302. Available at: https://doi.org/10.1177/0091552120909908.

Salmi, J. (2020) Paper commissioned for the 2020 Global Education Monitoring Report, Inclusion and education. UNESCO. Available at: https://unesdoc.unesco.org/ark:/48223/pf0000373689.

Tecilazić, A. (2021) 'Employability of Graduates: Effects of Internship on early career outcomes', in International scientific conference Employment, Education and Entrepreneurship (10; 2021; Beograd). Tenth International Scientific Conference Employment, Education and Entrepreneurship, Belgrade: Faculty of Business Economics and Entrepreneurship, Belgrade, Serbia, pp. 131–138.

Tecilazić, A. (2022) 'Student-centred learning: Effects on Graduates' Satisfaction with Study Programme', in EDULEARN22 Proceedings. 14th International Conference on Education and New Learning Technologies, Palma, Spain. 4-6 July, 2022.: EDULEARN22. Available at: https://doi.org/10.21125/edulearn.2022.

Tecilazić, A. (2023) 'Zapošljivost osoba s kvalifikacijama visokog obrazovanja: tranzicija iz visokog obrazovanja na tržište rada (Doktorska disertacija).

Tymon, A. (2013) 'The student perspective on employability', Studies in higher education, 38(6), pp. 841–856.

# GENERATIONAL PERSPECTIVE ON ENERGY-EFFICIENCY AWARENESS: A COMPARATIVE STUDY

Gabriella Francesca Amalia Pernice[1], Valeria Orso[1,2] and Luciano Gamberini[1,2]
*[1]Departments of General Psychology, University of Padova, Padova, Italy*
*[2]Human Inspired Technology Research Center University of Padova, Padova, Italy*

**ABSTRACT**

Recent advances in sustainable practices in all sectors, particularly energy consumption, are critical in addressing global challenges such as climate change and resource depletion. This paper explores the crucial role of engaging the youngest generations, who are poised to become the main market influencers and consumers. Despite its potential impact, existing literature reveals a gap in integrating the perspectives and contributions of these generations to sustainable energy practices. This oversight offers a significant opportunity for further investigation and integration of their preferences and behaviours into broader sustainability frameworks. The present study collected the responses of a survey of 128 participants of different generations in Italy on their attitude and trust in energy consumption and high-energy-efficient household appliances. The results present a substantial challenge for policy makers and businesses aiming to align their strategies with the values and expectations of this key demographic. The study calls for a more nuanced understanding of these generational differences to develop effective and sustainable energy strategies that resonate with the values of the youngest demographics, aligning these efforts with Sustainable Development Goal 7 (SDG 7) to ensure access to affordable, reliable, sustainable, and modern energy for all.

**KEYWORDS**

Energy-Consumption, Consumer Behaviour, SDG7, Generation Z, Millennials, Multigeneration

## 1. INTRODUCTION

The urgent need to adopt sustainable practices across various sectors, particularly in energy consumption, has become critically important in addressing global challenges such as climate change and resource scarcity (Bensouda and Benali, 2023). As the world transitions towards more sustainable methods, the role of young consumers, such as Generation Z and Millennials, gains increasing prominence. Characterised as digital natives with a strong inclination towards environmental activism, their environmental awareness significantly shapes their consumer behaviours and lifestyle choices (Fan et al., 2022). Although these generations are emerging as major market influencers and are poised to become the main consumers in the near future, their limited financial resources often restrict their impact on the energy market. This discrepancy between their green intentions and actual purchasing power is a critical oversight in the energy sector. Despite the potential of these generations to profoundly influence sustainability, there remains a significant void in research that focuses on their specific behaviours and contributions, particularly in the context of sustainable energy consumption. This deficiency is noteworthy, as daily life and consumption patterns are crucial to shaping future energy solutions. Understanding and integrating the environmental concerns, behaviours, and preferences of the youngest generations in energy consumption strategies is a requirement to develop effective and lasting sustainability policies (Yang, Solgaard, & Haider, 2016). This paper aims to provide a comparative analysis across different generations, with a specific focus on the youngest response on energy consumption. More specifically, it explores the preferences of five generations regarding high-efficiency household appliances and their purchasing intentions. Aligning these insights with the goal of Sustainable Development Goal 7 (SDG 7), which aims to ensure access to affordable, reliable, sustainable, and modern energy for all, this research seeks to enhance our understanding of how to effectively engage multiaged consumers in the energy transition (Walesiak and Dehnel, 2024). The findings of this study are intended to guide policymakers and stakeholders in developing strategies that address the environmental impacts of energy consumption while aligning with the values and expectations of key demographics who are set to inherit an environmentally constrained world. Furthermore, the development of specific tools and customised communication strategies to meet the diverse needs of different generations could be important to effectively engaging these groups in sustainable practices. This approach ensures that environmental policies and initiatives resonate more deeply and lead to more impactful and lasting changes in behaviour.

## 1.1 Related Work

Investigating sustainability through the lens of generational differences provides critical insights due to notable variations in behaviours and attitudes across ages. Based on the Generational Cohort Theory introduced by Inglehart in 1977, each generational cohort shares significant historical experiences and societal influences that shape consistent values, behaviours, and attitudes throughout their lives (Inglehart, 2020). Generally, the main adult consumer groups include Baby Boomers (born 1946–1964), Generation X (born 1965–1976), Millennials (born 1977–1995), and Generation Z (born 1996-present). A segment known as Xennials lies between the late Millennials and early Generation X, bridging the characteristics of both groups. These cohorts each show unique consumer behaviour and environmental attitudes, providing a significant basis for analysing patterns of sustainable consumption (Popa et al., 2018). Amid escalating global challenges such as climate change and the need for sustainable energy solutions, the focus has increasingly shifted toward the youngest generations, Millennials and Generation Z, who have not only witnessed the direct impacts of climate change, but have also become deeply involved in environmental advocacy (Inglehart, 2020). Some authors, such as Hickman (2020), underscored the rise of ecoanxiety among these youngest cohorts, as their awareness of ecological crises deepens, leading to active participation in movements like those led by Greta Thunberg. These movements are often amplified through social media, which serves not only to raise awareness but also to mobilise substantial policy and lifestyle changes (Boulianne et al., 2020). Likewise, social media deeply influenced the purchasing habits of the youngest generation, especially in fashion and food, not only enhancing their awareness but also motivating them to make sustainable purchase choices (Djafarova and Foots, 2022). Recent studies have highlighted the consumer behaviours of Generation Z and Millennials related to their environmental awareness. Their attitudes and perceived behavioural control significantly impact their decisions to purchase green products, both in fashion retail (Armutcu and Tan, 2022; Dabija and Chebeň Juraj, 2017) and in choosing sustainable and health-conscious food options (Lo et al., 2020; Kamenidou et al., 2019). Moreover, green marketing strategies that highlight the environmental benefits of products appear to be particularly effective among the youngest generations (Barska et al., 2023), who are generally highly receptive to these kinds of advertising. They tend to perceive information on sustainability as trustworthy and exhibit low green scepticism towards the sustainable information disseminated by brands (Pernice et al., 2023). However, while younger generations, such as Millennials and Generation Z, demonstrate a clear commitment to sustainability through their purchasing choices in retail sectors, a significant gap in research persists regarding their attitudes towards energy consumption (Yang, Solgaard, & Haider, 2016).

The present study seeks to bridge this gap by exploring and integrating the perspectives of these generations on purchasing high-energy-efficient household appliances and on their energy efficiency (EE) labels, where energy information is reported. This understanding is important for developing strategies that not only meet environmental goals but also resonate with the values of these consumer groups that would became the more influential on the market, thereby fostering a more sustainable energy landscape, starting with the household sector.

This study seeks to answer the following research question: How do different generations perceive, trust, and intend to purchase high-energy-efficient household appliances, and what factors influence these behaviours?

## 2. MATERIAL AND METHODS

### 2.1 Methods

The survey (30 items in total) used for this study was divided into three main parts: Demographic Information, Understanding and Perceptions of Energy Efficiency (EE) Labels, and Attitudes towards Refurbished Appliances. Demographic Information: This section collected basic information about the participants, including gender, age, educational background, occupation, monthly income, household size, and previous experience purchasing household appliances, through eight questions. Understanding and Perceptions of Energy Efficiency: This part, using a 7-point scale, assessed how participants perceive and understand energy efficiency, including 3 items on their general perceived knowledge of energy labels (PEK), adapted from Waris and Hameed (2020),. Additionally, it evaluated the trust participants place in these labels (ET) with three items, adapted from Hameed and Waris (2018), and measured their intentions to purchase energy-efficient appliances (PI) with 3 items, based on a study by Nguyen et al. (2017). Furthermore, participants were asked how much they would pay for a high energy-efficient household appliance in Class A (3 items).

Attitudes Towards Refurbished Appliances: The final segment explored the knowledge of and attitudes towards refurbished appliances using a 7-point scale. It included 3 items assessing perceived knowledge of refurbishment (REK) and purchase intentions (RPI; 3 items), both adapted from Waris and Hameed (2020). Finally, participants' environmental concerns (EC; 3 items) were assessed through an item adapted from Ward, Clark, Jensen, & Yen (2011).

### 2.2 Participants

The sample consisted of 128 participants. The age of participants was distributed among several generational cohorts, including Generation Z (18–24 years) at 25.78% and Millennials (25–40 years) at 24.22%. Xennials (41–44 years) represented 13.28%, while Generation X (45–56 years) represented 17.19%, and Baby Boomers (57+ years) represented 19.53%. Demographic data are reported in Table 1.

Table 1. Demographic data split for generations (%)

| Demographic | Generation Z | Millennial | Xennials | Generation X | Baby Boomers |
|---|---|---|---|---|---|
| Gender | | | | | |
| Female | 45,5 | 29,0 | 41,2 | 72,7 | 60,0 |
| Male | 54,5 | 71,0 | 47,1 | 22,7 | 36,0 |
| Household size | | | | | |
| Small (1-2 members) | 18,2 | 54,8 | 35,3 | 31,8 | 40,0 |
| Medium (3-4 members) | 57,6 | 25,8 | 47,1 | 54,5 | 40,0 |
| Large (5+) | 24,2 | 19,4 | 17,6 | 13,6 | 20,0 |
| Income | | | | | |
| Low | 51,5 | 25,8 | 23,5 | 18,2 | 24 |
| Medium | 39,4 | 45,2 | 35,3 | 45,5 | 40 |
| Medium-high | 9,1 | 19,3 | 11,8 | 22,7 | 24 |
| High | 0 | 9,7 | 29,4 | 13,6 | 12 |
| Education | | | | | |
| Elementary | 3,0 | 6,5 | 5,9 | 4,5 | 12,0 |
| Middle school | 27,3 | 19,4 | 35,3 | 59,1 | 60,0 |
| High-school | 42,4 | 6,5 | 17,6 | 9,1 | 4,0 |
| Specialization | 24,2 | 58,1 | 29,4 | 18,2 | 20,0 |
| Doctoral | 3,0 | 9,7 | 11,8 | 9,1 | 4,0 |
| Job | | | | | |
| Blue collar | 9,1 | 6,5 | 17,6 | 13,6 | 4,0 |

| | | | | | |
|---|---|---|---|---|---|
| Employee private sector | 21,2 | 41,9 | 35,3 | 36,4 | 4,0 |
| Employee public sector | 27,3 | 25,8 | 17,6 | 18,2 | 12,0 |
| Freelancer | 3,0 | 9,7 | 17,6 | 9,1 | 0,0 |
| Retired | 0,0 | 0,0 | 0,0 | 4,5 | 72,0 |
| Small entrepreneur | 0,0 | 0,0 | 0,0 | 4,5 | 8,0 |
| Student | 39,4 | 16,1 | 0,0 | 4,5 | 0,0 |
| Trader | 0,0 | 0,0 | 5,9 | 4,5 | 0,0 |
| Unemployed | 0,0 | 0,0 | 5,9 | 4,5 | 0,0 |

## 2.3 Procedure

The study was conducted in Italy. The research received ethical approval from the University Ethics Committee. Data were collected through an online survey conducted on the Qualtrics platform, with participants providing informed consent prior to their participation. The survey was sent through a link to a wide range of participants to achieve a diverse sample representative of various age groups, income levels and educational levels throughout the country.

## 2.4 Analysis and Result

The demographic profile provides a foundation for understanding the factors that influence consumer attitudes and behaviours toward sustainable household appliance choices across different segments of the population, as reported in this section. Data collected using questionnaires were analysed with IBM's SPSS (Version 25). Nonparametric analyses were performed, and the Benjamini & Hochberg correction was applied. To analyse the attitudes towards the energy-saving of household appliances and their perceived scepticism towards sustainable information related to household appliances, a series of Wilcoxon signed-rank tests were conducted. In particular, we compared the scores that participants assigned to each dimension with the median value of the scale (Mdn= 4), indicating a neutral attitude. Table 2 reports the results of the analysis. Overall, environmental messaging and products seem to be best received by Generation Z, who consistently show high levels of perceived knowledge of energy labels, trust and attitude regarding energy efficiency, refurbished devices, and environmental concerns.

Table 2. Wilcoxon signed-rank test split for generations. Mean, standard deviation, median, Wilcoxon signed-rank test statistics (W), Z-scores (z) and p.values (p) are reported

| Constructs | Generation Z | Millennial | Xennials | Generation X | Baby Boomers |
|---|---|---|---|---|---|
| *Perceived Environmental Knowledge* | | | | | |
| Mean | 4.74 | 5.02 | 5.35 | 5.2 | 5.48 |
| Std. Deviation | 1.22 | 1.12 | 1.14 | 1.22 | 0.85 |
| Median | 5 | 5 | 5 | 5.5 | 5.5 |
| W | 312.5 | 389 | 103.5 | 188 | 276 |
| z | 2.98 | 3.73 | 3.21 | 3.11 | 4.21 |
| p | .003* | .000* | .001* | .002* | .000* |
| *Environmental Trust* | | | | | |
| Mean | 4.87 | 4.80 | 5.44 | 4.52 | 5.2 |
| Std. Deviation | 1 | 1.09 | 0.76 | 1.28 | .99 |
| Median | 5 | 5 | 5.5 | 4.5 | 5.5 |
| W | 391 | 245 | 120 | 155 | 208 |
| z | 3.77 | 3.27 | 3.43 | 1.89 | 3.86 |
| p | .000* | .001* | .001* | .58 | .000* |
| *Purchase Intention* | | | | | |
| Mean | 5.83 | 6.25 | 5.67 | 5.7 | 5.38 |
| Std. Deviation | 1.10 | 0.7 | 1.02 | 1.46 | 1.04 |

| | | | | | |
|---|---|---|---|---|---|
| Median | 6 | 6 | 6 | 6 | 5.5 |
| W | 473.5 | 465 | 105 | 230 | 270 |
| z | 4.46 | 4.83 | 3.34 | 3.40 | 4.03 |
| p | .000* | .000* | .001* | .001* | .000* |
| *Refurbished Environmental Knowledge* | | | | | |
| Mean | 4.61 | 4.34 | 4.6 | 4.57 | 4.48 |
| Std. Deviation | 1.47 | 1.24 | 1.15 | 1.19 | 1.03 |
| Median | 4.66 | 4.33 | 4.66 | 4.5 | 4.33 |
| W | 383.5 | 238.5 | 102.5 | 115 | 195 |
| z | 2.24 | 1.19 | 1.79 | 1.83 | 1.73 |
| p | .025* | .232 | .072 | .067 | .082 |
| *Refurbished Purchase Intention* | | | | | |
| Mean | 4.78 | 4.66 | 4.6 | 4.2 | 3.86 |
| Std. Deviation | 1.71 | 2.07 | 1.2 | 1.5 | 2.20 |
| Median | 5 | 5.5 | 4.5 | 4 | 4 |
| W | 394 | 294.5 | 50.5 | 70 | 123.5 |
| z | 2.44 | 1.67 | .351 | .571 | -.445 |
| p | .014* | .093 | .726 | .568 | .656 |
| *Environmental Concern* | | | | | |
| Mean | 5.47 | 5.82 | 4.48 | 5.27 | 5.22 |
| Std. Deviation | 1.31 | 1.05 | 1.3 | .94 | 1.41 |
| Median | 6 | 6 | 4.33 | 5 | 6 |
| W | 452.5 | 377 | 131.5 | 223.5 | 248.5 |
| z | 4.04 | 4.54 | 2.63 | 3.8 | 3.38 |
| p | .000* | .000* | .008* | .000* | .001* |

*The significance level is 0.05

A series of Mann-Whitney tests were also run to compare the generation's scores toward energy efficiency and the perceived attitude towards refurbished appliances. For the sake of brevity, only significant results are provided. The analysis shows that Baby Boomers perceived to know more about the energy label than Generation Z (W= 264.5, p.value =.002). However, Generation Z and Millennials trusted more in the energy information reported on the energy label compared to the Xennials (GenZ vs. Xen; W= 185, p.value= .005; Mill vs. Xen; W= 168, p. value =.04). Although Xennials have more trust in this information than Generation X (W= 269, p.value = 0.02). Regarding refurbished appliances, it emerged that Millennials trust more in the appliances of reconditioned phenomenon compared to Xennials (W=3 58.5, p.value= .04), and Generation X (W= 461, p. value =.03). Finally, Generation Z and Millennials have more intention to buy high energy efficient household appliances compared to the Baby Boomers (GenZ vs. Old; W= 540, p.value=.04; Mill vs. Old; W= 583, p.value= .04).

In the last section related to the willingness to pay for a high-energy-efficient household appliance in energetic class A, the Xennials are the group that would pay on average more for a high-energy-efficient household appliance (646.8 €), followed by Gen Z (630.6 €). While Baby Boomers would pay on average 617.7 €, Millennials would pay 561 €, and Gen X would pay 547.7 €.

## 3. DISCUSSION

This study investigated the environmental consciousness of multiple generations, emphasising their commitment to sustainable practices and their complex responses to energy consumption. Previous research has extensively explored generational differences, providing information on consumer behaviour related to the selection of eco-friendly and ethically produced products, particularly in the retail and food and beverage sectors (Han and Lee, 2016).

Despite a broad focus on general sustainable practices, there remains a significant research gap on attitudes towards energy consumption and the adoption of high-energy-efficient household appliances. Our findings indicate that energy-related decisions are influenced by a range of factors, including socio-demographic variables, and that pro-environmental attitudes significantly vary across different generations. Although all generations display different behaviours and attitudes towards the environment, consistent with previous research (Paço and Lavrador, 2017), Millennials and Generation Z exhibit a deeper understanding of environmental crises. These youngest cohorts are more environmentally conscious, show greater concern about energy consumption than Baby Boomers, and are more likely to engage in energy-saving behaviours. They not only trust energy-related information, but also show a willingness to purchase high-energy-efficient household appliances and refurbished devices (Wijayaningtyas et al., 2021).

In exploring the economic aspects of sustainable consumption, particularly the willingness to invest in high-energy-efficient appliances, additional generational differences emerge. Specifically, when considering the purchase of appliances rated in the highest energy efficiency class (Class A), Generation Z, despite having the highest proportion of students and a considerable segment in low-income brackets, shows a strong inclination towards investing in high-energy-efficient appliances. This indicates a robust commitment to sustainability that transcends economic constraints, possibly driven by increased environmental awareness and societal values that favour green consumption (Kim and Lee, 2023). In contrast, Baby Boomers, despite having the financial means, exhibit a cautious approach to spending on such appliances. This could be attributed to a potential mistrust in the energy efficiency claims of newer appliances or a less acute perception of environmental urgency (Tomita and Akutagawa, 2009). Millennials, though displaying significant environmental concerns, do not invest in energy-efficient products as much as might be expected, possibly due to different priorities such as housing and career development, which might allocate their financial resources elsewhere. Finally, Generation X shows a more conservative approach, with a lower willingness to invest in high-energy-efficient appliances on average. Their lower investment could also reflect a generation that has not been deeply influenced by the current environmental discourse as younger generations.

## 4. CONCLUSION

The study confirms the need for more targeted research into the intersection of generational attributes and energy consumption behaviours; however, a limitation of this research lies in the exclusive use of quantitative data.

In fact, environmentally friendly attitudes do not always translate into concrete energy-saving actions, potentially contributing to environmentally harmful consumer-driven social trends (Gamberini et al., 2012). A notable discrepancy exists in the literature between people's declared intentions and actual behaviours, as situational factors significantly influence their sustainability practices (Balinska, Jaska & Werenowska, 2023). Therefore, future research would benefit from integrating qualitative data to provide a more comprehensive understanding of the factors driving sustainable energy practices.

The lack of focused research on energy consumption behaviour, particularly among the youngest generations, poses a significant challenge in aligning future energy strategies with their preferences and expectations. They are poised to become the main consumers in the energy market. Whatever proactive behaviour is driven by psychological and emotional factors that motivate engagement in sustainability initiatives (Verdugo, 2012), it is essential to understand and integrate their perspectives and translate them into actions to develop energy strategies. Aligning the sustainable actions with Sustainable Development Goal 7 (SDG 7), which aims to ensure access to affordable, reliable, sustainable and modern energy for all, is crucial. Without this alignment, achieving the objectives of the 2030 Agenda may not be feasible (Walesiak and Dehnel, 2024). Policymakers and stakeholders can develop more effective sustainability frameworks that resonate with and actively engage this critical demographic, ensuring that long-term sustainability goals are met. Critically, developing specific tools and tailoring communication strategies to meet the diverse needs of different generations could effectively engage these groups in sustainable practices, fostering lasting and impactful behavioral changes.

## 4.1 Limitations and Future Perspectives

The exclusive use of quantitative data adopted in this study might not fully capture the nuanced behaviors and attitudes toward energy-efficient appliances. Future research should integrate qualitative methods to provide a more comprehensive understanding of these behaviors. Additionally, this study focused on participants in Italy and could benefit from a broader geographic scope to compare generational attitudes toward energy efficiency across different cultures and regions.

# REFERENCES


Armutcu, B., & Tan, M. (2022). Generation Z Young Consumers' Sustainable Purchasing Behaviors and Intent to Purchase Recyclable Products: A Comprehensive Study in Turkey using SEM. *Business Research Proceedings*. https://doi.org/10.51300/brp-2023-72.

Balinska, A., Jaska, E., & Werenowska, A. (2023). Environmentally and Socially Sustainable Behavior of Generation Z in the Context of Tourist Trips. *EUROPEAN RESEARCH STUDIES JOURNAL*. https://doi.org/10.35808/ersj/3338.

Barska, A., Wojciechowska-Solis, J., Wyrwa, J., & Jędrzejczak-Gas, J. (2023). Practical Implications of the Millennial Generation's Consumer Behaviour in the Food Market. *International Journal of Environmental Research and Public Health*, 20. https://doi.org/10.3390/ijerph20032341.

Bensouda, M., & Benali, M. (2023). From "Fairly Good" to "Optimal" Energy Efficiency Practices within the Moroccan Manufacturing Sector: Are Financial Resources Sufficient?. *International Journal of Energy Economics and Policy*. https://doi.org/10.32479/ijeep.14046.

Boulianne, S., Lalancette, M., & Ilkiw, D. (2020). "School Strike 4 Climate": Social Media and the International Youth Protest on Climate Change. *Media and Communication*, 8, 208-218. https://doi.org/10.17645/mac.v8i2.2768.

Dabija, D. C., Chebeň, J., & Lančarič, D. (2017). Cross-cultural investigation of consumers' generations attitudes towards purchase of environmentally friendly products in apparel retail. *Studies in Business and Economics*, 12(3), 27-42.

Djafarova, E., & Foots, S. (2022). Exploring ethical consumption of generation Z: theory of planned behaviour. *Young Consumers*. https://doi.org/10.1108/yc-10-2021-1405. *Efficiency, 13*(8), 1653-1664.

Fan, A., Shin, H., Shi, J., & Wu, L. (2022). Young People Share, But Do So Differently: An Empirical Comparison of Peer-to-Peer Accommodation Consumption Between Millennials and Generation Z. *Cornell Hospitality Quarterly*, 64, 322 - 337. https://doi.org/10.1177/19389655221119463.

Gamberini, L., Spagnolli, A., Corradi, N., Jacucci, G., Tusa, G., Mikkola, T., ... & Hoggan, E. (2012, June). Tailoring feedback to users' actions in a persuasive game for household electricity conservation. In *International conference on persuasive technology* (pp. 100-111). Springer, Berlin, Heidelberg.

Gamberini, L., Spagnolli, A., Corradi, N., Jacucci, G., Tusa, G., Mikkola, T., ... & Hoggan, E. (2012, June). Tailoring feedback to users' actions in a persuasive game for household electricity conservation. *In International conference on persuasive technology* (pp. 100-111). Springer, Berlin, Heidelberg.

Hameed, D., & Waris, I. (2018). Eco labels and eco conscious consumer behavior: The mediating effect of green trust and environmental concern. Hameed, Irfan and Waris, Idrees (2018): Eco Labels and Eco Conscious Consumer Behavior: The Mediating Effect of Green Trust and Environmental Concern. *Journal of Management Sciences*, 5(2), 86-105.

Han, D., & Lee, J. (2016). An Effect of the Eco-friendly Consciousness of the Undergraduate by the Lifestyle on the Purchase Intention of the Eco-friendly Products. Journal of Digital Convergence, 14, 83-91.

Hickman, C. (2020). We need to (find a way to) talk about … Eco-anxiety. *Journal of Social Work Practice*, 34, 411 - 424. https://doi.org/10.1080/02650533.2020.1844166.

Inglehart, R. (2020). *Modernization and postmodernization: Cultural, economic, and political change in 43 societies*. Princeton university press.

Kamenidou, I., Mamalis, S., Pavlidis, S., & Bara, E. (2019). Segmenting the Generation Z Cohort University Students Based on Sustainable Food Consumption Behavior: A Preliminary Study. *Sustainability*. https://doi.org/10.3390/SU11030837.

Kim, N., & Lee, K. (2023). Environmental Consciousness, Purchase Intention, and Actual Purchase Behavior of Eco-Friendly Products: The Moderating Impact of Situational Context. *International Journal of Environmental Research and Public Health*, 20. https://doi.org/10.3390/ijerph20075312.

Lo, A., King, B., & Mackenzie, M. (2020). Segmenting Chinese Millennial Restaurant Customers: A Lifestyle and Health and Environmental Consciousness Approach. *Journal of China Tourism Research*, 16, 183 - 213. https://doi.org/10.1080/19388160.2019.1604281.

Nguyen, T. N., Lobo, A., & Greenland, S. (2017). Energy efficient household appliances in emerging markets: the influence of consumers' values and knowledge on their attitudes and purchase behaviour. *International journal of consumer studies, 41*(2), 167-177.

Paço A, Lavrador T. Environmental knowledge and attitudes and behaviours towards energy consumption. *J Environ Manage.* 2017 Jul 15;197:384-392. doi: 10.1016/j.jenvman.2017.03.100. Epub 2017 Apr 11. PMID: 28410516.

Pernice, G. F. A., Orso, V., & Gamberini, L. (2023). Consumer perspectives towards sustainability information in the household appliance marketing: An exploratory study. *IADIS International Journal on Computer Science and Information Systems*, 18 (2), 76-92.

Pernice, G. F. A., Orso, V., & Gamberini, L. (2023a). Is This Right for You?: The Key Role of Shop Assistants in Promoting Energy-Efficient Household Appliances. *Sustainability*, 15(24), 16630. https://doi.org/10.3390/su152416630

Popa, I., Dabija, D., & Grant, D. (2018). Exploring Omnichannel Retailing Differences and Preferences Among Consumer Generations. Applied Ethics for Entrepreneurial Success: Recommendations for the Developing World. https://doi.org/10.1007/978-3-030-17215-2_9.

Tomita, Y., & Akutagawa, Y. (2009). A Method for Measuring Environmental Value Consciousness by Factor Analysis. *International Journal of Urban Sciences*, 13, 172 - 186. https://doi.org/10.1080/12265934.2009.9693656.

Verdugo, V. (2012). The positive psychology of sustainability. *Environment, Development and Sustainability*, 14, 651-666. https://doi.org/10.1007/S10668-012-9346-8.

Walesiak, M., & Dehnel, G. (2024). Progress on SDG 7 achieved by EU countries in relation to the target year 2030: A multidimensional indicator analysis using dynamic relative taxonomy. *PLoS ONE*, 19(2), https://doi.org/10.1371/journal.pone.0297856

Ward, D. O., Clark, C. D., Jensen, K. L., & Yen, S. T. (2011). Consumer willingness to pay for appliances produced by Green Power Partners. *Energy Economics, 33*(6), 1095-1102.

Waris, I., & Hameed, I. (2020). Promoting environmentally sustainable consumption behavior: an empirical evaluation of purchase intention of energy-efficient appliances. *Energy*

Wijayaningtyas, M., Redjo, R., Handoko, F., Lukiyanto, K., & J., W. (2021). The Millennials' Energy Efficiency Behaviour towards Eco-friendly Home. *Civil Engineering and Architecture*. https://doi.org/10.13189/CEA.2021.090212.

Yang, Y., Solgaard, H., & Haider, W. (2016). Wind, hydro or mixed renewable energy source: Preference for electricity products when the share of renewable energy increases. *Energy Policy*, 97, 521-531. https://doi.org/10.1016/J.ENPOL.2016.07.030.

# A SUSTAINABILITY CAMPUS WALK AS A GAMIFIED TYPE OF BLENDED LEARNING

Anna-Marie Tuomala and Alexandra Maksheeva
*LAB University of Applied Sciences*
*Finland*

**ABSTRACT**

Education for sustainable development (ESD) provides a strong basis for the implementation of sustainability in universities. Using Bryant's rationale of space agnostic learning (SAL) and Carman's blended learning elements, a team at a Finnish higher education institution (HEI) came up with the idea of a Sustainability Campus Walk. The Sustainability Campus Walk is a blended learning approach combining elements of a physical walk with online learning materials. The aim was to introduce the HEI's work in sustainability for students of bachelor's and master's programmes. The project was completed using a novel approach, in a multi-stakeholder planning group, which enable the enrichment of the study materials with practical examples. Gamification elements were used to create checkpoints on the walk and make it more engaging for the participating students.

The methodology used to record the progress and results of the project was Kananen's observation diary model, which allowed a meticulous record of the activities. An analysis of the records of the coordinating member of the creation team and notes of the managerial members, provided the answers to two research questions about the involvement of the stakeholders in the planning process of gamified blended learning projects. We found that the stakeholders play a key role in planning learning projects, therefore using a multi-stakeholder team is recommended in further iterations of future courses.

**KEYWORDS**

Blended Learning, Gamification, Sustainability Learning

## 1. INTRODUCTION

## 1.1 Background and Rationale

Lozano (2010, 643) notes that if sustainability is integrated as a concept, in and among the different disciplines and tailored to their specific nature, universities could become more balanced, synergetic, trans-disciplinary and holistic. This enables students to become more proficient in sustainability. This idea is supported by Fischer & Isenmann (2023, 147) who underline that education for sustainable development (ESD) undeniably represents the core of implementing sustainability at universities. In their research, Hueske et al. (2022, 202) found several e-learning opportunities supporting education for sustainability development (ESD) and in terms of blended learning types, each HEI in their research applied several different formats.

### 1.1.1 Learning Design

Bryant (2021) introduces the space agnostic learning (SAL) concept in which mixed modes of participation allow students to undertake learning activities face-to-face or online, sometimes for the duration of the teaching period and other times varying. SAL involves designing teaching and learning activities so that *where* they occur does not limit the engagement and participation of students. SAL involves the design of teaching activities for implementation of different spaces that can be shared to enhance learning between students.

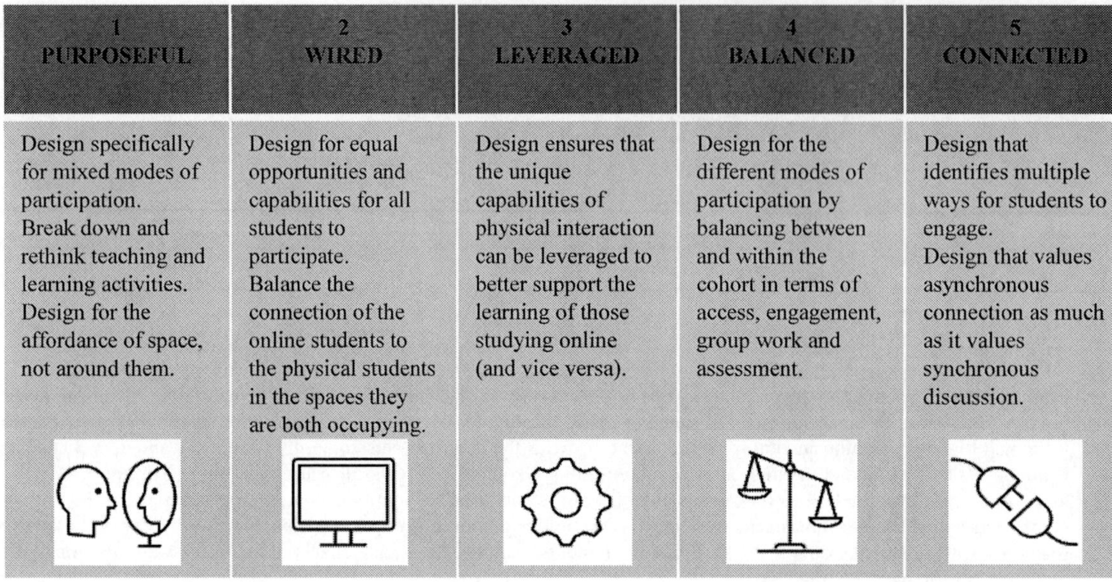


Figure 1. Diagram of space agnostic learning design principles (applied from Bryant, 2021)

Bryant underlines important planning principles within the SAL. The main idea is to put yourself in the position of a student, feeling empathy, and design learning activities by making the most of the spaces we have available for learning. To make it work, we must also 1) communicate clearly with students, thinking through the activity and how it works in entire learning process, 2) keep technical solutions simple because overly complex solutions easily lead to failure, 3) support ways to participate collectively and ensure the equivalence of experience and 4) design a framework of evaluation, because it is critical to improving the educational experience.

### 1.1.2 Blended Learning

There are many frameworks available, which conceptualize the relationship of blended learning elements. Carman's (2002, 2) framework has five key elements, which are referred to as ingredients: live events, online content, collaboration, assessment and reference materials. In addition, if we also adopt a learner-centred approach, focusing on the relationship between designs of blended learning environments and factors within the learner, empirical blended learning research might identify core attributes that influence learning, which can then be tested and better understood. Additionally, blended teaching and learning are central to enabling us to better manage complex contemporary information flows and integrating these flows into effective teaching (Ololube et al., 2015, 233). We see that because blended learning environments differ nowadays widely in terms of the technologies and methods used and the learning takes places in various physical environments, that blended learning is no longer learning that happens in an instructional context characterized by a deliberate combination of online and classroom-based interventions.

Sivapalan's (2017, 10, 21) findings combine blended learning and sustainability learning needs and possibilities. Some indicators have been found for higher educational institutions (HEIs) and sustainability educators planning to use blended learning to develop sustainability literacy and sustainability communication skills amongst the undergraduate engineering students. To add, blended learning provides flexible teaching and learning options for teachers and students – in comparison to conventional face-to-face studies.

### 1.1.3 Learning by Gamification

Dieleman & Huisingh (2005, 838-847) combine games in experiential learning for sustainability based on the theory of experiential learning, developed by David Kolb in the 1980s. According to that, the experience grasped through apprehension relies on the tangible and felt qualities of immediate experiences. This refers to the kind of learning processes one goes through while exploring things and looking at them. Dieleman & Huisingh see that the beauty of playing games is about 'learning by doing' and 'learning by failing', but without negative consequences in the real world. They emphasize that the theory of experiential learning provides numerous opportunities for teachers and learners to obtain valuable insights in learning processes and games could and should be added to the curricula in ESD.

## 1.2 Research Context

According to Pretorius et al. (2021, 1318) research in a case study format, which is presented as a reflective narrative with a focus on links with sustainability learning allows a contextualized description and assessment of the selected pedagogies, with inclusion of critical reflection.

A course titled the Sustainability Campus Walk was designed according to the SAL principles to promote sustainability literacy by immersing students in practical sustainability initiatives on campus and providing them with supplementary learning materials and reflections in a virtual environment. The course integrates physical and experiential learning with virtual, online components. This blended approach aims to enhance student engagement and understanding of sustainability concepts through real-world experiences. The learning objectives were established as the following for the students:

- to improve identification of possibilities and functions of sustainability in LAB Lahti campus
- to observe practical applications of the sustainable development in the operational environment
- to consider himself/herself better as part of the sustainable community and its different functions.

The research context also involves an educational approach where sustainability education is integrated into a course designed as a "hop on - hop off" experience, combining physical campus activities with virtual course components. In practice a student can enrol on the course at any time and when all the assignments are completed, and course result pass/not passed recorded, (s)he "hops off." Students attend physically according to their own individual schedules at specific locations on campus, each representing a sustainability checkpoint. These checkpoints are real-world settings of the university, where sustainable practices or initiatives are showcased or implemented. Each checkpoint corresponds to an assignment or activity with gamification elements on the virtual course platform. Here, students access additional resources, instructions, or reflections related to the physical activities they have completed.

## 1.3 Purpose, Objectives and Research Questions

### 1.3.1 Research Purpose

The LAB university of applied sciences has taken an ambitious goal to show and develop practical capabilities in the key areas of sustainability education and sustainability change management. It aims to follow global learning needs, defined by the OECD: a) to create self-directed learning options for the development of competencies, b) to apply transdisciplinary settings to make knowledge applicable and effective in real-life situations, and c) to consider overlapping the boundaries of academic versus nonacademic learning to achieve a holistic approach to lifelong learning (Tuomala 2022, 33). As a result, the purpose of this study is to investigate innovative approaches to sustainability education that empower individuals to autonomously develop competencies, apply knowledge effectively in real-life contexts through transdisciplinary settings, and integrate academic and nonacademic learning into a holistic framework aimed at advancing sustainability education and change management.

### 1.3.2 Research Objectives

The first objective delves into the practical application of sustainability knowledge acquired through SAL. It seeks to explore whether SAL is applicable to actionable study forms of sustainability. This objective is rooted in the idea that SAL can transcend traditional learning environments, allowing students to engage with sustainability concepts in a manner that is flexible and contextually relevant. By examining how SAL facilitates the acquisition and application of sustainability knowledge, the study aims to determine the effectiveness of this learning model in fostering real-world sustainability competencies. This involves analyzing student engagement, understanding, and the ability to apply learned concepts in practical settings. The findings from this objective will contribute to the broader discourse on innovative educational models that support the development of sustainability competencies in higher education.

The second objective is to evaluate if the gamification elements are efficient in promoting sustainability literacy and engagement. Gamification, which involves game design elements in non-game contexts, can make learning more interactive and enjoyable. This objective focuses on understanding how gamification can be integrated into sustainability education to enhance student motivation, participation, and retention of knowledge. By incorporating game-based activities into the curriculum, the study aims to determine the impact of these elements on students' sustainability literacy. This includes assessing the effectiveness of gamified assignments and activities in helping students grasp complex sustainability concepts, fostering a deeper understanding, and encouraging proactive engagement with sustainability practices. The research will analyze the extent to which gamification can transform traditional educational approaches, making learning about sustainability more engaging and impactful.

### 1.3.3 Research Questions

We formulated two research questions that address identified gap in gamified type of blended ESD courses in the literature. Recent research by Bryant (2023) was lacking an explanation on how co-design and co-generative dialogue could be supported when planning SAL at the higher education institution. It was noted that students require a more diverse amount of information to be included in the gamified process. A participatory approach involving multiple stakeholders could help to bring fresh perspectives to the quality of knowledge students would have after experiencing learning packages under SAL principles (Bakhanova et al. 2020).

To sum up, we noticed that there is a lack of information about the stakeholder's participation in course ideation and planning or involvement in general. We assume that the given context of stakeholders could also give rise to new knowledge that may benefit the field beyond the context of practice in which the knowledge was generated. When establishing clear and concise research questions, the context of the stakeholders guided our investigation towards meaningful and beneficial findings for the future. The resulting research questions are:

- RQ1 How does the implementation of SAL with a stakeholder group help impact students' understanding and application of sustainability concepts across diverse educational settings?
- RQ2 How could a collaborative approach to gamification strategies with stakeholder involvement be used to improve sustainability literacy and engagement among students?

These research questions serve as the foundation for our investigation, helping us to explore the research gaps identified.

## 2. METHODOLOGY

We created an intervention that consisted of blended methods of action research and SAL design thinking concerning teaching and learning. The aim was to apply Wetzler's research (2013, 4), which concluded that a collaborative organizational innovation methodology - that combines elements from action research and design thinking - has the potential to produce positive perceived outcomes related to novel solutions, personal commitment among organizational members, individual learning, and people's experience and relationship to the organization. The SAL concept of design thinking is introduced above in Chapter 1.1.1.

Argyris summarized the main principles of Lewinian action research into five theses (Argyris et al.1985, 8-9).

1. Operations research looks for solutions to concrete operational problems of organizations.
2. The research progresses cyclically, from identifying problems to planning the activity, to the activity itself, and to evaluating the results of the activity.
3. In action research, priority is given to educational measures aimed at changing the behavior and ways of thinking of the participants.
4. Action research questions the existing status quo from the perspective of democratic values.
5. Operational research simultaneously aims to promote the theoretical understanding of the phenomena it studies and their practical management.

Operational research progresses in cycles where planning, implementation, observation of operations and reflection are repeated. In this research theory has guided the search for new information and structured and systematized the process. The research progressed alternately as a dialogue between theory and empirics.

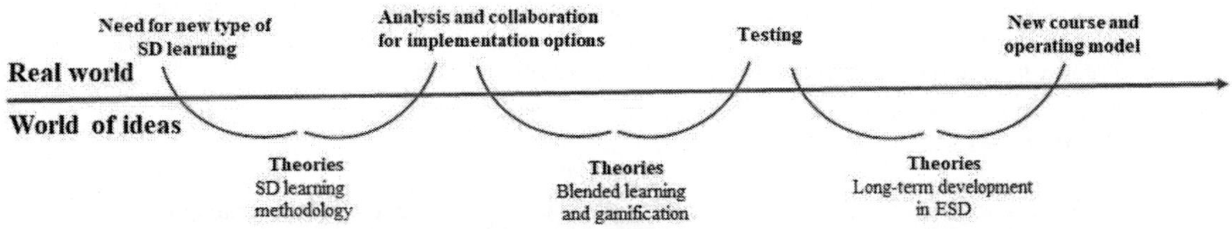


Figure 2. Stages of the dialogue cycle of the Sustainability Campus Walk between the real world and theory (applied from Saarinen, 2022)

In action research, the researcher's role is different from traditional research methods, because the researcher works as a group member throughout the research. The members participating in the activities agree on an equal basis throughout the research process. Action research fits with learning organizations, where the aim is both to train together and to develop the organization's operations to better meet its goals (Suojanen, 2014). From a triangulation point of view, this research also included applying 1) direct observations in action by the researchers, and 2) an analysis of enrolments and assignments occurring on an e-course platform and course management systems. A mixed-methods research approach was used to document the experiment. The documentation was carried out by the two action researchers/authors. Both researchers kept their own diaries for independent work. One kept a diary about the action in real life with the stakeholders and the other followed the management issues as a part of a larger RDI project. In the diary, the major steps and meetings with contributing stakeholders were documented. The reason for keeping the diary was two-fold: to record all the needs of the stakeholder group, which had to be highlighted in the Campus Walk and to make a log of the activities for the further analysis and presentation of the gamification method in sustainability education.

Table 1. Observation diary model for the Sustainability Campus Walk (applied from Kananen 2014, 69)

| Time | Place | Situation | Object | Activity | Description |
|---|---|---|---|---|---|
| | | | | | |
| | | | | | |
| | | | | | |

## 3. RESEARCH IMPLEMENTATION

Lahti Science Park is seen as a platform for education, research, development, and cooperation. It is owned by the City of Lahti, LAB University of Applied Sciences and the University of Helsinki and its facilities and operating environments are suitable for hybrid work and studies. An expert from Lahti Science Park also participated in planning the Sustainability Campus Walk.

### 3.1 Pre-Walk Preparation

The preparation process started in spring 2023 by contacting relevant stakeholders who would be potentially interested in contributing their material for the walk. The idea was to recognize points which would need to be highlighted for the check points of the campus walk. Relevant stakeholders who participated were Lahti Science Park, Lahti University Campus, chief research experts of LAB, and the City of Lahti and Compass Group. As the checkpoint material was collected and places to locate them on campus were identified, the next step was to create the demo and pre-test it. 7 checkpoints were identified at different locations of campus.

The checkpoints covered such themes as: sustainability in food (hosted by the campus restaurant), sustainability in design (hosted by the Institute of Design), sustainability in mobility (hosted by Lahti City), the sustainable future of the campus area (hosted by Lahti Science Park and Lahti University Campus), circular economy projects of LAB (hosted by the chief research expert), sustainable greening (hosted by university campus management) and sustainable laboratories (hosted by the Faculty of Technology). It was recognized that the possibility to do the campus walk had to be provided in two languages: Finnish and English, so that it could be inclusive for local degree students as well as for international ones. Therefore, the continuous planning process was carried out while keeping in mind this principle.

## 3.2 Demo Version Creation

During Summer School 2023, students were given a task to use the collected materials to organize the check points, create a story line and run through the walk. Two groups of international bachelor's and master's students completed the task with note of relevant aspects, like timing (they highlighted the average time spent on each checkpoint), flexibility (the route should be done so that is can be completed starting from any checkpoint in free order) as well as the necessity for incentives (it was needed to create a reward system for making walk more engaging). One of the important issues to consider when planning routes is to remember about the physical accessibility and to mark out the elevators or special elevating devices.

## 3.3 Building the Actual Walk

During August 2023 a cooperation team for LAB and the Lahti Science Park had built an online environment on the Moodle platform as an engaging virtual course with tasks in the form of games. H5P plug-in solutions were used to create various playful activities such as memory games, crosswords, drag and drop cards etc. As a practical example H5P plugins were used such as: "Multiple Choice", "Fill the Blanks", "Drag the Words", "Image Hotspots", "Memory Game", "Find the words" and "Crossword". Additionally, the material that the quizzes were based on was created in an engaging way via Canva platform and hidden under QR codes to be located at the 7 checkpoints during the physical walk. In addition, visitors from outside the university were also considered. Together with LUT Junior University, a Google game was created using interactive features from Google Forms, where the participants could answer a short set of 12 questions, connected to all the checkpoints on campus. This was done to promote the visibility of the programme.

An incentive programme was implemented in cooperation with the Compass Group, which is a chain of campus restaurants located on the premises of the university campus. It was agreed that after students pass the checkpoints and receive an overall badge with the project logo, they could get filter coffee at 50% reduction from the restaurant. In Finnish culture, coffee is the foundation of the daily diet, therefore this incentive was chosen.

One of the crucial issues, acknowledged by the coordinating team member, was to create a clear set of instructions connecting all the dots of the walk. It was an important step, as without the instructions, the walk would lose its user-friendliness, and it could create confusion. Additionally, some promotional materials were created, consisting of a poster and a brochure, to explain the concept of the walk and invite students to participate. The idea was to present the walk during the campus fair as a part of the orientation week starting on the 1st of September 2023.

## 3.4 Start of the Walk

The walk commenced on 1st of September 2023 during the orientation week. At the campus fair, the coordinating team member was present to promote the walk and to engage the students to participate in the fair. The walk sparked curiosity among the students and the discount on coffee created extra value as well.

It is important to note that the walk aimed to be active between 1st of September until 20th of October, i.e. half the semester until the academic break. The reason for this was to let students absorb the knowledge and reduce the time pressure. Additionally, hiding the checkpoints among the campus buildings created the the curiosity to randomly learn new knowledge.

## 3.5 Results, Management and Follow-Up

The Moodle platform allowed the monitoring of progress, and the coordinating team member could follow the progress and address troubleshooting problems. The troubleshooting was mostly related to one of the tasks which did not allow a restart if it failed. Therefore, some of the students could not finish the walk due to this issue. Surprisingly, most of the tasks had no technical problems and were easy to follow. In total, nearly 50 participants were recorded who had completed the walk. 35 were from the Moodle platform and 15 were from the Google survey for visitors.

#### 3.5.1 Presentation of First Results

At the Lahti Science Day 2023, the coordinating team member presented the walk concept to showcase the example of the sustainable campus walk on the local level and describe the results. The presentation was perceived with interest among the audience, with questions asked especially about the planning process of the campus walk.

#### 3.5.2 Creation of The Course Based on The Walk

In late November 2023, the creation team of the project met with the teaching support specialist as well as degree manager to discuss the formulation of a 1 ECT course for bachelor's degree students. The content of the Moodle page was copied to the course page and reported issues from students were fixed. Additionally, the instructions were slightly modified as enrolment on the course required application via the study systems. We used a normal study registration and study progress monitoring system for this course.

#### 3.5.3 Course Status and Follow-Up

Currently, the course is still running as a hop-on/hop-off course, offering 1 ECT. On average, the course has 20 active students enrolled in the study system. This course serves as preparation for transition to on-site teaching. Currently the course is part of Sustainable Solutions, a bachelor's programme in environmental engineering, which is entirely offered online. However, it is expected that the programme will be offered on campus starting from autumn 2025. Therefore, this form of blended learning is expected to be useful for the international students as well as Finnish students, during the transition period.

### 3.6 Further Development Work

We acknowledged the fact that our Campus Walk requires development work for full inclusivity. Therefore, soon, we plan to arrange a meeting with the audial team, which is responsible for providing the podcast and other audial content to the faculties' research teams. In addition, a run-through check and analysis would be organized with a special needs teaching specialist, to ensure the inclusivity of the materials.

## 4. RESULTS AND DISCUSSIONS

### 4.1 Assessing Usefulness of Gamification Elements in ESD Learning

The tasks of the Sustainability Campus Walk were formulated as a board game (Severengiz et al. 2018), which was presented as a map of the campus with marked checkpoints, where students could find the materials. A badge (Rohan et al. 2020) was awarded after the completion of quizzes which could be exchanged for a physical overall badge to receive further discounts. Below is a quick list of the gamification elements used in the campus walk, taken from Rohan et al. (2020) research with the quick explanation how they were used:

- Quests: when connecting project acronyms with a full description in the case of circular economy projects or matching a pair in a memory game about sustainable food.
- Clues: finding answers to fill in the gaps after reading the materials in sustainable laboratories.
- Crosswords and word trivia quizzes: searching for answers in the materials to solve the word puzzles on the future of the campus area or sustainable campus greening.
- Fill in quizzes: about sustainable mobility to strengthen information recall about modes of sustainable transport.

All these elements allowed for the creation of interactive content that students could engage with while learning about various sustainability initiatives the university has been pioneering. This approach made the learning process more enjoyable and helped reinforce the importance of sustainability in everyday campus life.

By integrating gamification into the educational experience, students were able to better retain information and feel more connected to the university's commitment to sustainability.

## 4.2 Assessing Learning Outcomes

The learning outcomes of students showed that the project led to increased awareness and understanding of sustainability issues within the study environment and practical applications of sustainable practices in daily life. It also enhanced students' critical thinking and problem-solving skills regarding sustainability challenges.

Our learning outcomes showed that the project still requires full development to serve a wider audience. As noted by Severengiz et al. (2018), careful planning of time use is crucial when building gamified solutions for learning, as they take longer to complete than multiple-choice tests. To make the content more engaging, even more stakeholders must be involved to provide novel and diverse information. Additionally, having several team members can help diversify the tasks and boost creativity.

## 4.3 Discussions

The results provide valuable insights into best practices for incorporating gamification in educational settings, significantly enhancing the effectiveness of sustainability education.

To answer the first research question, the implementation of SAL with a stakeholder group brings some benefits. It helps to provide more context-rich material for the study content, as various stakeholders have different ways to address sustainability problems. Additionally, creativity in achieving an interactive gamified learning experience is provided to students, as multiple stakeholders can bring more diverse solutions to be used.

To answer the second research question, the collaborative approach can be used when the university wants to show the diversity of solutions present in various streams of science. It provides more thorough planning and provides more perspectives.

Our research provides a robust framework for future iterations of similar courses in various scientific fields, particularly in environmental engineering and sustainability, ensuring these programs are more effective and engaging. Because Moodle is widely used in educational institutions worldwide (Aida 2023) and universities have their own compulsory platforms to follow the progress of studies, this example of gamified learning can be easily adapted as a common practice in other higher education institutions.

# DISCLOSURE STATEMENT

The authors declared no conflicts of interest regarding the research, authorship, and/or publication of this article. The authors declare that they have no relevant or material financial interests that relate to the research described in this paper. This research is funded by the Horizon Europe programme. With this statement we want to ensure clarity about the funding source and confirm that there are no conflicts of interest or relevant financial ties affecting the research.

# REFERENCES

Aida, S. (2023) Impact of E-Learning Orientation, Moodle Usage, and Learning Planning on Learning Outcomes in On-Demand Lectures. *Education Sciences.* 2023, 13, 1005.

Argyris, C., Putman, R, & McLain Smith, D. (1985). Action Science. Concepts, Methods and Skills for Research and Intervention. San Francisco: Jossey-Bass Publishers.

Bakhanova, E., Garcia, J. A., Raffe, W. L., & Voinov, A. (2020). Targeting social learning and engagement: what serious games and gamification can offer to participatory modeling. *Environmental Modelling & Software*, 134, 104846.

Bryant, P. (2021). Making the most of the spaces we have: Design principles for successful hybrid and hyflex learning. Co-Design Research Group. October 8, 2021. [Online]. Available: https://diberg.blog/2021/10/08/making-the-most-of-the-spaces-we-have-design-principles-for-successful-hybrid-and-hyflex-learning/

Bryant, P. (2023) Student experience and digital storytelling: Integrating the authentic interaction of students work, life, play and learning into the co-design of university teaching practices. *Education and Information Technologies*, 28(11), 14051-14069.

Carman, J. M. (2002). Blended learning design: Five key ingredients. *KnowledgeNet.* [Online]. Available http://facilitateadultlearning.pbworks.com/f/Blended_20Learning_20Design_1028.pdf

Dieleman, H. & Huisingh, D. (2006). Games by Which to Learn and Teach about Sustainable Development: Exploring the Relevance of Games and Experiential Learning for Sustainability. *Journal of cleaner production* 14.9 (2006): 837–847.

Fischer, K. & Isenmann, R. (2023). *Education for Sustainability at Distance and Online Learning Universities: Methodologies and Good Practices for Educating Sustainability Experts and Leaders of the Future.* In: Leal Filho, W., Lange Salvia, A., Pallant, E., Choate, B., Pearce, K. (eds) Educating the Sustainability Leaders of the Future. World Sustainability Series. Springer, Cham.

Hueske, A.-K., Aggestam Pontoppidan, C. and Iosif-Lazar, L.-C. (2022). Sustainable development in higher education in Nordic countries: exploring E-Learning mechanisms and SDG coverage in MOOCs". *International Journal of Sustainability in Higher Education*, Vol. 23 No. 1, pp. 196-211. https://doi.org/10.1108/IJSHE-07-2020-0276

Kananen, J. (2014). Laadullinen tutkimus opinnäytetyönä. Miten kirjoitan kvalitatiiviseopinnäytetyön vaihe vaiheelta. Jyväskylän ammattikorkeakoulun julkaisuja. Suomen Yliopistopaino. Oy – Juvenes Print, 2014.

Lozano, R. (2010). Diffusion of sustainable development in universities' curricula: an empirical example from Cardiff University, *Journal of Cleaner Production*, Vol. 18, pp. 637-644

Ololube, N. P., Umunadi, K. E., & Kpolovie, P. J. (2015). Barriers to Blended Teaching and Learning in Sub-Saharan Africa: Challenges for the next decade and beyond Curriculum Design and Classroom Management: Concepts, Methodologies, Tools, and Applications (pp. 846-861): IGI Global.

Pretorius, R. W. et al. (2021). Using Real-World Engagements for Sustainability Learning in ODeL in the Global South: Challenges and Opportunities. *International Journal of Sustainability in Higher Education* 22.6 (2021): 1316–1335.

Rohan, R., Pal, D., & Funilkul, S. (2020). Mapping Gaming Elements with Gamification Categories: Immersion, Achievement, and Social in a MOOC Setting. 2020 14th International Conference on Innovations in Information Technology (IIT), 63–68. https://doi.org/10.1109/IIT50501.2020.9299047

Saarinen, H. (2022). Toimintatutkimuksella ratkaisu käytännön ongelmaan. Väitöskuiskaaja-blogi / Thesis Whisperer blog. https://blogs.uwasa.fi/thesis/toimintatutkimuksella-ratkaisu-kaytannon-ongelmaan/

Severengiz, M., Roeder, I., Schindler, K., & Seliger, G. (2018). Influence of gaming elements on summative assessment in engineering education for sustainable manufacturing. *Procedia Manufacturing*, 21, 429-437.

Sivapalan, S. (2017). Sustainability, Blended Learning and the Undergraduate Communication Skills Classroom: Negotiating Engineering Undergraduates' Expectations and Perceptions. *On the Horizon* 25.1 (2017): 7–23.

Suojanen, U. (2014). Toimintatutkimus ammatillisen kehittymisen välineenä. Metodix –Metoditietämystä kaikille. https://metodix.fi/2014/05/19/suojanen-toimintatutkimus/

Tuomala, A-M. (2022). Virtual Studies in Learning of Solving Sustainability Challenges. *IADIS International Journal.* Vol. 21, No. 1, pp. 32-45. [Online]. Available: https://www.iadisportal.org/ijwi/papers/202321103.pdf

Wetzler, J.R. (2013): A case study of a "collaborative organizational innovation" intervention, combining action research and design-thinking methodologies. Teachers College, Columbia University ProQuest Dissertations Publishing, 2013.

## APPENDIX

# Map of the Campus Walk

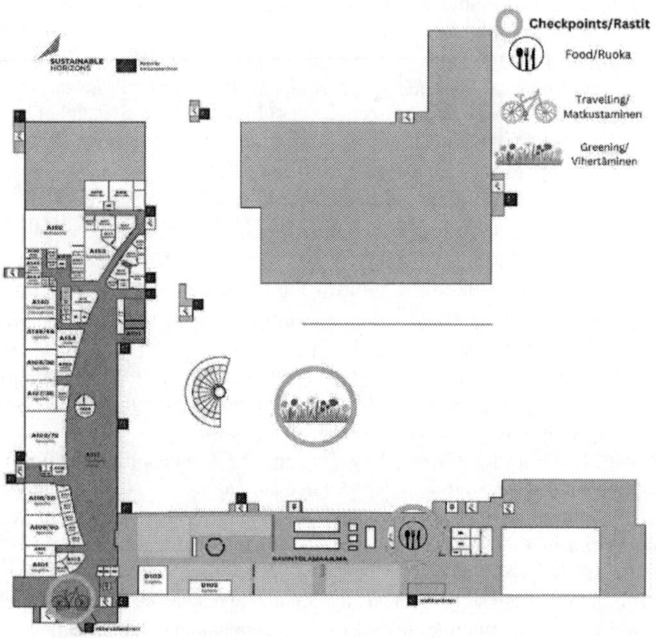


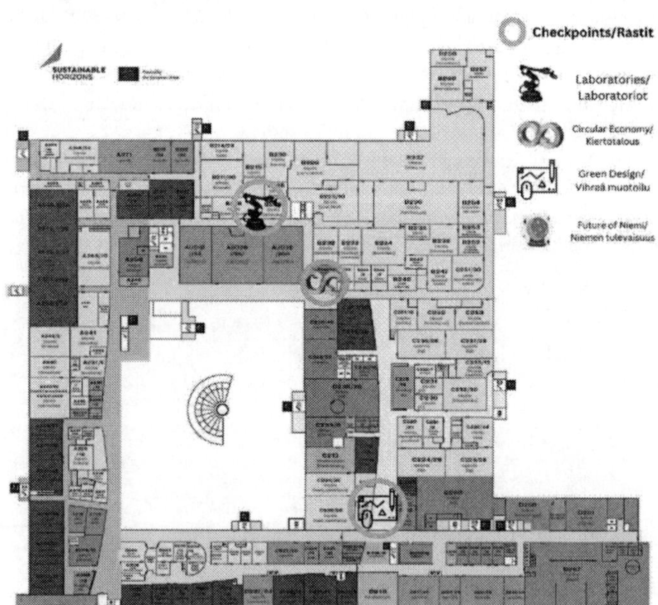

# Short Papers

# THE IMPACT OF THE MAKER MOVEMENT ON ENTREPRENEURSHIP FROM A RESOURCE PERSPECTIVE

Tsai-Hsin Chen*, Tsz-Yu Huang* and Chun-Wang Wei†
*Department of Healthcare Administration and Medical Informatics, Kaohsiung Medical University, Taiwan*

**ABSTRACT**

The maker movement provides opportunities for students to develop diverse interests, self-identity, and domain knowledge. Entrepreneurship is a mindset and a set of behaviors rather than a personality trait. Developing entrepreneurial skills greatly benefits students in bridging the gap between the workplace and their career advancement. This study aims to investigate the relationship between the maker movement and entrepreneurship. SmartPLS was employed to validate the proposed causal model. A total of 220 students participated in this survey research. The study's findings revealed that social, knowledge, and technology resources shape the features of the maker movement, which subsequently impact project-based outcomes. This study contributes to our understanding of students' involvement in the maker movement and its impact on entrepreneurship development.

**KEYWORDS**

Entrepreneurship, Knowledge Creation and Sharing, Maker Education, Maker Movement, Resource Perspective

## 1. INTRODUCTION

The maker movement refers to the growing trend of individuals embracing their creativity and becoming active creators, driven by the philosophy of experiential learning through hands-on building (Tabarés & Boni, 2023). Participation in the maker movement is driven by intrinsic interests and hobbies, with its real added value lying within the learning and enjoyment derived from "making." Makers are often active in academic circles or research institutions, and the success and dissemination of the maker movement might stimulate or strengthen traditional scientific approaches (Leavy et al., 2023). This study is grounded in the resource perspective, with the core concept that resource heterogeneity is the main driver of performance differences. Resources encompass all tangible and intangible assets possessed, including physical assets (such as facilities and equipment), knowledge assets (like patents and technology), and human resources (such as employee skills and experience). Efficient resource utilization fosters innovation and value creation (Pereira & Bamel, 2021).

Maker education is profoundly influenced by pragmatism and constructivism. Dr. Dewey believed genuine learning extends beyond acquiring knowledge in the classroom, emphasizing the importance of learning through engagement in complex social practices. This approach enables students to acquire more valuable knowledge through meaningful tasks and activities. From a constructivist perspective, students are seen as active knowledge constructors, where learning is a process of sharing and constructing knowledge. However, it also underscores the significance of constructing one's knowledge base through hands-on creation. The fundamental idea of constructivism revolves around student-centered teaching, shifting students from passive recipients of external stimuli to active constructors of knowledge. It emphasizes that with adequate resources to support student learning, they can create more authentic learning situations (Niiranen, 2021).

*Contributed equally
†Corresponding author: cwwei@kmu.edu.tw

Scholars have proposed a hypothetical model linking the three dimensions (social, knowledge, and technology) of the maker movement with entrepreneurship. However, validation of this model has not yet been conducted (Browder et al., 2019). The objective of this study is to construct and validate a theoretical model for the maker movement and entrepreneurship from a resource perspective. Miller's perspective on entrepreneurship, emphasizing innovativeness, risk-taking, and proactivity, stands as a mainstream viewpoint, identifying these three elements as the fundamental components of entrepreneurship (Miller, 1983). Innovativeness can be regarded as the critical factor in entrepreneurship. When traditional organizational and cultural management face bottlenecks, innovative ideas and behaviors aid in overcoming the challenges of business operations. Risk-taking refers to the proactive pursuit and acceptance of risks when seeking opportunities in uncertain environments, a decision-making behavior driven by a preference for risk. Entrepreneurs inclined towards low-risk-taking propensity tend to avoid risks and make cautious decisions, while those with a high-risk propensity tend to make quicker decisions to seize opportunities. Proactivity is a forward-looking response pattern wherein entrepreneurs, upon predicting future opportunities and market demands, take proactive actions to capture market opportunities or establish a leading position. Proactive entrepreneurs tend to gain a better competitive edge by continuously introducing innovative products, services, and management techniques into their companies to secure an absolute competitive advantage (Miller, 1983).

Although maker education doesn't necessarily direct students towards an entrepreneurial path, equipping students with an entrepreneurial mindset undoubtedly aids in creating a competitive workforce and adapting to the diverse environments of the workplace. Therefore, this study employed Miller's perspective as a measure of entrepreneurship.

## 2. METHODOLOGY

The primary resource dimensions facilitating the maker movement include social exchange, technology resources, and knowledge creation and sharing. Makers engage in social exchange with other participants, interact and collaborate within physical or virtual spaces dedicated to knowledge creation and sharing, and collaborate using technology and resources for joint creations (Browder et al., 2019). From the social exchange perspective, students' expertise, beliefs, subjective norms, values, and motivations influence their involvement in the maker movement. The technology resources dimension pertains to the hardware and software tools used; different creations require varying software and hardware needs, and access to suitable tools can impact the smooth completion of creations. Hardware equipment might encompass traditional woodworking and metalworking tools, alongside newer advanced programmable control apparatus. Software tools usually complement hardware to facilitate product design and hardware operation. The adequacy of technological resources provided by schools can influence students' willingness to engage in creation.

The knowledge creation and sharing space are crucial dimensions of the maker movement, fostering interaction, resource sharing, and experience exchange among participants. This space transcends departmental and professional boundaries, existing as physical maker spaces or virtual maker platforms. When participants from diverse backgrounds and expertise converge in one space, opportunities for knowledge sharing emerge, sparking creative inspirations and collaborations to develop products. Through these resources, makers develop communities and networks, engage in professional learning, and contribute to product design. As students actively participate in the maker movement, it reinforces their entrepreneurship in innovativeness, risk-taking, and proactivity. Based on the research objective and framework (see Figure 1), this study proposes the following hypotheses:

H1: Social Exchange has a significantly positive impact on Knowledge Creation and Sharing.
H2: Technological Resources have a significantly positive impact on Knowledge Creation and Sharing.
H3a: Knowledge Creation and Sharing have a significantly positive impact on Social Networks.
H3b: Knowledge Creation and Sharing have a significantly positive impact on Professional Learning.
H3c: Knowledge Creation and Sharing have a significantly positive impact on Product Design.
H4: Community Networks have a significantly positive impact on Entrepreneurship.
H5: Professional Learning has a significantly positive impact on Entrepreneurship.
H6: Product Design has a significantly positive impact on Entrepreneurship.

The subjects of this study are students aged eighteen or above who have previously participated in innovation and entrepreneurship competitions. A questionnaire survey method was employed to measure nine constructs. This questionnaire was designed based on the conceptual models of the maker movement by scholars Browder, Aldrich, and Bradley (2019), and the entrepreneurship concept by Miller (1983). Based on the comments of three domain experts, the questionnaire items were revised, and content validity was confirmed. The questionnaire (31 items) was distributed using an online survey method, disseminating the questionnaire link through the campus platform or the maker space at the institution. In this study, respondents' self-assessment of maker activities was measured using a Likert seven-point scale to understand maker resources' influence on entrepreneurship.

## 3. RESULTS AND DISCUSSION

A total of 220 valid samples were collected and analyzed. According to the descriptive statistics, the male participants slightly outnumbered the female participants, accounting for 58% and 42% of the total, respectively. The majority consists of seniors, accounting for 30%, followed by juniors (27.5%) and graduate-level students (27.5%), with the last category being underclassmen (24.5%). Higher-level students tend to amass more knowledge, experience, and resources, particularly in specialized fields and have established certain social networks, which may enhance their confidence and willingness to engage in entrepreneurial competitions.

The factor loadings for each item are all greater than 0.8, and the variance inflation factor (VIF) for each item is below 3, indicating the absence of collinearity issues among the indicators. The Cronbach's alpha values for each construct exceed 0.7, and the composite reliability (CR) values are all above 0.8, demonstrating strong internal consistency among the constructs. The average variance explained (AVE) for each construct ranges from 0.689 to 0.767. All AVE values for the constructs exceed the threshold of 0.5, indicating that the questionnaire possesses convergent validity. The square root of the AVE is greater than the inter-construct correlations, indicating discriminant validity among the constructs (Hair et al., 2017).

In the path analysis of the research model (see Figure 1), each path was significant, indicating support for all six hypotheses. Entrepreneurship is not only directly influenced by social networks, professional learning, and product design but also indirectly influenced by social exchange, technological resources, and knowledge creation and sharing.

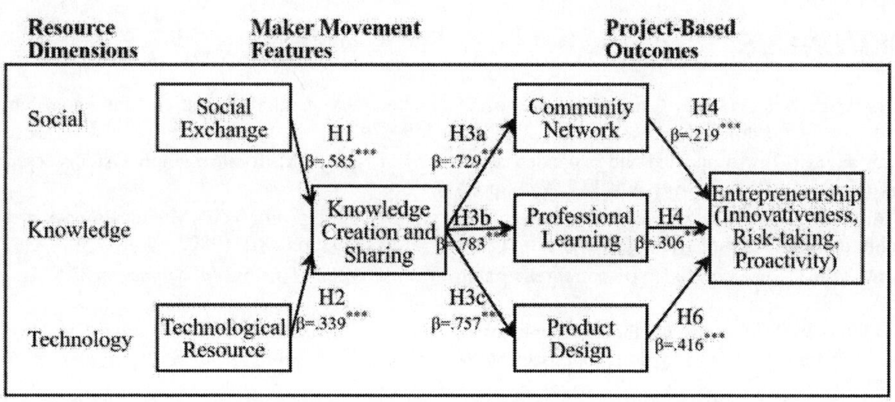


Note: *p < 0.05, **p < 0.01, *** p < 0.001.

Figure 1. Research framework and path analysis

Regarding the explanatory power of the model, the $R^2$ values for knowledge creation and sharing, social networks, professional learning, product design, and entrepreneurship were all above 0.5, indicating that the research model has good explanatory power. Furthermore, all the concepts in the research model have prediction effect values ($q^2$) greater than 0.35, indicating that the model exhibits good predictive ability (Hair et al., 2017). Therefore, the results of this study can provide reliable and effective predictive capability for future behaviors, events, or outcomes related to entrepreneurship explained by entrepreneurial resources. These findings can be applied to guide decision-making and implementation in entrepreneurial education.

## 4. CONCLUSION

Browder, Aldrich, and Bradley proposed a conceptual model of the maker movement and entrepreneurship. They proposed that resource dimensions shape the characteristics of the maker movement, influencing both non-commercial and commercial project-based outcomes (Browder et al., 2019). However, their conceptual model aimed at explaining the relationship between the maker movement and entrepreneurship without empirical validation. This study's model verification revealed that resource dimensions significantly impact the development of the maker movement. These dimensions provide the necessary foundation and support for innovation, practice, and collaboration, fostering a conducive environment for innovation and entrepreneurship.

The core concept of the resource perspective highlights that resource heterogeneity is a primary factor influencing performance differences. Thus, possessing resource diversity and innovative management can favor survival and development (Pereira & Bamel, 2021). Developing the maker movement based on the resource perspective enables educational institutions to leverage and integrate their internal resources and capabilities. This, in turn, encourages greater participation in the maker movement and affects students' cultivation of entrepreneurship. Through causal model analysis, this study examined students' engagement in the maker movement and entrepreneurship contests to understand student traits and needs, providing a basis for enhancing maker spaces and organizing maker movement initiatives within schools. Learning environments influence students' attitudes and perspectives, subsequently shaping their perceptions of concepts.

This study constructed a theoretical model beneficial for students' engagement in the maker movement and the development of entrepreneurship. It offers educational institutions insights into their supporting resources and understanding of students' characteristics, enabling personalized assistance to nurture industry professionals with entrepreneurship and competence. The primary objective of this study is to analyze students' perspectives on maker education. In the future, scholars can build upon this research foundation by employing artificial intelligence methods to analyze the actual resources invested by schools in maker education and its impact on the development of entrepreneurship among students.

## REFERENCES

Browder, R. E. et al., (2019). The emergence of the maker movement: Implications for entrepreneurship research. *In Journal of Business Venturing*, Vol. 34, No. 3, pp. 459-476.

Hair, J. et al., (2017). An updated and expanded assessment of PLS-SEM in information systems research. *In Industrial Management & Data Systems*, Vol 117, No 3, pp. 442-458.

Leavy, A. et al., (2023). The prevalence and use of emerging technologies in STEAM education: A systematic review of the literature. *In Journal of Computer Assisted Learning*, Vol. 39, pp. 1061-1082.

Miller, D., (1983). The correlates of entrepreneurship in three types of firms. *In Management Science*, Vol 29, No 7, pp. 770-791.

Niiranen, S., (2021). Supporting the development of students' technological understanding in craft and technology education via the learning-by-doing approach. *In International Journal of Technology and Design Education*, Vol. 31, pp. 81-93.

Pereira, V. and Bamel, U., (2021). Extending the resource and knowledge based view: A critical analysis into its theoretical evolution and future research directions. *In Journal of Business Research*, Vol 132, pp. 557-570.

Rouse, R. and Rouse, A. G., (2022). Taking the maker movement to school: A systematic review of preK-12 school-based makerspace research. *In Educational Research Review*, Vol. 35, 100413.

Tabarés, R. and Boni, A., (2023). Maker culture and its potential for STEM education. *In International Journal of Technology and Design Education*, Vol. 33, No. 1, pp. 241-260.

# THE PDSA APPROACH TO UNDERSTANDING SELF-REGULATED LEARNING STRATEGIES

Yassine Safsouf
*Laboratory LAMIGEP, EMSI Marrakech, Morocco*

**ABSTRACT**

Considering individualized learning strategies and fostering learners' self-regulated learning can enhance academic success and retention at all levels. While all learners exhibit some degree of self-regulation, not all of them are adept self-regulated learners. Moreover, these proficient self-regulated learners do not universally employ the same strategies. This paper introduces a novel approach to comprehending and summarizing the relationship between self-regulated learning strategies (SRLS), the phases and steps of self-regulated learning, and the empirical support for this potential correlation.

**KEYWORDS**

Self-Regulated Learning, Self-Regulated Learning Strategies, Self-Regulated Learning Models

## 1. INTRODUCTION

Self-regulated learning (SRL) is linked to how individuals control their emotions, cognitions, behaviors, and aspects of the context during a learning experience. It encompasses the processes through which individuals activate and maintain cognitions, affects, and behaviors systematically oriented toward the achievement of a goal. Recent studies have consistently demonstrated a robust connection between SRL and learners' educational success (Matzat and Vrieling, 2016), (Safsouf et al., 2020).

Although self-regulated learning skills are essential tools for lifelong learning, they are often not explicitly learned. Thus, many learners lack independence, organization, motivation, persistence, and a positive sense of wellness. It is therefore necessary to have an in-depth knowledge of the most important strategies to follow.

In this regard, we present a new approach in this paper through an emergent model to elucidate the phases, strategies, and steps of self-regulated learning. The rest of this paper is organized as follows: Section II delves into the self-regulated learning process, its strategies, and well-known models. Section III offers a perspective on the proposed model, and the conclusion in Section IV also outlines future work.

## 2. LITERATURE REVIEW

### 2.1 Self-Regulated Learning (SRL)

Self-regulation involves the control of one's mental processes, often employed to manage the mind's internal functions, states, and processes, as well as actions. In the field of learning, whether offline (traditional learning) or online (e-learning), a learner is deemed self-regulated when utilizing metacognitive, motivational, and behavioral processes to attain specific learning goals and performance outcomes (Zimmerman and Moylan, 2009). These processes encompass activities such as goal setting, metacognitive monitoring, self-evaluation, and seeking assistance. Metacognition enables self-regulation by allowing the learner to adapt the context of task completion based on cognitive or affective experiences and prior knowledge. Consequently, a self-regulated learner engages in metacognitive processes by reflectively thinking about and evaluating the cognitive strategies employed.

This involves selecting and employing strategies for learning tasks, assessing their effectiveness, monitoring progress, and adjusting strategies as needed.

## 2.2 Self-Regulated Learning Strategies (SRLS)

The self-regulatory process, often viewed as motivational, guides learners in regulating their behaviors, emotions, and thoughts, enabling full immersion in their developmental and learning experiences, while disregarding distractions or setbacks (H. et al., 2014). A self-regulated learner actively employs these processes through suitable strategies, also known as self-regulated learning strategies (SRLS) (Karapanos et al., 2009; Perry and Rahim, 2011). Individuals utilizing these strategies are recognized as being involved in what Zimmerman and Schunk refer to as self-regulated learning (SRL). These SRLS typically fall into three categories: personal, behavioral, and environmental:

- **Personal strategies** pertain to how the learner organizes and interprets information. These include the organization and transformation of information (summarizing, highlighting, use of diagrams, charts, etc.), goal setting and planning, record keeping and control, and rehearsing and memorizing.
- **Behavioral strategies** encompass actions that learners can apply, including self-assessment (checking quality or progress), self-reinforcement, and self-monitoring.
- **Environmental strategies** involve seeking help when needed and structuring the physical study environment. This includes searching for information (in books or on the internet), selecting a study environment (choosing secluded areas, eliminating or minimizing distractions), and seeking social support (from peers, teachers, or other adults).

All learners possess inherent self-regulation abilities to some extent, yet not all of them excel as strong self-regulated learners. Moreover, self-regulated learners may differ in their choice of strategies. For instance, one learner may opt for self-assessment and tracking of course progress, while another may choose not to engage in such practices. However, not all strategies contribute effectively to the processes in which self-regulated learners engage.

## 2.3 Self-Regulated Learning Models

Zimmerman defines SRL as a dynamic process in which the individual plans, monitors, and evaluates their learning, applying appropriate strategies to achieve goals. It comprises activities that individuals undertake proactively on their own (Zimmerman, 2013).

A recently published article, presents a review of the six most popular models of self-regulated learning (Panadero, 2017), These include Zimmerman's cyclical phases model of self-regulation (Zimmerman and Campillo, 2003), Boekaerts' dual process model of self-regulation (Boekaerts and Cascallar, 2006), Winne and Hadwin's model of exploring self-regulation from a metacognitive perspective (Winne, 2011), Pintrich's model based on the role of motivation in self-regulation (Pintrich and De Groot, 2003), Efklides' metacognitive and affective model of self-regulated learning (Efklides, 2011), and Hadwin, Järvelä, and Miller's model of self-regulation in the context of collaborative learning (Järvelä et al., 2013).

The article concludes that most of these models are composed of three essential phases, the forethought phase, the performance phase, and the self-reflection phase:

- The **forethought phase** of self-regulated learning encompasses the processes through which learners initiate the formulation of plans to accomplish tasks, ultimately leading to goal achievement. This may involve setting short-term goals as proximal pathways to longer-term objectives, utilizing strategies to guide cognition, control affect, and direct actions.
- The **performance phase** delineates the processes that learners employ to execute a task, encompassing its actual performance while monitoring and controlling progress. This involves selecting specific strategies to accomplish the task, utilizing mental images, managing time and monitoring thought processes while working towards goal attainment.
- In the **self-reflection phase**, the learner evaluates oneself, reflects on progress toward the task or goal, and reacts and adapts for future performance. Learners can utilize these insights to make informed decisions about completing the current task or embarking on a new one.

# 3. PROPOSED NEW SRLS MODEL

## 3.1 Deming Wheel and PDSA Cycle

The Deming wheel is a graphical representation of the PDCA (Plan-Do-Check-Act) quality management method (Deming, 1993). This method was originally developed by the statistician Walter A. Shewhart, but it was William Edwards Deming who introduced it to Japanese industrialists in the 1950s under the name of the Shewhart cycle. During the 1980s, Deming partially modified this concept to PDSA (Plan-Do-Study-Act) (Moen, 2009). In The New Economics, Deming wrote that the PDSA cycle is a flow diagram that facilitates learning and improving a product or process (McNary and Deane., n.d.).

This cycle aims to transform an idea into action and action into knowledge. It consists of four steps, each leading to the next, and aims to establish a virtuous circle. The first step in this cycle is **planning**. It involves identifying an objective, preparing, formulating a theory, and selecting the parameters necessary for the success of the implemented plan. This is followed by the **Do step**, during which the elements of the plan are implemented to proceed with the construction, development, and execution of the work.

These two steps are followed by the **Study step**, where the results are monitored, analyzed to check the validity of the plan, and to identify problems and areas for improvement. Monitoring tools such as performance indicators and dashboards are often used for this purpose. Finally, the **Act step** closes the cycle by acting, which involves implementing the studied changes, adjusting the objective, or restarting the cycle at the first step to modify methods or reformulate a theory based on the knowledge acquired during the previous cycles. These steps can be repeated over and over as part of a continuous improvement learning cycle.

## 3.2 PDSA Cycle of SRLS

Based on our research and proposed model, we suggest an emergent model for integrating SRLS and SRL phases into PDSA to achieve a PDSA cycle of SRLS (see Figure 1).

Figure 1. PDSA cycle of self-regulated learning strategies

We focus on the link between SRL strategies and phases, as well as the components of each step of the PDSA cycle. We propose representing the personal strategy by the forethought phase of Zimmerman and Campillo's model, as this strategy generally involves how the learner organizes and interprets information and begins to formulate plans to accomplish tasks. In the context of the PDSA cycle, this represents the first two steps: PLAN and DO.

The environmental strategy is associated with the performance phase, during which learners manage their time appropriately to accomplish tasks by avoiding distractions, modifying their physical environment to be more conducive to completing tasks, seeking help from others who are more knowledgeable, establishing clear positive or negative consequences for themselves, monitoring their thought processes as they work toward their goals, and engaging in an appropriate amount of self-recording. This phase corresponds to the STUDY step presented in Figure 1.

Finally, the behavioral strategies are described in the self-reflection phase, during which learners self-assess, note what they have learned (whether the implementation worked or not), check the quality or progress of the tasks accomplished, react, and adapt for future performance (if it did not work, what they could do in the next cycle to remedy it). This last phase represents the ACT step where the learner seeks to analyze what can still be improved, compared to the expected result.

The PDSA cycle of self-regulated learning strategies is continuous, which means that if a new action is taken or a previous action requires adjustment in the ACT step, the entire process must be repeated.

## 4. CONCLUSION AND FUTURE WORK

In this work, we have introduced an approach to represent self-regulated learning strategies and to comprehend their components, aligning them with the phases and processes of the self-regulation theory. To the best of our knowledge, the emerging model we propose is unique in its category. It offers an opportunity to explore the phases and steps necessary to establish a learning environment conducive to the implementation of self-regulated learning strategies by learners.

In our future work, we intend to validate this new approach through a study conducted in the context of higher education in Morocco.

## REFERENCES


Boekaerts, M., Cascallar, E., (2006). How far have we moved toward the integration of theory and practice in self-regulation? Educ. Psychol. Rev. 18, 199–210.

Deming, W.. E., (1993). Biographie of W . Edwards Deming. Knowl. Creat. Diffus. Util. 1–18.

Efklides, A., (2011). Interactions of metacognition with motivation and affect in self-regulated learning: The MASRL model. Educ. Psychol. 46, 6–25.

H., D.S., Meece, J.L., Pintrich, P.R., (2014). Motivation in Education: Theory, Research, and Applications.

Järvelä, S., Järvenoja, H., Malmberg, J., Hadwin, A.F., (2013). Part 1: Underexplored Contexts and Populations in Self-Regulated Learning. J. Cogn. Educ. Psychol. 12, 267–286.

Karapanos, E., Zimmerman, J., Forlizzi, J., Martens, J.B., (2009). User experience over time: An initial framework. Conf. Hum. Factors Comput. Syst. - Proc. 729–738.

Matzat, U., Vrieling, E.M., (2016). Self-regulated learning and social media – a 'natural alliance'? Evidence on students' self-regulation of learning, social media use, and student–teacher relationship. Learn. Media Technol. 41, 73–99.

McNary, Deane., L., n.d. The Deming management theory: A managerial leadership profile for the new economic age.

Moen, R., (2009). Foundation and History of the PDSA Cycle. Assoc. Process Improv. 2–10.

Panadero, E., (2017). A Review of Self-regulated Learning: Six Models and Four Directions for Research. Front. Psychol. 8, 422.

Perry, N.E., Rahim, A., (2011). Studying self-regulated learning in classrooms. Handb. self-regulation Learn. Perform. 122–136.

Pintrich, P.R., De Groot, E. V., (2003). A Motivational Science Perspective on the Role of Student Motivation in Learning and Teaching Contexts. J. Educ. Psychol. 95, 667–686.

Safsouf, Y., Mansouri, K., Poirier, F., (2020). An analysis to understand the online learners' success in public higher education in morocco. J. Inf. Technol. Educ. Res. 19, 87–112.

Winne, P.H., (2011). A Cognitive and Metacognitive Analysis of Self-Regulated Learning. Handb. Self-Regulation Learn. Perform.

Zimmerman, B.J., (2013). From Cognitive Modeling to Self-Regulation: A Social Cognitive Career Path. Educ. Psychol. 48, 135–147.

Zimmerman, B.J., Campillo, M., (2003). Motivating Self-Regulated Problem Solvers 233–262.

Zimmerman, B.J., Moylan, A.R., (2009). Self-Regulation, Where Metacognition and Motivation Intersect. Handb. METACOGNITION Educ. 299–315.

# SHIFTING PERCEPTIONS OF AI: AN INTEGRATED LEARNING PROGRAM IN THE MIDDLE SCHOOL SETTING

Won Jung Kim[1] and Sangjin An[2]
[1]*Santa Clara University, Santa Clara, CA, USA*
[2]*Dongshin Middle School, Busan, South Korea*

**ABSTRACT**

This qualitative study examines how three Korean middle school teachers - teaching Informatics, English, and Arts respectively - developed their perceptions on Artificial Intelligence (AI) through designing and implementing an integrative project-based learning program that utilized AI software. This program aimed to enhance students' environmental awareness, AI literacy as well as subject-specific learning goals.

The teachers engaged in a community-driven approach, where the informatics teacher introduced the English and Art teachers to basic AI concepts and instructional uses during co-design sessions. The 12-hour program began with Informatics classes focused on students exploring AI concepts, principles and practical problem-solving applications. In English classes, the teacher facilitated students' use of generative AI to develop and revise text for infographic posters raising environmental awareness. Art classes had students use AI software to transform drawings and sculptures of endangered species, gaining insights into generative AI artwork creation.

Data collected from pre-program design meetings, in-between reflection meetings, and post-program interviews were analyzed using constant comparative analysis. This revealed three steps: 1) Curricular analysis and cross-subject topic identification; 2) Program design and implementation; 3) Evaluation of student learning outcomes and teacher reflection. Teacher perception shifts included: abstract understandings became more sophisticated with knowledge of how AI operates and criteria for selecting software; students and teachers voluntarily iterated for better artifacts; teachers gained instructional confidence and experiential knowledge integrating AI software into their subjects. The results offer insights into AI education through interdisciplinary project-based learning program design involving AI.

**KEYWORDS**

Artificial Intelligence (AI), Project-Based Learning, Interdisciplinary/Integrative, Teacher Perceptions

## 1. INTRODUCTION

Digital utilization in education has gained massive momentum post-COVID-19, creating urgency for Informatics education to adapt students to societal changes driven by AI integration (Schwab, 2017). AI, or artificial intelligence, refers to the simulation of human intelligence in machines that are programmed to think and learn like humans. Generative AI like ChatGPT has proven applicable across societal contexts including education (Montenegro-Rueda et al., 2023). Countries are promoting widespread AI adoption in education (Yue et al., 2022), with Korea enhancing digital literacy and integrating AI across revised national curricula (Korea Ministry of Education, 2020).

However, determining how, what aspects, and to what extent AI can be taught across subjects is challenging (Yue et al., 2022). Ensuring in-depth student learning on AI concepts, principles, applications and problem-solving processes is difficult even in Informatics. Non-Informatics teachers may lack concrete AI understanding to effectively integrate it. Integrated cross-subject AI education approaches risk teachers lacking requisite AI knowledge for meaningful instruction.

To address these challenges, an Informatics teacher collaborated with English and Art teachers to design and implement a project-based program with subject-aligned activities enhancing environmental awareness while commonly using AI. This study explores how this integrative AI education approach shifted teachers' perceptions by asking:

1. How did the teachers conduct their integrated AI education approach across subjects?
2. How did the program design and implementation shift teachers' perceptions on AI education?

## 2. BACKGROUND: ARTIFICIAL INTELLIGENCE EDUCATION

AI education is largely described in two ways: Education about AI as a disciplinary subject like computer science, and Education with AI as a learning tool (ACM & IEEE-CS, 2020; Chen et al., 2023). Education about AI involves understanding AI components, functioning, problem-solving stages, and outcome evaluation, typically conducted in computer science/informatics. Education with AI utilizes AI software as an educational tool to enhance learning, focusing more on how AI aids problem-solving rather than deep AI understanding across subjects.

This study conceptualizes an integrative approach to AI education building on these two approaches, where students explore AI as an academic discipline enabling technological advances, while utilizing AI tools across subjects (Van Brummelen & Lin, 2020). However, little is documented on how teachers from different subjects, besides informatics teachers, voluntarily co-design and implement such an integrated AI education program when fairly new to AI as a discipline and technology.

Our study addresses this gap by examining a co-designed AI education program involving informatics, English, and art teachers, and its impact on teachers' perceptions of AI education. This integrated approach enables students to learn about AI explicitly while utilizing AI tools across subjects.

## 3. METHODOLOGY

This study employed Community-Driven Design Research (CDR) to develop an integrative unit encompassing Informatics, English, and Arts (Wilson, 2018). CDR involves iterative design processes to develop knowledge aimed at enhancing educational practices, bringing immediate shifts toward transformative futures.

The study took place at a South Korean middle school participating in the government-supported AI Leading School project. Teachers sought meaningful AI learning opportunities across subjects, recognizing students' interest and competence in AI from the previous year. The informatics teacher collaborated with English and art teachers who were willing to integrate AI into their subjects. CDR fits this context where the need and willingness for a new instructional approach evolved.

The informatics teacher (20 years' experience, doctoral degree) has extensive experience in educational materials development and national curriculum design. The English teacher (15 years' experience) is proficient in educational software and learning management systems. The art teacher (25 years' experience) is highly specialized but faces challenges with digital devices. These 8th-grade teachers co-designed and implemented the integrative program.

Primary data includes program design meetings, post-class reflection meetings, and post-program individual interviews. Supplementary data includes teachers' instructional materials and student project artifacts.

Constant comparative analysis was conducted using Atlas.ti software. The second author performed initial coding, refined by the two research questions and the conceptual framework of AI education as discipline and technology. Open codes were categorized to describe the CDR process (RQ1) and features of teacher AI perceptions (RQ2). The researchers revisited primary and supplementary data to confirm and triangulate the analysis across data formats, resulting in the two findings.

## 4. FINDINGS

The findings revealed three distinct steps in the community-driven design research (CDR) process undertaken by the teachers, along with three corresponding patterns of shifting perceptions towards AI that emerged throughout this iterative journey.

## 4.1 Teacher Perceptions Revealed During CDR Step 1

Initial discussions exposed teachers' limited, vague understanding of AI, equating it with applications like ChatGPT - a surface-level awareness conflating AI with encountered tools. When asked about using AI in their subjects, teachers inquired about specific AI apps or programs to utilize, focusing on pragmatic tool adoption rather than grasping underlying AI concepts and capabilities. Their questions implied viewing AI akin to other software, concentrating solely on practical usage without deeper comprehension of its functioning.

To bridge this gap, the informatics teacher's workshop delved into AI's core concepts, problem-solving approaches, hands-on AI software experience, and evaluating AI outputs. Teachers gained insights into AI systems' inherent characteristics, first-hand experience with different AI tools' problem-solving processes, and awareness of AI's potential strengths and limitations in education.

Consequently, teachers posed more nuanced questions reflecting a perception shift (Pattern 1). They inquired about refining inputs for desired AI outputs, acknowledged training data's impact on output variability, and grappled with criteria for selecting appropriate outputs when multiple options existed. This questioning signaled a transition from initial abstract notions to more substantively recognizing AI as a powerful educational tool warranting technological engagement.

However, while crucial, researchers emphasized deeper immersion and continued professional development would foster truly sophisticated AI understanding among teachers. The workshop alone, though dispelling misconceptions and sparking curiosity, could not single-handedly equip educators with comprehensive knowledge required for purposeful cross-curricular AI integration.

## 4.2 Teacher Perceptions Revealed During CDR Step 2

During program design and implementation, the informatics teacher spearheaded an "Education about AI" module providing a comprehensive overview of AI concepts, practical problem-solving applications, and enabling students to articulate and apply AI's problem-solving processes across domains.

Concurrently, the English teacher needed an AI tool for students to independently identify and correct writing errors, while the art teacher sought an AI solution to digitally represent and transform handcrafted student artwork into various styles.

As they collaboratively planned AI integration, teachers expressed apprehensions about the substantial time required for AI application proficiency, doubts about whether students could comprehend processes for achieving desired outcomes, and grappled with the extent individual subjects could comprehensively cover AI education, recognizing a potential need for a more centralized informatics foundation.

These dialogues marked an awareness expansion beyond initial fixation on AI's superficial applications, appreciating the necessity of imparting conceptual understanding of AI's underlying principles and problem-solving methodologies to students (Perception Shift Pattern 1). This realization emanated from attempts to connect AI education to tangible classroom implementation while evaluating potential educational impacts.

## 4.3 Teacher Perceptions Revealed During CDR Step 3

The third phase centered on evaluating student learning outcomes and facilitating teacher reflection. Initially, students exhibited constrained AI perceptions, associating it solely with specific apps like ChatGPT or characterizing it as a system that "does something automatically" – mirroring teachers' initial struggles to grasp AI's true depth and scope.

However, through the informatics curriculum systematically exploring AI principles, problem-solving approaches, and hands-on AI software experience, students' comprehension notably evolved. They progressed from perceiving AI as omnipotent and infallible to a nuanced understanding of needing to iteratively refine inputs and critically evaluate outputs for desired results – mirroring teachers' progression toward sophisticated AI perceptions.

The subsequent English class AI implementation for writing and revision further challenged teacher assumptions. Some students readily accepted outputs without critique, expressing sentiments like "AI writes better than me" or questioning English learning's necessity if AI automates writing. This prompted re-evaluating initial impressions of AI as mere convenience, recognizing foundational subject knowledge's indispensable role in effectively harnessing, rather than blindly treating AI as a panacea.

Within art classes, integrating AI tools for transforming student artwork yielded insights – struggling artists embraced AI more readily, while skilled artists expressed frustrations with inability to achieve desired AI results alone. This underscored the need for thoughtful, sophisticated lesson design ensuring meaningful, effective cross-disciplinary AI integration.

The culminating wrap-up workshop consolidated three distinct perception shift patterns:

1) Abstract to concrete understanding (Pattern 1) - From vague notions to concretely grasping how AI operates, critically evaluating outputs, articulating tool selection criteria aligned with goals, and elucidating AI's potential and limitations.

2) Iterative refinement enhances perceptions (Pattern 2) - Teachers and students iteratively refined inputs/prompts while evaluating AI outputs, exemplifying educational value through active participation, repeated exposure, and deeper understanding of effective AI utilization nuances.

3) Enhanced instructional integration knowledge (Pattern 3) - Alongside theoretical understanding, practical, experiential knowledge about purposefully integrating AI tools emerged – accounting for output uncertainty, ensuring meaningful learning beyond blind acceptance, navigating variability based on precise inputs.

These highlighted teachers' transformative comprehension evolution from vague, abstract AI notions to nuanced understanding of its educational technology potential – predicated on immersive experiences intentionally integrating AI tools in classroom practices, illuminating capabilities, limitations, and requisite pedagogical strategies.

## 5. CONCLUSION AND DISCUSSION

We used a community-driven design research method to explore how teachers' perceptions change when a new technology called AI is introduced into school settings. Throughout the stages of CDR, there was a noticeable evolution in teachers' perspectives on AI, as their initial notions gradually transformed into more concrete and stabilized perceptions. Additionally, it was found that the teacher's educational capabilities also changed depending on the type of AI software and results the teacher used.

It is crucial to acknowledge the localized nature of our study, focusing on a specific school, as variations in results may arise across different educational institutions. Furthermore, recognizing the potential impact of subject-specific nuances on teachers' perceptions of AI is essential. Different subjects may elicit diverse responses, and the integration of AI technologies may influence educational capabilities differently based on the subject matter.

Moreover, our research highlights the nuanced relationship between the type of AI software employed and the outcomes achieved by teachers. Future studies could delve deeper into subject-specific implications and explore how the integration of AI across diverse subjects influences educational practices, ultimately contributing to a more comprehensive understanding of AI's role in shaping teaching methodologies.

## REFERENCES

ACM & IEEE-CS (2020). *Computing Curricula 2020: Paradigms for Future Computing Curricula*. ACM & IEEE-CS. https://www.acm.org/binaries/content/assets/education/curricula-recommendations/cc2020.pdf

Bryant, A., & Charmaz, K. (2007). *The SAGE Handbook of Grounded Theory*. United Kingdom: SAGE Publications.

Chen, Y., Jensen, S., Albert, L. J., Gupta, S., & Lee, T. (2023). Artificial intelligence (AI) student assistants in the classroom: Designing chatbots to support student success. *Information Systems Frontiers*, 25(1), 161-182.

Korea Ministry of Education (2020). *Education Policy Direction and Core Tasks in the Age of Artificial Intelligence*, Seoul, Korea. Retrieved January 15, 2022, from https://www.korea.kr/archive/expDocView.do?docId=39237

Montenegro-Rueda, M., et al. (2023). Impact of the implementation of ChatGPT in education: A systematic review. *Computers*, 12(8), 153.

Schwab, K. (2017). *The fourth industrial revolution*. London: Portfolio Penguin.

Van Brummelen, J., & Lin, P. (2020). *Engaging teachers to co-design integrated AI curriculum for K-12 classrooms*. arXiv preprint arXiv:2009.11100. Retrieved January 7, 2022, from https://arxiv.org/pdf/2009.11100.pdf

Wilson, B. B. (2018). *Resilience for all: Striving for equity through community-driven design*. Island Press.

Yue, M., Jong, M. S. Y., & Dai, Y. (2022). Pedagogical design of K-12 artificial intelligence education: A systematic review. *Sustainability*, 14(23), 15620.

# COLLABORATIVE ONLINE INTERNATIONAL LEARNING (COIL): SWOT ANALYSIS FROM A TECHNICAL ENGINEERING COURSE PERSPECTIVE

Jaime Garbanzo-León[1], Luís Jorge Gonçalves[2] and Jorge Medina[2]
[1]*MSc, University of Costa Rica, Costa Rica*
[2]*PhD, University of Aveiro, Portugal*

**ABSTRACT**

Collaborative Online International Learning (coil) activities have become a widespread approach to promote internationalization because it is resilient, as proven during the pandemic COVID-19. Also, coil activities have other advantages such as inexpensive, easy to implement and coordinate. However, the design of these activities depends highly on coil partners and subjects. The objective of this paper is to reflect on these activities from the technical perspective of an engineering course. To reflect on these activities, we performed a SWOT (Strength, Weaknesses, Opportunities and Threats) analysis. To build this analysis we mixed our own observation with a literature review, which provides completeness. The final matrix contains 4 points on average per category. While the least points are defined in Strength, these points can outweigh the other categories in terms of importance, but this remains to be consulted to students and instructors.

**KEYWORDS**

Teaching, Internationalization, Remote Sensing, LiDAR

## 1. INTRODUCTION

The aim of this paper is to reflect on our coil (Collaborative Online International Learning) activity from the technical perspective of an engineering course. To conduct this analysis, a SWOT (Strengths, Weaknesses, Opportunities, and Threats) assessment was utilized. In order to build this matrix, our own observations were combined with various published activities observation and reviews to increase comprehensiveness of this metric. To provide a smooth flow of information, we first start defining what coil activities are and how they are formulated. Then, we describe our coil analysis to give the contextual information to the reader. Afterwards, each point of the SWOT analysis is further explained, which then is summarized in a SWOT matrix.

### 1.1 What is Collaborative Online International Learning (coil)?

Collaborative Online International Learning (coil) is a learning strategy to integrate global perspectives in the course curriculum. This action involves identifying related course content, resources, and activities, which can be carried out remotely by students residing in different parts of the world (Johnston & López, 2022). Johnston and López further elaborate on importance of having globally connected courses, for example, there are global threats nowadays more than ever such as climate change, rising nationalism, economic crisis, and other critical topics, which will require international collaboration to address these issues.

# 2. METHODOLOGY

In this section, we describe the different steps taken in the making of a COIL activity such as finding an international partner, modifying the course activities, and implementing it. In the final and fourth step, we evaluate our approach based on our observation and look at literature to validate them.

The process in three phases can be summarized in 3 steps: find a partner, design the COIL project, and develop the COIL project (CoLLab, n.d.). The first part not only implies finding an international partner to implement this approach but also who is committed to modifying their course content to fit this new activity. This part is the most critical because without an international partner, it is impossible to create a globally connected activity. To perform this search, there are various social media possibilities such as Sunny COIL center, COIL connect, and COLLab. The latter was used to connect the Geoscience Department and Surveying Engineering departments of the University of Aveiro and the University of Costa Rica respectively. Both departments were interested in implementing this activity about Remote Sensing Data and specifically on LiDAR.

LiDAR stands for Light Detection and Ranging and, it is a mapping technique that uses light in the form of pulsed laser to measure ranges (distances) to the earth. These data are combined with other measurements recorded by the airborne system to generate precise information about the shape of the Earth and its characteristics (NOAA, 2023). Furthermore, this mapping technique acquires a lot of point coordinates (X, Y, Z) without the attribute description, leaving the challenge of classifying the datapoints to assign them a class such a building, vegetation, ground, etc.

Then, the following step was to design the activity and change the course contents to leave room to carry out the activity. In several meetings and discussions, we designed an activity that can provide students from these two universities useful abilities in the future, in addition to all other benefits from the international exchange. This activity is described in section 3.1.

After the step of implementation, we decided to reflect on the activity creating a SWOT analysis. This analysis is a simple and known framework to analyze strength, weaknesses, opportunities, and threats for business in relevant external environments. However, this analysis is also relevant to teaching since it can serve to assess critically a teaching activity (Thomas et al., 2014). To perform this analysis, we first investigate literature and evaluations of the coil activities, where some strong and weak points were assessed. Then, we gathered this information and combined it with our observations. Once we have a full picture, classified the points as internal and external. Finally, we assigned the strong and weak internal points to the strengths and weaknesses list respectively, likewise, the strong and weak external points where mapped to opportunities and threats.

# 3. RESULTS

## 3.1 Activity

To create our coil activity, we took some principles of practices from the best practices for COIL learning and global engagement (López & Johnston, 2022). First, we set our objective, which is to provide the students with the skill to process a LiDAR dataset and to produce maps with this information while simultaneously engaging students to share knowledge and connect internationally. Afterwards, we chose to create an applied project which also includes a joint presentation, considering the hierarchy of assignments to promote global learning (see López and Johnston, 2022) and our courses′ context of engineering. Consequently, we designed a 4-week activity where only the first one had a lecture and a lab on what LiDAR is and how to prepare the software and to process it. The second- and third-week students should organize meetings to produce the maps and the presentation. During the last week students were presenting their results and experience on the task.

Students from the Surveying Engineering program at the University of Costa Rica had more knowledge on how to create maps, while the students from the Geoscience course of the University of Aveiro were more "knowledge experts", as these students were enrolled in a non-core course. This main difference made the course more interesting because knowledge could be gained from both sides. Also, the time difference was a good factor because students learn how to coordinate working sessions finding a proper spots in their schedules, which is an imperative practice in international remote work in today's context. Lastly, the base language we used during the session was English, however, Students were allowed to communicate as they wanted using English, Spanish, Portuguese, or "Portuñol". Portuñol is an informal word for mixing the languages since Portuguese and Spanish share a lot of the grammar rules. In the end, students not only exchange technical work but also information about cultural practices and typical food.

## 3.2 Strengths, Weaknesses, Opportunities and Threats (SWOT) Analysis

In this section, we elaborate on each categorized characteristic. In total, we identify 18 relevant points, which give us a good insight into this kind of activity as can be seen next.

### 3.2.1 Strengths

- Equity in Education: Promoting equality in education involves ensuring that all students have access to valuable learning experiences. One way to achieve this is by providing international experiences, even for those who may face travel limitations (Marcilla-Gómez M., Desilus, B., 2016).
- Adapting to Pandemic Challenges: The COVID-19 pandemic has demonstrated the vulnerability of face-to-face internationalization activities. However, virtualization emerges as a powerful tool to continue educational endeavors in a safe environment.
- Benefits of Virtualization: By embracing virtual approaches, we inherit the advantages of online education such as comfortability, no travel required and no physical room arrangement.

### 3.2.2 Weaknesses

- Compulsory activities for Mitigating Disengagement: To address disengagement, educational institutions have implemented compulsory activities. These measures include defining clear roles and fostering effective leadership within the learning environment. (Munoz-Escalona et al., 2020)
- Challenges of Cultural Exchange: While virtual exchanges offer valuable opportunities for cultural interaction, they may not fully replicate the richness of face-to-face exchanges. The absence of physical presence can limit the depth of cultural immersion.
- Language Skills Consideration: Virtual settings might not sufficiently challenge students' foreign language skills. In face-to-face interactions, linguistic nuances and spontaneous conversations play a more significant role.
- Diverse Student Backgrounds: The diverse backgrounds and abilities of students can impact the depth of learning activities. Tailoring content to accommodate this diversity is essential but can be challenging.
- Time-Intensive Preparation: Organizing and preparing for virtual exchanges demands substantial time and effort. Effective planning and coordination are crucial for successful outcomes.

### 3.2.3 Opportunities

- Creating Collaborative Partnerships: Encourage partnerships not only among students but also across different departments. These collaborations can lead to richer learning experiences and innovative initiatives.
- Creating Student Networks: Encourage students to build networks with partners, both within and outside the university. These connections can enhance their academic and professional growth.
- Motivating International Experiences: International exposure can serve as a powerful motivator. Consider face-to-face experiences such as student exchanges or participation in master's programs to broaden horizons.
- Managing Time Differences: Recognize the impact of time zones. Munoz-Escalona et al. (2020) highlight the importance of understanding and managing time differences in global collaborations.
- Coil Redesign: To facilitate Collaborative Online International Learning (COIL), consider redesigning the course. Fung (2004) emphasizes the need for intentional design to promote effective cross-cultural interaction.

### 3.2.4 Threats

- Disengagement: lack of contribution, motivation, and commitment are common issues faced by students. These challenges can hinder their active participation in academic activities. (Munoz-Escalona et al., 2020)
- Academic Overload: Students may feel overwhelmed by the sheer volume of activities, especially when dealing with challenging subjects. This additional burden can impact their motivation and engagement (Munoz-Escalona et al., 2020).
- Language Barrier: Some students might hesitate to participate due to the fear of encountering another language during academic interactions. Overcoming this language barrier is crucial for promoting inclusive learning environments.
- Technical Constraints: Factors like an unstable internet connection can disrupt students' ability to fully engage in online activities. Addressing technical challenges is essential for promoting effective learning experiences (Munoz-Escalona et al., 2020).
- Digital Literacy Disparities: Inequalities in virtual environment literacy can affect students' confidence and competence in navigating online platforms. Educators should consider strategies to bridge this gap and ensure equitable access for all learners (Vasquez & Ramos, 2022).

Figure 1 shows the summary of the SWOT analysis in matrix form.

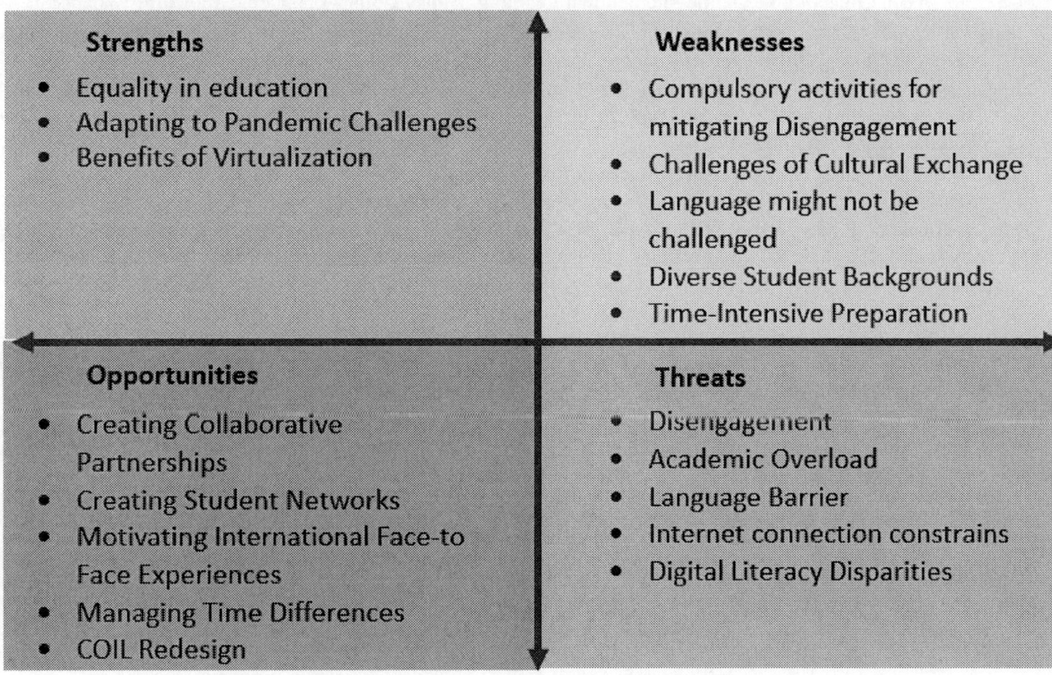


Figure 1. Strengths, Weaknesses, Opportunities, Threats (SWOT) matrix

## 4. CONCLUSIONS

In this short article, we shared the experience of designing, implementing, and reflecting on a Collaborative Online International Learning (coil) project. After the coil activity, we wanted to perform a critical analysis of our implementation. For this reason, we performed a SWOT analysis, which we then cross validated with the available literature. From this analysis, we identified 18 points which were placed in one of the SWOT categories. This was the first experience that the instructors of the courses had with a coil project. A second project is already planned in which we would like to systematically get the students SWOT analysis of the activity. However, how we can weigh each item in the SWOT matrix remains to be validated with students' and instructors' opinions.

## ACKNOWLEDGEMENT

I acknowledge the Department of Geoscience of the University of Aveiro, its collaboration to carry this activity out.

## NOTE

In the making of this article, we used Microsoft Copilot Chatbot to check the spelling and grammar.

## REFERENCES

CoLLab. (n.d.). Collaborative Online International Learning. Https://Collab-Edu.Com/Hub/Coil/p/Whatiscoil.

Fung, Y. Y. H. (2004). Collaborative online learning: Interaction patterns and limiting factors. *Open Learning: The Journal of Open, Distance and e-Learning*, Vol. 19, No. 2, pp 135-149.

Johnston, D., & López, I. (2022). Connecting Beyond the Classroom: Why You Should Be Globally Connecting Your Courses and How It Benefits You and Your Students. *The Wiley Handbook of Collaborative Online Learning and Global Engagement*, pp. 1–17.

López, I., & Johnston, D. (2022). Best Practices for Collaborative Online Learning and Global Engagement. *The Wiley Handbook of Collaborative Online Learning and Global Engagement*, pp. 247–263.

Munoz-Escalona, P., de Crespo, Z. C., Marin, M. O., & Dunn, M. (2020). Collaborative online international learning: A way to develop students' engineering capabilities and awareness to become global citizens. *International Journal of Mechanical Engineering Education*, 50(1), 89–104. https://doi.org/10.1177/0306419020934100

NOAA. (2023, January 20). What is lidar? Https://Oceanservice.Noaa.Gov/Facts/Lidar.Html.

Thomas, S., Chie, Q. T., Abraham, M., Jalarajan Raj, S., & Beh, L.-S. (2014). Education: An Application of the SWOT Framework A Qualitative Review of Literature on Peer Review of Teaching in Higher. Review of Educational Research, 84, 112–159. https://doi.org/10.3102/0034654313499617

Vasquez, E. and Ramos, E. (2022). Can the COVID-19 pandemic boost collaborative online international learning (COIL) in engineering education?–A review for potential implementations, *Procedings 2022 ASEE Annual Conference & Exposition*.

# COGNITIVE LOAD MEASUREMENT WITH ELECTROENCEPHALOGRAPHY BASED ON MACHINE LEARNING ALGORITHM

Yi-Hsuan Wang[1*], Hsing-Yi Wu[2*] and Chun-Wang Wei[1†]
[1]*Department of Healthcare Administration and Medical Informatics, Kaohsiung Medical University, Taiwan*
[2]*Institute of Medical Informatics, National Cheng Kung University, Taiwan*

**ABSTRACT**

Many scholars have pointed out an association between learning outcomes and cognitive load. In order to assist educators in adjusting their teaching methods timely and improving teaching effectiveness, it is crucial to measure learners' cognitive load promptly and effectively. In the field of psychology, cognitive load is defined as "subjective perception of the load on an individual's cognitive system." There is no explicit physiological measurement mechanism currently, but related research indicates a certain degree of correlation with brainwaves. Therefore, a supervised learning algorithm based on electroencephalography (EEG) is developed to construct a mechanism for measuring cognitive load. We design questions and scales, collecting participants' brainwave data and cognitive load questionnaires while they answer math questions of varying difficult levels. A supervised learning algorithm in machine learning is then applied to establish a cognitive load assessment model. Using easily accessible and reliable instruments for non-intrusive EEG detection, we explored five feature combinations and applied four supervised learning methods to identify the best algorithm and features for measuring cognitive load. In the end, Random Forest and K Nearest Neighbor algorithm models were constructed using feature sets of 14 dimensions and 7 dimensions respectively. The predictive performance achieved over 92%. The results of this study can be applied to assist educators in conducting "asynchronous follow-up tracking of learning outcomes" and "real-time monitoring of synchronization with the learning process." Through monitoring the cognitive load levels, educators can understand learners' actual learning situations, and make adjustments to the learning model as a supplementary tool to assist in the improvement of teaching and learning.

**KEYWORDS**

Cognitive Load, Electroencephalography, Supervised Machine Learning Methods, Artificial Intelligence

## 1. INTRODUCTION

There is a significant correlation between cognitive load and learning outcomes, moderate intrinsic and germane cognitive loads are helpful for learning, whereas extraneous cognitive loads are not. Therefore, effective measurement of learners' cognitive load will help educators to make timely adjustments to the teaching methods and contents of the teaching materials to prevent learners from developing excessive cognitive load, which will in turn affect the effectiveness of learning. Recently, most studies assess cognitive load by using self-reporting scales, which is not allowed for immediate measurement of cognitive load. On the other hand, some scholars have started utilizing physiological signal devices such as Electroencephalography (EEG), Magnetoencephalography (MEG), or functional Magnetic Resonance Imaging (fMRI) as non-intrusive neuroimaging techniques to assess learners' cognitive load in real-time (Havugimana et al., 2023).

Cognitive load is a theory in psychology that refers to a subjective load perceived by an individual on the cognitive system when engaging in a particular task (Siregar, 2023). The degree of cognitive load generated by an individual engaged in a particular task is influenced by the difficulty of the task and the individual's own abilities. Due to the lack of explicit physiological definition of cognitive load, we developed a supervised learning algorithm of machine learning to establish a cognitive measurement mechanism using EEG.

[*]Contributed equally
[†]Corresponding author: cwwei@kmu.edu.tw

The process of solving mathematical questions involves complex cognitive operations, using working memory resources for memorizing formulas, temporarily storing computations results, and connecting prior knowledge and experience. Without the working memory, individual isn't able to complete the complex cognitive operations. Thus, mathematical problem-solving is highly suitable for measuring an individual's perceived cognitive load. The purpose of this study is to design mathematical questions of different difficulty levels for participants to answer, and collect brainwave data and cognitive load questionnaires from participants to build a cognitive load assessment model using supervised learning algorithms, so as to monitor the cognitive load level of the learners in real time. This study will be helpful for the immediate assessment of the cognitive load level of the learners and the adjustment of the learning mode as well as serve as the basis for the subsequent related research.

# 2. METHODOLOGY

## 2.1 Experimental Tasks and Measurements

The experimental task in this study involved solving mathematical questions, with measured variables including EEG signals, subjective cognitive load, and correctness of responses. Throughout the task, participants wore a simple EEG device to collect brainwave signal. After completing each math question, participants used the Likert seven-point scale for self-assessment to measure their current cognitive load level. Higher scores indicate that participants perceived higher cognitive load. Meanwhile, the Neurosky MindWave Mobile 2 unipolar brainwave instrument was used as a brainwave collection tool to capture the EEG signals of the subjects during the experiment. The headband has a metal electrode that can be adjusted in height to collect electrical signals from the frontal lobe, the main cognitive area of the brain. In addition, the Brainwave Instrument is equipped with a clip-on ground electrode located on the left side of the headband, which contacts the subject's left earlobe (A1 region in Figure 1) when in use. This ground electrode is used to compensate for noise and interference that may exist in the brainwave signal, improving signal quality and reliability.

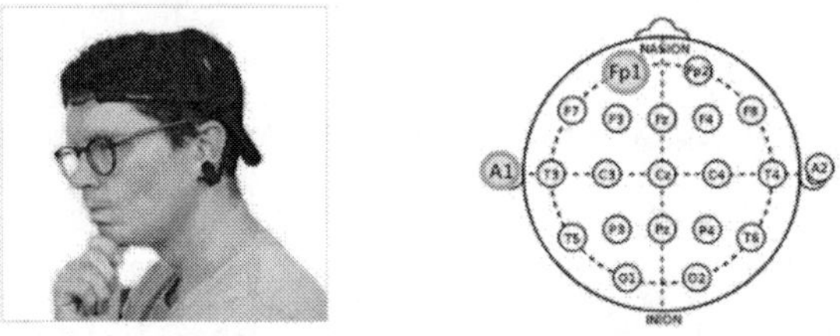


Figure 1. MindWave Mobile 2 Electrode Location (Source: NeroSky online store)

## 2.2 Data Processing and Feature Extraction

The purpose of feature extraction was to reflect cognitive load from brainwave power spectrum data for subsequent analysis and model building. The data generated by each subject for each question included the length of the question, the right and wrong answers, and the 12 brainwaves symptoms, etc., and were labeled according to the cognitive load scale values filled in by the subjects. The sampling frequency of the instrument is 0.25Hz, i.e., it outputs 8 data per second. The data was smoothed first, and the average of each second was calculated according to the time interval to obtain the data of each frequency per second. Then, according to the brain wave frequency band, the data was sorted and divided into five types of brain waves, which were divided as follows: Delta wave (0.5~4Hz); Theta wave (4~8Hz); Alpha wave (8~14Hz) was additionally divided into high Alpha (10~14Hz) and low Alpha (8~10Hz); Beta wave (14~30Hz) was divided into high Beta (20~30Hz), medium Beta (16~20Hz) and low Beta (14~16Hz); Gamma (30~80Hz) wave was divided into high Gamma (50~80Hz) and low gamma (30~50Hz) (Da Silva, 2023). The average value of the smoothed data was calculated from each brain wave band. Lastly, calculate the average of the data from the previous step according to the answering time of the question. The brainwave data of each segment of the subject answering the question was obtained, which was used as the 12 kinds of features data used in machine learning.

## 2.3 Machine Learning Algorithm Model Development

In this study, the cognitive load was categorized into seven categories, and the seven-point scale was used to derive the labels from low load to high load, ranging from 1 to 7. The categories were balanced by oversampling. Subsequently, on the basis of 80/20 rule dividing the data, which are used in model building: training phrase (80%) and testing phrase (20%). After that, 80% of data set in training phase divided "training set (64%)" and "validation set (16%)." Using stratified 10-fold cross-validation in model training phase, it is a stratified sampling method, which enables the categories in the training set and the validation set to be divided proportionally, and the seven categories of both data sets are equally accounted for.

We employed statistical methods to calculate Pearson correlation coefficients (PCCs) to assess the correlation between each feature. Ultimately, we selected a threshold based on the absolute value of the correlation coefficient as the criterion for feature selection. Then we used K-Star, KNN, Random Tree, and Random Forest algorithm classifiers to compare the prediction results and identified the most suitable model.

# 3. RESULTS

## 3.1 Feature Combination and Model Selection

In the feature selection section, C, D, E groups were filtered using a threshold of 0.02, 0.03, and 0.04, respectively. Considering the purpose of the study, the features that can be obtained at the moment of learning were included as one of the groups (Combination B). There were five sets of features for comparison, and the detailed descriptions of the feature sets are shown in Table 1. Four multi-classification modeling algorithms were used for training, and the results were obtained by cross-validation with 5-fold and 10-fold respectively. The average accuracy of each combination between the two by the average training accuracy was more than 90%, which was a stable and excellent training result. Combination A, which has the best average accuracy, and the Combination D, which has a good combination of dimensionality and accuracy, were put into the prediction and compared.

Table 1. Summary of Features Combination

| Combination | Filtering Criteria | Features | Dimension |
|---|---|---|---|
| A | No filtering | tf, sec, Delta, Theta, Alpha (total, low, high), Beta (total, low, middle, high), Gamma (total, low, high) | 14 |
| B | Brainwave Features | Delta, Theta, Alpha (total, low, high), Beta (total, low, middle, high), Gamma (total, low, high) | 12 |
| C | Threshold > 0.02 | Delta, Alpha (total, low, high), Beta (total, middle, high), Gamma (total, low) | 9 |
| D | Threshold > 0.03 | Alpha (total, low), Beta (total, middle, high), Gamma (total, low) | 7 |
| E | Threshold > 0.04 | Alpha (total, low), Beta (high), | 3 |

A comprehensive evaluation was conducted using various model evaluation metrics to choose the best model for feature combinations A and D. The experimental results of the Random Forest algorithm with 10-fold cross validation showed that it performed the best in the task of cognitive load level classification with an overall accuracy of 95.72%. Therefore, the 10-fold Random Forest model was chosen as the model for the A-feature group. In the model selection for combination D, the KNN model with 10-fold cross validation showed the best result. Thus, the KNN model was selected as the final predictive model for D-feature group.

### 3.2 Results of Prediction

The result showed that the Random Forest model, which was suitable for the A-feature combination, performed relatively poorly when the predicted cognitive load level was "2" or "5", but its performance was stable when predicting the other five levels (Table 2). On the other hand, the KNN model, which was the most suitable for D-feature combination, performs relatively weakly when the prediction level was "1" or "2", but still maintains a good prediction performance for the other levels (Table 2). Although the prediction ability of different levels varies slightly, according to the Kappa consistency coefficient, the prediction ability of the model is still excellent, and there is no excessive error due to the differences in levels. In this study, the supervised learning algorithm model developed in this study can accurately predict the level of cognitive load of learners and be used to monitor and analyze the cognitive load of learners in the learning process in real time.

Table 2. Results of prediction

| Model | Total | Correct | Incorrect | Mean accuracy | Kappa coefficient |
|---|---|---|---|---|---|
| Random Forest | 240 | 224 | 16 | 0.9333 | 0.9222 |
| K Nearest Neighbor | 240 | 222 | 18 | 0.9250 | 0.9125 |

## 4. CONCLUSION

In this study, a machine learning algorithm was established to assess cognitive load through real-time measurement of EEG, achieving an accuracy of more than 92%. Differing from previous studies, we used mathematical questions of different difficulty levels as a medium to induce cognitive load in learners. By extracting the characteristics of brain waveform changes during the learning process, we investigated how the brain waveforms reflect the level of cognitive load, and helped to complete the related data analysis and prediction model. The model developed is more applicable to real educational situations than results of previous research. Despite these advancements, it is essential to acknowledge certain limitations. The sample size was relatively limited, and the exclusive focus on mathematical questions may not fully capture the complexity of cognitive load in diverse learning contexts. In conclusion, this study enables real-time monitoring of learners' cognitive load, providing valuable insights for educator. However, future research should address these limitations by considering more diverse datasets and incorporating a broader range of learning scenarios to enhance the model's applicability and generalizability.

## ACKNOWLEDGEMENT

This study was supported in part by the National Science and Technology Council of Taiwan (MOST 112-2410-H-037-013), MOE Teaching Practice Research Program (PBM1120763) and Office of Global Affairs at Kaohsiung Medical University.

## REFERENCES

Da Silva, F. L., (2023). EEG: Origin and Measurement. *In EEG-fMRI: Physiological Basis, Technique, and Applications.* Cham: Springer International Publishing, pp. 23-48.

Havugimana, F. et al., (2023). Deep Learning Framework for Modeling Cognitive Load from Small and Noisy EEG data. *In IEEE Transactions on Cognitive and Developmental Systems*, doi: 10.1109/TCDS.2023.3319305.

Siregar, N. R., (2023). Explicit Instruction and Executive Functioning Capacity: A New Direction in Cognitive Load Theory. *In Journal of Education*, Vol. 203, No. 2, pp. 451-458.

# ZINES AND DIGITAL TASK-BASED PROJECTS: CONTENT-BASED INSTRUCTION AND TASK-BASED LEARNING

Peter Gobel and Makimi Kano
*Kyoto Sangyo University, Japan*

**ABSTRACT**

This paper discusses using project-based tasks to improve the understanding of cultural theory in a content-based language classroom. The course of concern is a university-level intercultural communication class, where students study a variety of cultural theories and then use these theories to explain cultural artifacts and cultural behavior in a variety of cultures. In previous years, various models of cultural theory, such as those of Hall (1976), Hofstede (2011), and Trompenaars (1997), were introduced to students through readings and lectures. These theories were then used as the basis for essays and final research papers. It was found that students only gained a shallow understanding of the models through readings and teacher-based instruction, resulting in very little thinking beyond the text. It was hoped that using digital projects, creating digital zines, would allow students to illustrate the theories, thus leading to deeper understanding of the models.
For each aspect of cultural theory presented by the instructor, students are given a set of digital tasks that help them create a zine. The tasks revolve around the reframing of the theoretical aspect from the readings, connecting the aspect to their own experience, and creating a digital representation of the theory. This story is then presented to their peers as an illustration of the complex cultural theory. It is hoped that by incorporating their own experiences and framing the theory as a story, students will more easily grasp these abstract concepts and incorporate them into their analysis.
The preliminary results of using digital project-based tasks in the classroom will be discussed in light of task efficacy, task completion, and the effect of the tasks on motivation and willingness to communicate. It is hoped that these results will provide evidence for the positive interface between CBI and digital task-based projects.

**KEYWORDS**

Content-Based Instruction, Task-Based Language Teaching, Zine, Project-Based Task, Cultural Theory

## 1. INTRODUCTION

The intersection of Content-Based Instruction (CBI) and digitally augmented Task-Based Language Teaching (TBLT) offers an approach to language education that integrates linguistic proficiency, cultural understanding, and digital literacy. This presentation explores the ongoing utilization of zine creation tasks. Zines are noncommercial, often homemade publications, usually devoted to specialized subject matter. They have often been used in areas of music fandom and subculture, where normal publications do not exist or are too general in nature. The presentation will explain how zines are used as a means to teach cultural theory within the frameworks of CBI and TBLT, emphasizing their synergistic benefits and implications.

## 2. UNDERSTANDING CULTURAL THEORY, CBI, AND TBLT

### 2.1 Content-Based Instruction

Content-based Instruction (CBI) integrates language learning with the study of subject matter content and uses authentic language and texts, providing meaningful contexts for language acquisition (Byram, 1997). In a CBI course, the focus of learning is not on learning of a language in isolation, but rather learning of language through the study of subject matter (Snow & Brinton, 2017). This means that the curriculum is based on a certain subject matter, and communicative competence is acquired in the context of learning about certain topics in that subject area. For the purposes of this presentation, the content concerned is the teaching of cultural theory. Cultural Theory delves into the exploration of societal norms, values, and practices within specific cultural contexts, essential for developing intercultural competence. Examples of cultural theory include those of Edward T. Hall (1976), Geert Hofstede (2011), and Franz Trompenaars (1997).

One of the problems with CBI is finding suitable material at an appropriate level. For the students in question, although the language level of the texts was generally suitable, there was a distinct lack of schema for the students. In other words, the examples and topics discussed in the text were unfamiliar to them, which made the readings more difficult to understand. This difficulty, compounded with the complexity of the models provided, resulted in a shallow understanding of the models and general concepts and a difficulty in connecting these cultural theories with their own lives. To resolve these problems, a series of task and task-based projects were designed, which would allow the students to reframe the theoretical concepts, connect these concepts to their own experience, and allow them to create personal narratives which would reflect the core concepts in each cultural theory.

## 2.2 Task-Based Language Teaching

Task-based Language Teaching (TBLT) prioritizes the use of tasks as the central unit of instruction, promoting authentic language use and communication. Tasks have a number of features. Among them are that a task should be communicative in nature (meaning-oriented and focusing on content); as authentic as possible; outcome oriented; and goal oriented (Nunan, 2004; González-Lloret & Ortega, 2014).

One of the perceived benefits of TBLT in a CBI curriculum is that the tasks can be designed to encourage authentic student output rather than the recitation of rote-memorized phrases and passages or the simple copy/paste of researched materials. This results in two additional benefits: it allows the students to create a product that is more readily accessible to their peers, and it allows the instructor to more accurately gauge student comprehension of the course material (the theoretical concepts involved). For the purposes of the course in question, students completed a series of tasks that resulted in a final product, a zine that was shared with the class in physical form.

## 2.3 Zine Production

Zine (short for "magazine") creation involves the creation of succinct explanations of events, objects, and theories (in this case, with the aid of digital technology), offering a creative platform to explore and express cultural experiences and perspectives (Bricks, 2017; Duncombe, 2008; Mullins 2018). Zines have developed as an important medium of communication in various subcultures and groups, and draw inspiration from a philosophy that is contrary to traditional conventions of design and magazine publishing. In general, zines are the product of either a single person or a discourse community and are usually printed or photocopied for distribution. Among the various reasons for creating a zine are: developing one's own identity; sharing a niche skill or interest in a particular genre; or developing or recounting narratives (Nilsen, 2019; Gu, 2020). A good example of zine production is the genre of the "fanzine," small publications created by devoted music fans about a group or individual artist. These are often sold or distributed at music stores, concert venues, and clubs.

# 3. INTEGRATING CBI AND TBLT

Integrating CBI, and digitally augmented TBLT in language education offers diverse opportunities for immersive learning experiences:

1. Digital Storytelling Projects on Cultural Themes: Students engage in digital storytelling projects that explore cultural theory concepts, creating narratives that reflect cultural perspectives, practices, and values within authentic contexts.
2. Task-Based Approach to Zine Creation: TBLT principles guide the task-based approach to digital story creation, emphasizing communication, collaboration, and negotiation of meaning among learners as they produce and share their narratives.

3. Reflection and Analysis of Cultural Narratives: Students reflect on and analyze cultural narratives created by their peers, engaging in critical discussion and inquiry to deepen their understanding of cultural theory concepts and perspectives.

This approach offers authentic opportunities for cultural engagement, allowing students to explore, express, and share their cultural experiences and perspectives. Integration of CBI and TBLT principles into zines also promotes language proficiency development by providing meaningful language input and opportunities for communicative language use within authentic contexts. In addition, through the exploration of cultural narratives, students develop cultural sensitivity and awareness, fostering respect and understanding for cultural perspectives and practices. Finally, the digitally augmented zine tasks enhance students' digital literacy skills, as they learn to navigate digital tools and platforms, create multimedia content, and effectively communicate and share their narratives with their peers.

## 3.1 Zine Creation

Zine creation involves a number of steps. The first of these is taking course material and limiting the focus of the project. This is done through a series of brainstorming tasks. In the case of this project, digital templates were used for the brainstorming process, allowing the students to brainstorm their zine, and narrow the topic. When the topic is reduced to a size appropriate for the final product (in this case, a zine booklet of 8-12 pages), students plan and reduce the information even further to fit the constraints of the final product. This task is done, again, using a digital template and a review checklist. This information is then input into a final template provided by the instructor which conforms to the unfinished zine format. In the case of the course in question, the template was provided in the form of a powerpoint, where the students would input text, pictures, and media. The students needed to learn various aspects of formatting to fit their media into the template and to make the final product as attractive as possible.

Although zines can be shared digitally, in this project, students are required to print out and fold their zines to create the final physical product. They then distribute the zines to their peers and explain the contents of the zine, page by page. An alternative is to create the zines and then explain them in a digital presentation. In either case, their peers are left with a physical representation of the student's understanding of a particular theoretical concept.

## 3.2 Preliminary Results, Challenges, and Considerations

The initial results of the zine project have been mixed. From a pedagogical perspective, the goals of the project are being met. That is, students are able to clearly and succinctly describe facets of culture that they have studied, and this success has transferred over to larger written works such as research papers and position pieces. On the other hand, it has been observed that a number of students were not ready to expand their technological proficiency beyond what they had already learned. Despite previous training in the use of digital tools such as PowerPoint, and plenty of "how to" videos that are available online, many students seemed to just struggle along with the knowledge they had, rather than attempt to improve their technological skills. Previous studies have reported on the reluctance of some Japanese students to embrace digital technology (Gobel & Kano, 2014, 2017; Cote & Milliner, 2017). This tendency might be specific to the Japanese learning environment, but it is worth considering when cretaing and implementing zine projects such as this one.

This is an ongoing project, and the process and final results are far from perfect or conclusive. To date, a number of challenges have arisen. Chief among them are the access to digital technology and resources, which is essential to the project. The instructor has to make sure that all students have access to the task materials outside of class. In this case, many students attempted all tasks on their mobile devices, which did not always work according to plan, since the tasks were designed for different operating systems. This is a common occurrence in tasks designed for one platform, and student platform usage should be taken into consideration when designing tasks (Godwin-Jones, 2015; Stockwell, 2017; Kukulska-Hulme & Viberg, 2018).

Another challenge is task design. Designing and implementing effective zine tasks requires careful consideration of task complexity, scaffolding, and learner proficiency levels to promote meaningful engagement and language learning outcomes. The anecdotal information provided here may not be sufficient to understand how students used the planning tools and input given to them. For future studies, the inclusion of a more in-depth analysis would be helpful in terms of finding out what the students are actually doing during the planning stages of the project. Qualitative data gathering, such as post-task interviews, retrospective interviews and think-aloud protocols may give a fuller understanding of actual student behavior during the planning stages.

Finally, assessing student learning outcomes in zine tasks requires the development of appropriate assessment criteria and rubrics that encompass linguistic proficiency, cultural understanding, and digital literacy skills. At present, the task efficacy has not been measured. For example, the percentage of effective task completion would be one good measure when evaluating task efficacy, but the effect of the task on motivation, willingness to communicate, or the development of academic skills (summary and communication skills) would also be helpful in determining the efficacy of the tasks and the total project. Related to this is the relationship between efficacy and presentation type. It would be interesting to see whether presentation type (one-way, interactive, face-to-face, or online), differing formats, or differing digital tools affect task completion, efficacy, or outcome.

## 4. CONCLUSION

The integration of CBI and digitally augmented TBLT offers a multifaceted approach to language education that fosters linguistic proficiency, cultural understanding, and digital literacy. By engaging students in the creation and exploration of multimedia narratives grounded in cultural theory, educators can promote meaningful language learning experiences that prepare students to navigate diverse linguistic and cultural landscapes effectively. While challenges exist, the transformative potential of this integrated approach underscores its significance in fostering intercultural competence and global citizenship skills essential for success in today's interconnected world. Future research will focus on a variety of factors that may affect task completion, such as topic familiarity and task complexity.

## REFERENCES

Brick, S. (2017). *Make a Zine!: Start Your Own Underground Publishing Revolution*. Microcosm Publishing.

Byram, M. (1997). *Teaching and assessing intercultural communicative competence*. Multilingual Matters.

Cote, T. J., & Milliner, B. (2017). Preparing Japanese Students' Digital Literacy for Study Abroad: Is More Training Needed?. *JALT CALL Journal*, *13*(3), 187-197.Duncombe, S. (2008). *Notes from Underground: Zines and the Politics of Alternative Culture*. Microcosm Publishing.

Dunn, P. (2014). *Make a Zine: When Words and Graphics Collide*. Beverly, MA: Quarry Books.

Frechette, D. (2018). *Zine DIY: A Guide to Zine Making*. New York, NY: Scholastic Inc.

Gobel, P., & Kano, M. (2014). Mobile natives: Japanese university students' use of digital technology. APACALL Book III. Cambridge Scholars Publishing.

Gobel, P., & Kano, M. (2017). The complexities of digital storytelling: Factors affecting performance, production, and project completion. In *Smart technology applications in business environments* (pp. 343-360). IGI Global.

Godwin-Jones, R. (2015). Emerging technologies: Mobile apps for language learning. *Language Learning and Technology, 19 (3)*, 8-20.

Gonzalez-Lloret, M. & Ortega, L. (2014). Technology-mediated TBLT. John Benjamins.

Gu, J. (2020). *Zines in Third Space: Radical Cooperation and Borderlands Rhetoric*. University of Pittsburgh Press.

Hall, E. T. (1976). *Beyond Culture*. Doubleday.

Harris, E. K., & Melamed, L. (2019). *Zine Making: From Idea to Illustration*. Cincinnati, OH: Impact.

Hofstede, G. (2011). Dimensionalizing Cultures: The Hofstede Model in Context. *Online Readings in Psychology and Culture, 2(1)*. https://doi.org/10.9707/2307-0919.1014

Kukulska-Hulme, A., & Viberg, O. (2018). Mobile collaborative language learning: State of the art. *British Journal of Educational Technology*, *49(2)*, 207–218.

Marx, W. D. (2023). *Status and Culture*. Viking Press.

Mullins, E. (2018). *Creating Zines: How to Make Collage Zines, Mini Zines, and Pamphlets*. New York, NY: Peter Pauper Press.

Nilsen, A. P. (2019). *So You Want to Publish a Zine?*. Microcosm Publishing.

Nunan, D. (2004). *Task-based Language Teaching*. Cambridge University Press.

Robin, B. (2006). *The educational uses of digital storytelling*. Association for Educational Communications and Technology.

Snow, M. A., & Brinton, D. M. (Eds.). (2017). *The content-based classroom: New perspectives on integrating language and content* (2nd ed.). Ann Arbor, MI: University of Michigan Press.

Stockwell, G. (2017). *Mobile-assisted language learning* (Vol. 21). Bloomsbury Publishing.

Trompenaars, F., & Hampden-Turner, C. (1997) *Riding the Waves of Culture*. Hachette.

# Reflection Papers

# ADAPTING ASSESSMENT DESIGN: NAVIGATING TECHNOLOGICAL ADVANCES AND ACCESS TO INFORMATION

Fran Greyling and Linda van Ryneveld
*University of Pretoria*
*University of Pretoria, South Africa*

**ABSTRACT**

The transformation of educational assessments in the digital age with AI advancements is crucial. This paper explores how these technologies necessitate evolving assessment methods. AI-driven tools can enhance assessments by automating grading, providing personalised feedback, and enabling adaptive testing. Challenges include ensuring equitable access, maintaining data privacy, and upholding academic integrity, requiring a multifaceted approach combining preventive measures with advanced detection technologies. Embedded professional development for educators is essential for effectively utilising AI in assessments. The paper advocates for a shift towards authentic assessments that evaluate higher-order thinking and real-world problem-solving abilities, aligning with modern workforce and societal demands. Leveraging AI in assessment design offers significant opportunities to enhance learning experiences, making them more meaningful, motivational, and memorable.

**KEYWORDS**

Educational Assessment, Artificial Intelligence, Online Learning, Academic Integrity, Authentic Assessment

## 1. INTRODUCTION

"Assessment is a moral activity" (Knight, 1995, p. 13). Knight's perspective, articulated before the ubiquity of information and communication technologies (ICT) and the rise of Artificial Intelligence (AI), highlights the importance of adapting assessment design more than ever.

Technological advances have significantly transformed educational practices, posing new opportunities and challenges in assessment design. Previous work highlights the importance of fair, valid, and reliable assessments, but traditional methods often fall short of these standards in the digital age (Mohammed & Mashkhal, 2023). The problem lies in adapting assessments to leverage AI while ensuring educators can proficiently use this technology and maintain equitable access, data privacy, and academic integrity.

This paper aims to reflect on these issues and consider strategies for effectively integrating AI into assessment practices. It contributes to the ongoing discourse on enhancing educational assessments by examining current technological capabilities and their implications. It provides insights into how AI can be harnessed to create more personalised, efficient, and relevant assessment experiences while also addressing the ethical and practical challenges that educators face. The purpose is to offer an approach to align assessment design with contemporary educational needs and technological advancements.

## 2. TECHNOLOGICAL ADVANCEMENT THAT ENABLES ASSESSMENT

The shift towards online learning, accelerated by the COVID-19 pandemic, has significant implications for traditional educational practices (UNESCO, 2020). Educators must rethink, amongst others, assessment design to ensure it aligns with the capabilities of digital platforms while also considering their limitations to develop a comprehensive and well-rounded assessment strategy. The AI-driven enhancements in online learning and assessment discussed next range from well-established tools to the latest innovations in AI technology.

AI accurately and automatically grades close-ended questions, providing instant feedback. Feedback can be personalised based on individual student responses. Recently, AI-driven Natural Language Processing tools have emerged to analyse essay-type assignments, offering automated grading and initial feedback immediately upon student submission. Similarly, Computer Vision AI tools now enable more frequent assessment and feedback of visual assignments. Additionally, AI technologies can assess spoken language skills in real time, evaluating aspects like pronunciation, fluency, and comprehension. These AI capabilities significantly reduce the grading workload for educators, all the while offering detailed and timely feedback that can assist students in refining their work.

AI offers additional avenues for educators to save time, such as automating administrative tasks related to assessments, including recording grades and generating reports. Furthermore, AI enhances assessment efficiency by analysing extensive datasets of student engagement and performance, identifying patterns in student performance, providing insights into areas where students may be struggling, and informing targeted interventions. This capability assists educators in identifying at-risk students and intervening proactively.

Educators can leverage AI capabilities to efficiently create high-quality, varied assessment content and scoring instruments. Based on provided instructional material, AI can generate a variety of question types tailored to different difficulty levels and learning objectives. In multiple-choice questions, AI can create plausible distractors based on common misconceptions, enhancing the utility of formative assessments. Additionally, AI can develop essay prompts and real-world scenarios for open-ended assessments, developing critical thinking and problem-solving skills.

AI significantly enhances the overall learning experience, fostering a more meaningful, motivational, and memorable learning environment. AI-powered adaptive tests automatically adjust the difficulty of questions in real time based on student performance, ensuring that assessment tasks are appropriately challenging. Furthermore, a competency-based assessment approach, which pinpoints specific knowledge and skill areas that students must master, enables educators to automatically release content according to predetermined criteria. This empowers students to advance at their own pace once they demonstrate achievement of these criteria. Gamified assessment incorporates game elements, such as badges and leaderboards, into the assessment process to make it more engaging and potentially increase student achievement.

These examples of AI-driven solutions demonstrate that AI streamlines the online assessment process for educators and enhances the overall learning experience. This is particularly important where large class sizes limit the opportunities for frequent assessments and effective feedback. However, the shift to online learning also introduces new challenges.

# 3. CHALLENGES AND CONSIDERATIONS

## 3.1 Addressing Equitable Access

To ensure equity and integrity, all assessments must uphold the principles of fairness, validity, and reliability (Miller, 2019), irrespective of the delivery mode. When utilising technology and AI tools, new concerns about fairness and bias may arise. Fairness in assessment can be affected by the digital divide, which is the gap between those who have access to Internet-enabled devices and those who do not. Promoting equity may involve providing resources or support for students who lack digital literacy, reliable Internet access or appropriate devices, as well as designing assessments that can be completed on a variety of devices and platforms and are text-to-speech or screen reader-compatible (Efe, 2022). Additionally, AI can be biased if it is trained on data that reflects existing inequalities and stereotypes.

## 3.2 Ensuring Data Privacy and Security

Protecting personal information is paramount as more learning activities move online. Strategies include selecting secure platforms, understanding privacy policies, and educating users about data privacy and security. Reliable online platforms ensure data privacy and security for educational institutions, organisations, and governments through features like privacy policies, regular security updates, secure authentication, data encryption, role-based access control, data anonymisation and deletion, activity logging and monitoring, backup and recovery, and secure file management with controlled access to sensitive documents (Moodle, 2024).

### 3.3 Upholding Academic Integrity

Academic integrity is a critical concern in online assessment due to the lack of physical supervision in online environments. Maintaining academic integrity requires a multi-faceted approach that includes preventive and detection measures.

The construction of assessment activities is an important preventative measure. Other preventative measures include the use of honour codes, fostering a culture of integrity by clearly communicating expectations and consequences of academic dishonesty, providing guidance about the importance of original work, and creating an environment where students feel supported and valued (Gynnild & Gotschalk, 2008). Detection methods can be managed by technology. The proposals from OpenAI (2024) focus on upholding academic integrity in online assessments by using secure login procedures, biometric verification, AI-based proctoring tools, advanced plagiarism detection, regularly updated question banks, and learning analytics to detect unusual patterns and prevent cheating.

### 3.4 Fostering Faculty Readiness

While emerging technologies may be exciting, acquiring them without proper consideration of their role in improving practice will likely result in mixed outcomes. The optimal use of technology in a modern learning environment seems to remain a challenge for educators (Hassan, 2021; Newton, 2020).

Na and Jung (2021) propose professional support and development to grow the capacity for using educational technologies. Practitioners and researchers recommend alternative approaches to traditional one-off training opportunities to shift practice (Darling-Hammond et al., 2017; OECD, 2023). These recommendations align with Greyling's (2007) proposed professional development approach, which is characterised by extended interventions that use practitioners' needs as a point of departure, promote community building, weave pedagogy and technical skills, prioritise authentic tasks, and create a safe space for learning and reflection.

## 4. ASSESSMENT APPROACH

The digital age necessitates a re-evaluation of traditional assessment methods. Traditional assessments may no longer be the most efficient or effective way to measure student learning. These methods often emphasise memorisation and rote learning, which are less relevant in a world where information is readily accessible. Such a world demands assessments that evaluate higher-order thinking skills, such as critical thinking, problem-solving, and creativity. (Thornhill et al., 2023).

Authentic assessment tasks within a problem-based learning framework offer an effective strategy for addressing the challenges posed by AI and access to information. These tasks require students to apply their knowledge in unique and meaningful ways to solve real-world problems, which reduces their susceptibility to the unauthorised use of AI and plagiarism. Tasks can include projects, presentations, simulations, and other activities that encourage deep understanding and application of knowledge. Additionally, authentic tasks can be tailored to individual students' interests, fostering deeper engagement and relevance.

## 5. CONCLUSION

The landscape of assessment is rapidly evolving with AI and technological advancements. These tools can enhance efficiency and effectiveness, offering personalised assessment activities and insights for targeted interventions. However, challenges such as equitable access, biases in AI systems, data privacy, and academic integrity require careful assessment design and robust protection measures. Continuous professional development for educators is crucial. The future of online assessment will likely combine advanced technology, innovative assessment design, and ethical standards to create meaningful and motivational learning experiences.

## REFERENCES


Darling-Hammond, L., Hyler, M. E., & Gardner, M. (2017). Effective teacher professional development. Palo Alto, CA: Learning Policy Institute. Retrieved from https://learningpolicyinstitute.org/sites/default/files/product-files/Effective_Teacher_Professional_Development_REPORT.pdf

Efe, A. (2022). The impact of artificial intelligence on social problems and solutions: an analysis on the context of digital divide and exploitation. Yeni Medya, 2022(13), 247-264.

Greyling, F. C. (2007). The Why and How of Technology-Assisted Learning: Authentic Professional Development for Higher Education Practitioners [Unpublished doctoral thesis]. University of Johannesburg, South Africa.

Gynnild, V., & Gotschalk, P. (2008). Promoting academic integrity at a Midwestern University: Critical review and current challenges. International Journal for Educational Integrity, 4(2).

Hassan, M. (2021). Online Teaching Challenges during COVID-19 Pandemic. International Journal of Information and Education Technology, 11(1), 1-6. Retrieved from http://www.ijiet.org/vol11/1487-AE0010.pdf

Knight, P. (Ed.). (1995). Assessment for Learning in Higher Education. Introduction by Peter Knight. London: Biddles Ltd.

Miller, D. (2019). Reliability in Educational Assessments. Oxford Bibliographies. DOI: 10.1093/obo/9780199756810-0228

Mohammed, A. & Mashkhal, S. (2023). Evaluating Online Assessment Strategies: A Systematic Review of Reliability and Validity in E-Learning Environments. North American Academic Research, 6(12), 1-18. https://doi.org/10.5281/zenodo.10407361

Moodle. (2024). Moodle Security and Privacy Compliance. Retrieved from https://moodle.com/security-privacy/

Na, S., & Jung, H. (2021). Exploring University Instructors' Challenges in Online Teaching and Design Opportunities during the COVID-19 Pandemic: A Systematic Review. International Journal of Learning, Teaching and Educational Research, 20(9), 308-327. https://doi.org/10.26803/ijlter.20.9.18

Newton, D. (2020, March 26). Most teachers say they are "not prepared" to teach online. Forbes. Retrieved from https://www.forbes.com/sites/dereknewton/2020/03/26/most-teachers-say-they-are-not-prepared-to-teach-online/?sh=a3150357f2cf

OECD. (2023). Lessons for Education from COVID-19. Retrieved from https://www.oecd-ilibrary.org/sites/8283a5c2-en/index.html?itemId=/content/component/8283a5c2-en

OpenAI. (2024). ChatGPT [Large language model]. Retrieved from https://www.openai.com/

Thornhill-Miller, B., Camarda, A., Mercier, M., Burkhardt, J-M., Morisseau, T., Bourgeois-Bougrine, S., Vinchon, F., El Hayek, S., Augereau-Landais, M., Mourey, F., et al. (2023). Creativity, critical thinking, communication, and collaboration: assessment, certification, and promotion of 21st century skills for the future of work and education. Journal of Intelligence, 11(3), 54.

UNESCO. 2020. Policy Brief: Education during COVID-19 and beyond. Retrieved from https://www.un.org/development/desa/dspd/wp-content/uploads/sites/22/2020/08/sg_policy_brief_covid-19_and_education_august_2020.pdf

# ADVANCING EARLY CHILDHOOD EDUCATION: A CONCEPTUAL FRAMEWORK FOR ARTIFICIAL INTELLIGENCE APPLICATIONS IN EARLY CHILDHOOD EDUCATION

Hesham Alomyan
*University of Petra, Jordan*

**ABSTRACT**

Artificial intelligence (henceforth AI) is one of the latest innovations in education, which provides new ways of improving learning in different educational settings. In particular, owning to their ability, the adoption of AI technologies at the preschool level can enhance and transform the teaching-learning process in as far as young students are concerned. The present paper introduces a conceptual framework of integrating AI into early childhood education. It is based on analyzing recent studies and literature that have discussed various components and considerations that should be taken into account when AI tools are used in education. The framework encompasses a wide range of practical applications such as AI-based adaptive learning systems, real-time assessment tools, tutoring systems, interactive activities, language learning applications, assessment and progress monitoring tools, assistive technologies, professional development resources, recommendation systems, and communication platforms. The suggested framework herein seeks to help instructional designers and teachers create inclusive and personalized learning experiences for young children.

**KEYWORDS**

Artificial Intelligence, Early Childhood Education, AI Technology Integration

## 1. INTRODUCTION

AI has the potential to bring immense change to many areas of life. Education is seen as an area whose recognition has grown in recent years. Among all levels of education, it is believed that early childhood education might benefit most from the introduction of AI since it can offer better personalized learning experiences for children thereby improving their educational outcomes. However, educators need more than just an understanding of these new tools; they need to be provided with a solid theoretical framework which guides their line of thought on how best they should use the AI tools in teaching and learning environments. This paper thus sets out such conceptualization by relying on current literature and research findings on the one hand and at the same time by offering practical suggestions of successful integration of AI systems into preschool settings on the other hand.

The paper is intended to give teachers and other interested parties a framework of using AI in early childhood education when investigating programs that have employed AI algorithms to provide children with personalized learning experiences. The paper is also hoped to look at real-time assessment tools, tutoring systems, interactive activities, language learning applications, assessment and progress monitoring tools, assistive technologies, professional development resources, recommendation systems, and communication platforms. These are all examples of AI applications that can optimize the learning experiences and foster the holistic development of young learners.

## 2. LITERATURE REVIEW

Recent studies have found that AI has an impact on young children's educational experiences. Alghamdi et al. (2020); Fu et al. (2020); Kumar and Boulanger (2020) examined the use of AI-powered adaptive learning platforms and found that these platforms provide personalized lessons tailored to each child's needs and adjust the difficulty of the exercises based on the student's performance, ensuring a customized learning experience. An example of such platforms is DreamBox Learning, which focuses on mathematics education for students in kindergarten through eighth grade. Robert et al. (2024) also reported that real-time assessment tools such as Edulastic, which use AI algorithms to collect data on young children's performance on different subjects on time, enable educators to identify areas requiring additional support or enrichment on students' part.

Moreover, Mollick and Mollick (2023) emphasized the provision of personalized feedback and guidance to young students using AI-driven tutoring systems. For example, MATHia by Carnegie Learning provides personalized tutoring support, which allows children to understand concepts at their pace, help them understand better the topic and foster independent learning. In relation to interactive and engaging activities, Dai and Ke (2022) discussed the combination of AI-powered games and simulations to offer challenges, which adjust to the children's learning progression and level. Code Combat was one of the tools that can be pointed out in regards to making learning more fun and interactive. In addition, Rizvi (2023) highlighted the importance of AI-driven virtual tutors, which provide instant feedback and stimulate activity among children. For example, Squirrel AI Learning furnishes a virtual assistant that follows the children's activities and offers guidance in case of enhancing subsequently student engagement and motivation.

Language acquisition and literacy are other areas in which AI-based applications and tools have proven to work. According to Creely (2023), AI-based language learning applications such as Duolingo have been beneficial, particularly in personalizing lesson plans and exercises depending on the proficiency of the student's level. On his part, Kim et al. (2021) also discussed the use of AI-driven speech recognition tools, such as Speechify, in examining children's oral language proficiency and personalizing the intervention to help enhance language skills.

AI technologies have also contributed to improve assessment and progress monitoring. According to Westera et al. (2020), AI algorithms are involved in automating processes including grading platforms like Grades cope. On the other hand, Robert et al. (2024) proposed the employment of AI data analytics tools such as Bright Bytes Clarity Analytics, which analyze students' data and display visualization depicting learning curves at a glance. These tools allow teachers to make sound decisions concerning teaching activities and to evaluate children's progress appropriately.

In addition to what has been stated above, the AI-empowered assistive technologies and translation tools address diversity of children's needs promoting inclusivity and accessibility. Alghamdi et al. (2020); Kumar and Boulanger (2020) are of the opinion that AI-enabled supportive tools like Read & Write are essential, because they are personalized and assist children to tackle reading and writing problems. Similarly, Ahmad et al. (2023) talked about the role of AI-based translation systems like Google Translate in providing equal access to educational materials for children whose cultural and linguistic backgrounds are so diverse.

Contemplating the role played by AI in professional development as well as pedagogic support, Gunawan et al. (2021); Lampos et al. (2021) revealed that AI-powered advanced learning tools can be utilized to provide professional development resources, such as Coursera. These tools, in order to improve instructional practices, assist educators to select and conduct courses and online training sessions, Moreover, Villegas-Ch et al. (2021) highlighted the importance of recommendation systems like Teach FX, which analyzes classroom environment for personalized recommendations of how teachers can improve their instructional strategies so that young children and preferences are met

Finally, AI-supported platforms for communication as well as chatbots make the interaction between teachers and parents easy. Lee and Kim (2021) emphasized the importance of communication platforms such as Class Dojo-like in providing real-time updates on students' progress and behavior to strengthen home-school partnerships. Lee et al. (2020) also stressed that parents chatbot like Remind can help parents get information about their children education and the resources needed to support their children school learning. In sum, AI can be successfully integrated into the early childhood education. It thus makes classrooms more inclusive and personalized, and enables children to thrive and reach their full potential.

# 3. FRAMEWORK FOR AI APPLICATIONS IN EARLY CHILDHOOD EDUCATION

Based chiefly on the aforementioned analysis of the literature regarding the incorporation of AI into early childhood education, a conceptual framework is built and presented in Figure 1. This framework outlines numerous possible uses and applications of AI in the context of early childhood education. The main objective behind this framework is to help instructional designers, teachers and other parties create AI-powered learning environments and programs that adequately cater for young learners' diverse needs.

As can be seen in Figure 1, the proposed framework covers main areas where AI can considerably improve educational experience of young children. These areas include real-time assessment, tutoring and feedback, interactive and engaging activities, language acquisition and literacy development, assessment and progress monitoring, assistive technologies, professional development, and parents-teachers' communication.

The particular feature of the framework denotes a separate area of early childhood education where AI can be employed to make children's learning outcomes have great advantages. For example, personalized learning, via AI algorithms, analyzes the learning styles and preferences to generate customized educational material and activities unique to each child. One of the key features of AI in the educational system is the real-time assessment, which utilizes AI to promptly get data on student performance to determine areas that need more support or enrichment in a timely manner.

Furthermore, AI-embedded tutoring systems furnish students with personalized feedback and tutoring, increasing their comprehension and promoting self-studying. Interactive and engaging activities comprise AI-enabled educational games and simulations that respond to children's level of skills and that make use of the best learning methods helping to make learning more attractive, efficient, and enjoyable.

AI-based applications and tools are also used to help language acquisition and reading literacy, including language learning apps and speech recognition technology. The AI algorithms can be used for automation of the grading processes and above all are able to analyze the student data to produce summaries and visualizations of the learning trajectories.

Besides, AI-powered assistive technologies are designed to provide differentiated learning for students who are struggling with reading and writing and to ensure that equal accessibility to education resources is available to students of different linguistic backgrounds. While communication between educators and parents is facilitated by AI-driven communication platforms and chatbots, professional development for teachers is enhanced by means of AI-driven resources, such as online training courses and recommendation systems that supports teachers' instructional practices and enhance their pedagogical skills.

# 4. CONCLUSION

In conclusion, when implementing the proposed framework, instructional designers and teachers can improve personalization by adapting educational experiences to specific student needs, which consequently reflect increased engagement and learning outcomes. They can also improve efficiency in a way of automating routine tasks such as grading and progress tracking and thus saving time for more effective instructional activities. The framework also assists in addressing diverse learners' needs when offering a range of accommodations and resources that makes learning accessible and inclusive for all young students. Furthermore, the use of the framework enables ongoing improvement as it provides data driven insights to refine and enhance instructional practices and curriculum design and thus makes young children able to thrive and reach their full potential. Lastly, the framework encourages partnership through improved communication and involvement with parents as well as other members within the school community, and this, in turn creates a learner' supportive culture.

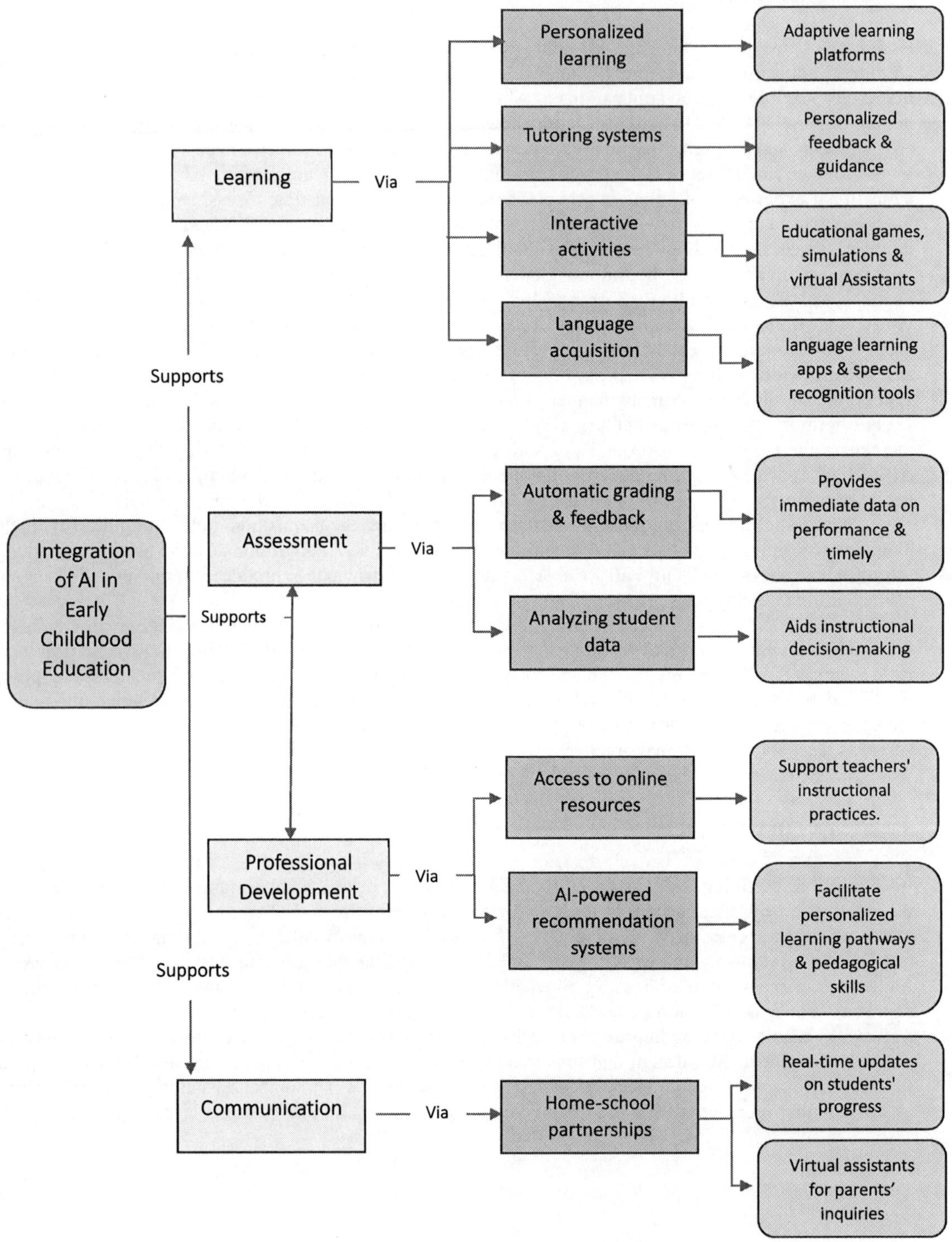


Figure 1. A Proposed framework for AI applications in early childhood education

## REFERENCES

Ahmad, M., Khasawneh, S., & Khasawneh, M. (2023). The potential of AI in facilitating cross-cultural communication through translation, *Journal of Namibian Studies,* 37, pp. 107-130.

Alghamdi, A. A., Alanezi, M. A., & Khan, Z. F. (2020). Design and implementation of a computer-aided intelligent examination system, *International Journal of Emerging Technologies in Learning*, 15(1), pp. 30-44. https://doi.org/10.3991/ijet.v15i01.11102

Creely, E. (2023). The possibilities, limitations, and dangers of generative AI in language learning and literacy practices. https://www.researchgate.net/publication/375558249_The_possibilities_limitations_and_dangers_of_generative_AI_in_language_learning_and_literacy_practices

Dai, C., & Ke, F. (2022). Educational applications of artificial intelligence in simulation-based learning: A systematic mapping review, *Computers and Education: Artificial Intelligence*, 3, 100087.

Fu, S., Gu, H., & Yang, B. (2020). The affordances of AI-enabled automatic scoring applications on learners' continuous learning intention: An empirical study in China, *British Journal of Educational Technology*, 51(5), pp. 1674-1692. https://doi.org/10.1111/bjet.12995

Gunawan, K. D. H. et al. (2021). Implementation of competency enhancement program for science teachers assisted by artificial intelligence in designing HOTS-based integrated science learning, *Jurnal Penelitian dan Pembelajaran IPA,* 7(1), 55–65. https://doi.org/10.30870/jppi.v7i1.8655

Kim, H. S., Kim, N. Y., & Cha, Y. (2021). Is it beneficial to use AI chatbots to improve learners' speaking performance?, *Journal of Asia TEFL*, 18(1), pp. 161-178. https://doi.org/10.18823/asiatefl.2021.18.1.10.161

Kumar, V., & Boulanger, D. (2020). Explainable automated essay scoring: Deep learning really has pedagogical value, *Frontiers in Education*, 5. https://doi.org/10.3389/feduc.2020.572367

Lampos, V., Mintz, J., & Qu, X. (2021). An artificial intelligence approach for selecting effective teacher communication strategies in autism education. *NPJ Science of Learning*, 6(1). https://doi.org/10.1038/s41539-021-00102-x

Lee, S. et al. (2020). The Role of AI-Driven Communication Platforms in Facilitating Parent-Teacher Communication in Early Childhood, *Journal of Early Childhood Education*, 45(3), pp. 123-145.

Lee, T., & Kim, J. (2021). Enhancing home-school partnerships through AI-driven communication platforms. *Journal of Educational Technology & Society,* 24(2), pp. 178-189.

Li, W., & Lei, C. (2021). The role of AI-driven virtual assistants in enhancing student engagement in early childhood education, *Journal of Interactive Learning Research*, 32(2), pp. 203-217.

Lin, C. C., Huang, A. Y. Q., & Lu, O. H. T. (2023). Artificial intelligence in intelligent tutoring systems toward sustainable education: A systematic review, *Smart Learning Environments*, 10, 41. https://doi.org/10.1186/s40561-023-00260-y

McCarthy, T., et al. (2016). An artificial intelligence tutor: A supplementary tool for teaching and practicing braille, *Journal of Visual Impairment & Blindness*, 110(5), pp. 309-322. https://search.ebscohost.com/login.aspx?direct=true&db=eric&AN=EJ1114767&site=ehost-live&scope=site

Mollick, E., & Mollick, L. (2023). Using AI to implement effective teaching strategies in classrooms: Five strategies, including prompts. Wharton School of the University of Pennsylvania & Wharton Interactive. https://ssrn.com/abstract=4391243

Rizvi, M. (2023). Investigating AI-powered tutoring systems that adapt to individual student needs, providing personalized guidance and assessments, *The Eurasia Proceedings of Educational and Social Sciences*, 31, pp 67-73. https://doi.org/10.55549/epess.1381518

Robert, A., & Frank, L. (2024). The impact of artificial intelligence on students' learning experience, *Wiley Interdisciplinary Reviews: Computational Statistics*. https://doi.org/10.2139/ssrn.4716747

Villegas-Ch, W., S´anchez-Viteri, S., & Rom´ an-Ca˜ nizares, M. (2021). Academic activities recommendation system for sustainable education in the age of COVID-19, *Informatics*, 8(2), pp. 29; https://doi.org/10.3390/informatics8020029

Westera, W. et al. (2020). Artificial intelligence moving serious gaming: Presenting reusable game AI components. *Education and Information Technologies*, 25(1), pp. 351–380. https://doi.org/10.1007/s10639-019-09968-2

# Posters

# APPLYING VIRTUAL LEARNING ENVIRONMENTS ASSISTS LANGUAGE LEARNERS TO STUDY ENGLISH

Pei-Lin Liu and Chiu-Jung Chen
*National ChiaYi University*
*No. 85, Wenlong Village, Minxiong Township, Chiayi County 621302, Taiwan*

**ABSTRACT**

Since the 1990s, teaching topic content in a foreign language through Content and Language Integrated Learning (CLIL) has grown in popularity. However, it's sometimes difficult for teachers to integrate language and subject matter effectively, which makes adopting CLIL difficult. These challenges include not having enough CLIL teaching tools, not having much experience teaching non-language courses, not comprehending the CLIL structure, and being unable to teach non-language subjects in English. A Taiwanese study highlighted these issues, particularly with science education.

**KEYWORDS**

Immersive Virtual Reality, Content and Language Integrated Learning, Total Physical Response

## 1. INTRODUCTION

Due to globalization, studying a language is becoming a typical way for students to absorb academic content. Introduced in the 1990s, Content and Language Integrated Learning, or CLIL, is an educational technique rapidly spreading throughout Europe (Tagnin & Ríordáin, 2021). The term CLIL describes the teaching of any non-linguistic subject in a foreign language at the primary and/or secondary education levels. According to some accounts, the two primary educational goals of CLIL are language acquisition and subject understanding.

## 2. LITERATURE REVIEW

Since the relationship between language and content is important to any implementation, this integration is one of the main characteristics of the CLIL approach. Teachers find it difficult to integrate language and subject, as numerous studies have shown (Liu & Chen, 2023). Research in various contexts has consistently shown that CLIL teachers typically see themselves more as content teachers than as language teachers. Taiwanese CLIL teachers were surveyed and interviewed for the study, which reveals the challenges these educators faced, such as their lack of CLIL teaching resources, their inexperience with the CLIL framework, and their incapacity to teach non-language subjects in English (Liu & Chen, 2024). The last two considerations may help to explain why content teachers are unable to use language as much as they would have liked. Since subject matter specialists usually instruct subjects, creating CLIL instructional resources for content teachers is crucial.

One example of an advanced visualization technology that has become more and more popular in recent years is immersion virtual reality (IVR). When utilizing IVR, customers don a headset to completely shut out the outside world and immerse themselves in a stereoscopic, computer-generated 3D display of pictures. Users can act and experience the consequences of their actions thanks to IVR technology's real-time recording of their body movements. To increase student involvement, these IVR technology components give them the impression that they are physically present in the new virtual environment and that the virtual activities are actually taking place. An intentional movement or action is required when applying the action approach to physical reactions. Learners might physically perform a new expression to aid in their retention of new vocabulary in the language of instruction.

Because of its distinct focus on pictures, interactivity, and embodied movement, virtual reality (VR) offers unique learning opportunities (Radianti et al., 2020). These qualities have attracted additional studies to investigate IVR's educational advantages. IVR's educational advantages have been thoroughly examined, particularly in the fields of science and engineering education. Researchers have assessed the literature on IVR's efficacy for engrossing students and accomplishing learning objectives in light of the growing interest in incorporating it into education (Liu, 2022). Even though science education is an important sector of educational IVR research, previous studies have generally addressed broad educational problems without specifically addressing the particular needs of science education.

## 3. METHOD

The value of in-person interactions, creations, and experiences in education has increased. When students use the physical exercise method, they can retain new material more effectively (Liu & Chen, 2014). IVR has been used to improve language learning in a number of settings, and there appears to be proof that it is superior to conventional teaching techniques for language acquisition. There is currently little research on using IVR technology to learn foreign languages.

The primary objective of this study was to investigate the use of physical activity and the CLIL approach in an IVR learning environment for English language training. As a result, this study recommends using the CLIL approach to create IVR scientific learning environments. Total physical reaction (TPR) was incorporated into the study's educational activities. The "Happy Stone Face" scientific learning simulation was developed specifically for this study, which enables the researchers to introduce the benefits of IVR into the classroom without incurring costs associated with complex equipment or setting up a school lab. The students were given three different lab tasks to see firsthand: examining rocks, utilizing the Mohs hardness scale, and observing the flow of water.

The main educational goal of the simulation is to improve understanding of the effects of various rock kinds and water movement. The goal of the researchers' virtual environment immersion was to encourage innovative approaches to learning. They provide students with the course material in an enjoyable and stimulating environment.

In this work, the researcher developed the following hypothesis to assist determine the study's goals and objectives: People who employ immersive virtual reality significantly outperform those who use the traditional way regarding language acquisition success.

## 4. CONCLUSION

Science is a popular CLIL subject that requires students to actively participate in class discussions as well as understand scientific concepts. Despite the fact that hands-on learning is an important part of science education, safety concerns, equipment costs, resource accessibility challenges, and time constraints can make it difficult for teachers to encourage their students' interest in and performance in these activities.

To address these issues, a study proposes integrating immersive virtual reality (IVR) into CLIL for science instruction. Augmented reality (AR), which is characterized by immersive 3D images and real-time body movement monitoring, increases student engagement by creating a sense of physical presence in virtual environments. This is in line with the 1989 recommendation of the American Association for the Advancement of Science that students take an active role in their science education.

## ACKNOWLEDGEMENT

The work was supported by the Taiwan National Science and Technology Council (NSTC) of the NSTC 112-2410-H-415-028.

# REFERENCE


Liu, P. L., & Chen, C.-J. . (2023). Using an AI-Based object detection translation application for English vocabulary learning. Educational Technology & Society, 26(3), 5-20. https://doi.org/10.30191/ETS.202307_26(3).0002 (Q1 SSCI, IF: 4.358)

Liu, P. L., & Chen, C.-J. ., (2024). Teaching young EFL learners new vocabulary: A comparison of the efficiency of traditional and personalized TPR strategies. Sage Open, Accepted. (Q2 SSCI, IF: 2.272)

Liu, P.-L. & Chen, C.-J. (2014). Learning English through actions: a study of mobile-assisted language. Learning. Interactive Learning Environments, 23(2), 158-171. https://doi.org/10.1080/10494820.2014.959976

Liu, P.-L. (2022). Online English direction phases learning through a virtual reality tour during the COVID-19 pandemic. Research of Educational Communications and Technology, 130, 23-45. https://doi.org/10.6137/RECT.202212_(130).0002

Radianti, J., Majchrzak, T. A., Fromm, J., & Wohlgenannt, I. (2020). A systematic review of immersive virtual reality applications for higher education: Design elements, lessons learned, and research agenda. Computers & Education, 147, 103778. https://doi.org/10.1016/j.compedu.2019.103778

Tagnin, L. & Ríordáin, N. M. (2021). Building science through questions in Content and Language Integrated Learning (CLIL) classrooms. *International Journal of STEM Education, 8*(34), 1-14. https://doi.org/10.1186/s40594-021-00293-0

# DESIGN OF STACKABLE CERTIFICATE COURSE MODULES IN AN ONLINE PROGRAM

Parviz Ghandforoush
*Virginia Tech*
*7054 Haycock Road,*
*Falls Church, VA 22043, USA*

**ABSTRACT**

In this paper we present a case study involving the design of stackable certificates focused on cybersecurity within an information technology master's degree program offered fully online. Although online graduate degree programs are readily available to professional learners, their semester/term long course structure (duration) prohibits a timely completion of the degree for this group of learners. In this paper we present an approach involving various stackable certificates that can complement the master's degree and allow the student the opportunity to modularize their studies, earn the appropriate credential while also pursuing the full graduate degree. We provide information on three stackable cybersecurity certificates that can be taken as part of a 33-credit master's degree in information technology allowing students to demonstrate progress and earn recognition to benefit their careers while also earning the degree.

**KEYWORDS**

Online Course, Stackable, Certificates

## 1. INTRODUCTION

Online accessibility to higher education has allowed working professionals, military personnel in isolated locations and those living in rural areas to earn degrees that were wholly inaccessible just a few years ago (Grenzky and Maitland, 2001; Li, 2007). Furthermore a number of traditional institutions report that residential students, seeking to increase their course load, or take advantage of the flexible scheduling and convenience of online programs, account for a significant portion of their online sections (Trentin, G., 2001 and Rungtusanatham, et al., 2004). The objective of this paper is to present a case study of using stackable certificates within a degree program to expedite the completion of relevant modules that will allow students to earn certificate(s) to demonstrate their mastery of the subject matter sooner than waiting to earn the full degree in two or more years. In the following sections we will describe a case involving an existing degree program structure followed by a discussion of stackable certificates and how may be redesigned and implemented.

## 2. PROBLEM DEFINITION

This case involves an interdisciplinary degree program consisting of courses in business, computer science, and electrical engineering. The curriculum design for such a program requires consideration given to variation in content and format among the three disciplines. For example, courses in computer science and engineering typically involve higher degree of quantitative and scientific content than those taught in business. Often computer science and engineering courses rely heavily on lecture-based presentations and hands-on laboratory experiments. Whereas, business courses involve greater level of faculty-student interactions on such things as group presentations and case analyses (Brower, H. H., 2003).

Courses in the program are designed to meet synchronously on Zoom but are fully recorded and available asynchronously for those who may not be able to participate in real-time. They can play and replay the lecture sessions as needed. Course materials are posted on software like Canvas. To create a classroom "feel" the online technology used for instruction enables student teams to set up discussion forums. It permits texting between the instructor and students during class meetings, and enables case and project presentations in real-time by student groups using PowerPoint slides or dynamic web pages.

Most students earn their degree on a part-time basis and finish in 2.5 to 3 years by taking two courses per semester. As a degree candidate, students also have the opportunity to earn graduate certificates. The graduate certificates are officially conferred by the university as separate credentials in addition to student's diploma. Students may opt to only earn the certificate(s) as stand-alone credentials if not seeking a full master's degree at this time.

## 3. STACKABLE CERTIFICATES

Each certificate within the program comprises three courses and can be completed in a year or less. These courses are the same as those taken by degree-seeking students, providing an equal academic experience. Certificate-only students can later apply their earned credits toward the full degree program if they choose to pursue it. Three certificates are available within the cybersecurity discipline, the Graduate Certificate in Cybersecurity Management, the Graduate Certificate in Cybersecurity Policy, the Graduate Certificate in Cybersecurity Technologies. A brief description of each certificate is provided below.

The Graduate Certificate in Cybersecurity Management is designed to teach students fundamentals of this field, which includes administration and coordination of all activities and personnel required to protect an organization's information systems and computer networks from cyber-threats (e.g., attacks, intrusions, malware) and data breaches. Students will learn the fundamentals of cybersecurity governance and risk management to evaluate and apply security frameworks that align with an organization's objectives and goals. Students will also learn how to use various business and cybersecurity tools and strategies (e.g., security analytics, risk assessment) needed to plan, manage, and assess cyber technologies to secure corporate data and information.

The Graduate Certificate in Cybersecurity Policy is designed to teach students how to develop policy to protect organizational information from external threat. Students will learn how to use cyber threat research to write policy that supports organizational business information and transactions across the internet in a secure manner. Students will also learn how to incorporate national and international legal, ethical, and privacy standards in policy and procedures (e.g., standard operating procedures, intervention plans). Graduates will be able to create action plans, write policy, and implement policies and procedures governing the protection of organizational operations from cyber threat.

The Graduate Certificate in Cybersecurity Technologies is designed to teach students about the technology necessary to collect and assess threat data to protect computers, networks, and organizational information from external intrusion. Cybersecurity technologies include techniques and tools such as digital logic (e.g., software to identify abnormal file movements, bioprinting), smart grid technologies (e.g., encryption gateways), and system application hardening (i.e., reduces vulnerability of computer data by removing redundant or unnecessary account functions, access, ports). Students will learn how to assess and protect computer systems with data elements (e.g., data that includes personal information stored on a cloud) that are vulnerable to a threat.

## 4. IMPLEMENTATION

For students pursuing the certificate-only option courses in these three certificates can be designed in a modular fashion such that each course may be taken as 1, 2, or 3 credits during a given semester. A one credit asynchronous course maybe completed in two weeks and the student may return at another time to complete one or more remaining credits from the same course. In the course of a semester a student may complete three courses and earn one of the above certificates. Given this arrangement it is possible to complete up to three certificates in a 12-month cycle. Of course, courses within each certificate taken in a modular fashion will count toward the full master's degree if the student decides to complete it. It's also possible to earn less than 3 credit hours for each course and not fully complete the 3-credit course. However, in this case the remaining credits, out of the nine required, maybe taken from another certificate provided course prerequisites are met.

In this case a student will receive a general cybersecurity certificate rather than one of the above specialized certificates.

## 5. CONCLUSIONS

Currently over 600 professional learners are enrolled in the program on a part-time basis with an average completion time of 2.5 -3 years. The current approach for offering certificate courses to earn the credential does not appeal to those who wish to earn their certificate in a timelier manner. We plan to implement the proposed modular stackable approach in a limited number of courses in the next academic year for its operational feasibility and to assess students' interest and will report the results. The proposed course design maybe implemented broadly to certificate topics other than cybersecurity. We plan to test the application of this method to other topics in the future.

## REFERENCES

Brower, H. H., 2003. On emulating classroom discussion in a distance-delivered OBHR course: Creating an on-line learning community. *In Academy of Management Learning and Education,* Vol 2, No. 1, pp. 22–36.

Rungtusanatham M. et al, 2004. Toward a typology of business education in the internet age. *In Decision Sciences Journal of Innovative Education,* Vol. 2, No. 2, pp. 101-120.

Trentin, G., 2001. Designing online education courses. *In Computers in the Schools,* Vol 17, No. 3, pp. 47–66.

# LEARNING ANALYTICS WITH MAPPER: INVESTIGATING PARAMETER CONFIGURATION

Hitoshi Inoue[1], Koichi Yasutake[2], Osamu Yamakawa[3], Takahiro Tagawa[4], Takahiro Sumiya[2] and Kiyomi Okamoto[5]
[1]*Nakamura Gakuen University, Japan*
[2]*Hiroshima University, Japan*
[3]*Fukui Prefecture University, Japan*
[4]*Chikushi Jogakuen University, Japan*
[5]*Osaka University, Japan*

**ABSTRACT**

This study aims to clarify the multidimensional data structure in learning processes using a topological data analysis method, Mapper. Unlike conventional dimensionality reduction techniques such as Principal Component Analysis, Mapper visualizes data while preserving its inherent topological characteristics. The outcomes from Mapper significantly depend on selecting filter functions, covering parameter settings, and clustering algorithms. We conducted a series of experiments to evaluate these factors systematically using a comprehensive educational dataset. We present case studies demonstrating how different parameter configurations impact the analysis results, highlighting Mapper's potential in uncovering valuable insights from educational data.

**KEYWORDS**

Learning Analytics, Topological Data Analysis, Mapper

## 1. INTRODUCTION

Learning analytics often target data that spans a broad and complex range. As data volume increases, analysts find conducting detailed analysis more challenging, which can result in overlooking crucial information. For instance, analysts commonly use Principal Component Analysis (PCA) to analyze multidimensional data. However, ensuring analysis and visualization remain actionable and insightful as dimensions increase may require more than just a few principal components.

To address these challenges, we apply topological data analysis (TDA) (Edelsbrunner, 2002), which conceptualizes data through a topological lens and interprets its shape algebraically. Mapper, a TDA method creates simplified network structures from data based on its geometric characteristics (Singh, 2007). Mapper's effectiveness depends on the choice of filter functions, covering parameters, and clustering algorithms.

This paper describes the methodology for selecting and optimizing Mapper parameters. We then present case studies using a dataset collected from a Learning Management System, demonstrating the impact of different parameter settings on the analysis results.

## 2. MAPPER

Topological data analysis applies to various fields, including protein structure analysis. Although only a few studies (Inoue, 2023; Ordóñez, 2021) have explored its use in learning analytics, its potential has already started to capture attention, especially following its discussion in a renowned journal (Munch, 2017).

Mapper is a topological data analysis method that creates a simplified network structure from a dataset's geometric characteristics (Singh, 2007). The Mapper methodology unfolds through these critical steps:

**Filter Function Selection** We project the dataset, seen as a high-dimensional point, onto a lower-dimensional point that reflects the dataset's properties.

**Cover Creation** We divide the filter-function-projected region into several overlapping areas.

**Clustering** In each region, we cluster points, transforming each cluster into a graph node.

**Graph Generation** This step involves establishing edges between clusters that share common points in overlapping regions, thereby constructing a comprehensive graph.

Optimizing Mapper's effectiveness in analyzing educational data requires selecting suitable parameters for each of Mapper's processing steps, including the filter function, the number of coverings, the overlap of coverings, and clustering.

# 3. EXPERIMENTATION

## 3.1 Dataset Description

We applied Mapper to a dataset collected by a Learning Management System's learner activity recording tool (Amrieh, 2015; Amrieh, 2016). This dataset includes 480 records featuring 16 characteristics and grades divided into three levels. We selected four features with integer values for the dataset. Our previous research (Inoue, 2023) shows that the top and bottom-grade groups have a similar topological structure, differing from the middle-grade group.

## 3.2 Parameter Selection

Optimizing Mapper's effectiveness in analyzing educational data requires selecting suitable parameters for each of Mapper's processing steps. To establish a reliable parameter configuration, we initially experimented with the "arm.stl" mesh data by Artec Group Inc. (https://www.artec3d.com/), from which we extracted 7,000 data points. We chose this dataset because its structure is well-known, allowing us to assess the correctness of the Mapper representation visually. We tested various configurations and found that using seven coverings with an overlap rate of 0.2 and Principal Component Analysis (PCA) as the filter function best represented the original point cloud data structure without redundancy. For clustering, the First Gap Heuristic was optimal as it reflected the underlying structure of the data well.

While these parameters were optimal for the "arm.stl" dataset, they may not necessarily be the best fit for the target dataset. However, using the "arm.stl" data provided a typical example to guide parameter selection.

# 4. RESULTS AND DISCUSSION

The Mapper graphs revealed distinct topological structures for the different grade groups:

**Top Grade Group**: The graph showed a single, significant connected component, indicating uniformity in learning behaviors. This suggests that high-performing students exhibit consistent study patterns.

**Middle Grade Group**: The graph displayed several smaller clusters and loops, reflecting diverse learning strategies and interactions. This group's complex structure suggests varying levels of engagement and performance.

**Bottom Grade Group**: Similar to the top-grade group, the bottom-grade group had a more unified structure but centered around different characteristics, indicating specific areas of struggle.

These findings support the hypothesis that high and low-performing students tend to cluster around specific behaviors or characteristics, while average students display more varied patterns. To further illustrate the application of Mapper, we conducted a detailed case study on the middle-grade group. By dividing this group into 15 coverings, we observed consistent loop formations in the Mapper graph, which aligned with the Persistent Diagram results. These loops represent recurring behavioral patterns, providing insights into the diverse strategies employed by middle-grade students.

The insights gained from this topological analysis have significant implications for educational practice. Understanding the distinct learning patterns of different student groups can help educators tailor their interventions more effectively. For instance, targeted support can be designed for middle-grade students to address their diverse needs. At the same time, specific strategies can be developed to reinforce the successful behaviors observed in top-grade students.

## 5. CONCLUSION AND FUTURE WORK

In this study, we demonstrated the utility of the Mapper method in analyzing educational data through topological data analysis. We uncovered meaningful patterns in student performance data by carefully selecting and optimizing parameters. Our findings show that different parameter settings can significantly influence the interpretation of Mapper results, emphasizing the importance of systematic parameter evaluation.

Moving forward, we plan to expand this analysis to larger and more diverse datasets, exploring Mapper's potential in various educational contexts. Additionally, we aim to develop automated tools for parameter selection to further streamline Mapper's application in educational research. By providing actionable guidelines and demonstrating practical applications, we hope to enhance the understanding and adoption of topological data analysis in learning analytics.

## ACKNOWLEDGEMENT

This work was supported by JSPS Grant-in-Aid for Scientific Research JP22K18614.

## REFERENCES

Amrieh, E. et al., (2015). Preprocessing and analyzing educational data set using X-API for improving student's performance. *Applied Electrical Engineering and Computing Technologies (AEECT)*, The Dead Sea, Jordan, pp. 1-5.

Amrieh, E. et al., (2016). Mining Educational Data to Predict Student's academic Performance using Ensemble Methods. *International Journal of Database Theory and Application*, Vol. 9, No. 8, pp. 119-136.

Edelsbrunner et al., (2002). Topological persistence and simplification. *Discrete & Computational Geometry,* Vol. 28, No. 4, pp. 511–533.

Inoue, H. et al., (2023). Learning Analytics through a Topological Data Analysis Approach. *Proceedings of EdMedia + Innovate Learning*, Vienna, Austria, pp. 445-450.

Lum, P. et al, (2013). Extracting insights from the shape of complex data using topology. *Scientific Report* 3, 1236.

Munch, E., (2017). A User's Guide to Topological Data Analysis. *Journal of Learning Analytics*, Vol. 4 No.2, pp. 47–6.

Ordóñez, S. J. et al., (2021). Sentiment Analysis to Assess Educational Methodologies in a Competency-Based Educational System. *Machine Learning-Driven Digital Technologies for Educational Innovation Workshop*, Online, pp. 1-8.

Singh, G. et al., (2007). Topological methods for the analysis of high dimensional data sets and 3d object recognition. *Eurographics Symposium on Point-Based Graphics*, Prague, Czech Republic, pp. 91-100.

**International Association for Development of the Information Society (IADIS)**
Rua Sao Sebastiao Da Pedreira
No. 100, 30, 1050-209
Lisbon, Portugal

ISBN 979-8-3313-0304-4
9 798331 303044

CURRAN ASSOCIATES INC.
proceedings.com